AN INTRODUCTION TO ETHOROBOTICS

This pioneering text explores the emerging discipline of ethorobotics which brings together the fields of animal behaviour and robotics. It encourages closer collaboration between behavioural scientists and engineers to facilitate the creation of robots with a higher degree of functionality in animal/human environments and to broaden understanding of animal behaviour in new and intriguing ways.

Utilising the knowledge of key ethologists and roboticists in the field today, this book is divided into four major parts. The first part is written for those with little or no background in the biology of animal behaviour, particularly for those coming from an engineering background seeking an accessible introduction to the field and how it can be applied to robotic behaviour. Topics include problem solving in animals, social cognition, and communication (visual, acoustic, olfactory, etc.). The second part is an introduction to the basic construction of robots for non-engineers, and the possibilities offered by current technical achievements and their limitations to the study of animal behaviour. The third part explores the core theme of ethorobotics, the basic framework of the discipline, the field's evolution, and current topics including ethical considerations, autonomy, to 'living' social robots. The fourth and final chapter looks at ethorobotics in practice through key research projects which have had the biggest impact.

This is a ground-breaking interdisciplinary text which will appeal to upper-level undergraduates, postgraduates, and researchers focusing on animal behaviour and cognition, as well as those undertaking courses in engineering, social robotics, biologically inspired robotics, AI, and human–robot and animal–robot interactions.

Judit Abdai, Department of Ethology, Eötvös Loránd University, Hungary.

Ádám Miklósi, Department of Ethology, Eötvös Loránd University, Hungary.

AN INTRODUCTION TO ETHOROBOTICS

Robotics and the Study of Animal Behaviour

Edited by
Judit Abdai and Ádám Miklósi

LONDON AND NEW YORK

Cover image: © Alexa Ivan

First published 2025
by Routledge
4 Park Square, Milton Park, Abingdon, Oxon OX14 4RN

and by Routledge
605 Third Avenue, New York, NY 10158

Routledge is an imprint of the Taylor & Francis Group, an informa business

British Library Cataloguing-in-Publication Data
A catalogue record for this book is available from the British Library

ISBN: 9781032023335 (hbk)
ISBN: 9781032023342 (pbk)
ISBN: 9781003182931 (ebk)

DOI: 10.4324/9781003182931

Typeset in Times New Roman
by codeMantra

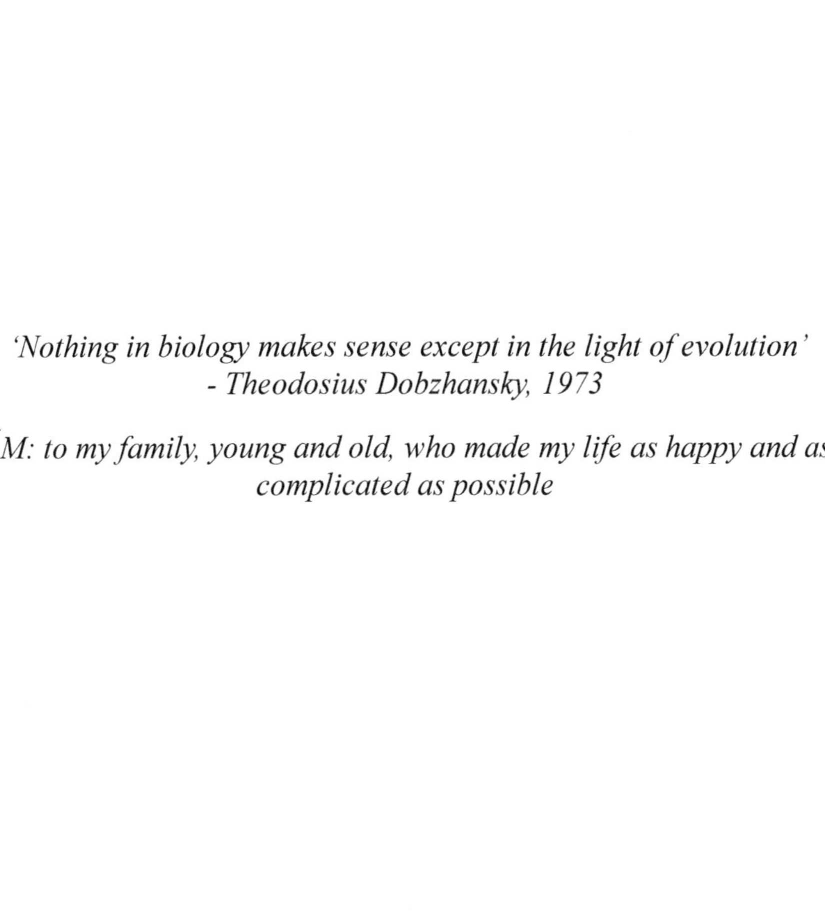

'Nothing in biology makes sense except in the light of evolution'
- Theodosius Dobzhansky, 1973

ÁM: to my family, young and old, who made my life as happy and as complicated as possible

CONTENTS

CONTRIBUTORS (EDITED BOOKS)

Judit Abdai

HUN-REN-ELTE Comparative Ethology Research Group
Budapest, Hungary
Department of Ethology, Eötvös Loránd University
Budapest, Hungary
Center for Mind/Brain Sciences (CIMeC), University of Trento
Rovereto, Italy

Bence Ferdinandy

Department of Ethology, Eötvös Loránd University
Budapest, Hungary

Beáta Korcsok

HUN-REN-ELTE Comparative Ethology Research Group
Budapest, Hungary

Ádám Miklósi

Department of Ethology, Eötvös Loránd University
Budapest, Hungary
HUN-REN-ELTE Comparative Ethology Research Group
Budapest, Hungary

Péter Telkes

Budapest University of Technology and Economics
Budapest, Hungary

PREFACE

The true story behind this book

This book would never have been written if it weren't for the advertising campaign that introduced the first pet robot: the AIBO. This occurred in the 1990s, and for a dog researcher, a robot behaving like a dog immediately sparked a plethora of experimental questions. How would a dog respond to such a creature? Would dogs perceive an AIBO as a fellow member of their species? Would they show interest in engaging in social interactions, such as playing? With limited funds to purchase a robot, we sent hundreds of emails to various researchers, and patiently awaited any response for many months. Finally, a letter arrived from the Sony Computer Laboratory (SCL).

This book would never have been written if Frédéric Kaplan, a senior researcher at the SCL at that time, had not been so curious and courageous as to bring the AIBO robot to our laboratory in Budapest. Together, we staged the first experiment observing interactions between AIBO and adult dogs as well as puppies. Following this initial testing, Enikő Kubinyi visited Paris for an extended period, during which time our observations on AIBO–dog interactions went viral on the internet, bringing us some dubious recognition in the form of being awarded the first place for the 'maddest scientific experiments' by the *New Scientist* journal in 2008.

This book would never have been written if Kerstin Dautenhahn had not invited us to participate in a large consortium supported by the European Union (LIREC), where we all aimed to develop agents that could live and interact as companions with humans. Her idea was that studying dogs could inform roboticists on how to design socially interactive robots. We still vividly remember the surprised facial expressions of the hard-core engineers when we first met at the joint workshop. Despite their initial shock upon learning that a large portion of the funding would be allocated to dog ethology rather than robot development, we became valuable partners and collaborated on many projects.

This book would never have been written if it were not for the immense interest and support shown by two engineers in mechatronics, Péter Korondi and Péter Baranyi, who offered collaboration and financial support right from the beginning. With no funds to purchase a real robot, students of Péter Korondi built our first, second, and third robots from scratch. We were thrilled if these robots worked for even just a few hours or days, learning that such challenges were quite typical for robots operating in research laboratories worldwide.

For biologists like us, '*nothing in biology makes sense except in the light of evolution*' is one of the fundamental statements made by the evolutionary biologist Theodosius Dobzhansky (1973). Robotics is no exception to this principle. Robots, with their embodiments and behaviours, are increasingly becoming subjects of the behavioural sciences, including ethology. While animals are not machines, machines may become animals of some kind.

From another perspective, all biologists are aware that the complexity observed in living beings emerged after a long evolutionary process. Life itself emerged between 4.2 and 3.5 billion years ago, followed by the appearance of eukaryotes around 1.8 billion years ago. The symbiotic relationship between cells and plastids and mitochondria occurred between 1.8 and 1.6 billion years ago. Approximately 870 million years ago, multicellular organisms emerged, with evidence for central neural structures ('brains'), appearing 100 million years later. The adaptive immune system is estimated to be about 500 million years old, while eusociality, as seen in termites, evolved around 200 million years ago. Finally, anatomically modern humans (*Homo sapiens*, the 'wise') began populating the Earth between 100,000 and 50,000 years ago, characterised by their linguistic abilities.

What one should learn from this timescale is that evolution spent the most time establishing the robust building blocks of life, with much less time needed to develop abilities such as language once the basic biological structure ('brain') was in place. From this perspective, it seems a bit naive and typically anthropocentric to expect that we can 'create humans' in the blink of an eye (in terms of evolutionary time) by reverse-engineering, starting from the current end of the evolutionary process.

As non-engineers, our goal is not to revolutionise robotics, although we certainly wish we could. Instead, our objective is to strengthen the bridge between robotics and biology, facilitating a meeting point at halfway for scientists from both fields. By taking the accomplishments of evolution more seriously, formulating realistic plans for technical development, and aiding in the development of a shared scientific language within this domain, we believe we can collectively take at least one significant step forward.

Finally, this book would have been never written if there were not many very curious and ingenious ethologists who could be convinced easily that mixing behaviour studies with social robotics, make a very challenging combination, leading to a new interdisciplinary science where we can not only learn about how to make better social robots but also what is specific in our furry companions, the dogs. This strong interest in ethorobotics led to the establishment of an independent research group in which now the third generation of young scientists are working on such projects including some of the present authors, Judit Abdai, Beáta Korcsok, and Bence Ferdinandy.

Numerous researchers have played a pivotal role, directly or indirectly, in the creation of this book. We extend our heartfelt gratitude, particularly to Márta Gácsi, whose inspirational contributions were instrumental in our exploration of human–dog interaction. Despite her primary focus on dog ethology, Márta has been deeply involved in all facets of developing the concept of ethorobotics. Enikő Kubinyi stands as one of the pioneers in this research domain, delving into the interactive capabilities of the AIBO robot. Péter Korondi has worked a lot to foster a strong bond between ethologists and roboticists.

We owe a debt of gratitude to numerous other researchers, including Péter Baranyi, Kerstin Dautenhahn, Tamás Faragó, Dario Floreano, Luca Gambardella, Frédéric Kaplan, Veronika Konok, Szilveszter Kovács, Gabriella Lakatos, András Lőrincz, Vicente Oliveira Matellán, Mihoko Niitsuma, Pierre-Yves Oudeyer, Péter Pongrácz, Csanád Szabó, and Dávid Vincze. Their invaluable contributions have provided us with a comprehensive understanding of this rapidly evolving field. Ádám Miklósi would like to thank Krisztina Soproni who was always very helpful when he struggled in the web of unwanted administrative duties, Krisztina Ajkay who inspired him many times of need, and Krisztina Sándor who has always been available for insightful discussions on science from the viewpoint of the younger generation. We are grateful to Gabriella Lakatos, Péter Pongrácz and Péter Löw for reading and commenting parts of the book.

Throughout our journey, our research has been supported by many grant organisations. We express our sincere appreciation to the Hungarian National Research, Development, and Innovation Fund (OTKA Research grants and fellowships; Cooperative Doctorate Program), the Hungarian Academy of Sciences (HUN-REN-ELTE Comparative Ethology Research Group and János Bolyai Research Scholarship), the European Research Council (LIREC, ID: 215554), the Swiss National Science Foundation (SWARMIX, CRSI22 133059), the Sony Computer Science Laboratory (Paris, France), the Hungarian State Fellowship Program, and MSCA-Seal of Excellence by the University of Trento.

Just as this book is going to press, we have heard news that some people are constructing “dog robots” equipped with guns to kill people in combat. While we cannot stop any army from producing these machines, we hope and rely on the goodwill of the public to stop labeling these agents as animals. Animals are a natural part of Earth, and no one should misuse this partnership for disgraceful purposes.

INTRODUCTION

This book represents a pioneering effort to explore the relationship between ethology and robotics. Over generations, inventors have found it challenging to build machines that not only replace human work but also that are able to behave somewhat similarly to animals. However, only in recent decades, it has become feasible to construct agents that can engage in meaningful interactions with humans and other animals. Many experts believe that the emergence of socially interactive robots is inevitable and will bring significant benefits to society. Although these technologies are still in their early stages and most social robots exist only in scientific laboratories, this presents a valuable opportunity to foster collaboration between ethology and robotics.

We propose that the interdisciplinary field of ethorobotics offers a promising framework for such collaboration. Ethorobotics broadly involves applying ethological knowledge to the construction of socially interactive robots. Ethology can contribute evolutionary, functional, mechanistic, and developmental insights about the behavioural phenotype in robotics. By building so-called ethorobots based on strong biological foundations, we may enhance their ability to function as effective partners for humans and other living creatures.

Chapter 1 presents a brief overview of essential ethological concepts that are important for both engineers and biologists. Unlike typical ethology books, this chapter provides concise descriptions of animal behaviour, with an emphasis on the underlying phenomena rather than exhaustive details. Thus, many specific references to animal behaviour are omitted, and the interested reader is requested to look for the relevant examples in textbooks (e.g., Barnard, 2004; Dugatkin, 2009; Goodenough et al., 2009; Shettleworth, 2010).

Chapter 2 delves into the world of ethorobotics by drawing on the key concepts from modern ethology that are crucial for constructing a new generation of robots. We propose that robots designed to be similar to living beings should represent novel species and we should not aim for mimicking existing ones. To differentiate

 DOI: 10.4324/9781003182931-1

this approach from others, we will refer to ethologically inspired artificial agents as ethorobots.

In *Chapter 3*, we examine how modern technology can contribute to the development of robots that are of service to humans (or animals), going beyond being 'only' socially interactive. This chapter is not about technology per se, but rather, we aim to emphasise the importance of reliable and robust technology for ethorobotics. In this field, technology should serve the function of the robot, and not the other way around. Thus, the function of the robot is paramount, although limitations in available technology inevitably constrain its usefulness.

Chapter 4 presents and evaluates a selection of important case studies, which provide a strong foundation for the future development of ethorobotics. We hope that in a few years, there will be even more adequate case studies to report. There is an increasing demand for socially engaging robots, but the delivery is lagging behind. These examples also aim to highlight problematic issues, showing that a collaborative effort of multiple disciplines is necessary to develop ethorobots that perform well in various settings, such as elderly homes, restaurants, and space travel.

References

Barnard, C. J. (2004). *Animal behaviour: Mechanism, development, function and evolution.* Pearson Education.

Dugatkin, L. A. (2009). *Principles of animal behavior* (2nd ed.). W.W. Norton.

Goodenough, J., McGuire, B., & Jakob, E. (2009). *Perspectives on animal behavior* (3rd ed.). John Wiley & Sons.

Shettleworth, S. J. (2010). *Cognition, evolution, and behavior* (2nd ed.). Oxford University Press.

1

ETHOLOGY

The science of animal behaviour

Judit Abdai and Ádám Miklósi

1.1 Ethology as the study of behaviour

Ethology is the biological study of behaviour (Tinbergen, 1963). It emerged gradually at the end of the 19th and beginning of the 20th century based on the insight that behaviour of animals is a stable and measurable feature of the phenotype just like colour, shape, or other aspects of morphology. Most early ethologists had a background in zoology or biology, and since the name ethology was coined only in 1902 by William Morton Wheeler (1902), many of them did not know that their research was contributing to a new branch of science. Darwin is a fine example. Known for his work on evolution, he could as well be honoured as one of the first ethologists. His book *The Expression of Emotions in Man and Animals* should not only be regarded as the first book in ethology, but it is still one of the best in its area. It is no wonder that being an evolutionary biologist, Darwin saw no major divide between 'animals' and 'man'. The separate reference in the title was probably intended more for the readers. The treatment of humans as part of the animal kingdom regarding their behaviour and mental functioning was truly revolutionary. In the last few years, there seems to have emerged a consensus among researchers that human-like emotional states also exist in animals. This is the result of an intensive debate spanning ethology, psychology, physiology, and the philosophy of science.

Comparative psychology is another way of inquiry into the behaviour of animals and humans. However, its starting point was rather different. The animals under study were represented by less than a handful of species (mostly pigeons (*Columba livia*), rats (*Rattus norvegicus*), cats (*Felis sylvestris catus*), dogs (*Canis familiaris*), and rhesus monkeys (*Macaca mulatta*)), and most of the research was carried out inside bleak laboratory environments. Animals often served as a simplified model system to study phenomena, especially learning, from a human aspect, after realising that working with humans can be very subjective and experimental studies on humans have several, including ethical, limitations.

 DOI: 10.4324/9781003182931-2

It must be acknowledged that comparative psychologists have indisputable merits in improving the methodology of studying animal behaviour, and several ideas and concepts developed by them have found their way into ethological thinking. Psychology, with its wide-ranging interest in all aspects of human life, is by nature anthropocentric. Thus, it seems more strategically important to have a distinct knowledge hub for the biological study of behaviour as part of the phenotype of a living being, which at some point in the future may also include artificial agents. The psychological background, in addition to the subjective experience of scientists, provides a continuous anthropomorphic influence. It is still debated whether anthropomorphism is an interesting hypothesis-generating tool (Burghardt, 1991) or if it provides a rather biased influence on interpreting behaviour (Wynne, 2007). The idea of strengthening and maintaining an 'independent' science for the study of the behavioural phenotype seems reasonable.

Some argue that ethology should be replaced by behavioural biology due to its broader foundation. Indeed, several interconnected causal factors have to be considered to explain behaviour, and for this, acquaintance with ecology, physiology, genetics, etc. is indispensable. While such an interdisciplinary approach certainly has its merits, it is likely that this would result in a loss of focus on behaviour and its mentalistic explanations. Ethology has always maintained close relations with sciences investigating different levels of biological organisation. Neuroethology, behaviour genetics, behavioural ecology, and the latest, cognitive ethology, all have a long history and have established specific methodologies of their own.

1.1.1 Our vision

The very existence of ethology helps also to recognise that the scale of behaving non-biological agents is growing. Although these are not part of natural evolution and are emerging as part of a technological and cultural process, they also display various behaviours that can and should be studied by the ethological methodology. With improvements in technology, analogously to an evolutionary scenario, such artificial agents will start to display different kinds of and more and more complex behaviours. It seems to be warranted that ethology establishes a link towards a science of engineering, which constructs these agents, robots, especially in the case when they are supposed to exist in the same environments as 'living beings'. The knowledge and insights gathered by ethologists over the last 150 years have found their way into many disciplines of robotics. Behaviourally, or ethologically inspired theoretical constructs as well as practical applications can be found in many present-day robots (e.g., Arkin, 1998; Arkin et al., 2003; Cañas & Matellán, 2007; Miklósi & Gácsi, 2012), but there is still room to tighten these ties.

Our motivation for writing this book stems from our vision that ethorobotics has the potential to establish a general scientific underpinning for the development of novel types of social agents. A pivotal lesson gleaned from ethology is the absence of a singular supremely fit species for survival; there is no species atop the evolutionary hierarchy. Instead, an abundance of species has demonstrated their adeptness at enduring across millions of years. Embracing this perspective prompts us to temper our aspirations towards creating a 'perfect' artificial emulation of humans, given the extraordinary diversity of life.

Moreover, there is a growing imperative to establish a unified conceptual framework encompassing all entities that exhibit behaviour (non-human or human animals or artificial agents) alongside the implementation of standardised methodologies. In this context, ethorobotics should assume the role of forging a common ground, melding divergent yet related and interconnected concepts from psychology, ethology, and robotics.

1.1.2 A cautionary note on 'social robots'

Throughout the text, readers will encounter numerous references to robots, delineating their capabilities or limitations. The authors recognise that these observations may resonate differently with individuals whose experiences with robots span a broad spectrum. Many have been involved in constructing robots or engaging with them as researchers at cutting-edge tech enterprises. In contrast, some have gleaned their insights from films or various forms of traditional and digital media. A few may have interacted with robots in settings such as restaurants or assisted living facilities.

To mitigate potential misconceptions, it is important to clarify that when the authors make overarching statements about 'social robots', they specifically refer to those that are operational and efficient (!) and actively engage with people under real conditions. Additionally, it is worth noting that such robots are relatively scarce in kind and number.

1.2 The 'four questions' of ethology

Interestingly, ethologists agreed only lately on the definition of behaviour. According to Levitis et al. (2009), behaviours are 'internally coordinated responses (actions or inactions) of whole living organisms (individuals or groups) to internal and/or external stimuli, excluding responses more easily understood as developmental changes'.

A less emphasised insight of ethology was that behaviour is an important feature of the phenotype. It can not only be observed as, for example, morphological traits but despite being more variable and dynamic, it is still an important characteristic of individuals of any species. This may also explain why at early times ethologists focused more on behavioural traits having a more rigid nature, called fixed action patterns. These, often very conspicuous behaviours, could be typically observed during courtship, aggressive interactions, parenting, etc. Subsequent research revealed that it was a mistake to indicate that these actions are 'fixed'. They may be also subject to modifications through maturation and experience, the degree of which may however differ.

In his famous essay, Tinbergen (1963) asked the following simple question: 'Why do these animals behave as they do?'. For a biologist, this question can only be answered by observing animals in their natural environment. Although much of present-day ethological research takes place in laboratories, which may offer a relatively impoverished environment for the animals under study, the scientific questions should be based on observations, including systematic collection of behavioural data, conducted in nature.

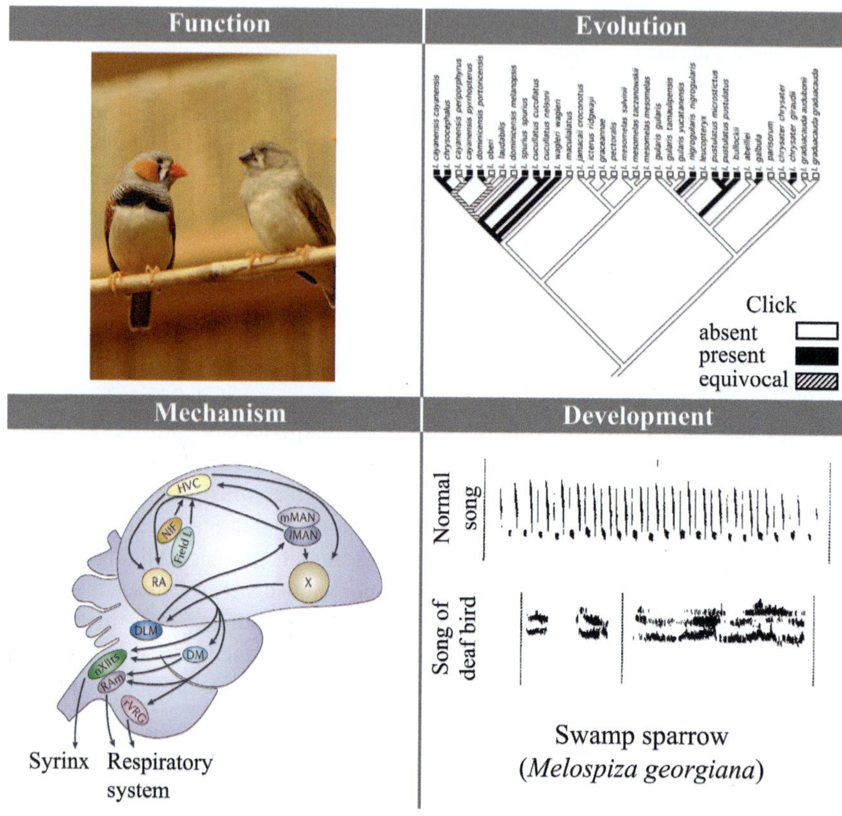

Figure 1.1 Song of songbirds from the viewpoint of Tinbergen's four questions. *Function*: attracting females and warning rivals by signalling their quality (photo by Balázs Rosivall). *Evolution*: characteristics of the songs vary between phylogenetic lines (presence and absence of clicks; figure from Price et al., 2007). *Mechanism*: neural circuitry of brain nuclei and their projections (schematic diagram of the parasagittal view of songbirds' brain; figure from Bolhuis & Gahr, 2006). *Development*: songs are usually learned from conspecifics during a sensitive period (Marler & Sherman, 1983, copyright 1983 Society for Neuroscience).

It has turned out that ethologists were interested in different aspects of this 'Why' question. These studies also utilised different methods, but often it was not clear in what way the results answered the actual question. It was argued that some 'Whys' are more important than others. Looking at the history of ethology provides clear evidence that different types of answers are equally important because they are closely interrelated. A complete answer to the above question can be only provided if the behavioural phenomenon is investigated from different perspectives in parallel. In the following, we summarise the four questions, and as an example use the learned song of songbirds as a specific trait (Catchpole & Slater, 2008; Dugatkin, 2009) (see also Figure 1.1).

1.2.1 Function of behaviour

In contrast to other meanings of 'function', in ethology, function of a behavioural trait is to increase the fitness of the individual, in other words, function refers to the adaptive significance or current utility at the present evolutionary state of the species (Bateson & Laland, 2013). In ethology, a somewhat separate branch of research is devoted to establishing the function of specific phenotypic traits: behavioural ecology. This is a quite formidable and complicated task because animals have a huge number of traits, the measure of which can also vary, and there are no general rules to predict whether a trait defined by specific measures provides the animals with some positive functional consequences. In addition, fitness is often estimated by the number of offspring reaching maturity, but such data are often difficult to get, so researchers often settle for proximate variables, for example, the number of eggs in the nest (but it is easy to see that these are not the same measures).

There is a consensus that the song of male songbirds may contribute to fitness. One typical assumption is that the main function of song is to attract females. Males' attractiveness should vary according to their song quality, so ethologists need to find out experimentally what characteristics of the song have the best influence on female birds. Note that attraction is only a very proximate measure of fitness (but it is more difficult and timelier to estimate the number of adult offspring), and song quality (or quantity) can be estimated by hundreds of parameters.

Importantly, there is often some trade-off between phenotypic traits. While songs (c.f., time spent with singing) may attract the best females, predators may also be alerted, so this behaviour also carries some risks. This means that the singing behaviour that appeals to females could be subject of another selective factor (predators). Thus, it is likely that the physical features of the male song may not reach maximal efficiency but rather arrive at an optimal (the best possible) state. Trade-off between selective forces may be different in various environments that can have a species-specific influence on the characteristics of the songs. Thus, species may differ in which traits are selected for affecting fitness.

Functionality in this biological sense should also apply to robots. This is trivial for robots working in the industry but has been often neglected for robots in the human environment. Robots providing no benefit for humans have a reduced fitness and should not survive. In contrast to living agents, functionality can be defined and tested in advance. Engineers have developed tools, called benchmarking, which provide a measure for usefulness (see Section 2.2.4) that can be also used to compare effectiveness among different robots. It is unfortunate that such testing is only rarely performed on social robots, and functionality is only of secondary importance. The understanding of trade-offs must also stand out as a clear lesson form functionality in robotics. Any specific skill or ability may come at a cost to have other important features. Evolution had time to find some optimal or balanced solutions while robotics often suffers from designers' optimistic aspirations.

1.2.2 Evolution of behaviour

Behaviour of animals can also be explained in relation to their evolution. Biological continuity is reflected in traits that have their antecedence, and there is sometimes good evidence for this phenomenon in the fossil record; however, behaviour does not fossilise. Continuity in living agents results in constrains that strongly influence their evolvability and adaptability. Biologists are also keen on restricting the use of adaptation to evolutionary processes. A trait is considered to be adapted if there is evidence that it conveys fitness benefits to the organism by being specifically selected for surviving in that environment (niche). Importantly, traits may also provide the organism with selective advantage if they have not been selected specifically. This evolutionary process is called exaptation (Gould & Vrba, 1982), that is, when traits evolving for some other functions are co-opted for a new utility. This phenomenon also shows that evolution is a rather conservative process, and it seems to be easier to reuse and accommodate existing traits than inventing something new. Exaptive processes are often involved in ritualisation when one trait with some related function evolves into a communicative signal during evolution (see Section 1.8). For example, it is assumed that tails of many mammals acted typically as a balancing organ but gradually they gained a signalling function.

Mechanisms of the behaviour are not independent from the environment in which these evolve. The same function can be achieved by different mechanisms. For example, using a simple tool can be a genetically pre-programmed behaviour, or it can be learned either individually (by trial-and-error) or by observing others (see Table 1.1).

Thus, many traits share a common evolutionary origin (homology) but then diverge with respect to environmental challenges faced by descendant species. Limbs of vertebrates share a common basic morphological structure but are very different in shape and form. Animals with very different evolutionary histories may face very similar selective forces in their environment. Their similarity is a consequence of convergent processes. Convergent traits, e.g., song of songbirds and song in gibbons, may have a similar function (see above) but the mechanisms that regulate them

Table 1.1 Comparison of benefits and risks of generic transmission, social transmission, and individual learning in terms of flexibility and temporal window in the case of a specific trait (Boyd & Richerson, 1985)

Origin of the behaviour	*Ecological factors*	*Flexibility*	*Risks*	*Individual level*
Species typical behaviour	Stable environment	Low flexibility	Few negative consequences	Highly reliable
Individual learning	Rapidly changing environment	High flexibility	Negative consequences	Very variable
Social learning	Slowly changing environment	Medium flexibility	Few negative consequences	Variable

and their phenotypic development may be different. Birds with the capacity to learn songs emerged three times independently of each other (songbirds, parrots, and hummingbirds) during avian evolution. Songs have the same function in these clades of species, and the production of the song also shares many similarities.

Robots differ from living beings in that there is no inherent continuity between subsequent generations. Thus, new robots could be designed *de novo*, but apart from specific inventions that may occur (e.g., invention of the electric engine), engineers rely on the knowledge and habits of their predecessors. However, the pace of technological development and cultural changes, as well as the capacities of the engineers, may set a limit of rapid changes in robots. While the direct application of many evolutionary insights may not bear with much advantage for robotics, there are specific solutions that could be adopted (see Section 2.3).

1.2.3 Mechanisms of behaviour

Behavioural output of an organism is subject to a multi-level control system that functions on very different time scales. Molecular processes within and between cells, including protein synthesis and myriads of molecular interactions that have physiological consequences on local or global scale, are represented by the neural, hormonal, or immune system. These mechanisms ensure that the individual is able to adjust its behaviour and maintain integrity in the face of rapid and sudden environmental disturbances. All these mechanisms can be related directly or indirectly to some measurable chemical changes. However, from the behavioural perspective, there is also a need to refer to mechanisms that reveal the overall organisation of the controlling systems. Such mental or cognitive mechanisms provide a kind of algorithmic description of the working of the animal and help getting a grasp on how molecular systems are integrated. Mental mechanisms represent an independent level of behaviour control and should be studied on their own right.

Research during the last 50 years has revealed that songbirds possess a dedicated system that allows them to learn their father's song and produce a complex song by the end of maturation. In some species, song learning takes place only once in their life, when the chicks are in the nest, whereas others can learn a new song each year. Specific genes and a complex network of neural structures are involved in this process. It seems that the emergence of this capacity was a major evolutionary undertaking. A small number of comparative studies showed many parallels in the neural and hormonal mechanisms in species belonging to the independently evolving clades of singers (e.g., Gahr, 2000).

There is a specific research direction in robotics that implements biological mechanisms in robots. Such bio-inspired artificial agents show sometimes close similarities to their biological counterparts, but technology often does not reach the level of complexity and sophistication present in animals. This collaboration has huge potential and can mutually reinforce each other's efforts. Mental models developed by ethologists may also find their way into robotics, and in some cases, the structure of robot architecture is also based on insights in biology or more specifically, cognitive ethology (see Section 2.4).

1.2.4 Development of behaviour

The study of mechanisms looks at an organism at a specific time. Developmental investigations aim to find out how these processes change over time while maintaining the integrity of the organism. As development in living beings starts with a single cell and leads to a multicellular organism, it should also explain the rules of the emerging complexity. Another way of looking at development is to focus on the specific information that is directly transmitted from one generation to the next. At the individual level, this includes the nuclear or mitochondrial DNA, factors related to genomic imprinting, or many non-genetic parental effects. The basic question is: what is the underlying genetic contribution to a specific behavioural trait? Environmental effects, both social and non-social, can have a major influence on development depending on the plasticity of the organism. This process is often referred to as epigenesis during which temporally and spatially coordinated genetic and environmental effects interact to generate an adult organism. The lack of environmental input or feedback alters natural developmental processes. Many such inputs should occur within a specific time window (c.f., sensitive period) that ensures the most advantageous effect.

It is also often the case that animals pass through different developmental environments during their lifetime (e.g., mammals rely on a range of specific abilities when living on the mother's milk that differ from self-supporting eating). The ability of behavioural traits to adjust the organism to its actual environment is also supported by learning mechanisms that allow for changing the probability of behaviours in relation to environmental events. Learning processes are the basis for changes in the mental mechanisms of the organisms. Much of the plasticity is also achieved by learning, but in the same vein, failure to learn can have detrimental consequences. The ratio of the contribution of learning opposed to genetic influence in the case of a specific trait is a consequence of selective processes in evolution.

In songbirds, the interplay between genetic inheritance and environmental factors holds paramount significance in shaping species-specific vocalisations. Among others, cross-fostering experiments revealed that young birds can learn the song of their foster parents but if given a choice, they often show a preference for the songs of their own species. The father has an important role by exposing the offspring to the relevant songs. Nevertheless, as the fledglings mature, they may introduce novel elements into their songs, thereby upholding a degree of variation within the broader population.

Developmental processes are very rarely, if at all, part of the robotic design. Typically, robots do not develop over time (apart from some learning) because most of their features, including embodiment and behaving capacities are relatively rigid. Currently, there appears to be no technology capable of effecting changes on a scale comparable to those observed in the animal kingdom. This is unfortunate because developmental adjustments to the actual environment could be an important contribution to fitness (see Section 1.11). The only exception is learning: some robots utilise learning mechanisms and rely on experience to better prepare for the future.

1.2.5 Individuality in behaviour

Animals are characterised by two, somewhat paradoxical tendencies. They are robust to maintain stability over time, but at the same time, they also show significant variability in their behaviour in response to environmental events. In the case of behaviour, the robustness is now generally described as an aspect of personality (Section 2.4.10), but it does not explain the total measurable variability of the individual's phenotype. Thus, the remaining variability can manifest in unique differences among individuals. Such uniqueness is also subject to selective processes, but in addition, it also has other functions, especially in social animals. Unique features are the basis for individual recognition and the establishment of individualised relationships.

So far there has not been a particular effort in robotics to make robots differ on individual bases. Typically, engineers expect robots to show invariable, robust, and highly predictable behaviour that is a must for industrial applications but may be less helpful for social interactions.

1.3 Description and analysis of behaviour

One of the main contributors of ethology to the biological sciences was to provide a method for making behaviour measurable (for a short introduction on this topic, see Bateson & Martin, 2007). At that time, zoologists were used to measure mainly physical traits of animals in order to characterise and categorise them. Behaviour seemed to be not just too complex but also very variable. Ethologists had two important insights. First, they argued that despite this seemingly huge variability, many aspects of the behaviour are species specific which provide a strong base for stability (see 'modal action pattern' Section 1.4). Second, the continuous flow of behaviour can be decomposed to well-defined behaviour units (elements) (segmentation).

1.3.1 The ethogram

An ethogram provides a detailed and comprehensive catalogue on the behavioural repertoire of an animal species rest on the assumption that all individuals share the same general behavioural features (Lehner, 1998). These descriptors are based on human sensory and perceptual capacities because the observational data are collected by human observers. The detailed descriptions also serve to ensure that different observers have the same understanding of a specific behaviour unit, and measurements of inter-observer agreement should be provided for all ethological analyses to guarantee the reliability of the collected data.

The segmentation of the behaviour is a complex issue, and theoretical and practical questions should be considered:

1 Behaviour is hierarchically organised both at the level of the physical components involved (muscle, limbs, and body) and in terms of the temporal proceedings. The researcher has to decide in advance at what level of behaviour organisation units

should be defined. This also means that different types of ethograms could be invented for the same behavioural process (e.g., hunting);

2 The measurement of the described behaviour unit should be in line with the hypothesis tested;
3 It makes no sense defining behavioural units very specifically if they cannot be observed reliably. The description should be as detailed as possible and complemented by pictures or videos;
4 More detailed segmentation of the behaviour stream (splitting) can be advantageous if later some units need to be grouped together because the reverse possibility is not possible after data collection.

Ethograms are useful for describing the behaviour of any agent. Here are some criteria for constructing a useful ethogram (Abdai et al., 2018):

1 The position and form of the body and its parts should be described in relation to angles of articulations, directions of movements, and the necessary degrees of freedom;
2 Location and orientation of the element should be provided in relation to the animal's actual environment;
3 If possible, the typical duration and intensity (speed and amplitude) of the unit should be referred to if it can be judged by observation;
4 Possible variations should be mentioned which do not exclude the observed behaviour from that specific unit category;
5 Behaviour units should be named by a descriptive label that should not refer to any proposed or hypothesised function.

The unit of interaction: an action of one partner is followed by a specific reaction of the other where one can experimentally show by repeated measures that there is a temporal correlation between the two actions with a minimal delay, and changing the shape, context of the actions affects the reaction in some predictable way.

1.3.2 Structure of behavioural data

It is important to remember that the segmentation process to define measurable units disregards the temporal aspects of behaviour. Subsequent actions are not happening at random, but they are the outputs of a hierarchically organised control system involving (repeated and imbedded) cycles of various forms.

Quantitative description: behavioural units are regarded as being independent from each other, and the temporal aspects are not considered. Thus, for any specific observation period, the frequency and the total duration are calculated. Sometimes the time to the first occurrence (latency) and the mean duration of the unit (bout) are also noted. These measurements provide the basic data for any further statistical analyses.

Temporal description: the analysis of the temporal aspects of behaviour raises several problems, and there are many different approaches to deal with these issues.

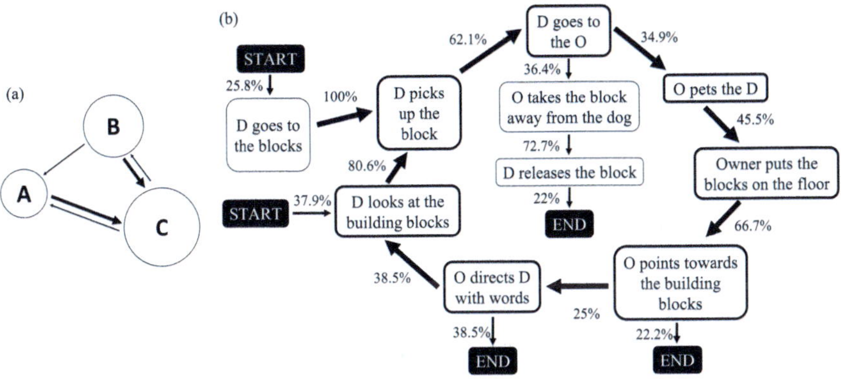

Figure 1.2 Two methods for the temporal analysis of behaviour. (a) Hypothetical scenario with A, B, and C behaviours following each other; the size of the circles indicates the frequency of the occurrence of the specific behaviour, and the thickness of the arrows indicate the probability of the two actions following each other in a sequence (see Table 1.2). (b) T-pattern analysis of a behaviour sequence (Magnusson, 2020) in which an owner and a dog interacts with each other when manipulating a building block (redrawn based on Kerepesi et al., 2005). Describing temporal pattern of interspecific interaction can be advantageous for developing robot behaviour for complex interactions.

Table 1.2 Example of scenarios in which A, B, and C behaviours follow each (see Figure 1.2a)

		Follow			
		A	*B*	*C*	*Overall*
Precede	A	7	0	20	27
	B	9	10	18	37
	C	8	11	3	22
	Overall	24	21	41	86

In a behaviour sequence, units do not follow each other randomly. The probability of one unit following the other can be calculated when all occurrences in the sequence are known.

Markov chains provided the earliest and simplest method for such analyses (e.g., Bateson & Martin, 2007). A first-order Markov chain calculates the probability of two behavioural units following each other in a behavioural sequence. For this calculation, occurrences of all observed units are put into a table, called transition matrix (Figure 1.2a, Table 1.2). Other similar methods have been also developed, but they have not become popular because there was a lack of long and detailed behaviour recordings, the calculations were complicated, and the results were more qualitative than quantitative.

The T-pattern analysis offers a different approach to the problem (Magnusson, 2020). The basic assumption of this method is that patterns in a behaviour sequence may stretch over non-adjacent units and occur over a longer time period like seconds or minutes. Thus, two or more units can still form a pattern (where the temporal dependency is determined by statistical methods) independently from what is happening between these two or more actions (Figure 1.2b, based on Kerepesi et al., 2005).

The analysis of sequences of events has become a field of statistics on its own, but these statistical methods are still not common in ethology. The main reason is that for robust analysis, there is a need for long behavioural recordings and the calculation gets very complicated as the number of units increases.

1.3.3 Automatisation of behaviour data collection

At the beginning, behavioural data were collected by direct observation and making notes on paper. Video recordings and the utilisation of computers have speeded up the process and made also more detailed descriptions possible although videos often miss some details of the behaviour that could be observed on site (Bateson & Martin, 2007). This activity is called behaviour coding or annotation.

The utilisation of data provided by wearable sensors and analysis of video images has the potential to change the way how behaviour is quantified. Especially software based on machine learning procedures can provide a big advantage (e.g., Bohnslav et al., 2021; Datta et al., 2019; Valletta et al., 2017). There are two main approaches (for details see Section 3.7):

1 *Supervised learning*: these applications may make human observers redundant by aiming for replicating the behavioural data produced by them. In this case, the software is trained by a human observer until there is a close agreement between them. All further analysis is done by the software and thus, massive amount of data can be collected. However, some further quality control by the researchers is still needed. It may also happen that the recognition capacities of the software are still limited, and the affected behaviour units have to be coded by the experimenter. Note that the ability of software to recognise behaviour units depends crucially on the recognition capacity and accuracy of the human.

2 *Unsupervised learning*: the task of the software is to find regularities in the data without any human influence, often based on a type of statistical clustering. The relationship among these groups can vary depending on the clustering method used. As an outcome, exclusive, hierarchical, overlapping, etc. groupings can emerge. This approach offers the possibility to detect patterns that had not been recognisable for the human eye, but precisely because of this, often it becomes difficult to make sense of these newly discovered patterns.

Probably some kind of combination of these two basic approaches leads to a robust solution to collect large amount of behavioural data replacing human workload, but also to reveal underlying structure of the data not accessible for human observers.

It is also important to note that the data collected by sensors and analysed in the above ways could also be used to enrich behavioural skills of ethorobots. The relationships between behaviour units or the hidden regularities detected in the data can act as rules in the programming of artificial agents. In the case of ethorobots, collection of behavioural data can be automatised because the agent can provide an output about its own actions. Although this could make human observations unnecessary, ethorobots may display also emergent behaviours for which a third-person perspective is useful.

1.4 Modelling of behaviour systems in ethology

Ethologists observe the environment and the individuals' behaviour in it. An answer to the mechanistic interpretation of Tinbergen's question (Section 1.2), why the animal does what it does, can only be answered by developing mental models on mental functioning. The aim of these models is to provide a realistic, but mostly hypothetical connection between environmental events and inner machinery that produces the expected behavioural response.

Mental models have always played an essential role in conceptualising behaviour, but the task seems to be immense partly because the models depend very much on our scientific preparedness. The human mind was often visualised as a kind of clockwork, phone exchange, or a computer. It seems that our imagination about the mind depends on the technical achievements of the human society. Thus, it is not surprising that the inspiration of mental models comes from mechanics, electric science, cybernetics, fuzzy theory, and many other sources. While so far, there was no preference for any kind of mental models, this may change when one wants any of these to be tested in an embodied robot under real circumstances in the field. Importantly, these ethologically inspired mental models are too simple and rely on several assumptions that make their direct implementation difficult:

1 It is assumed that the individual has other mechanisms to maintain homeostasis in general;
2 It is assumed that the individual is able to always maintain its physical integrity, even if challenged in unexpected situations;
3 The models focus only on the core mental processes, without many details about perceptual or motor aspects.

Despite these limitations, it may be worthwhile to provide a short summary of these models that had significant influence on scientific thinking and may provide a basis for the next generation of more sophisticated approaches (Section 3.6).

1.4.1 Early models of animal behaviour

Ethologists struggled a lot with the concept of 'instinct' because they felt that despite its widespread use, it has little explanatory power. Instead, the idea was to focus on the observable aspects of the behaviour. The model has specific inputs and outputs that are connected by a simple hypothetical central mechanism (innate releasing mechanism; Goodenough et al., 2009; Lorenz, 1981) (Figure 1.3).

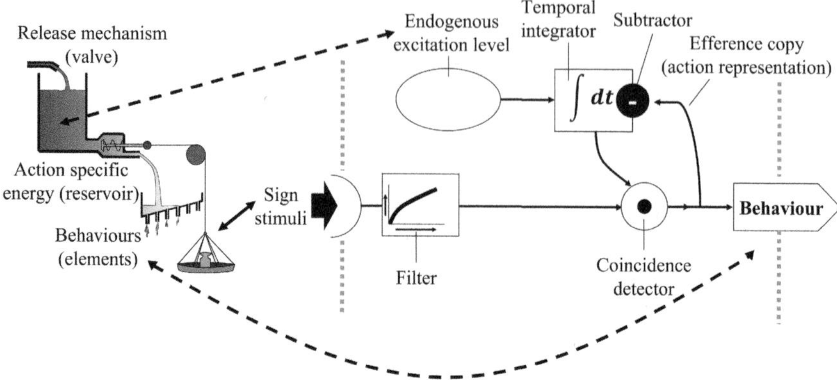

Figure 1.3 A comparison of two conceptually different models on the generation of Modular (Fixed) Action patterns. (a) 'Psychohydralic model' (Lorenz, 1981): action-specific energy gathers within a reservoir. The water is released by weights on a scale pan, which represent specific sign stimulus. The water purring out of the system (through a valve) represents the action pattern: the stronger the pulling effect of the weights, the more water is released, illustrating the richness of the action pattern. In some situations, when the water content builds up on the valve, it can also activate the system in the absence of any sign stimulus (pulling) (vacuum activity). (b) Biocybernetic model (Hassenstein, 1965): the effect of a complex pattern of stimuli (sign stimulus) combined with the strength of motivation (endogenous excitation level) in the coincidence detector. If the sum of both elements is sufficiently high, then the specific behaviour is triggered. A feedback loop reports back the changes caused by the action (efference copy signal), which temporarily reduces motivation (subtraction). After some time, the animal becomes hungry again, the state of supply deteriorates, and motivation rises once more. Note that the two models have some overlapping features (e.g., Action-specific energy = endogenous excitation level) but they rely on very different mechanistic interactions among the components of the model.

Sign stimulus as input: specific contextual features of the environment that release a specific type of action or actions in the absence of experience. Sign stimuli differ from typical stimuli in that their effect is behaviour specific, often they function only under specific circumstances (e.g., mating period), and they are multi-featured, involving various physical and/or chemical parameters.

Fixed (modal) action pattern as output: well-defined, morphologically distinct actions or action sequences, which are displayed upon sensing the relevant sign stimulus or stimuli.

However, there seems to be a tautology here, because the sign stimulus and the fixed action pattern determine each other mutually. In practise, both the input and the output are determined experimentally, that is, specific tests are conducted to find out the exact configuration of the effective sign stimulus, and observations on many individuals support the relative stable form and sequential pattern of the action.

The model was also criticised because it did not incorporate room for any learning. This shortcoming was soon corrected, because it was well-known for observers of animal behaviour that even those 'fixed' patterns of behaviour are subject to change after experience. Animals also learn about other features that are typically associated with the sign stimuli. Accordingly, we should abandon any reference to 'fixed action pattern' and the use of modal action pattern is preferred because it implies the possibility for flexible changes induced by learning (Goodenough et al., 2009).

Conceptually, sign stimuli should be seen also as 'attention grabbers' which direct the sensory system towards this input. This canalising effect allows the individual to learn about other significant aspects of the environment. For example, the moving body, neck, and head with eyes are strong releasers (sign stimuli) for following behaviour in one-day-old chicks. While they follow the hen, they also learn the individual features of the parent, so that they can discriminate her from others in the flock (Andrew, 1991). Nest building in bird relies on a series of complex modal action patterns, and accordingly nest building improves a lot over subsequent years revealing the effect of experience.

This simple model of behaviour provides a basic framework for mental organisation. The main hypothesis is that individuals are relatively predetermined to react to a range of specific stimuli in a prearranged way. Thus, from early on there is a crudely functional system in place, which shows some preference for stimuli and actions. Improvement and fine-tuning of both the sensory-perceptual and motor systems take place via learning mechanisms (which may be complemented by further maturational processes during development).

Interestingly, the controlling system of most presently functioning social robots is very close to this basic ethological model. These robots react to specific environmental cues (e.g., human faces) and have a collection of fixed actions to be performed in specific situations. Thus, these robots adjust their behaviour to the environment by selecting the most appropriate action in their repertoire based on inputs they receive, and no or little learning takes place 'on the fly'. Robots' capacities could be enhanced by allowing them to learn about the environmental inputs that correlate with the appearance of sign stimuli. For example, the recognition of a human face, as a sign stimulus, could automatically activate a learning mechanism that collects information about visual and acoustic features associated with that face.

1.4.2 The triadic division of mental systems

Most simple models of behaviour consist of three main modules following the computer metaphor: perception, central mechanisms (often used synonyms: representations, referential system, and concepts), and action. The central mechanisms are the most critical components because their existence differentiates living and non-living systems (see Box 1.1). Importantly, Shettleworth (2010) defines cognition as a set of interconnected mechanisms by which animals 'acquire, process, store and act on' input from the environment, which includes perception, learning and memory, decision making and action. According to this definition, behaviour systems are cognitive systems, and the difference is mainly in the computing power at each module. A behaviour system may or may not have central mechanisms. Cnidarians are an example of the latter case. In specific situations, behavioural systems may not involve central mechanisms in the execution of an action upon stimulation. Reflexive actions (e.g., patella reflex) provide an illustration of a direct input-output coupling. Note, that in contrast to reflexes, innate releasing mechanisms always involve central processing.

The modular organisation has the following main features (Box 1.1; see also Section 2.5). It is worth noting that although the distinct separation of these modules at the mental, anatomical, and physiological levels is not straightforward, the triadic differentiation offers conceptual benefits that aid in directing research efforts.

Perception: this module is responsible for relying, processing, and combining input from the sensory systems. Processing may involve very different, and on their own very complex mental calculations, which allows the individual to track, remember, and recognise different events/objects even if only partial information is available. Perception can be biased to recognise sign stimuli, and this may cause exaggerated versions of the sign stimuli (supernormal stimulus) elicit stronger responses. Features of sign stimuli may have an additive effect, that is, the reaction of individuals may be more intense (heterogenous summation).

Central mechanisms: by using the very general terminology of 'central mechanisms' one may avoid the problem of anthropomorphism, and there is also less need to be specific since these mental processes are of different natures. Some of these are responsible for representing the actual state of the individual: motivational, emotional,

Box 1.1 A simple overview of the mind's structure

General models of the mind have three main parts (see Shettleworth, 2010). (1) Perception deals with the preprocessing of sensory input encompassing active search for sources (attention and expectation) and relying on past experience (learning and memory). (2) Central mechanisms are involved in organising multiple and parallel processing of perceptual and motor input and integrating them with past and current states of the system. The outcome of central functioning are decisions regarding output actions. (3) Action mechanisms execute central decisions considering the physical state of the system and providing feedback on the alterations that arise as a result of the activity (see Figure 1.4).

Male three-spined sticklebacks (*Gasterosteus aculeatus*) develop a red patch on their belly during the breeding season, serving as a sign stimulus for other males with territories. When these males approach an already occupied territory, the owners respond by attacking to safeguard their property. Sign stimuli are identified based on inner 'templates' that form without prior experience (genetic disposition). Typically, these templates are simplified representations of actual stimuli. For instance, the 'red patch' signifies an intruder for the territory owner, prompting them to attack any smaller objects displaying a red colour during the breeding season. Learning plays an important role in such systems. During interactions, animals learn both about the stimuli and also about the way to interact with them.

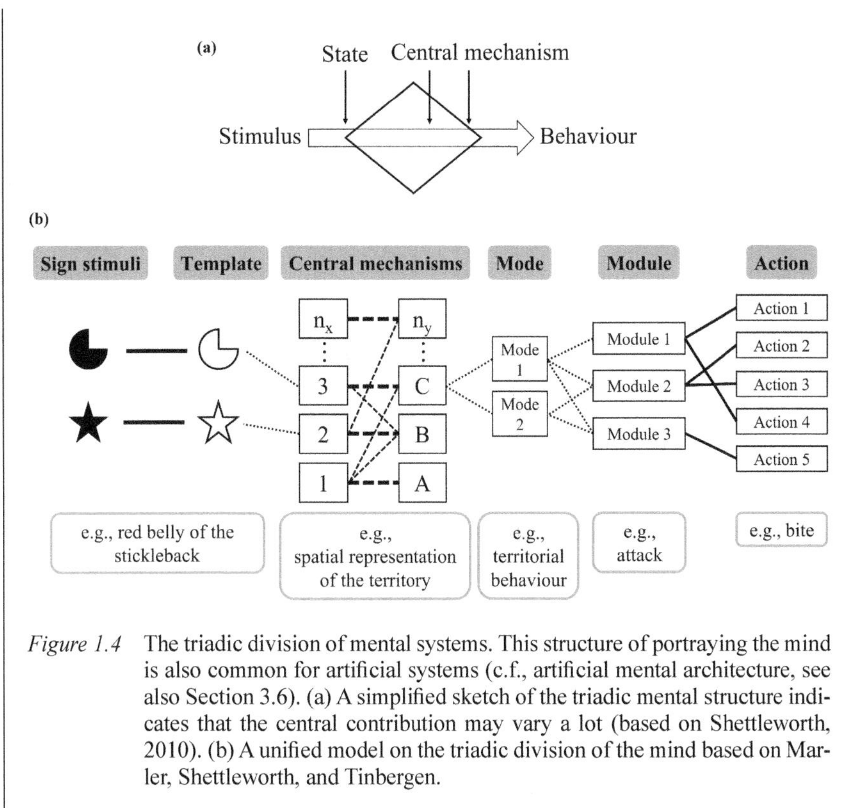

Figure 1.4 The triadic division of mental systems. This structure of portraying the mind is also common for artificial systems (c.f., artificial mental architecture, see also Section 3.6). (a) A simplified sketch of the triadic mental structure indicates that the central contribution may vary a lot (based on Shettleworth, 2010). (b) A unified model on the triadic division of the mind based on Marler, Shettleworth, and Tinbergen.

and cognitive. Others act as a kind of memory store making experience available upon demand. Finally, other processes make specific decisions by taking into account current and past states of the mind. Both the absolute and relative size of the nervous system devoted to central mechanisms determine the sophistication of these processes. Probably there is also a trade-off between fast reactions, involving limited number of central mechanisms, and most efficient reactions when there is a need to rely on information provided by different sources. Processes of the latter type may be more common in more complex minds. Ashby's law (1958) provides an important insight for modelling the mind. Accordingly, a system can function successfully only if it possesses at least as much complexity as the environment it is operating in. This also means that more complex systems are expected to respond to environmental challenges in a more flexible manner. Since the natural evolution tends to produce novel entities over time, these living beings have to operate in an increasingly complex environment which in turn may select for more complex mental operations.

Action mechanisms: the ability of animals to perform context-dependent skilled actions is often taken for granted, even though such actions depend on complex low-level feedback processes. These processes involve sensory-motor coordination as well as manoeuvres that ensure a smooth transition between actions. Behavioural modelling of action mechanism aims to describe a higher level that determines which actions are displayed at a time, and how actions follow each other. The behaviour of individuals can be modelled by a hierarchical control system, assuming that the emergence of actions is governed by a series of inhibitory control mechanisms that operate at different levels of the animal's mental (and neural) system. The main idea is that during the resting state all actions are inhibited, and the less inhibited (most activated) action has the highest chance of being executed. This ensures that the individual's behaviour is contextually appropriate and functional (Tinbergen, 1951).

Behavioural research also discovered specific action mechanisms that are responsible for generating and maintaining rhythmic activities. These central pattern generators (CPGs) play a fundamental role in controlling walking, swimming, or flying, but depending on the species, they may be involved also in generating the rhythms of feeding behaviour. Although CPGs are investigated at the neural level, it is likely that these processes affect mental control of the behaviour, for example, they may interfere with decision making or are involved in the generation of stereotypic behaviours (Wagenaar et al., 2010) (see also Box 1.2).

Behavioural models developed by ethologists serve only as orienting anchors for planning and building the minds of ethorobots (Arkin et al., 2003), despite the fact that the level of complexity shown by these models and the control system of present-day robots are comparable (see Section 3.6.1). Most of the computing in robots concerns very basic perceptual and motor skills that are below the level of the ethological modelling. In ethology, it is implicitly assumed that the animal has basic capabilities to sense and act functionally, but obviously this is not the case in robots.

Box 1.2 Central pattern generator

The CPG is a network of neurons that produces the basic locomotor rhythm, independently from any sensory input or feedback. Their tightly coupled patterns of neural activity are responsible for driving rhythmic and stereotypical motor actions such as walking, swimming, breathing, or chewing. Although they possess the ability to function independently of inputs from higher brain regions, their operation still relies on modulatory inputs, and their outputs remain adaptable. To be categorised as a CPG, systems have to consist of at least two sequentially interacting processes, causing each process to cyclically increase and decrease, and a consequence CPG consistently returns to its initial state. CPGs have been identified in invertebrates and vertebrates, including humans. CPGs allow rapid reaction and also smooth functioning; it may be useful to be implemented also in ethorobots (see Figure 1.5).

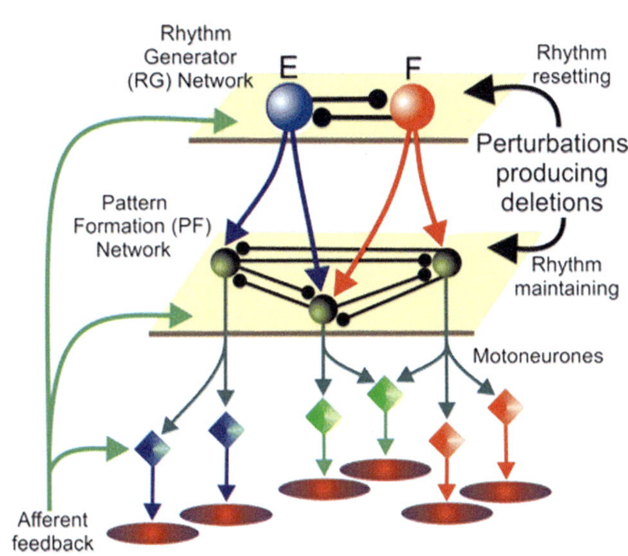

Figure 1.5 The rhythm generator (RG) network generates the rhythmic signals which are mediated by the pattern formation (PF) to the motoneurons. The interneuron populations of the PF are connected to each other with inhibitory connections. They synchronise the activities of the motoneurons and time the phase transitions without shifting the phase of resetting the rhythm. The motoneurons integrate the output of the PF network and the muscle sensory feedback; the output of the muscles is determined by the extensor and flexor motoneurons. (Figure from Rybak et al., 2006).

1.5 Cognitive ethology: analysis of problem-solving behaviour

In the early years of ethology, the focus was primarily on the observable behaviours of animals, and there was little interest in studying the underlying cognitive processes. Some ethologists even argued against studying cognitive processes, citing concerns that existing approaches were either too reductionist, relying solely on laboratory observations with a narrow focus on learning, or too anecdotal, relying too heavily on anthropomorphic interpretations (Kamil, 1998). However, some researchers successfully combined the study of animal behaviour under both field and laboratory conditions with the search for underlying mental processes, particularly in the study of navigation in animals. The debate surrounding whether animals possess a 'cognitive map' (see Section 1.6.1) has been particularly fruitful in this regard. Shettleworth (2010) provided a crucial synthesis of the ethological and cognitive approaches, which laid the groundwork for the emerging field of cognitive ethology. Although there are numerous labels for this interdisciplinary field, such as 'comparative cognition', 'cognitive ecology', or 'cognitive biology', as Kamil (1998) argued, the original term proposed by Griffin (1981) remains the most fitting one.

However, when an ethologist studies nest building behaviour of blackbirds then he is actually studying the problem of building a nest given the bird's behavioural skills and specific environmental conditions for carrying out the task. Accordingly, the study is on the behaviour per se but on the problem of the bird of how to build a nest. Thus, cognitive ethology is concerned with providing mental explanations in terms of perception, central mechanisms, and action for the problem-solving behaviour of animals from a biological perspective. For the present purpose, we define problems as regular/specific states/events in the environment of a species that generate alternative behavioural decisions at specific frequencies that differ in fitness consequences. Problem-solving behaviour reflects changes (decisions) when an animal responds (behaves/acts) to regular or specific states of its environment in a way to change it or its state to its advantage. The focus on problem-solving behaviour is important for several reasons (Abdai & Miklósi, 2022):

1 Behaviour/performance can be measured directly by relatively objective tools (Section 1.3) and can provide the basis for comparison of species;
2 The determination of the problem animal face in nature should precede the hypothesis about the mental processes (which is often done inversely);
3 Most debates focus on the cognitive interpretations rather than the validity of the problem-solving behaviour from an ethological and ecological perspective;
4 Multi-perspective investigations of the problem-solving behaviour may provide a rich source of behavioural input that could facilitate more objective hypotheses on mental processes.

Scientists often make important simplifications when talking about the abilities of animals. The implied common-sense concepts, however, do not provide any tangible scientific explanations. For example, saying that 'the dog knows the way home' (instead of saying, 'the dog is able to find a way home', that is, is able to solve a problem) does not implicate any specific cognitive mechanism despite the reference to a mental state ('knows'). This rather anthropomorphic view goes against the recommended ethological perspective. Nevertheless, it is still very prevalent to attribute complex anthropomorphic mental processes and concepts to the animal mind without knowing too much about their functioning. We also know relatively little about how these concepts come to existence in our mind (Shettleworth, 2010).

It may be useful to separate solving problems of physical nature from problems that have to be tackled in the social world. These notions may rely on the assumption that there are dedicated mental mechanisms (and respectively different neural structures) that deal with these different types of problems. This division is also practical from a research perspective, but it may not reflect the organisation of the mind, that is, mental structures involved in navigation in the physical world may overlap with those that are utilised in navigation in the social network (Peer et al., 2021; Schafer & Schiller, 2018).

An important strength of cognitive ethology is to widen the horizon of research and avoid the anthropocentric fallacy (Kamil, 1998). Comparing the problem-solving abilities of species on a wider spectrum can offer insights about possible mechanisms shared by them, or some species-specific solutions and how these depend on the evolutionary or ecological constraints. The cognitive ethological approach also ensures that these investigations are carried out under behaviourally and ecologically valid conditions, avoiding the generation of artefacts.

Wilson and Golonka (2013) suggested that the behavioural analysis of problem-solving behaviour should aim to answer four questions.

1. What is the task at hand? It is crucial to evaluate the complexity of the problem, considering factors such as its hierarchical structure, the various approaches to overcoming it, and whether there are particular solutions favoured by the animals.
2. What resources are available for accomplishing the task? This examination should encompass the potential sensory inputs from both the physical and social environments, the physical and motor components of the executed behaviour. Additionally, hypotheses regarding cognitive abilities (incl. perception) may be proposed.
3. How can these resources be orchestrated to solve the task? A comprehensive list of alternative methods for task resolution, involving different facets of sensing, acting, and mental skills, should be compiled. This should be correlated with observed behaviours during task-solving.
4. Does the organism assemble and employ these resources? Definitive answers should be derived from specific experiments designed to stage the task in various ways, including novel and unexpected scenarios based on the diverse hypotheses put forth.

Concentrating on problem-solving behaviour also moves researchers away from the old dichotomy of trying to distinguish between 'simple' and 'complex' mental mechanisms. This long ongoing (and not very fruitful) debate was also often formulated as a yes or no choice between 'associationism' and 'cognitivism' (e.g., Barrett, 2012; Shettleworth, 2010). It is likely that mental mechanisms utilise a wide range and combination of specific mechanisms that may reflect some simplicity or complexity.

Ethorobotics could capitalise from increased collaboration with cognitive ethology. Widening the range of well-understood animal systems could help to find the best solutions for building a bio-inspired agent with some expected function. In the long run, the performance of robots can also help in enciphering possible mental mechanisms in animals. Robots possessing and merging different kinds of bio-inspired mental architectures could show the advantages of one solution over the other, also in relation to different other (e.g., environmental) parameters.

1.6 Physical problem solving

Physical problem-solving behaviour in animals serves the purpose of managing their interaction with the non-living environment. This involves navigating various substrates (such as the ground, water, or air) and physically interacting with objects using their whole body or different body parts. Problem-solving abilities manifest in various degrees across diverse animal species, albeit with varying levels of complexity. Here, we only introduce some of the most important examples that may concern ethorobots. It should be emphasised that without having robust functioning in solving basic problems, for example, to navigate and avoid obstacles, it makes little sense to tackle complex problems offered by the social environment for a fully functional ethorobot. The history of social robotics seems often to show opposite trends: robots aiming to show complex social skill lack rudimentary abilities of object avoidance.

Human cultural and technological evolution has yielded an array of navigational methods that can be juxtaposed against those found in the animal kingdom. On the contrary, contemporary robots are notably less adept at object manipulation when compared to certain avian and mammalian counterparts. By advancing our cognitive and neural comprehension of object manipulation, we can potentially derive bio-inspired solutions to develop physically adept ethorobots.

1.6.1 Spatial problem solving

One of the remarkable abilities exhibited by some animals is their capacity to navigate across large stretches of land or water. Such skills support searching for specific locations, like food sources, patrolling a restricted territory to monitor potential threats, or finding home from large distance. Animals exhibit egocentric orientation when they rely solely on internal (bodily) information, while for allocentric orientation they utilise external sources of environmental information.

1.6.1.1 Egocentric orientation – path integration

Egocentric orientation is a common characteristic observed in many animal species. However, it is particularly useful for those living in environments lacking conspicuous environmental cues for orientation, such as deserts. Comparative studies conducted on golden hamsters (*Mesocricetus auratus*), dogs (*Canis familiaris*), and humans in a simple homing situation demonstrated similar performance levels (with similar error rate) across all three species, suggesting the presence of a comparable mental mechanism (Etienne et al., 1996). During the outward journey, animals have to mimic somehow the mathematical procedure that integrates measures on the distance travelled and changes in directions when calculating the homeward path (Figure 1.6). The capacity for relying on this mechanism can be very useful in the absence of other solutions, and when exploring new terrains. However, this mental mechanism is prone to error, and there is a need for regular recalibration based on external cues.

1.6.1.2 Allocentric orientation – beacons

Environmental cues that are in close proximity to, or at the goal, are referred to as beacons. Typically, beacons are visual cues, but certain odours or sounds can also produce a similar effect. These cues trigger an approach response in animals that is based on their perceived location. For instance, when an animal relies on visual orientation, it moves forward to increase the projected image of the beacon on its retina. Alternatively, an animal can follow an odour trail by detecting systematic changes or differences in the concentration. For example, honey bees (*Apis mellifera*) rely on beacons to locate their hive, and Atlantic salmons (*Salmo salar*) sense odour trails in the water to locate their spawning sites (Goodenough et al., 2009).

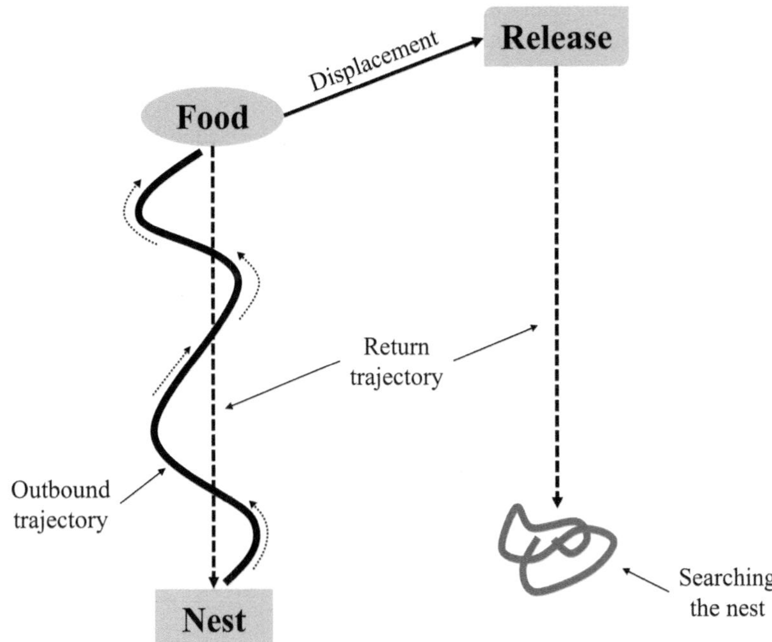

Figure 1.6 Egocentric orientation in ants. Desert ants embark on foraging journeys by exploring their surroundings in search of food (based on Srinivasan, 2015). (a) Once they locate a food item, they adopt a direct homeward path. This path's orientation and length are determined through an approximate calculation involving the average walking angle and the aggregated distances covered in the same direction while travelling from the nest to the food source. This computation is then employed in reverse to plan the return route. (b) Support for this mechanism arises from experiments involving the displacement of ants just before they begin their journey back home. The displaced ant follows a path parallel to the original trajectory, missing the nest by the distance of the initial displacement.

1.6.1.3 Allocentric orientation – landmarks

Often, there are no useful specific cues at or close to the target location, so the animal has to look for alternatives. Landmarks are distal environmental cues, and the animal has to make some kind of extrapolations to locate the goal. Tinbergen (1951) arranged pinecones in a circle around a nest of a digger wasp (*Philanthus triangulum*). When the wasp approached the pinecones, it flew straight to the centre and entered the nest. The wasp displayed the same behaviour when the cone circle was relocated to a new location without a nest in the middle. Tinbergen concluded that the wasps learned to recognise the spatial relationship between the pinecones and the entrance. Similarly, Norway rats (*Rattus norvegicus*) exhibit comparable behaviour when searching for food in an arena filled with water.

There are various considerations and hypotheses based on experiments regarding the mental mechanisms that aid animals in locating targets using landmarks. It is crucial to note that using a particular mental mechanism may depend on the species, context, and individual experience. Therefore, any observed differences may be due to ecological biases or experiential effects. Moreover, it is likely that many species rely on a combination of navigation mechanisms and deploy them based on the situation they face. However, researchers have identified specific mechanisms that utilise different methods of computation (Shettleworth, 2010). These models offer several significant advantages:

1 These mechanisms rely on mathematical calculations that make precise predictions about the navigation patterns exhibited by the animal;
2 They can be experimentally tested to determine the accuracy of these solutions under natural conditions;
3 Transfer experiments can also be conducted, in which the subject gains experience in one condition and is then tested in a series of systematically altered situations. This helps to determine the representational aspects of the actual mental model. To exclude the effect of other mechanisms, the local cues from the goals or targets, the goals should not be perceivable during transfer experiments;
4 These models can also be implemented in artificial agents, which provide independent validation of the concept and support their navigation.

1.6.1.4 Template matching

Cartwright and Collett (1983) proposed that bees, and possibly other species, navigate by using 'snapshots' of their environment. They suggested that bees could create a series of 2-D mental images of their surroundings, which could be compared to the actual image perceived during their flight. By matching these images and identifying differences, the bees could determine the direction of their flight until they reached their destination. For example, bees were trained to search for food at a specific location using one landmark. After this training, the landmark was reduced by 50%, and as a consequence, the bees were searching for food 50% closer to the landmark. It should be noted that these snapshots are likely different from the mental images humans create when observing the same scenario due to differences in the perceptual system. However, storing these snapshots may require a significant amount of memory, and researchers believe that bees perform specific transformations to compress these images (Meyer et al., 2020).

1.6.1.5 Vector sum model

Cheng (1989) tested pigeons (*Columba livia*) in a simple arena with a few conspicuous cues to find experimental support for spatial navigation based on summing of vectors. An animal may realise that a specific location of food is at certain distance from, for example, two conspicuous landmarks. The measure of this distance can be used on the next occasion when the animal approaches the same location form a specific distance. According to the model, the animal has to carry out a summation of two vectors: its own distance from the landmark and the distance of the food location from the landmark (based on its memory), and the resultant determines the actual vector which has to be followed to the goal (see Figure 1.7a, b). This calculation should be carried out a few times until the food location is reached.

In the experiment, pigeons showed evidence that the vector sum model provides an accurate prediction of their behaviour. After some experience in a specific scenario, in a transfer test, one of the two landmarks was shifted along the wall to one side for a test. In this case, the summation of the two vectors points to a location that is also shifted to the side parallel with the wall. As expected, the pigeons looked for the food along this virtual line, showing a shift to the same side. The observations have also shown that pigeons tend to exhibit a bias towards a nearer landmark, and they never modified their search in the orthogonal direction to the landmark shift (Cheng, 1989).

This vector summation model is effective at short distances from the target, as the estimation of distance becomes less accurate at increasing range. Pigeons may use their precise depth vision to calculate distance vectors, and they may also learn which landmarks are more reliable as a reference point. This is suggested by the search behaviour of the pigeons, as they gave more weight to the nearby wall of the arena than expected (Cheng, 1989).

1.6.1.6 Multiple bearings model

Complex scenarios offer a variety of landmarks for navigation. Some kind of landmarks may be better than others, but in parallel, the more landmarks are utilised, the more accurate could be the estimation about the goal. The unexpectedly skilful localisation in Clark's nutcrackers (*Nucifraga columbiana*) prompted the suggestion by Kamil and Cheng (2001) that using their vision these birds may rely on multiple bearings for finding their caches. Their hypothesis makes the following assumptions based also on some experimental evidence:

1 Nutcrackers are able to calculate the target locations by estimating the vectors between goals and multiple landmarks;
2 This encoding relies preferentially on the directional information;
3 Involvement of multiple landmarks increases the precision of searching.

Thus, nutcrackers need to encode the bearings between the target location and a landmark. In almost all cases, coding another similar bearing (Figure 1.7c) provides an intersection close to the target, and a third bearing closes a triangle within which the target is located. Any further bearings help to make the search effort more focused. According to Kamil and Jones (2000), error rate of nutcrackers is lowered in the presence of more landmarks. This mechanism can be also advantageous if there are many potential landmarks, and the subject may forget some bearings by the next visit, or some landmarks disappear. Both situations have some relevance because nutcrackers often re-visit their caches after many weeks or months.

These and other models of navigation are very interesting both from a theoretical and practical point of view. Despite a relatively large amount of experimental

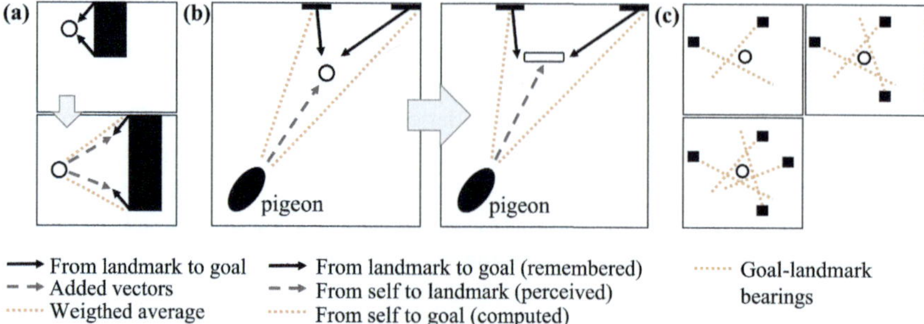

Figure 1.7 Using landmarks to determine a specific location based on the (a) and (b) vector sum model (based on Cheng, 1989; Shettleworth, 2010), and (c) multiple bearings model (based on Kamil & Cheng, 2001). Circles indicate the goal and filled rectangles the landmarks in all panels (see text for details).

data involving many species, there are still a lot of open questions. One assumption is that a wide range of animals share many of these skills and deploy them in a situation-specific way. Alternatively, evolutionary history and ecological challenges may have provided a selective environment for one to excel in one or the other form of navigation. Consider the difference between pigeons and nutcrackers. The former is searching for small, scattered food while for the latter it is extremely important to hide their caches because no other food source is available during winter. One may suppose that under these circumstances, the pigeons may survive at a larger error rate than nutcrackers. Although there is some experimental evidence in favour of this assumption, differences in the methods applied in these experiments prevent making an unambiguous statement (Shettleworth, 2010).

1.6.1.7 Cognitive maps

Researchers studying the mechanisms of navigation were often fascinated by the possibility that animals (and humans) may have a global representation of their surroundings. This hypothesis has been always treated with a lot of caution, and results in favour of a cognitive map met scepticism. Also, ethologists could not come up with a good and testable model (Bennett, 1996). Some maintained that the cognitive map is basically analogous to any kind of mental representation (Gallistel, 1990), whereas others argued that it is an allometric representation of environmental features (landmarks) (Thinus-Blanc, 1988). Cognitive maps were often referred to as global representations of the area where the animal lives, and local changes in the map do not affect other parts of the representation. Thus, it is also assumed that a cognitive map functions much like a visualised geographic map.

Most authors agree that the ability to make a novel shortcut on the terrain could be a proof for utilisation of a cognitive map. However, the problem is actually more complex than previously thought because it has to be ensured that the animal has not explored the area involving the shortcut, and that there are no cues (landmarks) which may provide a common reference point for navigation in the area involved.

In bees, Dyer (1996) conducted a very carefully designed experiment to see whether they utilise a cognitive map (Figure 1.8). He looked for a particular terrain where there was an asymmetry with respect to the visual perspective from the feeding sites. The feeding site A was at an open space and could be seen from multiple positions, including the route that the bees flew from the hive to the feeding site B. In contrast, the feeding site B was behind a group of trees, so it was invisible when the bees flew from their hive to the other feeding site. Initially, bees were trained to fly solely on the routes from the hive either to feeding site A or B on different times of the day. During the test, the bees were caught shortly after heading out on one of the routes, and they were transferred to the opposite feeding location where no food was available at that time. So generally, the bees had two options: fly home or take a shortcut to the other feeding location, which was their original goal. Bees transferred to the feeding site A typically flew towards the other location, taking a new path that they had not previously travelled. In contrast, most bees taken to feeding site B chose to fly home instead.

According to Dyer (1996), this asymmetry in the behaviour of the bees arose because the bees could see feeding site A when flying towards the location B, and thus they could integrate this experience with their actual view when they found themselves (after the transport) at feeding site A. This was, however, not the case in the opposite situation. As the feeding site B remained hidden during their travel to feeding site A, there was no possibility to connect directly the current view (which was rather restricted) with the experience collected during the outward flight. This means that during the training, when visiting the two feeding locations from the hive, the bees did not form an integrated map containing hypothetical bearings between feeding sites A and B. Dyer (1996) concluded that bees do not construct a cognitive map (in the classical sense), but their navigation is based on many route maps (learning about landmarks along the route) that 'cover' the respective area (see also Bennett, 1996).

If animals possess multiple mechanisms for navigation, it is possible that they may not need to rely on a cognitive map. In environments with familiar, large, and conspicuous landmarks, simpler egocentric or allometric orientation mechanisms may take priority, especially in small laboratory settings. Rats have a home range that can span from 20 to 100 m in diameter, which is at least one order of magnitude larger than their typical testing environment. As a result, they are confined to small cages and have no opportunity to practice large-scale mapping.

During navigation, it appears that animals activate multiple mechanisms simultaneously, gathering both egocentric and allocentric information. Cognitive maps are advantageous when alternative routes represent long detours. However, it is unclear if animals such as wolves (*Canis lupus*), who need to travel tens of kilometres, may take shortcuts through unfamiliar areas to reach their destination.

1.6.2 Time and rhythms

The rhythmic changes in the environment, most of which are directly or indirectly linked to the rotation of the Earth, have a profound impact on animal behaviour. The cycles of light and dark periods, as well as changes in tides, can either facilitate or hinder certain activities. Animals have evolved specific mechanisms not only to keep

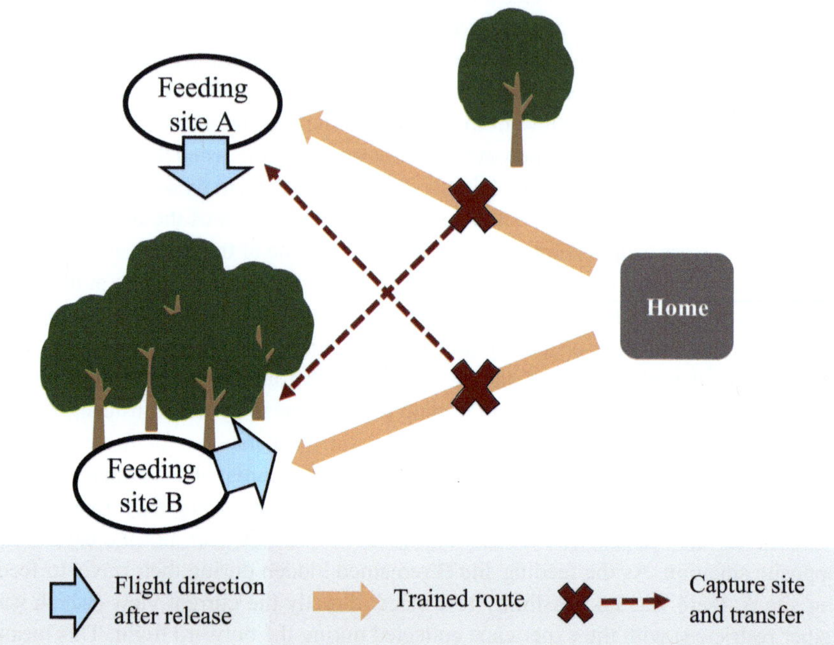

Figure 1.8 Depiction of the study about cognitive maps in bees by Dyer (1996). A and B are the two feeding sites, blue arrows indicate the trained route, red Xs and dashed arrows indicate the capture site and transfer; and orange short arrows indicate the flight direction of bees after release (see text for details) (based on Dyer, 1996).

track of time but also to predict specific events that are related to these rhythmic changes (Häfker & Tessmar-Raible, 2020). These biological clocks have the ability to measure absolute and relative time, similar to human-made clocks.

When animals are kept in constant light conditions, they display a rhythmic activity level that roughly corresponds to a 24-hour cycle. Although, as time passes, they slowly deviate from the typical day-night cycle, it suggests the existence of a clock-like mechanism. Such a deviation is significant because maintaining a precise 24-hour cycle could be difficult for a biological system. However, it also means that these clocks can be calibrated by environmental cues. Light is a common cue, known as a 'zeitgeber', that can synchronise the animal's activity with its environment, a process known as entrainment. In fact, animals kept in the dark can quickly synchronise their activity to a 24-hour cycle with even a brief exposure to light (Shettleworth, 2010).

Biological clock-like mechanisms provide a major advantage in that they enable the animals to organise their activities and prepare for future events. For example, bees are able to learn the dynamic pattern of when flowers offer nectar at different times of the day and optimise their foraging behaviour accordingly. Time-based optimisation of behaviour is observed in a wide range of species. Additionally, animals that experience predictable exposure to food often exhibit increased activity before the actual feeding time, a phenomenon known as anticipatory behaviour. In natural circumstances, this behaviour is crucial for animals to arrive on time at the expected location for feeding.

Carrying out such experiments under controlled conditions, Bolles and Moot (1973) also showed that learning about the time of the day and not hunger was controlling the behaviour. After exposing rats to two meals per day, they omitted one or the other meal on specific days. The anticipatory activity increased before each meal, as expected, but it also decreased both after a realised feeding session and when they did not get any food (and stayed hungry).

There is also evidence that some animals are able to make time-based decisions about events that happened in the past. Scrub jays (*Aphelocoma coerulescens*) prefer moth larvae over peanuts if given a choice. Moth larvae perish much faster after a few hours and become inedible for the birds. Scrub jays were offered both types of food for caching in a large arena, and then they were taken back to their homes. If the birds were allowed to search for the cached food within a few hours, then they preferentially looked for the moth larvae. If much more time (e.g., days) had passed, they chose the less preferred peanuts. Thus, scrub jays can take track of the time passed since the caching (Clayton & Dickinson, 1998).

1.6.3 Mental architectures for objects

In the case of navigation, the challenges for non-human animals and humans are much the same, despite species may differ regarding the actual environment in which they are navigating (ground, water, or air) and the sensory and mental skills involved. When it comes to interactions with objects, the situation is very different. Humans excel in using objects in myriad of ways, including simple actions like hammering, to more complex ones like knitting or playing a violin. Humans possess a range of sensors (for optic and haptic input) and manipulators, such as hands and fingers, that allow them to gather a wide variety of information about objects. Objects also play an important role in social interactions among humans, and knowledge about them can be acquired through the experiences of others. However, outside of a few specific species, such as monkeys and elephants (e.g., *Loxodonta africana*), objects play a relatively limited role in the lives of animals. As a result, our understanding of how animals handle physical problems involving objects is limited, and existing research is often influenced by an anthropocentric perspective.

Behavioural observations have revealed numerous examples of tool use in animals. However, it remains unclear what kind of mental representations are employed in this behaviour. The following short section is based on (folk psychological) concepts about objects that appear as logical when dealing with problem-solving behaviour related to object use. The central question is whether animals utilise general mental representations about objects or rely on more specific mechanisms.

1.6.3.1 Stationary features

All objects share several fundamental attributes, such as their enduring presence over time (permanence), consistent dimensions (constant size), and inherent solidity in the absence of external forces. From these three features, object permanence received the most interest in comparative studies (Doré & Dumas, 1987), partly because of the heritage of Jean Piaget (1936/1963) who was the first to carry out systematic developmental research in children on this topic.

Thus, object permanence is both a feature of an object, and also a mental skill that ensures that the animal is able to mentally manipulate representations of an object without seeing it. In the typical experimental arrangement, the subjects are trained to search for objects (often food) at various locations. Next, they are exposed to different scenarios to see whether they are able to successfully recover the hidden targets. In the standard condition, there are three hiding locations at the same distance from the subject (e.g., Doré et al., 1996) (Figure 1.9). The target object is moved from one side to the other passing behind the hiding locations. Depending on the condition, the target object remains behind a hiding location or re-emerges at the other end and continues its move towards the next location. Just as if a small prey (mouse) would run behind three wide tree trunks and decide at each location whether to stay or move to the next. Would a potential predator (cat) be able to find it?

This task may appear to be straightforward for a predator, as cats and dogs have been able to successfully locate the target under these conditions. However, a more complex version of the task involves placing the target inside an opaque box that is then moved in a similar manner (invisible displacement). In this version, the box is opened after emerging from behind the last occluder to determine whether the target is still inside or was left behind at the previous location. Until now, dogs and cats

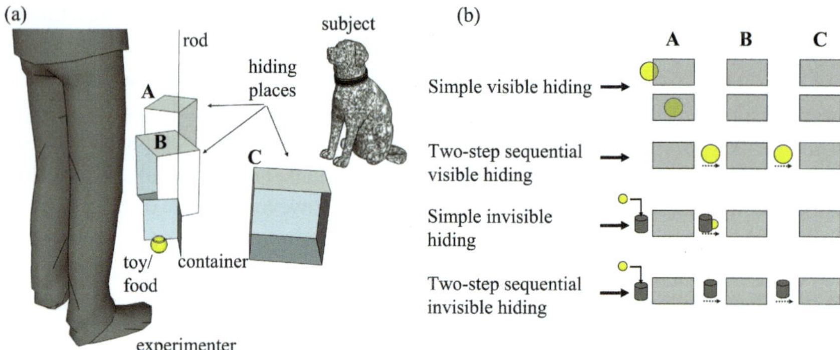

Figure 1.9 Illustration of testing object permanence. (a) Hiding an object in one of three hiding places. The object (food or ball) is moved by means of a specific carrier by which the experimenter can reveal whether the object is still there or was left behind the last hiding place. (b) It has been assumed that the mental representation of the object depends on its visibility during the hiding. The different tasks represent increased challenge from top to bottom. Human infants are able to master more complex tasks as they get older. Based on their ecological status animal species may also differ in their skills to represent hidden objects. During visible hidings the subject can directly follow the object visually, but in the case of invisible hiding, the object is placed inside a container first and subjects can only follow the movements of the container. Black circles represent hidden items; grey squares represent opaque screens as hiding locations.

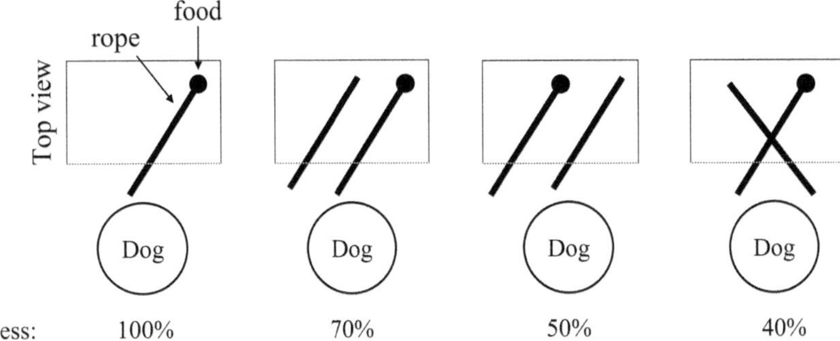

Figure 1.10 Dogs' success in retrieving food by the means of a rope in different arrangements (based on Osthaus et al., 2005). Dogs typically fail when from their perspective the bait (black circle) is perceived to be closer to the end of the rope near their paw.

failed in this rather unnatural task, while apes (Call, 2001) and corvids (Ujfalussy et al., 2013) showed some evidence of being able to deal with invisible displacements.

Another similar object-related concept is connectedness. The problem is that connection between objects can take many physical forms depending on the context. Consider a simple case with two strings, each of which is connected to a piece of food at the end. The food can be obtained by pulling the other end of the string (Figure 1.10). When the two strings were put close to each other in a parallel arrangement, and only one of them had the bait attached to its end, the tested animals (e.g., dogs, wolves; Osthaus et al., 2005) performed well. However, the same individuals made many mistakes when the strings crossed each other. Although the connection between one rope and the food was clearly visible, often the subject pulled the rope whose end was closer to the food.

Interestingly, ravens can solve such problems (Jacobs & Osvath, 2022), suggesting that the skill may depend on the ecology of the species. Although there is a large variation in tool use in corvids, members of some species have displayed complex interactions with objects. These abilities may be rooted in their foraging behaviour, and as a result, they may possess a more advanced representation of connectedness.

One specific case with the New Caledonian crows (*Corvus moneduloides*) also shows that even relatively restricted manipulative skills can support complex use of tools. These crows make different types of tools from twigs or stiff plant leaves to pick out insects from below the tree bark or small holes (Hunt, 1996). Although these animals show various forms of tool use in nature, they seem to solve even more complex tasks under laboratory conditions, often on the first attempt or after few trials (Bluff et al., 2007). New Caledonian crows were observed to solve the following tool-related problems with reference to the physical aspects of the tool (Hunt, 2021):

1 Bending an aluminium stick to pull out a small container from a tube;
2 Connecting two sticks to get to a target;
3 Adjusting stick length to the distance to the target; and
4 Choosing the stick of the right length for reaching the target.

This species is regarded as one of the most skilful tool-user among birds, and they may also rival the performance of apes (McGrew, 2013). Researchers are debating what kind of mental representations control this ability. Such sophisticated tool use may require complex planning in advance, but it may also rely on a set of optimised rules deployed flexibly in a specific scenario (Hunt, 2021).

1.6.3.2 Estimation of amounts and quantities

There has been a relatively large interest in researchers to study how animals deal with the problem of quantities, or as it is often referred to, whether non-human beings possess some forms of numerical competence. As far as it can be judged, there are typically rare occasions in nature when animals have to base their decisions on such skills. However, from an anthropocentric view, the study of numerical competence may shed some light on the evolutionary antecedents of the corresponding human ability (Shettleworth, 2010). Under controlled laboratory conditions, individuals of many species display remarkable numerical skills. For example, Alex, a grey parrot (*Psittacus erithacus*), was able to count objects of different colour, shape, and size up to nine items, perform addition of small numbers, and by learning a vocal label for zero and numbers smaller than nine, he also demonstrated some level of symbolic number representation (Pepperberg, 2006).

Studies conducted in more naturalistic settings have shown that many species are capable of making food choices based on quantity. They consistently exhibit a preference for larger quantities, such as two food items over one, but this preference decreases as the quantity increases, as predicted by Weber's law (as explained in Box 1.3) (e.g., Jordan & Brannon, 2006). Interestingly, this preference shift occurs even at relatively low quantities, suggesting that animals tend to rely on the overall amount (surface area covered) rather than counting the number of items. This strategy may make sense, as the difference between choosing five or six food items may not be critical for survival. Therefore, it is intriguing to consider why many animals are able to exceed their natural abilities when presented with specific learning experiences.

Box 1.3 Weber's law

Weber's law states that perception of a change is not constant, but rather the increase or decrease of a stimulus (e.g., intensity of light) is proportional to the pre-existing stimuli. The higher the intensity of the perceived stimulus, the larger the change needed for it to be perceivable. The Weber law applies well to the success of subjects discriminating quantities (see Figure 1.11).

It is relatively easy to discriminate two quantities when their ratio is 1:2, 2:4, or 3:6, but the discrimination becomes more difficult if the ratio is changed to 2:3, 4:5, or 5:6. Based on the discovery of Weber, his student Fechner described the phenomenon:

$$k = \Delta I / I,$$

where I is the original intensity of the stimulus, ΔI is the minimal addition in the intensity that makes the difference perceivable (also called just-noticeable difference, JND), and k is (the Weber) constant. It indicates the degree of sensitivity or discrimination ability of a sensory system. A smaller Weber constant implies greater sensitivity, as it suggests that small changes in stimulus intensity are more easily detectable. Different sensory modalities (e.g., vision, hearing) and individuals may have different Weber constants, reflecting variations in sensitivity. For example, if the Weber constant for a particular sensory modality is 0.05, it means that a change in stimulus intensity needs to be at least 5% of the original intensity for a person to notice the difference reliably. Importantly, Weber's law plays an important role in signal perception in communicative interactions as well.

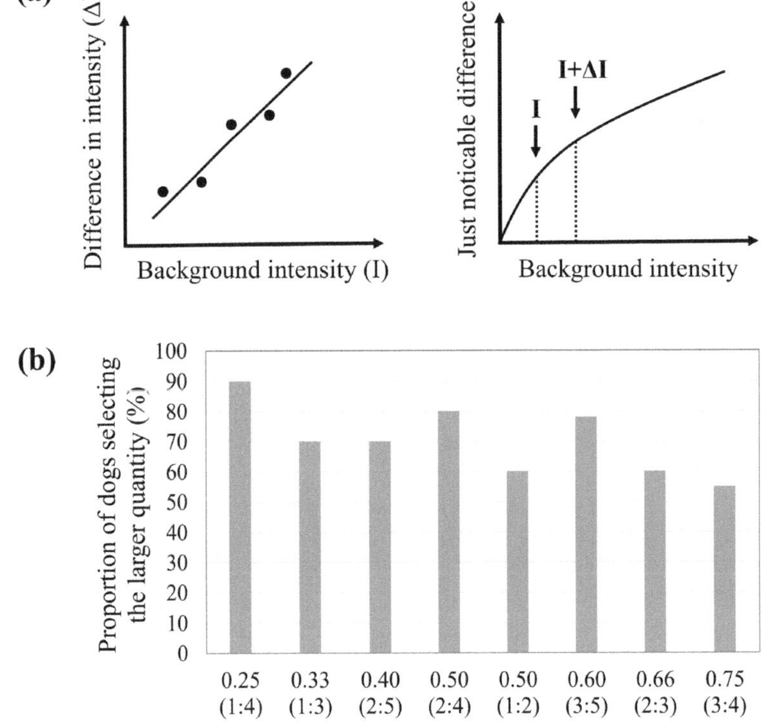

Figure 1.11 (a) Visual depiction of Weber's law. (b) Quantity judgement in dogs may follow Weber's law (Ward & Smuts, 2007). The dogs were offered a choice between two sets of food items, e.g., 1 vs 2 (ratio: 0.5) or 2 vs 3 (ratio: 0.66). Dogs show significantly higher performance at lower quantities, especially when the difference between the two quantities is larger.

1.6.3.3 Dynamic features

In certain situations, objects may also move along a path, but in contrast to animate entities that can move by themselves, the motion of inanimate objects requires an external force (e.g., collision with another moving object). Due to this, based on the observation of the object's motion, its future trajectory can be determined. For example, Völter et al. (2020) tracked the eye movement of dogs when they observed on a screen two humans throwing a Frisbee. Dogs did not simply follow the motion of the Frisbee, but they looked at the catcher before it arrived indicating that they anticipated the future position of the object. In a second experiment, the video was stopped while the Frisbee was mid-air and then it was rewound. Dogs here also followed the movement of the object, and they looked at the catcher more rapidly when its motion was predictable (video moving forward), compared to when it was unpredictable (rewound video). Following the movement of objects in space can be advantageous for both to avoid or to prepare for collisions, and to find (currently non-visible) resources (e.g., watching a fruit falling down to the bushes). Many predators have to catch living entities (prey), even though their behaviour tends to be less foreseeable.

Typical tracking behaviour relies on olfactory cues but in some cases complex mental efforts are needed when only visual information is available. For example, the banded archer fish (*Toxotes jaculatrix*) can bring down flying insects over substantial distances by firing an aimed shot of water at them. Notably, since the fish cannot adjust their shot after it is unleashed, they must pre-emptively factor in the spatial trajectory of both the intended target and their expelled 'projectile' to successfully intercept moving prey (Schuster et al., 2006).

Objects exposed to gravity also follow a specific trajectory down to the ground. American crows (*Corvus brachyrynchos*) often fly above concrete surfaces (e.g., roads) with a nut in their beak and release it from 8 to 10 m high (Cristol & Switzer, 1999). These manoeuvres offer a good way to crack open hard-shelled nuts, and birds tend to fly higher if the shell is thicker. This behaviour may arise through trial-and-error or by observing others. However, this skill also suggests that some animals may develop a more general representation of 'gravity', where objects follow a typical trajectory towards the ground in the absence of support. Human infants show gravity bias, that is, they search for a falling object along its vertical trajectory, even if it enters an opaque tube that diverts its route diagonally. Eventually, children learn that such physical constrains defy gravity, but this initial bias is interpreted as a mental representation of gravity (Tecwyn & Buchsbaum, 2018). Despite many efforts, no such gravity bias was found in multiple species, including chimpanzees (*Pan troglodytes*), marmosets (*Callithrix jacchus*), and dogs. However, Bliss et al. (2023) found that newly hatched chicks already prefer upward-moving stimuli compared to downward-moving ones. In adult humans, moving against gravity has been linked to animacy perception (internal energy source is needed for such movement), indicating that chicks may also have some representation of gravity.

1.6.3.4 Rules for predicting events or relationships

Inferences are essential for survival and are based on experience and learning. However, gaining experience can be costly, so organisms may have evolved to use as little experience as possible for making robust inferences. Transitive inference is a good example of this. For instance, members of a group with pronounced hierarchical relationships may observe that animal A wins over animal B, and B wins over animal C. Different species, including cichlid fish, domestic hens (*Gallus gallus*), and pinyon jays (*Gymnorhinus cyanocephalus*), provide evidence that after such experience, they recognise that A is superior to C in the absence of direct information. Transitive inference has also been observed in relation to physical qualities, such as colour or odour (Vasconcelos, 2008).

Object permanence provides another example of how inferences are made about the location of objects. Suppose an individual knows that an object can be at one of two hidden locations. The actual location of the object can be determined either directly by observing it at one of those hiding places, or indirectly by inferring its location from its absence at one of the other locations. When exposed to such an experimental arrangement, dogs were able to locate a ball both based on direct observation and by finding out its absence at the other possible hiding place (Erdőhegyi et al., 2007). Thus, dogs behaved as if they were able to make inference about the location of the ball. Chimpanzees (*Pan troglodytes*) are similarly skilful in such tasks, but in contrast to dogs, they are also able to infer the presence of an object in a box by hearing its sound when the experimenter shakes it (Call 2004).

1.6.4 Ethorobotic perspective

Navigation is a fundamental skill for ethorobots. Endowing embodied robots with robust navigational skills is an intensive field of research (e.g., Gul et al., 2019). However, unlike animals, most of these robots currently operate in specific and familiar locations with limited travelling range. As a result, it is simpler and more efficient to provide these robots with a pre-existing ‘cognitive map’, eliminating the need for them to construct a map based on idiosyncratic sensory information. As a consequence, depending on the presence or absence of other navigation skills, these robots may not be capable of recognising changes in their environment, such as the repositioning of a door or the construction of a wall, and they require re-programming when deployed in new locations.

The invention of the clock and its electronic counterparts has made it relatively easy for current ethorobots to represent time. This presence is a nice case of convergence between a functional trait in animals and human technology. However, despite the availability of time as a reference, this feature does not play a prominent role in the control of robot behaviour, and it could be still advantageous to introduce specific zeitgebers in some situations. Nevertheless, noting and remembering when events occur could be a valuable additional environmental input for robots to make appropriate decisions.

There is still much to be learned about how animal species solve physical problems, and it is not entirely clear how comparative research can aid in the design of better ethorobots. However, regardless of the underlying mental mechanisms, living organisms display robust solutions to cope with environmental challenges that have evolved over millions of years and are crucial for everyday survival. Ethorobotics has the potential to acquire and implement fundamental physical problem-solving skills observed in animals, uncovering how seemingly complex tasks can be solved by simple means. Animals offer a compelling example that simple skills can yield effective outcomes across diverse environments. The utilisation of a 'proximity rule' to deal with connectedness could prove advantageous for an array of ethorobots, obviating the need for the development of more advanced visual capabilities. Furthermore, ethorobots possess the unique advantage for combining diverse and complementary skills in ways not feasible for living organisms, which have undergone divergent evolutionary histories.

To evaluate the performance of any mental model, the actual solution of a physical problem can be measured without much bias. In a recent study, Höfer et al. (2018) investigated the mental models proposed to explain ball-catching actions. Generalist models rely on general decision-making processes based on the position, velocity, and predicted trajectory of the ball. Specialist models emphasise the direct perception of the ball from the subject's own point of view. A mathematical analysis revealed advantages for both models depending on the actual environment and the subject's ability to obtain information about the event. Therefore, the study concluded that the best results could be achieved by a combination of the two different approaches.

1.7 Social problem solving

The study of social problem solving is central to cognitive ethology. To facilitate efficient group living, animals have evolved a variety of behaviours that promote cohesion, synchronicity, and various forms of interaction that can be utilised in dyadic or more complex inter-individual formulations. The decision to live a solitary life or form a group involves a trade-off, which is determined by the benefits and costs associated with each lifestyle during a species' evolution. Benefits of group living may include protection from predators, increased efficiency in finding food and potential mates, and the opportunity to learn from one another. However, there may also be increased competition and a greater risk of infection, making group living too costly. An animal group can be defined as an enduring structured network of individuals engaging in common activities through dyadic or multiple interactions. This definition excludes animals that are present together in a particular location only because they are attracted by the same environmental conditions, such as lizards sunbathing.

As ethorobots are designed for integration within social groups, one can draw valuable insights from gregarious animal species that have evolved traits advantageous for communal living. Notably, many contemporary social robots are developed primarily as 'solitary' agents, with a few supplementary social behaviours incorporated. A paradigm shift could occur if their design embraced their inherently gregarious nature, reflected not only in their physical appearance and behaviour but also in their mental architecture. A pertinent lesson from animal social behaviour research is that seemingly simple yet specific skills can wield a significant impact on the quality of social interactions. Leveraging ethological insights about sign stimuli, for instance, could enhance the receptiveness towards artificial agents.

1.7.1 Group structure

Animal group structure can be mathematically modelled using a simple framework that considers the relationships between a set of individuals. The hierarchical relationships among group members determine the nature of the network. Inter-individual relationships can range from symmetric to asymmetric and vary in strength (see Box 1.4). Traditional methods of measuring relationships involve observing physical or communicative interactions over a prolonged period of time. Individuals who consistently win these dyadic interactions are deemed dominant, while those who lose are considered subordinate (Goodenough et al., 2009). A linear hierarchy can be formed if all group members can be ranked from the most dominant (i.e., 'alpha') to the least dominant (i.e., 'omega'), with the steepness of the hierarchy reflected in a value between 0 and 1 (Landau index for the group). Additionally, individuals can receive a score based on their position in the rank order (Davis index for each group member) (Bayly et al., 2006).

Measuring hierarchical relationships in animal groups can be challenging, as overt fights are rare, and individuals may rely on subtle signals to indicate dominance or submission. Counting signals of dominance may not accurately capture the hierarchy established based on submissive signalling, and in large groups, some animals may not interact or have context-dependent dominance relationships. Despite the difficulty in measuring hierarchy, it provides several benefits for group functioning. The hierarchical structure determines the division of resources, reducing time-consuming disputes. The dominant individual also plays a leading role in group actions, facilitating decision making. As the strongest and often one of the oldest members of the group, the dominant can maintain its status over a long period and use its experience for the benefit of the group. Hierarchies provide a degree of stability to the network, and the position of the dominant individual often remains unchallenged, even when younger and stronger followers are present, who could potentially take over the leadership (social inertia).

Box 1.4 Flack and de Waal's (2004) social relationship model (based on the social structure of macaque species)

The overall use of the agonistic and formal (status) interactions decides the nature of the social system:

- *Despotic:* strongly asymmetric relationships maintained by harsh aggression; all relationships are precisely defined (formalised: precise place in the hierarchy);

- *Tolerant:* highly asymmetric relationships where the level of aggression is low (some relationships are not formalised);
- *Relaxed:* there are asymmetrical (formalised) relationships, but most are not precisely defined; aggressive poses are present;
- *Egalitarian:* asymmetrical relationships between couples are rare; the relationships are not formalised, and many relationships are based on equality (Table 1.3).

Table 1.3 Hierarchy rank and behavioural interactions in wolves and dogs, based on Flack and de Waal (2004) and Cafazzo et al. (2010). Agonistic interactions are initiated by the higher-ranking individual, whereas formal (status) interactions by either party. Agonistic interactions initiated by a high-rank individual can be ritualised or non-ritualised threats, but their formal actions are always ritualised

Type of the interaction	*Frequency of conflicts*	*Higher ranking*	*Lower ranking*
Agonistic	Often	**Initiator ('agonistic', 'violent')** *Ritualised Threat:* threatening position, showing teeth, pulling up hackles, growling, barking *Non-Ritualised Threat:* chase, fight, bite	**Recipient ('submissive behaviour')** Avoiding eye contact, head down, flat ears, tail down or between legs, crawling, laying on floor, showing chest and stomach, avoiding, stepping back
Formal (Status signals)	Rarely	**Initiator ('formal', 'assertive')** *Ritualised threat:* upright, stiff posture, head and tail high, ears up, snout or paws wagging behind back, tail high	**Initiator ('submissive', 'affiliative behaviour')** Slightly stooped, flat ears, tail down and wagging, quickly licking each other's snout

Despite living in a group, animals remain subject to individual selection, with their fitness determined by the benefits gained and costs incurred. However, being members of the same group, individuals are often each other's relatives, which can lead to alternative behavioural tactics. The kin selection theory suggests that genetic relatives may display more altruistic behaviour towards each other, even at some cost (Birch & Okasha, 2015). Closer relationships predict stronger altruism, and competition for resources is reduced within families. In some cases, increased cooperation among individuals can lead to a reduction in strong asymmetric relationships. Moreover, increased sociality, combined with a long lifespan, may lead to individualised personal relationships, including friendships, in certain animal species, such as chimpanzees (Silk, 2002).

1.7.2 A behavioural model on describing social complexity

The complexity of social behaviour systems makes it challenging to simplify the many interconnected components. Many textbooks focus on specific social skills, without discussing the generality of social behaviour.

Csányi (2000) proposed a theoretical three-dimensional space for representing social behaviour, based on sociality, synchronicity, and construction ability. By placing specific social skills in this virtual space, their relationship (distance) can be visualised. The *sociality* dimension generally involves interactions that increase or decrease the distance between group members, while *synchronicity* reflects the ability of group members to act in concert and present themselves as a unified entity. *Construction* ability refers to the collaborative creation of physical or social performances, such as tools or languages. Any specific social skill, such as cooperative hunting, attachment, or synchronised migration, involves all three dimensions to varying degrees. For instance, although tool making is primarily a construction ability, it also relies on learning from others and synchronising with the group's activities.

1.7.2.1 Sociality

Interactions between individuals and their surroundings play a crucial role in shaping social relationships. In mammals, close proximity between mother and offspring is essential for survival, but this dynamic changes, as the young mature (Section 2.6.2).

In animals, the use of space varies widely depending on the species and ecology. Home range is the area where animals spend most of their time, and it usually includes a core area for resting, hiding, and raising offspring. Home ranges of different individuals may overlap, in contrast, territories are exclusive areas actively defended by placing visual and olfactory markings at borders or attacking intruders, which allows territory holders to safeguard valuable resources such as food or mates.

In groups, the protection of individual space can be more complex, with success depending on an individual's rank in the hierarchy when competing for resources such as food, resting places, or mates. However, some rights to possession and social space are usually respected to some degree. For example, food is typically not taken from another individual if they are holding it in their mouth, beak, or hand.

Social tolerance is a measure of the typical proximity between individuals in a group. Short distances can be perceived as threats to possessions or resources, or as physical obstacles when escaping danger. Selection for tameness in domesticated animals has increased social tolerance by decreasing acceptable inter-individual distances and increasing acceptance of physical restraint.

Altruism, or behaviour that benefits others at a cost to oneself, can be explained by genetic relationships between partners or mutual/group-level shared activity to maintain cohesion (Goodenough et al., 2009). In a group, unselfish behaviour can be maintained if individuals recognise each other and punish selfishness by excluding the individual from mutual sharing. For example, group-living vampire bats (*Desmodus rotundus*) beg others for food after an unsuccessful foraging session but do not provide food to individuals who were selfish on other occasions (Wilkinson, 1984).

Sharing of resources can be passive or active. In passive sharing, the recipient initiates the interaction repeatedly until it achieves its goal, or it may give it up after a while. This is relatively common in many species, and it helps donors to establish and maintain relationships with other group members as potential allies. In contrast, active sharing with non-relatives is very rare among animals, and humans are outliers in this respect.

1.7.2.2 Synchronisation

Several different mental mechanisms ensure that the group can behave as a unit if needed. There are different contexts in which synchronised behaviour may be advantageous, and synchronicity may affect only small groups (dyads), larger sub-groups, or the group as a whole.

Movement synchronisation: individuals in groups often have to move together at lower or higher speeds. To achieve this, group members must behave in a non-independent manner, taking into account the position and path of others in order to avoid collisions (Herbert-Read, 2016). The level of synchronisation can be characterised by the strength of correlations among the members within a specific timeframe. Several factors influence the achieved synchronicity: (1) sensory-motor systems used for the coupling (e.g., vision or olfaction); (2) substrate of movement (e.g., air or water); (3) species differences; (4) speed of movement; (5) proximity preferences/tolerance; (6) group composition; and (7) individual preferences for positions within the group (Herbert-Read, 2016).

Emotional synchronisation: the mental states of group members can be synchronised through communicative signals, which can also affect the movement dynamics of the group. Thus, emotional synchronisation or empathy occurs when one animal's mental state becomes similar to another's after exposure or perception of a different mental state (Hoffmann, 2000; Preston & de Waal, 2002). Empathy can promote group cohesion and may have evolved as a mechanism for establishing closer relationships, such as friendship, and promoting reconciliation. It also plays a role in altruistic behaviours, such as helping (Hoffmann, 2000; Preston & de Waal, 2002).

Preston and de Waal (2002) introduced the Perception-Action Model of empathy, which operationalises the phenomenon in three phases: (1) observing the behaviour of the target individual; (2) activation or priming of a similar mental state in the observer, accompanied by autonomic and somatic responses; and (3) display of similar behaviour by the observer. Notably, this process entails the suspension of the observer's own mental state, including the focus and direction of its ongoing activities. Empathy is primarily based on emotional mental states, but it may also involve a cognitive component such as perspective taking.

Social referencing and empathy have some overlapping features, although they are typically discussed separately. When faced with a new object or situation, a naïve observer may gather information by observing the behaviour or reaction of an experienced individual and modifying its behaviour accordingly. For instance, if the experienced individual displays fear signals, the observer may retreat (Merola et al., 2012).

This scenario can also be understood through the Perception-Action Model, considering how the interaction begins and its consequences. The key difference is that empathy can occur spontaneously after encountering a companion, whereas social referencing is triggered by a specific environmental event or situation.

According to Preston and de Waal (2002), empathy should be distinguished from sympathy. The key difference is that empathy involves the observer experiencing a similar emotional state as the target individual, whereas sympathy does not require such a shared experience. Instead, sympathy aims to alleviate the suffering of the target individual without necessarily experiencing their emotions. An example of sympathy may include a non-involved third-party attempting to console a distressed individual without experiencing the same emotional state.

1.7.2.3 Construction

Collaboration among individuals within a group can yield a diverse array of outcomes or products that would be unattainable through solitary existence. In humans, many of these outcomes are integral to the group's culture. Consequently, the process of construction not only encompasses the collective effort invested in the actual realisation of these phenomena but also encompasses their transmission across generations. Language serves as a prime example of social construction, albeit one exclusive to humans. While less pronounced in animals, constructive activities are still evident in forms of communication and collaboration. Thus, complex communicative signalling systems, joint hunting endeavours, and instances of teaching are recognised as instances of construction, each of which holds considerable importance within social dynamics. Due to their paramount importance within the realm of social interactions, these aspects will be addressed in Sections 1.8– 1.10.

1.7.3 Social skills facilitating group living

Individual recognition: it is a complex hierarchical process that plays a fundamental role in shaping inter-individual interactions. Individuals can share morphological or behavioural traits due to their membership in the same group, familial ties (are relatives), or social class. Discrimination of these features or their combinations can aid in identifying individuals, but the most sophisticated level of social recognition occurs when individuals recognise each other (Ward et al., 2009). Therefore, the target set for individual recognition contains only one item in terms of set theory. Animals are also capable of remembering individuals and adjusting their memory in response to potential changes in their morphology or behaviour.

Social monitoring: following events within a group can be essential for all members. In addition to direct interactions, gaining a third-party perspective can also provide advantages for the observer. Social monitoring enables individuals to remain vigilant to any occurrences in the group's life. Eavesdropping (or overhearing) occurs when the observed interactions directly influence the observer's behaviour. For example, if an observer male guppy (*Poecilia reticulata*) witnesses that most males prefer a specific female, then it may also approach that individual given the opportunity. Similarly, observing one partner winning over another can encourage the observer to show a preference towards the winner to gain an advantage in the rank order (Peake, 2005).

Social gaze: the head, face, gaze, and eyes serve as important means for conveying signals that reflect an individual's inner state (see Section 1.8). Facial cues are also significant in affiliative and cooperative situations, so individuals forming complex, individualised groups possess a range of social attention skills that manage facial interactions and process the gathered information (Emery, 2000). Although social attention primarily relies on visual cues, auditory input can also play a role in some cases.

Facial interactions exhibit varying levels of complexity (Figure 1.12). The simplest form is mutual gazing, which can be followed by either attraction or avoidance. In many animal species, enduring and stiff gazing at another individual is a form of threat, but a relaxation or alteration of this gaze pattern can indicate affiliative tendencies (Emery, 2000). Many species demonstrate the ability to follow another's gaze and orient towards a point in space. Others exhibit joint attention, where members of a dyad share an object of interest. Finally, shared attention occurs when participants

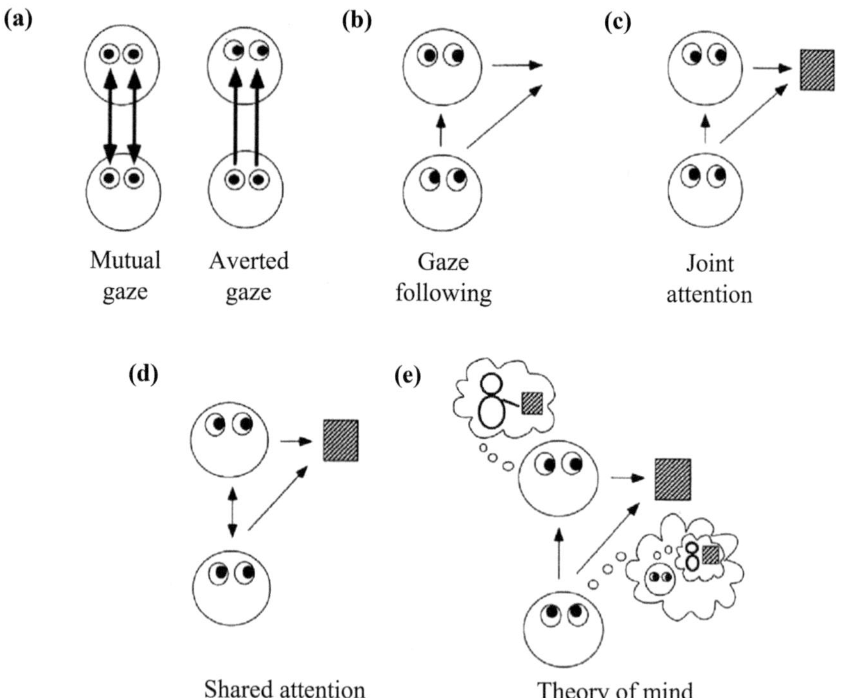

Figure 1.12 The direction of the other's gaze may provide information about the internal or external state of the individual. Thus, this skill has important role in social cognition, including perspective taking or empathy. For the description of the facial interactions see the main text (from Emery, 2000).

mutually acknowledge their shared interest in a particular target. From an anthropomorphic perspective, one may also question what the other is thinking about the object of mutual interest. Mental attribution (theory of mind, see Section 2.5.7) may come into play if an individual takes a third-person view of the mental representations in another's mind. Evidence suggests that most types of social attention occur in the majority of vertebrates, with the exception of mental attribution (for a review, see Zeiträg et al., 2022).

Different types of social attention can aid individuals in predicting the behaviour of others. Competitiveness among group members may have led to the development of abilities that leverage these possibilities for the observer's advantage. It is well-established that attention often precedes goal-directed actions (e.g., anticipatory look; see Section 2.6); for instance, an individual may briefly look at a mug on a table before picking it up, which is necessary for proper sensory-motor coordination when executing the grabbing action. From the perspective of an observer, this type of attention can be used to anticipate the other's next move. There is evidence that animals also attend to others to learn more about their goal-directed behaviour or to discern what type of behavioural cues predict the other's actions in specific contexts. The mental representation of the other's observable goal is the first step towards attributing intentions to the other (Dennett, 1978; see also Section 2.5.7), which may also involve the attribution of mental states like desires or beliefs. However, there is limited evidence of this type of attribution in animals thus far.

Self-recognition: ability to recognise oneself as a distinct individual separate from others is referred to a self-recognition. This ability is a subcategory of individual recognition, but it differs in that the representation of the self is treated as if it were another individual, distinct from all others. This ability rests on a hierarchical mental representation, similar to individual recognition, because different features of the self can be recognised independently.

In the famous 'mark test' (Gallup, 1970), researchers put a coloured mark on the forehead of sleeping children, chimpanzees, elephants, etc. When the subject wakes up and sees its reflection in a mirror, any attempts to remove the mark, such as by washing or rubbing it off, are taken often as evidence of self-recognition or even self-consciousness. However, some researchers, such as Heyes (1995), have argued that this experiment only measures the ability for 'body self-recognition' rather than true self-recognition or self-consciousness. It is also likely that due to the different ecological background, representation of the self manifests differently in different species. Bekoff and Sherman (2004) argued that representation of the self should be considered as a continuum with increasing complexity. Also, Lenkei et al. (2020) proposed a modular approach to self-representation, that is, the representation of the self is an array of interconnected cognitive skills, and the building blocks present varies between the different species (see also Box 1.5).

Groups as sources of innovation: in group settings, individuals may exhibit proactive behaviours tailored to specific situations, or they might strategically seek to benefit from the actions or accomplishments of others (Giraldeau & Dubois, 2008). For instance, producers actively search for food and even innovate new methods of acquiring it. Scroungers opt to wait and opportunistically reap a portion of the

Box 1.5 Alternative models: do animals know their own size?

Lenkei et al. (2020) investigated a specific aspect of body self-representation in dogs. In their study, dogs had to go through different opening sizes in an experimental setting (see Figure 1.13): (1) changing large enough and too small openings, with a last trial being medium sized; (2) changing large enough to smaller and smaller opening until the dog could not fit through, ending with a step back to the last size the dog could fit through. Dogs approached the large enough openings sooner than the too small to fit through openings (the latency of going through the mid-size opening fell in-between), and dogs were more likely to try the large enough opening than the too small one before it. By relying on the latency to start moving and arriving authors attempted to measure differences in the a priori decision making of their subjects, instead of the result of a direct trial-and-error strategy.

There are at least two possible explanations for these observations. Firstly, dogs might, through extensive experience with passing through openings, develop an abstract

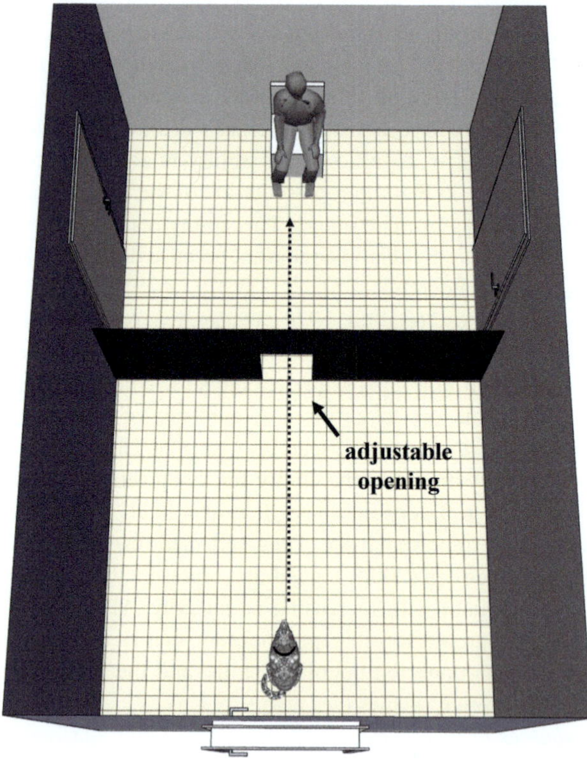

Figure 1.13 Illustration of the experimental setup used by Lenkei et al. (2020). Dogs were encouraged to go through a door the size of which varied between large enough to cross easily to too small to fit through.

representation of their own body. Input from both vision and haptic sensors could contribute to this representation. This body representation might be utilised flexibly, independent of the situations where the sensory input was collected. For instance, it could provide crucial input for deciding whether to engage with an opponent of a specific size. Secondly, dogs may learn, through experience, the visual perspectives of openings from a typical distance they can pass through, or the size of adversaries they can overcome. In this case, the mental representation controlling the behaviour does not reflect body size but rather specific states of the environment.

In a subsequent experiment, researchers found that dogs recognise their bodies as obstacles when attempting to pick up a toy attached to the rug they are standing on (Lenkei et al., 2021). This simple issue highlights the difficulty of revealing the nature of mental representations, necessitating carefully controlled experiments and multiple approaches to form stronger hypotheses.

Social robots encounter analogous challenges when navigating the human environment. Skills such as passing through doors, entering elevators, or moving through crowds of people require abilities similar to those of dogs. However, unlike dogs, engineers set the robot's size as a fixed parameter for the algorithm dealing with object avoidance. While this might seem reasonable initially, it limits the robot's flexibility. Dogs, for example, may force their way through a slightly smaller opening, but a robot programmed as suggested would be unable to do so.

rewards obtained by the producers. Novel resource acquisition demands innovation, for example, venturing into uncharted territories or finding out new ways of food acquisition. Evidence suggests that larger groups tend to excel in solving novel challenges. Liker and Bókony (2009) found that house sparrows *(Passer domesticus)* in larger groups outperformed their smaller-group counterparts in opening a problem-box. This advantage is attributed to the increased phenotypic diversity in larger groups, which increases the likelihood of having innovative individuals present. With reduced individual-level demands for vigilance and food search, innovative members within larger groups may have more time to explore new solutions. While scroungers may opportunistically capitalise on gains, other individuals can copy the innovators, thereby disseminating beneficial new behaviours throughout the group.

1.7.4 Ethorobotic perspective

Ethorobots currently face a major challenge in displaying social skills that are on par with those of gregarious animal species. They lack the necessary perceptual abilities to process complex information related to first- or third-person interactions and do not have mechanisms in place to evaluate inputs and display context-dependent actions. While existing social robots may perform adequately in specific tasks, such as acting as body trainers for stroke patients (Dembovski et al., 2022), they are still far from being socially competent agents. For instance, they can display certain movements for humans to imitate and may also recognise the success of imitation, but this falls short of the complex and nuanced social skills that are required for effective interaction with other beings.

1.8 Communication

When asked to define communication, the common answer centres around information transfer. However, human communication is not limited to just exchanging factual knowledge and experiences through verbal or written means in everyday life and work. This superficial understanding of communication is often apparent in human–robot interaction (HRI) research, where the focus is primarily on linguistic communication, particularly in light of the commonplace use of conversational software tools (e.g., chatGPT). Non-human animals have a rich array of communicative behaviours that is also present in humans, and that goes beyond verbal exchanges. While the conversational abilities of computers and mobile devices are sometimes enhanced by basic non-verbal signalling (such as emojis), embodied ethorobots possess an even greater potential for engaging in natural communication with humans.

Ethological research has not only uncovered the richness of animal communication but has also provided evidence about some rules that appear to operate across a diverse array of species, including humans. Animal communication also provides examples of how even subtle modifications in signalling can improve communication. Studying the evolution of communication systems has revealed the importance of signal accuracy, clarity, and redundancy. The ethorobotic approach, centred on the development of non-human embodied agents, can gain valuable insights by drawing inspiration from animal communication.

1.8.1 Definition of communication and the use of terms

Based on evolutionary considerations, communication is defined as the display of specifically shaped morphological traits or behaviours by the sender, which have a specific effect on the receiver by modifying its behaviour to the advantage of the sender, without relying on significant physical forces (Goodenough et al., 2009). The communicative interaction can also be beneficial for the receiver, and any advantage should be calculated for a longer duration and not necessarily for each act. The sender's behaviour was also often interpreted as a kind of manipulation of the receiver. This concept is, however, not useful because 'manipulation' has a negative connotation, and equating all communicative interactions with manipulation hinders to specify those cases when the sender actually deceives the receiver (see below).

While the terms 'signals' and 'cues' are often used interchangeably, it is important to differentiate between them. Signals are specific evolved traits that serve a clear function in the communication between sender and receiver, while cues can be any other features that gain a communicative function during a series of interactions, and thus also involve learning. For example, growling is a signal used to threaten an opponent before an attack, while saying 'Stop it!' to a dog is a communicative cue because the dogs have to learn it and actually, any other words had the same effect if used in a similar fashion. Note that when the person's voice contains acoustic features that correspond to growling (e.g., low frequencies with harsh tone), vocalisation represents a mix of a signal and a cue. One can also envisage that signals and cues represent two ends of a spectrum indicating the arbitrariness. Evolutionarily selected signals are non-arbitrary (e.g., growling), while cues based only on learning (e.g., name of an object) are arbitrary. Although learning plays a role in the sending and receiving of communicative means, interactions based on signals have strong evolutionary roots. For example, signals like growling are sent or perceived without the need for much experience. But senders may need to learn how and when to use the signals, and receivers should come to know the optimal reaction to signals based on the actual circumstances (Hauser, 1996).

From the sender's perspective, there are static signals (or cues), which are always 'on' (displayed by the signaller), while others are dynamic and can be turned on or off in line with the senders' actual preference. A red mark on the adult male robin's plumage is always 'on' and it signals its 'maleness', while the bird can always decide to sing or not to sing when protecting its territory against intruders.

There is an ongoing debate about the 'information' that is 'exchanged' during communicative interactions. Many scholars are in favour of using concepts derived from human linguistic communication models, while others are against it (see Font & Carazo, 2010; Rendall et al., 2009). From a mental perspective, communication can be successful only if the participants share the same general framework of using a specific set of signals (and/or cues), and at the same time, they have the potential to rely on a shared repertoire of actions in response. This is also likely as it is expected, that depending on the situation, the same individual is either the sender or the receiver. For example, when a sender growls in an agonistic interaction, the vocalisation serves to achieve a goal of making the other leave. This outcome is perceived and recognised by the sender, and thus it could be formulated as the message of the signal.

In an analogous way, upon facing a growling opponent the receiver may decide to step back and leave by processing the meaning of the signal (but see Font & Carazo, 2010). It is important to distinguish the mental representation for the sender and the receiver because their overlap or agreement is the function of both evolutionary and developmental factors, and it depends on the actual communication system. Selection acts on both aspects of communication (Figure 1.14), and there is a trend to narrow the gap between the mental referents of the message and its meaning.

One may wonder how certain traits acquire signalling functions. There are two crucial factors to consider: first, signals must possess distinct features that differentiate them from other behaviours, and second, the receiver should be able to recognise them and respond appropriately. Ethologists propose an evolutionary process called ritualisation, to explain this phenomenon, during which the function of a specific behaviour evolves over many generations. For instance, the wing movements of birds, which are crucial for relaxing muscles and arranging feathers before flight, have evolved into a signal for taking off, aiding in the coordination of group behaviour. Although providing concrete evidence for this idea is challenging, the concept of evolutionary ritualisation has gained wide acceptance (Goodenough et al., 2009).

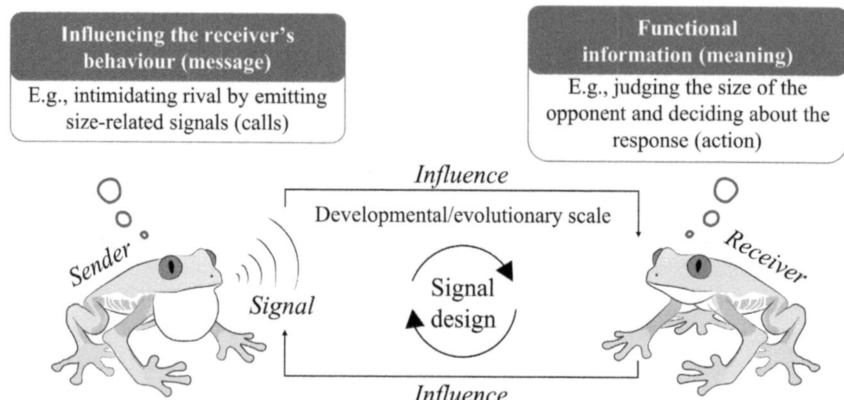

Figure 1.14 Schematic illustration of the communicative interaction (based on Font & Carazo, 2010). The sender emits a signal (message) with a predetermined structure that undergoes modifications based on the sender's immediate internal state and environmental conditions. The receiver detects the signal through its senses, initiating a process of perception and central processing to formulate a decision (meaning) regarding the appropriate behavioural response. Throughout evolution, signalling behaviours might exploit receiver biases, but ultimately, signals evolve to exert an impact on the receiver. Receivers could develop mechanisms to withstand unreliable signals. During development, actions (cues) can transform into signals if they fulfil the same functions expected from a signal (as discussed further in the text). The signal's persistence, whether during evolution or development (note the varying time scales), depends on its effectiveness in regulating communicative interactions. The message does not correspond to the signal and similarly, the receiver's action is not a direct consequence of the decoded meaning. Both aspects of the communication are influenced by inner state of the sender/receiver as well, how they evaluate the actual state of the environment. Drawings by Zsófia Bognár (see also Section 2.6.3).

Evolutionarily ritualised signals have a strong genetic component. Dogs do not need to learn how to growl, bare their teeth, or perform a sophisticated courtship dance. At the same time, learning may be needed to improve the effectiveness of signalling particularly in terms of when and how to display them. Thus, most animal signals are to be considered as modal action patterns (see Goodenough et al., 2009). Warning calls seem to have a similar origin, but by the means of learning, animals increase the efficiency of signalling. For example, they learn to use the calls selectively for one specific type of predator. This means that senders also learn about the predators, about their specific features, and as a result, they form a specific mental category. Over time, senders develop a complex mental representation of potential predators that is linked to a specific call type.

An analogue process is assumed to take place on a developmental timescale. Individuals may display different, sometimes arbitrary, actions (cues) during their interactions with others, which may gain an analogous function of a signal, as a result of mutual learning (developmental ritualisation) (Tomasello, 2008). The sender may learn that its action has a reliable effect on the receiver, and the receiver learns about the consequences of its reactions. For example, a chimpanzee may hit the ground repeatedly and then initiate a playful interaction with a companion. Over time, the hitting may become a specific action for an invitation to play. As specific behavioural cues become means to communicate, they obtain a similar function as signals, thus there is some logic in referring all behaviours as signals if they have a communicative function, independently of their origin.

However, significant distinctions exist between signals that arise through evolutionary or developmental ritualisation. Evolutionary signals are typically shared by all members of a species due to their substantial genetic basis (e.g., growling in wolves). In contrast, signals resulting from developmental ritualisation are shared solely in idiosyncratic communicative interactions within dyads or possible groups, but do not achieve a communicative function for the species as a whole (e.g., human language).

1.8.2 Form of signals

Communication is inherently uncertain, and signals (messages) may not always be detected or interpreted (meaning) accurately by the receiver. As a result, senders have an interest in increasing the probability of successful communication by refining the signalling action, that is also supported by ritualisation processes.

In general, signal detection theory deals with these issues, and sender's aim is to increase the signal-to-noise ratio when the noise is generated by the environment that includes all actions, events, etc. in parallel to the signalling incident (Hauser, 1996). It should be also noted that improving signalling efficiency is always associated with some cost (e.g., the sender uses energy for producing the signal), and it can also have negative consequences on the sender. For example, an effective warning signal by a prey animal can also become a useful behavioural cue (and not signal!) for the predator about its location. Thus, there is typically a trade-off between the effectiveness of signals and the potential risks and costs associated with signalling. This trade-off is influenced by both evolutionary and ecological factors and may vary depending on the specific communication system and context (Johnstone, 1997). Ritualisation processes form the signals in the following ways:

- *Conspicuousness*: for effective communication a signal should stand out from the surrounding environment. Signalling actions may include the use of body parts that become more colourful or vocalisations that utilise sound frequencies distinct from background noises (Halfwerk & Slabbekoorn, 2009).
- *Stereotypy*: signals are easier recognised if the receiver's perceptual apparatus is repeatedly activated in the same way. Stereotypy is achieved by making the signal simpler, predictable, and repetitive.
- *Redundancy*: signals may activate different senses of the receiver. This increases the chance that the signal is processed and reacted to. This explains why many signals have visual, acoustic, and also chemical components.
- *Sensory exploitation*: animal species may display biases in processing environmental inputs due to their specific sensory receptor structure or ecological challenges.

Table 1.4 Different characteristics of the four main modes of signal transmission (modified after Goodenough et al., 2009). These features play an important role in relying on specific types of signals or their combination under different ecological conditions, predetermined by the evolutionary history of the species

Characteristics	*Mode of signal transmission*			
	Olfactory	Auditory	Visual	Tactile
Range	Long	Long	Medium	Short
Transmission rate	Slow	Fast	Fast	Fast
Travel around objects	Yes	Yes	No	No
Night use	Yes	Yes	No	Yes
Fade-out time	Slow	Fast	Fast	Fast
Locate sender	Difficult	Varies	Easy	Easy
Cost to send signal	Low	High	Medium	Low
Effective distance	Long	Long	Medium	Short
Exchange	Slow	Rapid	Rapid	Rapid
Durability	Long	Short	Rapid/medium	Short

For example, some populations of guppies *(Poecilia reticulata),* exhibit a preference for food with an orange colour. The males' integument in this population is dominated by spots of the same colour. It is assumed that as a result of a selection process males exploit the females' sensory preference for their own interest (Rodd et al., 2002).

1.8.2.1 Channels of communication

Evolutionary and ecological constraints limit or may extend the capacity of animals' sensory systems. Apart from tactile (haptic) signals, during typical communicative interactions the sender does not get into physical contact with the receiver, and thus different mediums (channels) are utilised, such as light, sound, molecules, and electricity. The sensory organs and the medium also determine many important aspects of communication, including signal form, communication distance, and dynamics (see Table 1.4). Animals utilise many composite signals that can be sensed by different types of sensors. This also ensures that signals are not only redundant but can be sent under different conditions.

1.8.2.2 What are signals about?

Signal messages can be interpreted in a variety of ways, and their content can be subject to uncertainty due to the complex mechanisms involved in their generation. Researchers differentiate between two basic types of signal messages (Goodenough et al., 2009):

Analogue signal systems: for some researchers, communication represents a way to regulate the behaviour of individuals. Accordingly, the message of signals is based on the inner state of the sender. A dog growling at its partner reveals its inner state of anger elicited by the presence of a strange conspecific individual. Similar associations exist between signals that reveal fear, affection, and other inner states. It follows that the number of affective signals a species uses is closely tied to the number of specific inner states and their possible combinations it experiences (see Section 2.5.5). This may also explain why species with more complex social interactions, and therefore more diverse inner states, tend to have a richer signal repertoire. Inner states are typically graded, meaning their intensity can vary from 0 to some maximum level of excitation (e.g., annoyance, anger, or fury provide a linguistic example for increased activation). As the intensity of the inner state changes, signals tend to mirror those changes, often with increased signalling complexity, motor activity, and energy, involved as the signal intensity increases.

Digital signal systems: some signals may not directly reflect the inner state of the sender, but instead indicate changes in the environment. Warning signals (mobbing) may provide an example (e.g., Curio et al., 1978; see Box 1.6). Diana monkeys (*Cercopithecus diana*) emit different calls upon noticing aerial (e.g., eagle) or ground predators (e.g., leopard). These signals help group members to seek suitable protection. They run up the trees when leopards approaching or climb down the tree if an attack from the bird of prey is expected. They do the same if hearing the respective warning signals. It can be said that the signal indicates a specific environmental event, that is, it has an external referent (referential signals; Zuberbühler et al., 1997). While this seems to be a plausible argument, most researchers are more cautious because they see no unequivocal proof that these signals function as more typical referential signals, such as words in human language (Font & Carazo, 2010). It is important to distinguish between referential signals (which refer only to an external event) and functionally referential signals, since the latter may still involve changes in the internal state of the sender. Nevertheless, the notion on functionally referential signals highlights the role of the environment in shaping animal communication and underscores the complexity of the signalling systems used by different species.

1.8.3 Intentionality in communication

From an anthropomorphic and anthropocentric perspective, communicative interactions seem to be intentional, while even in human communication many signalling activity has no intentional component (e.g., blushing when caught cheating). To capture this distinction, some researchers differentiate between voluntary and involuntary signalling. In cognitive ethology, much effort has been made to identify the intentional component of animal communication. This is important because the ability to engage in intentional signalling is thought to increase the flexibility of the sender's behaviour (e.g., Townsend et al., 2017). Specific external or internal conditions, for example, lengthening of the light periods and increasing testosterone levels may elicit courtship behaviour in males of many species (e.g., fishes). This kind of signals are not typically regarded as intentional because they are under strict neurohormonal control, and there seems to be little room for any flexibility. In contrast, signals that invite a partner to play may be intentional because the sender could have various alternative goals regarding the playful interactions (e.g., play-fighting or object play).

Based on earlier discussions, Townsend et al. (2017) suggested three criteria to account for intentionality in specific communicative interactions (see also Leavens et al., 2005):

1. The sender should act with a clear goal representation in mind. It shows goal-directed signalling towards the receiver, persists in signalling, and/or elaborate the signal until the desired goal is reached;
2. The sender's signalling behaviour is flexible (may be withheld or not), and must be directed at the receiver, including sensitivity to the other's attention and other social factors;
3. The senders' signalling changes the behaviour of the receiver in a way that it becomes closer to realise the goal.

1.8.4 The versatility of communication systems

In their analysis of linguistic and non-linguistic communication, Hauser et al. (2002) argued that certain features of communication are shared by both humans and other species, adding to the flexibility of their signalling systems and allowing for more complex forms of interaction.

Regarding the sensory-motor system involved in communication, the ability to invent, learn, and imitate signals, having sophisticated discriminatory ability, and possessing cross-modal perception is especially important. The flexibility of communicative systems can be increased by the large number of signal units that can be combined freely, and which allow also for the emergence of metacommunication (the potential of a signal to modify the message and perhaps also the meaning of a subsequent signal). The presence of a conceptual-intentional system is manifested by having voluntary control over signal production, the utilisation of signals with a referential component, and the development of non-linguistic conceptual representations (Hauser et al., 2002). Many of these attributes can be observed in African grey parrots (*Psittacus erithacus*), dolphins, apes, dogs, or bees among other species (Shettleworth, 2010).

1.8.5 The communicative cycle

While simple communicative exchanges involve sending a signal that elicits a response in the receiver, more complex interactions involve a process known as the communicative cycle. This process involves several steps, depending on the communication channel involved, as the sender seeks to ensure that the signal is received and has the desired effect on the receiver.

1. The sender first aims to direct the receiver's attention onto itself by emitting attention-getting signals or ensuring that the receiver is in a state to potentially sense

the signal. For example, a courting male guppy may move and display various behaviours around the head of a female to catch her attention;
2 The sender sends the signal, repeating or elaborating it until it achieves the desired effect. The male guppy continues to signal until the female signals willingness to mate;
3 The receiver female assesses the male's signalling and other qualities before emitting signals for readiness to mate or rejection;
4 The sender evaluates the effect of its signals on the receiver. The male guppy recognises the female's signalling and proceeds to mate with her or leaves.

The communicative cycle highlights the importance of mutual attention during communication, particularly when it is a means of increasing collaborative interactions (see also Section 2.6.3 and Box 2.21 for details).

1.8.6 Intra vs interspecific communication

Communication is typically taking place among members of the same species, with conspecific individuals playing both the roles of sender and receiver. The evolution of signalling behaviour is closely tied to the evolution of receptive sensory systems, ensuring that signals are effective in achieving their intended function.

There is a tendency for signals used for similar functions to share structural similarities. This may be due to behavioural homologies among related species that persist despite divergent selective pressures, or general social rules that favour convergent similarities. For instance, in many mammalian species, low-frequency and harsh growling vocalisations are used to deter opponents in conflict. The structural-motivational hypothesis (Morton, 1977) proposes that low-frequency and harsh vocalisations are evolutionary ritualised signals that stand for 'larger bodies' that provide advantage in physical combat (Figure 1.15). This evolved relationship between the growling sound and the body size is based on the acoustic rule that large-bodied objects tend to produce harsher and deeper sounds in general (c.f., compare double bass vs violine).

The generality of some signal features allows for the development of interspecific communication, particularly in situations where two species cohabitate over extended periods, such as in course if domestication (Miklósi, 2007). Learning mechanisms can also support the mutual exchange of signals, as senders may learn about the most effective ways and times to signal, and receivers may learn from the consequences of the signalling event.

1.8.7 Linguistic vs non-linguistic communication

Language is a unique and highly complex form of human communication that is based on a set of rules that govern the pattern of signalling (Hauser et al. 2002). The use of language facilitates the rapid synchronisation of mental states between conversational partners, resulting in the creation of shared knowledge that develops over time through interactions. Human language is especially capable of dealing with

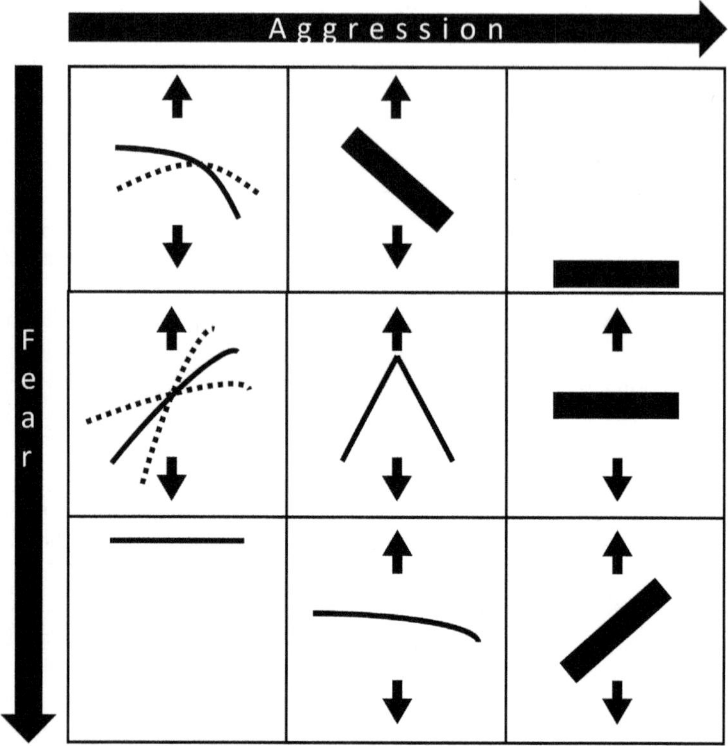

Figure 1.15 A simplified diagram depicts alterations in tonality and pitch of vocalisations based on the levels of fear and aggression. The vertical position within the boxes represents frequency, while rectangles signify harshness. Thin lines denote tonality, and dotted lines indicate possible shifts in slope. Arrows illustrate potential frequency adjustments within the boxes (adapted from Morton, 1977). This model finds support across various species, with birds and mammals largely aligning with the predictions put forth.

immense numbers of signals most of which are referential, that is, they are referring to states of the environment.

As far as we currently know, no other animal species has a communication system with a comparable level of complexity as human language. It is essential to emphasise that the terms 'body language' (often referred to in social robotics) or 'dance language' (when bees communicate about distant food sources) are misapplications of the concept. Non-human animals have limited ability to develop shared mental content, which means that their minds remain isolated from each other in comparison to humans who use language.

In language, specific rules assure the sensory-motor functioning of the signalling system (e.g., cross-modal abilities), the emergence of signals (e.g., learning and teaching), the links to other representation systems (e.g., categorisation), and social

Table 1.5 Comparison of some key features of acoustic communication in learned birdsong and human language (from Berwick et al., 2012). The mechanisms of singing behaviour in many bird species are studied because it has many shared features with human language. This similarity involves behavioural, cognitive, and neural features (Berwick et al., 2012; Doupe & Kuhl, 1999; Zhang et al., 2023)

	Birdsong	*Human language syntax*
Precedence-based dependencies (first-order Markov)	Yes	Yes, but in sound system only
Adjacency-based dependencies	Yes	Yes
Non-adjacent dependencies	In some cases	Yes
Unbounded non-adjacent dependencies	Not known	Yes
Describable by (restricted) finite-state transition network	Yes (k-reversible)	No
Grouping: elements combined into 'chunks' (phrases)	Yes	Yes
Phrases 'labelled' by element features	No	Yes (words)
Hierarchical phrases	Limited	Yes, unlimited
Asymmetrical hierarchical phrases	No	Yes
Hierarchical self-embedding of phrases of the same type	No	Yes
Hierarchical embedding of phrases of different type	No	Yes
Phonologically null chunks	No	Yes
Displacement of phrases	No	Yes
Duality of phrase interpretation	No	Yes
Crossed-serial dependencies	No	Yes
Productive link to 'concepts'	No	Yes

functions (e.g., managing social relations and modelling other individuals' mental content) (Hauser et al., 2002; Pinker & Bloom, 1990).

It is worth noting that non-linguistic communication systems can also achieve or maintain complex interactions, and animals may possess one or more cognitive skills that are fundamental components of human language. For example, the songbirds share many features with language, including early learning, imitative abilities, and hierarchical use of signalling units (see Table 1.5).

1.8.8 Honesty and deception

In the context of communication, signals are honest if they correspond to the sender's actual inner state, true qualities, or to an environmental event. In evolutionary biology, the question of honesty is mostly raised in connection with the relationship between the genetic quality of the individual and the attractiveness of the signals, produced for finding the best match. As cheating is always a possible alternative tactic, but at high frequencies, it may destabilise the communication system, one should be able to explain what kind of processes maintain honest communication (Getty, 2006; Számadó et al., 2023).

The production of signals is mostly costly, and its consequences can be risky (e.g., being noticed by a predator). Partly this investment by the sender ensures that it should benefit from the interaction, especially because similar signals are less likely produced by individuals in poorer condition (cheaters). It is also the sender's interest to make the signalling as effective as possible, that is, by gaining the most with least expenditure. Although dishonest signalling may occur in short term, the need for efficiency supported by Darwinian selection maintains honest communication in animals.

Tactical deception occurs when a sender emits a signal in the absence of corresponding environmental events or inner states, with the goal of gaining an advantage through the receiver's misinterpretation of the signal (Byrne & Whiten, 1991). The term 'tactical' refers to the flexibility of this behaviour, and that the signals are part of the sender's honest communicative repertoire. For example, tufted capuchin monkeys (*Sapajus nigritus*) use alarm calls in a tactical deceptive manner to monopolise food sources (i.e., scaring away their competitors from the food by convincing them that a predator is in the vicinity), and they adjust their deceptive behaviour based on the availability of food and their spatial relationships with others in the group (Wheeler, 2009).

1.8.9 Ethorobotic perspective

Animal communication systems have evolved over millions of years, allowing ample time for the optimisation tailored to diverse environments. These solutions encompass various modes of transfer, taking into consideration signalling structures (including body and behaviour) and sensory systems. Humans also adhere to specific rules of communication, enabling them to establish communicative relationships not only with other species (such as dogs and cats) but also with individuals of diverse species (e.g., Alex the parrot). This suggests that the development of artificial agents should not be constrained by our humanness. While existing social robots often exhibit a range of non-verbal, sometimes animal-like communicative signals, their signalling systems are not coherent and frequently lack the capacity to serve as recipients of similar human signals.

One could hypothesise that enhancing the non-verbal signalling capabilities of ethorobots could potentially foster stronger connections with humans. It is crucial that these ethorobots do not just act happy because they have a machinery to do so, but genuinely manifest positive states in functional contexts. The communicative proficiencies of ethorobots should ideally contribute to trust and believability, essential elements for cultivating meaningful human–robot interactions. Humans possess the ability to identify 'cheaters' due to evolutionary adaptations, implying that unrealistic communication from ethorobots might jeopardise the integrity of the human-robot relationship.

Ethorobots have the potential to verbally communicate with humans if equipped with a natural language processing system (Nadkarni et al., 2011), but for this, no robotic embodiment would be needed. However, any addition of verbal communication should only take place if it can be made believable from the receiver's perspective.

At present, ethorobots lack the versatility of non-verbal skills exhibited by both animals and humans, which hinders their effectiveness in genuine social contexts. The intricate nature of human communication, characterised by nuanced and situation-specific signals, highlights the importance for ethorobots to exhibit a more streamlined array of non-verbal cues akin to those observed in a conceivable animal species, rather than striving to emulate human behaviour.

1.9 Social information gathering

Groups can efficiently disseminate information among their members. In many cases, animals acquire knowledge by observing others, which can be advantageous as it allows individuals to avoid potential risks of trial-and-error learning (Section 1.11; Boyd & Richerson, 1988). There are arguments that sociable animals have an advantage over solitary ones in this scenario. Animals that live alone can only rely on information they have personally gained from experience and interaction with their hazardous environment. A lone animal must come into contact with a predator to gain experience, but members of a group can also learn by watching how others respond.

Humans heavily rely on social information, a trait that sets them apart from all other non-human beings. This perspective has also left its mark on early social robotics, with significant emphasis placed on equipping these agents with the ability to learn from humans. However, social robots have an alternative avenue for learning. Rather than allowing them to learn from humans, these robots can be trained in a non-social context. Depending on the robot's underlying cognitive structure, the outcomes of social and non-social learning might diverge. Currently, creating social robots having robust social learning skills, remains a challenge. Nonetheless, the diverse scenarios of social learning in the animal kingdom, along with the types of knowledge acquired, could offer insights for the development of ethorobots. It may also emerge that complex information acquisition via social learning is not always necessary; even little socially gained experience can enhance an observer's problem-solving skills.

1.9.1 Social influence

Social influence includes all scenarios in which the behaviour of one partner activates some actions on the part of the other, typically in the absence of learning. Members of a group typically spend a lot of time in close proximity. Depending on the species or ongoing activity (and many other factors), this can fall in the range of millimetres to metres. Group members need to coordinate their activities and actions to avoid chaotic situations. The coordination may take place at different levels in different situations, and social in influence contributes also to regulate and synchronise group activities.

Action coordination during group travel: there are several complex sensorimotor mechanisms in place that minimise the collisions of the group members, for example, during coordinated swimming or fleeing from predator attacks.

Activity coordination: there are several external (e.g., light cycle) and internal cues (e.g., level of blood sugar) that result in coordinated activities, but activities shown by group members also have a facilitating effect on others, that is, specific actions act as a releaser for similar actions in the observer. In dog puppies or chicks, eating behaviour can be elicited in satiated individuals when they are exposed to hungry companion consuming food. Attacking behaviour observed towards individual singled out, increases the chance that others also join in.

Social monitoring: members of a group observe each other continuously. This mutual awareness is advantageous if there is a need for rapid (coordinated) reaction, for example when the group is about to venture out for a longer travel. Social monitoring can also reveal what the partner is up to, so that the observer has a better chance to join in or intervene.

1.9.2 Social learning

Social learning is a central process for obtaining information from other group members, which the observer can use to its advantage at some point later in time (Whiten & Ham, 1992). In social groups, some individuals (demonstrators) could be more knowledgeable, experienced or skilful and this situation provides an opportunity for the others (observers) to learn. Such social information is very valuable because it contributes to the fitness of the (less experienced) observer at a lesser cost than individual learning. This is especially important for young and unexperienced offspring which is often at greater risk due to their constrained sensory and motor skills.

Generally, it is assumed that observers can pick out from whom it pays to learn because social learning often emerges when the individual is facing a problem. For example, having nothing to eat, the animal is attracted by feeding individuals and this can lead it to a potential food source. Social learning can also lead to problems if the observer obtains less valuable or outdated information from the other. Thus, the individual may need to recognise its own inability to solve a problem and find and recognise a partner who shows evidence of such capacity (Kendal et al., 2018).

In animals, most of the research has been focused on isolating the different learning mechanisms that work during this process. Researchers carried out many carefully designed experiments aimed at separating one kind of mechanism from the other but still could not avoid heated debates on the interpretation of the observations. The main issue is centred on the exact content of the learning process: did learning take place, and what did the animal learn?

Research in recent years has led to the characterisation of processes that correspond to the two questions above. However, it is not clear whether these categories indeed reflect completely separate mental processes or, rather, utilise overlapping structures when activated (Heyes, 1994). It is likely that, similarly to other cases, these processes take place within a hierarchical system, the components of which are part of the same network. This interwoven feature of different mechanisms is also clear from the difficulty of providing explicit mutually exclusive verbal descriptions for most of them. Despite the above, there seems to be a consensus among researchers in discriminating four basic mechanisms of social learning (Galef, 2012; Heyes, 1994; Whiten, 2005).

Social enhancement: any activity of a group member may raise interest in the others. Being at or going to a specific location, interacting with objects directs the attention of observers to that aspect of the environment. The observed behaviour acts as a trigger to do the same. As a consequence of social enhancement, the observer learns about the significance of specific places (e.g., locations of food) or objects (e.g., consumes the same type of food). For example, a naïve rat (observer), after interacting briefly with a previously fed conspecific (demonstrator), exhibits an enhanced preference for the diet its demonstrator had been fed (Galef, 2012), providing a case for stimulus enhancement.

Observational learning: the activity of the demonstrator may lead the observer to learn about a relationship between a specific action and an outcome (Box 1.6). In principle, the knowledge gained is analogous to what would have been learnt as a result of an own action (operant conditioning). The advantage of such learning is especially clear in case of potentially dangerous situations when the observer can obtain the same knowledge in the absence of negative experience. Rhesus monkeys (*Macaca mulatta*) acquire fear of snakes through observational conditioning (Mineka et al., 1984) when they witness the escape behaviour of experienced wild-born companions in the presence of snakes. Such learning is facilitated when the demonstrators also display specific communicative signals, such as high-pitched sounds emitted during the escape which are usually associated with a state of fear. This suggests that learning takes place if the situation elicits analogous inner state in the demonstrator and the observer. Eavesdropping seems to be also a form of observational learning (see Section 1.7.3).

Box 1.6 Young birds learn about predators by social learning

In an experimental setup, blackbirds (*Turdus merula*) captured from the wild are introduced to a simulated threat by a stuffed owl (stimulus). Simultaneously, young and inexperienced birds are positioned on the opposite side of the apparatus, where they

Figure 1.16 The arrangement of the experiment investigating how young birds learn about their predators (based on Shettleworth, 2010).

encounter an unfamiliar object, a bottle (novel stimulus) (see Figure 1.16). As the adults emit alarm calls (sign stimulus for the young) in response to the stuffed owl, the fledglings associate the bottle with a predatory presence. Subsequent to this encounter, the observer or listener naïve bird displays mobbing behaviour (learnt behaviour) in response to the bottle, prompted by the learned association (Curio et al., 1978) (see Table 1.6).

Table 1.6 Overview of the learning process (based on Shettleworth, 2010)

Earlier time (T1)		**Later time (T2)**	
Sign stimulus →	Modal action pattern	Learned stimulus →	Modal action pattern
Mobbing calls and behaviour by demonstrator	*Mobbing responses by the observer*	*Bottle*	*Mobbing*
Novel stimulus ↗ *Bottle*			

Imitation: the demonstrator exposes the observer to an action that has some novel components in itself or in relation to the environment. Imitation occurs if the observer repeats this action topographically after some delay in a similar situation. Imitation allows the observer to learn about the organisation (e.g., shape and dynamics) of the action or sequences of actions that lead to improving motor skills (Heyes, 1994; Whiten & Ham, 1992) (see also Box 1.7).

In animals, imitation of actions most often takes place in relation to acting on an object than copying extreme forms of movement. The existing range of actions and the relationship of the novel observation to a specific action may also determine whether imitation takes place. If the demonstrated action is very different from any action in the repertoire of the observer, then it is less likely that copying is successful. Matching is achieved step by step approximations, and several attempts have to be made until the observer displays a rather similar action. Voelkl and Huber (2007) exposed marmoset monkeys (*Callithrix jacchus*) to a conspecific demonstrator that was previously trained to open a plastic box in a specific manner by using its teeth. Detailed analysis showed that the observers precisely copied the movement patterns of the novel action demonstrated by the demonstrator.

Copying the sequence or the hierarchical (nested) structure of actions has been also regarded as a form of imitation (Byrne & Russon, 1998). While the logic of this argument is acceptable, the underlying mental structures for action imitation and action sequence imitation are probably different.

Box 1.7 Imitative learning may be more efficient than trial-and-error learning

Dogs can acquire new behaviours through various methods, with operant conditioning being the traditional and widely adopted approach. In this method, the dog is motivated to perform various actions, and the trainer rewards the behaviour only if it aligns with

Figure 1.17 (a) In Do as I Do training, dogs learn to imitate actions demonstrated by the human partner upon hearing a common command (b) In trial-and-error learning, dogs have to 'figure out' what the partner wants from them. They should try displaying various behaviours and then determine whether they receive a reward for that particular action or not. (Photos by Claudia Fugazza).

the target action envisioned by the trainer or closely resembles it. The most popular form of the trial-and-error procedure is called Clicker Training, which involves using a clicking device to indicate the correct action for the dog. An alternative approach known as the Do as I Do procedure aims to teach the dog to imitate actions demonstrated by the trainer. In this case, the trainer executes a specific action and instructs the dog to replicate it. Dogs need prior training for such performance to understand the meaning of the command (for a detailed description of the two procedures, see Fugazza & Miklósi, 2014).

In an experiment conducted by Fugazza and Miklósi (2014), the performance of dogs certified in either Clicker Training or the Do as I Do procedure was compared. Dog owners were instructed to train their dogs for various novel tasks using either of the two methods. Tasks were categorised as simple (e.g., ring a bell), complex (e.g., pull a drawer), or action sequences (e.g., place a purse into a basket). While there was no difference in the speed of learning simple tasks between the two groups, dogs trained through the Do as I Do method demonstrated faster mastery of complex tasks and action sequences (Figure 1.17).

Emulation: there are cases when observers show no tendency to copy the demonstrated behaviour, rather they try to solve the demonstrated problem on their own way, relying on their habitual action set. It seems that it may be situation dependent whether observers copy the demonstrated action or aim to achieve the goal based on their own experiences. For example, if the demonstrated behavioural solution is too difficult, complex, or very different from the action set used, or if it cannot be perceived well, observers may be inclined to go their own way. Naïve chimpanzees could observe trained companion to use a rake for pulling food into their cage. When given the chance, they also used the rake with similar aims but did not succeed, because they used the rake inappropriately (with upward spikes) (Tomasello et al., 1987).

As indicated above, social learning is only one of several mechanisms for adapting to ecological challenges (see also Box 1.8). Utilisation of modal action patterns and trial-and-error learning can also yield similar results. Laland (2004) proposed a hierarchical relationship among different forms of behaviour control (Figure 1.18).

1.9.3 Learning from whom?

There are situations where information obtained from others can be outdated, inaccurate, or flawed. Therefore, indiscriminate reliance on social learning may have costs that outweigh the disadvantages of individual learning.

How should an observer know from whom to learn? Laland (2004) investigated various 'when' and 'who' strategies that may contribute to the success of social learning. Learning by observation can be particularly valuable when alternative forms of learning are costly or the individual lacks the knowledge to solve the problem independently.

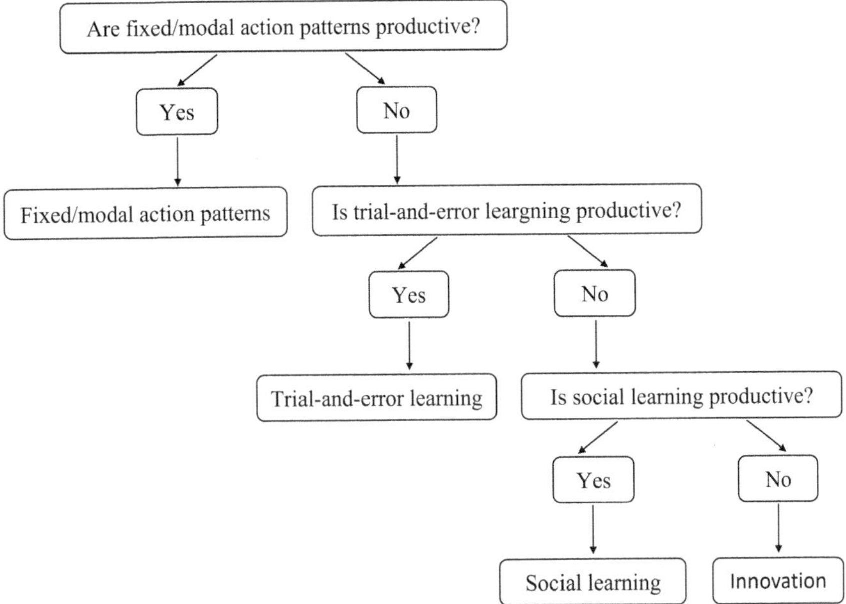

Figure 1.18 Hierarchical control of behavioural strategies (based on Laland, 2004). Fixed or modal action patterns play a predominant role in early development and are also crucial in novel situations, such as the first mating season. The learning capacity rapidly develops, enhancing and expanding the repertoire of actions grounded in fixed or modal patterns. While trial-and-error learning significantly contributes to an individual's possibility to accommodate to its environment, it also entails certain risks. Therefore, learning from experienced others may be a preferable solution when circumstances permit, such as when parents are present.

Box 1.8 The effect of social learning on problem-solving behaviour

Social learning in dogs can both help them overcome their own tendencies and potentially lead to the adoption of disadvantageous habits. Researchers, led by Pongrácz and colleagues, conducted several experiments to investigate the impact of social learning on problem-solving behaviour in dogs.

1 Dogs require only 2–3 observations of a demonstrator to independently solve a detour problem. In contrast, using trial-and-error learning, they need at least 7–8 attempts to solve accomplish a detour around the fence on their own (Pongrácz et al., 2001);

2 When dogs have the opportunity to cross a fence through an opening at its apex, they tend to develop a preference for this route. However, social learning accelerates the process of reversing this tendency if this opening is closed. Two or three observations of an experienced partner are sufficient for the dog to change its tactics and follow the route demonstrated (Pongrácz et al., 2003); in contrast, dogs without such experience continue to make attempts to pass through the closed opening. This demonstrates that social learning can mitigate the persistence of disadvantageous preferences;
3 While dogs quickly learn that detouring is an efficient way to navigate a V-shaped fence, repeated observations may lead them to miss opportunities for shorter routes

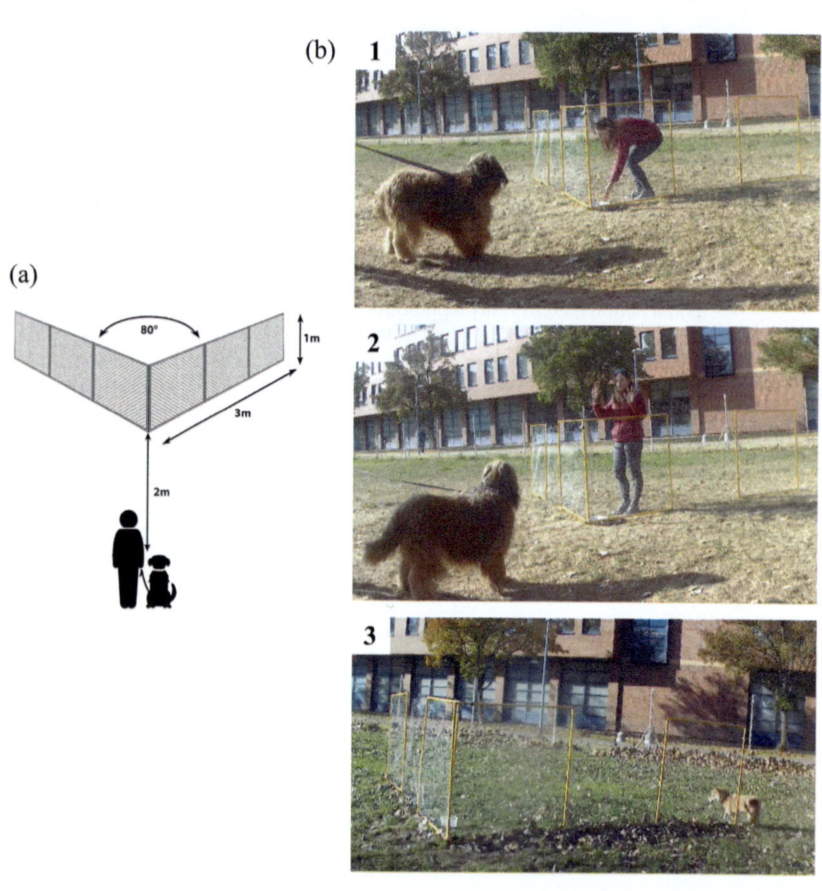

Figure 1.19 (a) The V-shaped fence used for the detour experiments (figure from Pongrácz et al., 2021); (b) (1) the experimenter shows the dog the reward; (2) the experimenter places the reward on the inner edge; and (3) a dog detours around the fence (photos by Csenge Lugosi and Péter Pongrácz).

through just-opened openings (Pongrácz et al., 2003). In some cases, social learning may initiate or maintain disadvantageous behaviours.

These experiments illustrate that the ability to learn from experienced partners plays a vital role in an individual dog's life, providing flexible ways to accommodate to its environment. Importantly, the utility of socially obtained information depends on the learner's experience and various contextual factors (Figure 1.19).

Observers can make informed decisions by copying behaviours that are (1) commonly performed, (2) demonstrated by successful individuals, (3) superior to their own solutions, or (4) executed by kin, friends, or more experienced group members.

1.9.4 Ethorobotic perspective

Social learning serves as a valuable source of insights for designing ethorobots, revealing the technical hurdles that must be solved before this mechanism can be harnessed effectively by artificial agents. Even within ethology, little attention has been devoted to social monitoring, which underpins many processes of social learning. While observing others, the observer's focus extends beyond anticipating the partner's potential future actions, it also encompasses a curiosity about novel facets in their activities. Thus, the observer has to distinguish between past and present occurrences and evaluate whether this novelty holds significant informational value, prompting decisions on how to respond or engage with it. This may involve approaching the object or location, or attempting to replicate the observed action. Responding to such events in a spontaneous manner necessitates a complex mental analysis of the scenario, coupled with memory recall of past events.

1.10 Cooperation

The concept of cooperation among living organisms has often been met with incomprehension, as it appears to defy the principles of Darwinian natural selection. Animals' interaction to achieve a common goal that could not be achieved through solitary effort, is a common definition of cooperation (Dugatkin, 2009). Joint actions result in a net benefit for the participants calculated over a longer period. This helps to explain why partners are willing to invest in cooperation that may also involve some risk.

The utility of ethorobots hinges critically upon their capacity for effective collaboration. In the animal kingdom, the decision to engage in an ongoing or future collaborative interaction is a matter of choice. For instance, a female great tit (*Parus major*) has to make a serious decision considering multifaceted factors before she joins (chooses) or not a male to establish a family and collaborate with him in parenting. A great tit has little chance to bring up the chicks alone but choosing a bad partner could be even worth (e.g., Norris, 1990).

In contrast to animals, ethorobots do not have real alternatives because they are programmed to work with people, and there is no need for any return of the investments. Nonetheless, ethorobots should learn to assimilate knowledge about their prospective human partners, thereby gaining insights into their behaviours and capabilities. Additionally, ethorobots could be equipped with the capacity to assess the efficacy of their interactions, empowering them to decide about altering partners or adjusting their own behaviour accordingly.

1.10.1 Cooperation and defection

From a different perspective, some individuals may be viewed as altruistic, as they are willing to incur costs in order to provide reproductive advantages to their potential competitors. Animals that exhibit altruistic behaviour may be at risk if their partners act selfishly, because cooperative interactions offer many opportunities for cheating. Prey fish, such as sticklebacks (*Gasterosteus aculeatus*), need to collect information on nearby predators. They may do this by approaching potential predators in small groups to minimise the risk of being attacked (and potentially being eaten). In this type of collaboration, the leader fish is followed by the others. However, if the follower(s) decide to abandon the approach, the leader may find itself alone and vulnerable to a predator attack (Milinski, 1987).

This behaviour of the sticklebacks is known as 'tit for tat' behaviour strategy, in which the leader's cooperative move is followed by similar actions by others in the group. This strategy enables the interaction to be sustained if cooperative acts are reciprocated. According to Axelrod (1984), the 'tit for tat' strategy is based on a simple rule: the initiator must act cooperatively first, and subsequent actions by either party should mimic the other's previous behaviour. This cooperation strategy may evolve in situations where the likelihood of encountering the same partner is high, such as in groups. In such cases, individuals are more likely to adopt the 'tit for tat' strategy, where they start by cooperating, but will cheat if their partners respond by cheating. Additionally, the success of this strategy relies on individuals not holding onto memories of their partner's past actions for too long (Dugatkin, 2009).

1.10.2 Contexts for cooperation

Evolutionary biologists have proposed a few scenarios that could serve the basis for the emergence of cooperation. First, partners may cooperate if they are genetic relatives. In this case, altruism increases the fitness of its own kind. This happens when unselfish great tit (*Parus major*) parents feed their offspring, or when the family of related animals go out for a joint hunt, like in wolves. Another common mechanism for cooperation is reciprocal altruism in which the partners are members of the same group, recognise each other on the level of individuals, and possess a behavioural mechanism that maintains cooperation through some form of punishment for those who cheat (Dugatkin, 2009).

The vampire bat (*Desmodus rotundus*) serves as a prime example of reciprocal altruism (Wilkinson, 1984). Vampire bats require a regular intake of blood to avoid starvation, as they cannot survive more than one to two days without food. When a bat has had an unsuccessful hunt and is in danger of starving, it begs for regurgitated food from other bats, typically receiving only a small amount, just enough to survive the next day. Interestingly, it has been observed that in the future, when the roles are reversed and former recipients have become donors, those who had previously received help are more likely to return the favour by providing regurgitated blood to their former recipients. This system of reciprocal altruism is also supported by the high chance for being unsuccessful for any bat in the colony, that is, today's winners can easily become tomorrow's losers.

Cooperation can sometimes be enforced by a dominant animal in certain situations, even when a partner has no gains from the interaction and attempts to defect. For example, keas (*Nestor notabilis*) were trained to operate a seesaw cooperatively to access food. Dominant keas were able to monopolise the food and they forced subordinates to activate the apparatus so that they could eat (Tebbich et al., 1996) (see Figure 1.20a). While this observation highlights an alternative mode of collaboration, such interactions may not be stable under natural conditions.

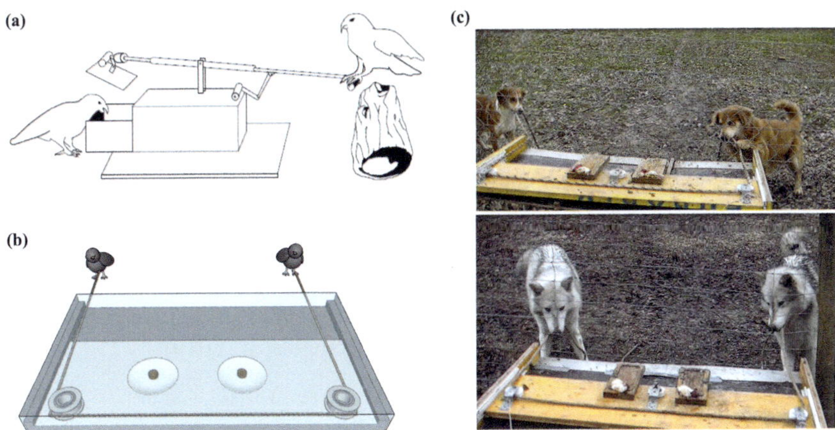

Figure 1.20 (a) Asymmetric (forced) collaboration: one kea, typically the subordinate individual in the dyad, sits in the handle of the seesaw, while the dominant is feeding from the container. This situation does not represent a voluntary choice by the subordinate. The dominant chases the other until it flies to the handle, this is the only 'safe' place for that bird (from Tebbich et al., 1996). (b) Symmetric collaboration: in the loose-string-pulling task, the two ravens need to coordinate their behaviour with each other in order to be successful, and the reward is divided equally among them (e.g., Asakawa-Haas et al., 2016). (c) The string-pulling task performed by dogs (upper) and wolves (below) (figure from Marshall-Pescini et al., 2018).

1.10.3 Behavioural synchronisation

The coordinated behaviour of joint actions is an essential aspect of successful cooperation. Such coordination may involve spatial and temporal aspects that can make the task challenging. However, various solutions have evolved to reduce the perceptual and motor load of mental processing, making coordination of actions achievable through the use of behavioural rules or contextual constraints. Cooperative mound building in termites (*fam.* Termitidea) is an example of the former case where individuals follow a sequence of rules. They first dig for soil and then deposit it near larger deposits made by others. As these small hills grow, termites connect them at certain heights, forming arches. Although a specific pheromone released by the builders may also influence this collective behaviour, the individual termites work independently of each other (Green et al., 2017).

Hunting wolves align themselves in a particular pattern when chasing a large prey, such as a white-tailed deer (*Odocoileus virginianus*). This configuration, however, may not result from specific planning. Instead, each wolf aims to get as close to the prey as possible while also maintaining a safe distance from the other chasers. These two rules alone can explain why chasing wolves may form a semi-circle behind the escaping deer, and why some wolves need to run ahead of the deer to surround it from the front (Mech, 1970).

Generally, laboratory studies do not challenge the participants with complex cooperative tasks, although some skills for coaction are needed to solve these problems. The so-called loose-string-pulling task is a popular approach for studying cooperation (Hirata, 2003). This consists of a rolling wooden board placed out of reach from the subjects with two ring bolts on both ends. A long rope is driven through both bolts ending close to the subjects' area (Figure 1.20b, c). To move the board, both ends of the rope must be pulled simultaneously. If only one end is pulled, the rope slips out of the ring bolts without moving the board. Successful completion of this task requires partners to wait for each other to approach the ends of the rope and perform a coordinated pull. For example, capuchin monkeys (*Sapajus apella*) and chimpanzees have successfully learned to solve this task in pairs (Silk, 2007), while wolves and dogs have also been observed collaborating with humans. However, in joint pulling tasks with conspecifics, only wolves were successful (Range et al., 2019).

1.10.4 Recruitment and choosing partners

Recruiting partners is an important aspect of cooperation because group members may be engaged in different tasks at the same time. Thus, the initiator may either have to wait for others to volunteer or actively modify their behaviour through communication. Alternatively, harassment may also be a possible tactic in a dominant-subordinate relationship (see above).

Studies have shown that specific signals for recruiting conspecific partners were rarely observed in chimpanzees and wolves when they faced the string-pulling task. Typically, they waited until the other approached one end of the rope. However, communication did take place with human partners. In contrast, cliff swallows (*Hirundo pyrrhonota*) circle around in the air and emit alarm-call-like sounds to recruit partners for their next trip to a newly discovered food source upon returning to their colony (Brown et al., 1991).

Partner selection is an important factor for successful cooperation, and individual attributes such as a prosocial, supportive attitude or better skills can make a significant difference (Abdai & Miklósi, 2016). In most experiments, subjects were given the opportunity to observe two humans, one of whom shared food with a begging person while the other did not. While chimpanzees displayed some preference for the prosocial partner (Melis et al., 2006), in dogs, no clear preference was observed. Similarly, laboratory observations did not provide strong support for the idea that collaborators choose more skilful partners over clumsy ones. Thus, it remains unclear whether animals are capable of evaluating these characteristics and making choices accordingly. However, despite many experimental efforts, it remains unclear whether dogs or chimpanzees prefer prosocial partners over antisocial ones.

In some cases, smaller sub-groups or coalitions may form to take action against one or more individuals. Coalition formation is common among primates, but it has also been observed in wolves and dolphins (Harcourt & de Waal, 1992). To achieve their goals, individuals within these coalitions must act in concert. In natural conditions, coalitions are often involved in agonistic interactions (attacks or protective moves), or for example, male coalitions in bottlenose dolphins (*Tursiops aduncus*) are organised to gain control over females in order to mate with them (Connor et al., 2022).

1.10.5 Sensitivity for outcomes

The benefit of the cooperative action may play an important role in determining future interactions between the partners. Partners engaged in collaborative actions may share an asymmetric social relationship within the group, leading to a non-equal distribution of gains. For example, chimpanzees engaged in joint hunts on monkeys do not share the meat equally; dominants usually take the larger portions. Observations have also shown that meat can be used as an investment for gaining support from others. Male donors of meat were supported in group disputes by former recipients, and this relationship was typically mutual (Mitani & Watts, 2001).

While complex and compensatory mechanisms that maintain cooperation are often observed in nature, laboratory experiments have shown that individual-level partners may be sensitive to reward contingencies associated with collaboration and exhibit 'inequity aversion' (Brosnan & de Waal, 2003). In these experiments, animals are required to perform an action (such as handing over a token or pressing a lever) to receive food, with two animals present who can observe each other during the experiment. Both individuals readily cooperated with the experimenter while they were receiving the same quality reward. However, an animal partner may reduce its readiness to cooperate or stop acting altogether if it observes that the other animal is receiving a better-quality reward. The sensitivity to such asymmetry may be species- and context-specific. For instance, rhesus monkeys are more sensitive to reward quality than dogs in similar tests (Oberliessen & Kalenscher, 2019). Dogs stop cooperating only when they do not receive food while the other animal is rewarded, indicating that they are more susceptible to the absence of social interaction with humans (Range et al., 2009).

1.10.6 Teaching

Teaching represents an active form of social learning that involves the teacher initiating and establishing a collaborative interaction with the learner (Thornton & Raihani, 2008). It requires a significant investment of time and effort from the teacher, with the expectation of reaping benefits only later. Caro and Hauser (1992) proposed three criteria for effective teaching: (1) the teacher modifies its behaviour (only) in the presence of the pupil, (2) there is a cost associated with teaching and no immediate benefits involved, and (3) the pupil acquires knowledge and skills faster than in the absence of such experience, only through trial-and-error learning.

It is important to distinguish teaching from social learning, in which the knowledgeable individual (demonstrator) displays the activity regardless of the presence or absence of an observer. When teaching, the teacher's role is to provide the learner with an appropriate context for learning. This can be facilitated through various non-exclusive behavioural mechanisms. A classic example of teaching behaviour can be observed in many felids, where a mother cat teaches her cubs to hunt by gradually introducing them to more challenging tasks (Caro & Hauser, 1992). These are typical contexts for teaching (Thornton & Raihani, 2008):

1. *Attention getting*: the teacher directs the pupil's attention to some aspects of the environment and/or to itself;
2. *Physical support*: the teacher helps the pupil to achieve a goal by physical interaction;
3. *Occasion setting*: the teacher ensures and facilitates that the pupil has the means needed for learning;
4. *Task setting*: the teacher follows the advancement of the pupil, and this is reflected in the increased difficulty of the subsequent tasks.

Human-like teaching behaviour (Csibra & Gergely, 2009) is extremely rare among animals, although some elements of teaching can be found in various taxa, both in invertebrates and vertebrates (Kline, 2015). Dogs represent a specific case, because they are especially sensitive to human teaching (Topál et al., 2009). In their study, Topál and colleagues found that dogs show a specific sensitivity for human ostensive communication, when it was used to signify particular searching strategies demonstrated by the experimenter. Importantly, in that experiment, dogs behaved very similarly to human infants, while socialised (tame) wolves disregarded the ostensive communication.

1.10.7 Ethorobotic perspective

Collaboration is a fundamental feature in social animals probably selected for or emerged as a by-product of group living. These features often manifest in relatively simple phenomena, like feeding the offspring or going on a hunt together. However, looking more closely one can detect a complexity of tightly connected and coordinated actions that make the controlling of the collaborative activity complex.

Crucially, ethorobots are anticipated to possess the capacity to engage in collaborative endeavours by aiding humans. While this facet has received limited explicit attention, the potential for robots to deliver the greatest impact lies in their assistance with physical tasks. The adept manipulation of objects and the safe transport of heavy stuff by ethorobots could alleviate individuals from dangerous and tiring duties, enabling them to allocate more focus to the interpersonal facets of their human interactions (c.f., quality time, Broadbent et al., 2011).

Contemporary ethorobots still lack the intricate proficiency to engage with objects, particularly when juxtaposed with human capabilities. Consequently, the majority of these robots are unable to operate independently and necessitate human supervision. Consider a waiter robot, for instance, which can assist only in carrying dishes to and from tables but cannot fully assume the role of a human waiter. However, even in such scenarios, these robots could benefit from exhibiting elementary social skills as they pass by humans or navigate around them. It is conceivable that such attributes might heighten human awareness of the robots' presence and enhance their acceptance among both staff and guests.

1.11 Development of behaviour

Development is the intricate process by which a single-cell organism transforms into a complex multicellular body capable of organised behaviour and reproduction. It involves much more than just increasing the number of cells and establishing a modular body. Instead, developmental processes must follow a complex spatial and temporal pattern that is both robust and plastic, stretching over days, years, or sometimes even decades.

For many organisms, the developmental environment changes over time, leading to alterations in the phenotype. Evolutionary processes ensure that organisms adapt specifically to these temporal environments. In some organisms, developmental stages are less pronounced, while in others, such as frogs, the changes are more abrupt, involving three stages of metamorphosis from eggs to tadpoles to adult forms. In humans, distinct developmental stages include newborn, infancy, early childhood, middle childhood, adolescence, early adulthood, middle adulthood, and old age.

Today's ethorobots do not undergo development in the conventional sense as observed in living organisms. Their embodiment remains constant, and developmental changes primarily occur through the learning process. Nevertheless, even this learning process is subject to limitations, partly due to the high cost associated with storing vast amounts of information. Nonetheless, certain facets of development could hold valuable insights for the future of ethorobots, for instance, the utilisation of play as a means to facilitate interspecific interactions or the impact of environmental stimuli (or their absence) on behaviour.

1.11.1 Developmental plasticity

Developmental plasticity enables organisms to reach reproductive adulthood by continuously accommodating to their actual environment. This plasticity may be exhibited through an organism's sensitivity to its environment, such as growing larger in response to increased food availability. It may also result in the production of various phenotypes from the same genotype, such as adopting different reproductive tactics (e.g., becoming male or female). Additionally, some traits may be reversible depending on environmental conditions, such as the ability to revert to a different sex under environmental conditions (DeWitt & Scheiner, 2004; West-Eberhard, 2003).

The degree of plasticity in an organism's phenotypic trait can vary depending on the trait itself. Some traits may develop optimally during transitions from one environment to another, while further changes may not be possible due to prior developmental processes determining the possible future directions.

Waddington (1966) introduced a metaphor to aid in understanding development: the 'epigenetic landscape', represented by a hill down which a ball rolls. Balls released from the top of the hill would arrive in various valleys, illustrating how the environment affects organisms differently, resulting in unique individuals. However, even if the same ball were released repeatedly, it would likely land in different valleys. Waddington's idea is also useful to explain that development is more complex then suggested by the hill metaphor, as organisms represented by the ball are not rigid structures and can change in many ways (e.g., grow larger or smaller). The epigenetic landscape may have valleys connected to one another, representing alternative developmental routes that eventually lead to similar phenotypes. Additionally, there may be smaller hills on the way down, indicating the potential reversibility of some developmental processes. The behaviour of these organisms may also depend on their experiences while rolling down the hill (i.e., learning), as they may actively avoid certain valleys. Although the epigenetic landscape is a crude simplification, it provides the ground for thinking about specific developmental mechanisms that may be important for a plastic organism.

1.11.2 Maturation

Kingsley and Garry (1957) defined maturation as the process by which physical changes in the body, such as growth, affect behaviour. Maturation begins after fertilisation of the egg and ends before death, including changes that involve not only growth but also destructing physical, morphological, and physiological alterations, which result in compensatory behavioural modification associated with old age.

Maturation affects all aspects of structural organisation, including morphology, physiology, and behaviour. Some researchers differentiate between physical maturation, which mainly involves body morphology, and mental maturation, which includes the increasing complexity of thought processes. Importantly, there is an interaction between the structural maturation of the brain and the maturation of cognitive skills.

Maturation is a sequential process that must be coordinated both spatially and temporally. It interacts with other aspects of development, such as learning, but is

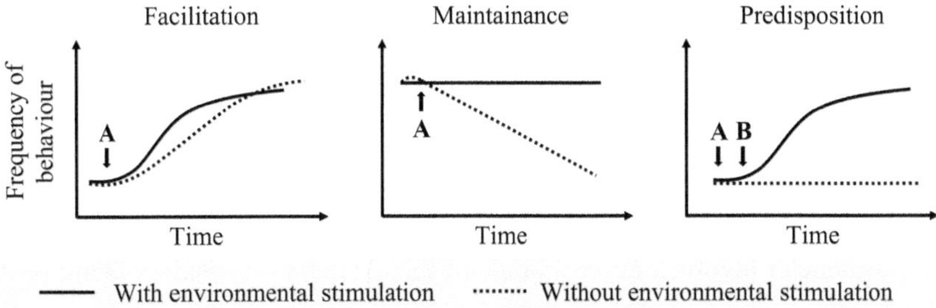

Figure 1.21 Influence of facilitating, maintaining, and predisposing environmental events on the developing organism. A and B indicate the specific environmental stimulus (based on Miklósi, 2007).

also sensitive to environmental influences. For example, children who are overfed may develop larger amounts of adipose tissue and become more likely to be obese as adults, with fewer chances to reduce their weight by fastening (Rodgers & Sferruzzi-Perri, 2021).

1.11.3 Environmental effects on development

Caro and Bateson (1986) provided a useful overview of how environmental effects can influence development (Figure 1.21). Facilitating effects can speed up the development of specific traits. For example, kittens may learn to hunt on their own, but they learn more quickly if their mother provides them with prey and opportunities to practice (Caro, 1980).

Maintaining effects are important for ensuring that a skill remains available to the individual. For example, living in social isolation for some time can lead to the malfunctioning of social skills (Weldon et al., 2016). Predisposing environmental effects may prepare an individual to gain an advantage from later experiences. For example, neonatal exposure of wolf pups to humans has a strong taming effect, which can increase their ability to learn about human communicative signals.

1.11.4 Sensitive periods

During development, there are time frames when the input from the environment is vital. These sensitive periods (referred to also as 'critical periods' in the earlier literature) are characterised by enhanced sensitivity and rapid learning (e.g., Knudsen, 2004). Domestic chicks are precocial, that is, they are able to run and feed on their own soon after hatching. Although they are mobile, without much experience their survival is at risk. The hen provides a secure environment for them but only if the chicks learn about her identity. As a result, chicks and many birds of similar lifestyle have a sensitive phase to obtain information about the visual and acoustic features of their mother (McCabe, 2019). If they do not have the chance for early learning (i.e., in the sensitive period), then they are unable to follow their mother because they do not know how she looks like.

Sensitive phases may also emerge as a result of hormonal changes that also prepare the neural structure for the new experience. It seems that oestrogen and oestradiol may play an important role during the sensitive phase for learning songs from the father in many species of songbirds (e.g., Knudsen, 2004). The sensitive period for song learning is also associated with local neural development in the brain nuclei involved in the storage of the obtained information (song of the father).

Sensitive periods have the important function to focus and speed up learning at a specific period in life. Their duration can vary a lot and has been optimised during evolution depending on the ecological situation. For example, while in chicks with relatively limited mobility the sensitive period lasts for a few days, in horses or sheep in which the newborn depends crucially on the milk of the mother, a similar sensitive period ends after a few hours (Goodenough et al., 2009).

1.11.5 The origin of problem-solving skills

When observing a bird constructing its nest, a cat successfully hunting a mouse, or termites constructing their mounds, one may wonder how they know what to do. Historically, the debate about the role of 'instinct' and 'learning' in these abilities has been futile. Many animal behaviour textbooks have a separate chapter on learning, but one could argue that learning, as a specific mechanism for accommodating local environmental changes, belongs to the topic of development. Learning is crucial at the beginning of life because failure to learn at an early age leads to a shorter lifespan. As a result, learning should be viewed as an opportunity to increase behavioural plasticity and improve survival chances during development.

Learning refers to the increase or decrease in the frequency of behaviour displayed, following either a single instance or repeated experiences, with either a short or extended delay, which typically benefits the individual. Conversely, behavioural changes resulting from maturation, sensory fatigue, or tiredness are not considered as learning (Shettleworth, 2010).

During development learning does not occur in a vacuum; many perceptual, mental, or motor aspects of behaviour have to be established aforehand, although some may not function correctly. Therefore, both maturation and learning are required for the system to achieve optimal functioning. For instance, newly hatched chicks frequently make mistakes when pecking at small seeds on the ground. However, even without additional experience, such as keeping them in the dark for 24 hours without food, their performance improves, indicating some evidence for maturation. Repeated exposure aids in this improvement by facilitating sensory-motor coordination through operant learning (Cruze, 1935). This simple example demonstrates that genetically predisposed mechanisms and maturation interact complexly with learning, which can also be conceptualised as a specific type of feedback (Shettleworth, 2010).

Table 1.7 A schematic overview of the advantages and disadvantages of genetically pre-programmed and learned behaviours. These concepts represent extreme scenarios, and animals typically exhibit a blend of genetically pre-programmed and learned behaviours in varying proportions. The superiority of either mode of determination hinges on the species' ecological context

	Genetically pre-programmed behaviour pattern	*Learned behaviour pattern*
Advantage	Fixed problem (e.g., sun)	Ecological context-dependent problem (e.g., food)
	Immediate pre-programmed response to a significant programme	Rapid reaction to environmental changes, variable behaviour patterns
Disadvantage	Limited and rough solutions	Delayed reaction
	Slow accommodation	Increased chance for error
	Less than optimal behaviour in changing environment	Increased risk during the learning process
Life course characteristics of species (trends)	Small size, short lifespan, no parental care	Large size, long lifespan, parental care

Although the strict differentiation between 'instinct' and 'learning' may be unnecessary, it remains a pertinent question to what extent a particular behavioural skill should rely on genetically predisposed mechanisms, and the level of flexibility that learning should allow. One could argue that in very stable environments where events, challenges, and solutions are highly predictable, evolution would favour the development of behaviour mechanisms without any reliance on learning. Such fixed and inflexible behaviour skills might be advantageous because learning is always accompanied by risk. This trade-off between genetically predisposed mechanisms and learning may alter if the environment changes rapidly, and the individual must adapt quickly (see Table 1.7).

The examination of learning in the context of development is beneficial because animals have the ability to learn how to learn. For instance, when a rat is required to learn to press a pedal to receive a small piece of food, it has extensive prior experience with learning, even though laboratory animals may have less exposure due to the limited conditions of their environment. Learning generally has two outcomes. First, the animal learns a new approach for addressing environmental challenges (problem-solving behaviour), and second, this experience alters its potential to participate in the subsequent learning experience. Learning is never entirely reversible (Laland et al., 2020).

1.11.6 Exploration, attention, and learning

It is crucial for individuals to maintain some level of exploration to accommodate to changing environments and unexpected events. By continuing to explore, individuals ensure that their mental representation of the environment remains up-to-date and relevant.

From a functional perspective, *attention* can be defined as the sensory and perceptual focus on a specific part or aspect of the environment, which involves excluding or reducing input from other sources (Oberauer, 2019). During attentional processes, there is a trade-off between gathering focused information and maintaining wide-ranging monitoring capabilities. For instance, when an ostrich (*Struthio camelus*) searches for food in the grass, its attention is primarily directed towards this task, making it less likely to notice approaching predators. Ostriches possess good vision, and they exhibit scanning behaviour by raising their necks and surveying the surroundings. However, searching for food and monitoring for predators are incompatible actions. In larger ostrich groups, this issue can be partially mitigated through task-sharing in monitoring activities. Observations in nature have supported this hypothesis, showing that individual ostriches in larger groups spend more time engaged in feeding behaviours (Bertram, 1980).

Attention behaviour can be explained, in part, also by the limitations of mental resources. However, evidence suggests that at least humans are capable of divided attention, which is connected to the concept of 'multi-tasking' (Herbranson, 2017). Functionally, the constraints on multi-tasking are not solely related to mental resources but also involve the compatibility of sensory capacities and limitations in action execution. Multi-tasking is more feasible when it involves different sensory and motor processes that do not interfere with each other. For instance, one can walk and read (using a mobile device) simultaneously, but running motion would make reading impossible.

Latent learning: Tolman (1949) has noted that rats exhibit faster learning when searching for a food source in a maze if they have been previously exposed to the maze. It was hypothesised that the individuals acquired knowledge of the maze's internal structure during the pre-exposure period and utilised this information to navigate when they were later searching for hidden food. Since the learning effect was not immediately apparent during the pre-exposure phase and there was no obvious reinforcement present to facilitate learning, this phenomenon was labelled as latent learning.

Perceptual learning is a significant outcome of environmental monitoring. Through experience, individuals enhance their ability to learn the significance of stimuli or objects (Goldstone, 1998). Familiarity, which arises from continuous monitoring, facilitates the subsequent discrimination of these stimuli. For instance, rats exposed to visually distinct objects in their cage are able to learn and differentiate them more quickly in subsequent learning tasks.

The presence of latent and perceptual learning strongly suggests that animals actively seek out information in their environment. Continuous monitoring not only allows for the recognition of new environmental input but also enables the accumulation of additional information about the unchanged aspects of the surroundings. Repeated encounters with the same stimulus or object may result in a more detailed mental representation, ultimately making subsequent discrimination tasks easier.

Search image: exploration can be also under more direct perceptual control. The concept of a search image has been used to explain patterns of searching behaviour in various species (Reid & Shettleworth, 1992). The term suggests that the decision-making process during search is influenced by a mental representation or image of the target (prey) (Langley, 1996). In many observations, predators, such as birds searching for insects or seeds, do not exhibit optimal behaviour. They may delay selecting a novel food type when it appears in the environment, resulting in a feeding frequency that is lower than expected (Blough, 1989). Despite a decline in the frequency of a preferred food type, predators often continue to favour it, even if this choice becomes less profitable. The generation of a mental search image is believed to be the underlying cause of this suboptimal behaviour.

1.11.7 Learning mechanisms for accommodation to environmental challenges

Extensive laboratory observations have led to the identification of several common types of learning in animals through research in animal learning. These mechanisms have garnered significant support from the study of a broad array of species, including humans. Moreover, many of these insights have been successfully utilised in artificial systems to simulate learning capacities, such as artificial neural networks (Zador, 2019).

It is worth mentioning that in real-world environments, these apparently distinct mechanisms do not operate in isolation, and the dynamics of events differ significantly from those encountered in controlled laboratory settings. As such, rather than detailing these mechanisms extensively (which can be found in any textbook on learning, such as the chapters in Shettleworth, 2010), this discussion will only focus on their role in promoting survival (see also Section 3.7.2 for discussions on learning mechanisms).

Attention to events: organisms exhibit a decreasing response rate to stimuli or events that do not pose a threat to their immediate survival. This ability is crucial as it enables individuals to focus their attention on a select portion of their environment. Such diminished responsiveness is referred to as habituation, and it can occur in response to specific stimuli, as well as in more complex scenarios (Shettleworth, 2010). For instance, repeatedly touching a particular area of a small snail's body results in a progressively reduced response (contraction). However, applying the stimulus at a different location causes the contraction to intensify once again (dishabituation).

Habituation can also occur in response to more complex situations, such as entering a new environment with many novel elements. In such cases, exploration behaviour is activated as the animal aims to map the novel features of the environment and learn how to navigate it. As the animal becomes more familiar with the environment, its exploratory activity decreases, and habituation takes place. This decrease in exploration behaviour is not due to sensory fatigue or motor exhaustion but rather reflects the animal's increasing familiarity with the environment. However, if a novel object is encountered on the familiar terrain, the animal's interest can be revived, and dishabituation may occur. Both habituation and dishabituation are central learning processes that occur in response to changes in the animal's environment. These processes play an essential role in an animal's ability to accommodate to its environment and respond appropriately to novel stimuli or ignore familiar ones (Rankin et al., 2009). A somewhat reverse process occurs when repeated encounter with the same stimuli increases the animal's reaction to it (sensitisation). Experiencing dangerous situations frequently may increase the sensitivity (the reaction shown) towards the specific stimuli.

Predicting events: a considerable proportion of events do not occur randomly, even though they are associated with significant uncertainty. For an individual, it is important to find out what is the chance that some noise in the grass indicate moving of a prey, or how likely it is that it finds some seeds on the junipers. From a different perspective, noise or junipers predict the occurrence of some kind of food. The mental mechanism that deals with this kind of problem is called Pavlovian or classic associative learning (Shettleworth, 2010).

The importance of Pavlovian learning becomes apparent when comparing individuals who have had the opportunity to learn with those who have not. Hollis et al. (1995) conducted an experiment in which three-spot gourami fish (*Trichogaster trichopterus*) were placed on two sides of a small tank separated by a fixed glass wall and a movable opaque wall. A light stimulus was presented to the fish, followed by raising the opaque wall, allowing the territorial fish to attack their partner through the glass. Over time, the latency for attack decreased, indicating that the fish had learned to associate the light stimulus with the appearance of the opponent. After this initial training, the trained fish were paired with naïve gouramis in similar tanks. Earlier, the naïve fish were exposed only to the light stimulus without any subsequent event.

In the experiment, the researchers assessed the behaviour of the fish when the opaque wall was removed, following the presentation of the light stimulus. They found that the trained fish were more successful in winning fights against the naïve fish. This may be because the trained fish were able to quickly change their inner state and behaviour in response to the light stimulus and launch a successful attack, while the naïve fish were unprepared. Competitive situations where speed is advantageous are common in nature, and individuals may compete to see who can better predict the spatial and/or temporal occurrence of an event.

Sign stimuli are crucial because they often facilitate learning by directing the attention of the subject. For example, the two horizontal eye spots on the head of a predator can capture the attention of a prey fish and prompt it to focus on the carrier. If this encounter is followed by an attack, the next time the fish encounters a similar sight or smell associated with the predator, it may activate avoidance behaviour as a learned response (Csányi, 1993).

Predicting action outcomes: every individual is born with a unique set of genetically predetermined actions that can have varying outcomes in different contexts. Some actions may have positive, neutral, or negative consequences for the animal. Learning enables animals to maximise the benefits of their actions by adjusting their behaviour in response to environmental feedback. This type of learning is known as

Skinnerian or operant learning, which relies on the consequences of actions to modify behaviour. Positive feedback ('reward') can increase the frequency of specific behaviours, while negative feedback ('punishment') can decrease their frequency (Shettleworth, 2010).

In many cases, genetically predetermined actions serve as the foundation for learning, where a rudimentary version of the action is initially carried out and then refined through feedback. This process is exemplified by the development of bite inhibition in dogs (Miklósi, 2007). Puppies initially rely on biting to 'grasp' or 'hold' body parts of others in their mouth, but their lack of control and sharp teeth can cause pain. Through feedback from their peers, which may include withdrawal of the body part, high-pitched cries, and counterattacks, the puppy learns to bite (grasp with their muzzle) without causing pain. The other's feedback is essential for operant learning, and the behaviour gradually takes its optimal shape when puppy is able to grab the others by a painless bite. Similarly, complex sequences of actions can also be acquired through operant learning.

From a cognitive perspective, learning about actions in a context-dependent manner forms the foundation of goal-directed behaviour. When faced with various possibilities in an environment, a specific internal state can activate actions with specific outcomes. Balleine and Dickinson (1998) demonstrated this concept by training rats to pull a rope when thirsty and press a lever when hungry. Satiated but thirsty rats pulled the rope repeatedly if given both options, in the absence of any food or water. This type of differential experience can lead to specific mental associations between actions and outcomes, which may form the basis of goal-directed behaviour. Although similar actions may lead to achieving different specific goals, the mind may develop a general representation of these goals, referred to as intentions.

1.11.8 Play

Most researchers generally agree that play is common among mammalian species, but it has also been observed in many species of birds. However, the existence of playful behaviour in other groups of animals, such as reptiles, fish, or cephalopods, remains a topic of debate. Because play or playful activities are difficult to define, some experts, such as Burghardt (2005) and Galpayage Dona et al. (2022), have suggested providing a list of characteristics that can be used to differentiate between play and non-play:

1 Playful behaviours are distinct from those displayed by adult animals and often involve interrupted or incomplete behaviour sequences;
2 Play occurs voluntarily, and is rewarding on its own;
3 Play activity is repeated but not stereotyped and differs from its functional form;
4 Play occurs when there is no immediate need for vital activity, and the animal is in a relaxed mental state;
5 Playful behaviours do not always have a specific goal or serve a practical purpose what would be typical for an adult.

Playful activities are often categorised as locomotor play, object play, and social play, as these types are believed to play a specific role in behavioural development (Held & Špinka, 2011). The frequency and types of play vary among species and throughout an animal's life. In most species, play emerges only within a certain time period during development, for example, in rats social play peaks at one month of age and gradually declines thereafter, a pattern also observed in many feline species. In social species, playful behaviour is often part of the adult repertoire and is displayed specifically when there are offspring present in the group or family (Špinka et al., 2001).

The ethological approach to animal behaviour is typically focused on understanding its function, but when it comes to play, a simple explanation remains elusive. Over the years, several non-exclusive hypotheses have been proposed to explain why play is prevalent in many mammals and birds but absent in most other species. Play may serve various functions in animals, including:

1 Enhancing physical and social skills, such as strength, speed, and coordination;
2 Providing opportunities for learning and gaining experience;
3 Facilitating the establishment of social partnerships, such as friendships;
4 Allowing for practice in collaboration with others;
5 Offering situations for experiencing different mental states; and
6 Preparing individuals to successfully face unexpected events.

Researchers generally agree that these features of play are best carried out under safe and relaxed conditions, when the individual is not under pressure from other needs, such as hunger. However, even under safe conditions, play activities can be risky and can contribute to fatal accidents. In fact, play is often the cause of death among young animals.

While play is often considered a foundation for the development of complex cognitive skills such as fairness, trust, and morality (Bekoff, 2001), there are few studies that could provide strong evidence that being more playful has a fitness advantage at the individual level for the adult life. One of the best-known examples comes from studies on kittens, which suggest that more playful kittens may have better-developed predatory skills (Caro, 1981). However, it is possible that individuals with strong predatory tendencies are simply more likely to play when they're young, rather than play itself having a direct causal effect on the development of those skills.

1.11.9 Ethorobotic perspective

While development in the biological sense may not be a feasible option for ethorobots due to the constraints imposed by their inorganic material, learning mechanisms remain the primary means for these agents to accommodate to changing environmental conditions. Despite this, those few robots, which had been sold and perform activities in real situations, still rely on a limited number of predetermined (fixed) action patterns, resulting in minimal flexibility.

Living organisms possess a remarkable ability to continuously sense, select, perceive, mentally process, and evaluate environmental information, enabling them to be in a constant state of alertness and readiness to respond to a variety of inputs. In contrast, ethorobots typically lack such monitoring capabilities, and while they may be able to perform certain tasks simultaneously, they are not well-equipped to handle unexpected events. This shortcoming is particularly problematic in social problem-solving situations, where nuance differences in reactions could derail the social interaction with others.

Development is crucial in creating distinct individual robots. While ethorobots can be programmed to exhibit personality traits advantageous for specific tasks, exposure to real-world work environments and interactions with people could modify these traits, giving the robot a sense of individuality.

Considering development as changes in behaviour over time, it can also strengthen social relationships based on the duration of time spent together. This may be influenced by the experiences one accumulates with a partner over time. Expressing interest through an open attitude, seeking help in necessary, and displaying receptiveness to learning may foster a positive response from human partners towards these ethorobots. Additionally, incorporating elements of play or playfulness could further facilitate engagement with partners, making ethorobots appear more social and acceptable.

References

Abdai, J., Korcsok, B., Korondi, P., & Miklósi, Á. (2018). Methodological challenges of the use of robots in ethological research. *Animal Behavior and Cognition*, *5*(4), 326–340. https://doi.org/10.26451/abc.05.04.02.2018

Abdai, J., & Miklósi, Á. (2016). The origin of social evaluation, social eavesdropping, reputation formation, image scoring or what you will. *Frontiers in Psychology*, *7*, 1772. https://doi.org/10.3389/fpsyg.2016.01772

Abdai, J., & Miklósi, Á. (2022). After 150 years of watching: Is there a need for synthetic ethology? *Animal Cognition*, *26*, 261–274. https://doi.org/10.1007/s10071-022-01719-0.

Andrew, R. J. (1991). *Neural and behavioural plasticity: The use of the domestic chick as a model*. Oxford University Press.

Arkin, R. C. (1998). *Behavior-based robotics*. MIT Press.

Arkin, R. C., Fujita, M., Takagi, T., & Hasegawa, R. (2003). An ethological and emotional basis for human-robot interaction. *Robotics and Autonomous Systems*, *42*(3–4), 191–201. https://doi.org/10.1016/S0921-8890(02)00375-5

Asakawa-Haas, K., Schiestl, M., Bugnyar, T., & Massen, J. J. M. (2016). Partner choice in raven (*Corvus corax*) cooperation. *PLoS One*, *11*(6), e0156962. https://doi.org/10.1371/journal.pone.0156962

Ashby, W. R. (1958). Requisite variety and its implications for the control of complex systems. *Cybernetica*, *1*(2), 83–99. https://doi.org/10.1007/978-1-4899-0718-9_28

Axelrod, R. (1984). *The evolution of cooperation*. Basic Books.

Balleine, B. W., & Dickinson, A. (1998). The role of incentive learning in instrumental outcome revaluation by sensory-specific satiety. *Animal Learning and Behavior*, *26*(1), 46–59. https://doi.org/10.3758/BF03199161

Barrett, L. (2012). Why behaviorism isn't satanism. In J. Vonk & T. K. Shackelford (Eds.), *The Oxford handbook of comparative evolutionary psychology* (pp. 17–38). Oxford University Press. https://doi.org/10.1093/oxfordhb/9780199738182.013.0002

Bateson, M., & Martin, P. (2007). *Measuring behaviour: An introductory behaviour* (3rd ed.). Cambridge University Press. https://doi.org/10.1017/CBO9780511810893

Bateson, P., & Laland, K. N. (2013). Tinbergen's four questions: An appreciation and an update. *Trends in Ecology and Evolution*, *28*(12), 712–718. https://doi.org/10.1016/j.tree.2013.09.013

Bayly, K. L., Evans, C. S., & Taylor, A. (2006). Measuring social structure: A comparison of eight dominance indices. *Behavioural Processes*, *73*(1), 1–12. https://doi.org/10.1016/j.beproc.2006.01.011

Bekoff, M. (2001). The evolution of animal play, emotions, and social morality: On science, theology, spirituality, personhood, and love. *Zygon*, *36*(4), 615–655. https://doi.org/10.1111/0591-2385.00388

Bekoff, M., & Sherman, P. W. (2004). Reflections on animal selves. *Trends in Ecology & Evolution*, *19*(4), 176–180. https://doi.org/10.1016/j.tree.2003.12.010

Bennett, A. T. D. (1996). Do animals have cognitive maps? *Journal of Experimental Biology*, *199*(1), 219–224. https://doi.org/10.1242/jeb.199.1.219

Bertram, B. C. R. (1980). Vigilance and group size in ostriches. *Animal Behaviour*, *28*, 278–286. https://doi.org/10.1016/S0003-3472(80)80030-3

Berwick, R. C., Beckers, G. J. L., Okanoya, K., & Bolhuis, J. J. (2012). A bird's eye view of human language evolution. *Frontiers in Evolutionary Neuroscience*, *4*. https://doi.org/10.3389/fnevo.2012.00005

Birch, J., & Okasha, S. (2015). Kin selection and its critics. *BioScience*, *65*(1), 22–32. https://doi.org/10.1093/biosci/biu196

Bliss, L., Vasas, V., Freeland, L., Roach, R., Ferrè, E. R., & Versace, E. (2023). A spontaneous gravity prior: Newborn chicks prefer stimuli that move against gravity. *Biology Letters*, *19*, 20220502. https://doi.org/10.1098/rsbl.2022.0502

Blough, P. M. (1989). Attentional priming and visual search in pigeons. *Journal of Experimental Psychology: Animal Behavior Processes*, *15*(4), 358–365. https://doi.org/10.1037/0097-7403.15.4.358

Bluff, L. A., Weir, A. A. S., Rutz, C., Wimpenny, J. H., & Kacelnik, A. (2007). Tool-related cognition in New Caledonian crows. *Comparative Cognition & Behavior Reviews*, *2*(1), 1–25.

Bohnslav, J. P., Wimalasena, N. K., Clausing, K. J., Dai, Y. Y., Yarmolinsky, D. A., Cruz, T., Kashlan, A. D., Chiappe, E., Orefice, L. L., Woolf, C. J., & Harvey, C. D. (2021). DeepEthogram, a machine learning pipeline for supervised behavior classification from raw pixels. *ELife*, *10*, e63377. https://doi.org/10. 7554/ eLife. 63377

Bolhuis, J. J., & Gahr, M. (2006). Neural mechanisms of birdsong memory. *Nature Reviews Neuroscience*, *7*(5), 347–357. https://doi.org/10.1038/nrn1904

Bolles, R. C., & Moot, S. A. (1973). The rat's anticipation of two meals a day. *Journal of Comparative and Physiological Psychology*, *83*(3), 510–514. https://doi.org/10.1037/h0034666

Boyd, R., & Richerson, P. J. (1985). *Culture and the evolutionary process*. University of Chicago Press.

Boyd, R., & Richerson, P. J. (1988). An evolutionary model of social learning: The effect of spatial and temporal variation. In T. R. Zentall & B. G. J. Galef (Eds.), *Social learning* (pp. 29–48). Lawrence Erlbaum Associates.

Broadbent, E., Jayawardena, C., Kerse, N. M., Stafford, R. Q., & MacDonald, B. A. (2011). Human-robot interaction research to improve quality of life in elder care - An approach and issues. *Human-Robot Interaction in Elder Care: Papers from the 2011 AAAI Workshop*, 13–19. www.aaai.org.

Brosnan, S. F., & de Waal, F. B. M. (2003). Monkeys reject unequal pay. *Nature*, *425*(6955), 297–299. https://doi.org/10.1038/nature01987.1.

Brown, C. R., Brown, M. B., & Shaffer, M. L. (1991). Food-sharing signals among socially foraging cliff swallows. *Animal Behaviour*, *42*(4), 551–564. https://doi.org/10.1016/S0003-3472(05)80239-8

Burghardt, G. M. (1991). Cognitive ethology and critical anthropomorphism: A snake with two heads and hognose snakes that play dead. In C. A. Ristau (Ed.), *Cognitive ethology: The minds of other animals: Essays in honor of Donald R. Griffin* (pp. 53–90). Lawrence Erlbaum Associates.

Burghardt, G. M. (2005). *The genesis of animal play: Testing the limits*. MIT Press.

Byrne, R. W., & Russon, A. E. (1998). Learning by imitation: A hierarchical approach. *Behavioral and Brain Sciences*, *21*(5), 667–721. https://doi.org/10.1017/S0140525X98001745

Byrne, R. W., & Whiten, A. (1991). Computation and mindreading in primate tactical deception. In A. Whiten (Ed.), *Natural theories of mind: Evolution, development and simulation of everyday mindreading* (pp. 127–141). Basil Blackwell.
Cafazzo, S., Valsecchi, P., Bonanni, R., & Natoli, E. (2010). Dominance in relation to age, sex, and competitive contexts in a group of free-ranging domestic dogs. *Behavioral Ecology, 21*(3), 443–455. https://doi.org/10.1093/beheco/arq001
Call, J. (2001). Object permanence in orangutans (*Pongo pygmaeus*), chimpanzees (*Pan troglodytes*), and children (*Homo sapiens*). *Journal of Comparative Psychology, 115*(2), 159–171. https://doi.org/10.1037/0735-7036.115.2.159
Call, J (2004) Inferences about the location of food in the great apes (Pan paniscus, Pan troglodytes, *Gorilla gorilla, and Pongo pygmaeus*). *Journal of Comparative Psychology*, 2004 June; *118*(2), 232–41. doi: 10.1037/0735-7036.118.2.232
Cañas, J. M., & Matellán, V. (2007). From bio-inspired vs. psycho-inspired to etho-inspired robots. *Robotics and Autonomous Systems, 55*(12), 841–850. https://doi.org/10.1016/j.robot.2007.07.010
Caro, T. M. (1980). Effects of the mother, object play, and adult experience on predation in cats. *Behavioral and Neural Biology, 29*(1), 29–51. https://doi.org/10.1016/S0163-1047(80)92456-5
Caro, T. M. (1981). Predatory behaviour and social play in kittens. *Behaviour, 76*(1–2), 1–24. https://doi.org/10.1163/156853981X00013
Caro, T. M., & Bateson, P. (1986). Organization and ontogeny of alternative tactics. *Animal Behaviour, 34*(5), 1483–1499. https://doi.org/10.1016/S0003-3472(86)80219-6
Caro, T. M., & Hauser, M. D. (1992). Is there teaching in nonhuman animals? *The Quarterly Review of Biology, 67*(2), 151–174. https://doi.org/10.1086/417553
Cartwright, B. A., & Collett, T. S. (1983). Landmark learning in bees - Experiments and models. *Journal of Comparative Physiology A, 151*(4), 521–543. https://doi.org/10.1007/BF00605469
Catchpole, C. K., & Slater, P. J. B. (2008). *Bird song: Biological themes and variations* (2nd ed.). Cambridge University Press. https://doi.org/10.1017/CBO9780511754791
Cheng, K. (1989). The vector sum model of pigeon landmark use. *Journal of Experimental Psychology: Animal Behavior Processes, 15*(4), 366–375. https://doi.org/10.1037/0097-7403.15.4.366
Clayton, N. S., & Dickinson, A. (1998). Episodic-like memory during cache recovery by scrub jays. *Nature, 395*, 272–274. https://doi.org/10.1038/26216
Connor, R. C., Krutzen, M., Allen, S. J., Sherwin, W. B., & King, S. L. (2022). Strategic intergroup alliances increase access to a contested resource in male bottlenose dolphins. *Proceedings of the National Academy of Sciences of the United States of America, 119*(36), e2121723119. https://doi.org/10.1073/pnas.2121723119
Cristol, D. A., & Switzer, P. V. (1999). Avian prey-dropping behavior: II. American crows and walnuts. *Behavioral Ecology, 10*(3), 220–226. https://doi.org/10.1093/beheco/10.3.220
Cruze, W. W. (1935). Maturation and learning in chicks. *Journal of Comparative Psychology, 19*(3), 371–409. https://doi.org/10.1037/h0063532
Csányi, V. (1993). Genetics and learning in individuality. In P. P. G. Bateson, P. H. Klopfer, & N. S. Thompson (Eds.), *Perspectives in ethology* (pp. 1–51). Plenum Press.
Csányi, V. (2000). The 'human behavior complex' and the compulsion of communication: Key factors of human evolution. *Semiotica, 128*(3–4), 243–258. https://doi.org/10.1515/semi.2000.128.3-4.243
Csibra, G., & Gergely, G. (2009). Natural pedagogy. *Trends in Cognitive Sciences, 13*(4), 148–153. https://doi.org/10.1016/j.tics.2009.01.005
Curio, E., Ernst, U., & Vieth, W. (1978). The adaptive significance of avian mobbing: II. Cultural transmission of enemy recognition in blackbirds: Effectiveness and some constraints. *Zeitschrift Für Tierpsychologie, 48*(2), 184–202. https://doi.org/10.1111/j.1439-0310.1978.tb00255.x

Datta, S. R., Anderson, D. J., Branson, K., Perona, P., & Leifer, A. (2019). Computational neuroethology: A call to action. *Neuron, 104*(1), 11–24. https://doi.org/10.1016/j.neuron.2019.09.038

Dembovski, A., Amitai, Y., & Levy-Tzedek, S. (2022). A socially assistive robot for stroke patients: Acceptance, needs, and concerns of patients and informal caregivers. *Frontiers in Rehabilitation Sciences, 2*, 793233. https://doi.org/10.3389/fresc.2021.793233

Dennett, D. C. (1978). Toward a cognitive theory of consciousness. In C. W. Savage (Ed.), *Perception and cognition: Issues in the foundations of psychology* (pp. 201–228). University of Minnesota Press.

DeWitt, T. J., & Scheiner, S. M. (2004). *Phenotypic plasticity: Functional and conceptual approaches*. Oxford University Press.

Doré, F. Y., & Dumas, C. (1987). Psychology of animal cognition: Piagetian studies. *Psychological Bulletin, 102*(2), 219–233. https://doi.org/10.1037/0033-2909.102.2.219

Doré, F. Y., Fiset, S., Goulet, S., Dumas, M. C., & Gagnon, S. (1996). Search behavior in cats and dogs: Interspecific differences in working memory and spatial cognition. *Animal Learning and Behavior, 24*(2), 142–149. https://doi.org/10.3758/bf03198962

Doupe, A. J., & Kuhl, P. K. (1999). Birdsong and human speech: Common themes and mechanisms. *Annual Review of Neuroscience, 22*, 567–631. https://doi.org/10.1146/annurev.neuro.22.1.567

Dugatkin, L. A. (2009). *Principles of animal behavior* (2nd ed.). W.W. Norton.

Dyer, F. C. (1996). Spatial memory and navigation by honeybees on the scale of the foraging range. *Journal of Experimental Biology, 199*(1), 147–154. https://doi.org/10.1242/jeb.199.1.147

Emery, N. J. (2000). The eyes have it: The neuroethology, function and evolution of social gaze. *Neuroscience and Biobehavioral Reviews, 24*, 581–604. https://doi.org/10.1016/S0149-7634(00)00025-7

Erdőhegyi, Á., Topál, J., Virányi, Z., & Miklósi, Á. (2007). Dog-logic: Inferential reasoning in a two-way choice task and its restricted use. *Animal Behaviour, 74*(4), 725–737. https://doi.org/10.1016/j.anbehav.2007.03.004

Etienne, A. S., Maurer, R., & Séguinot, V. (1996). Path integration in mammals and its interaction with visual landmarks. *Journal of Experimental Biology, 199*(1), 201–209. https://doi.org/10.1242/jeb.199.1.201

Flack, J. C., & de Waal, F. B. M. (2004). Dominance style, social power, and conflict management in macaque societies: A conceptual framework. In B. Thierry, M. Singh, & W. Kaumanns (Eds.), *Macaque societies: A model for the study of social organization* (pp. 157–181). Cambridge University Press.

Font, E., & Carazo, P. (2010). Animals in translation: Why there is meaning (but probably no message) in animal communication. *Animal Behaviour, 80*(2), e1–e6. https://doi.org/10.1016/j.anbehav.2010.05.015

Fugazza, C., & Miklósi, Á. (2014). Should old dog trainers learn new tricks? The efficiency of the Do as I do method and shaping/clicker training method to train dogs. *Applied Animal Behaviour Science, 153*, 53–61. https://doi.org/10.1016/j.applanim.2014.01.009

Gahr, M. (2000). Neural song control system of hummingbirds: Comparison to swifts, vocal learning (songbirds) and nonlearning (suboscines) passerines, and vocal learning (budgerigars) and nonlearning (dove, owl, gull, quail, chicken) nonpasserines. *Journal of Comparative Neurology, 426*(2), 182–196. https://doi.org/10.1002/1096-9861(20001016)426:2<182::AID-CNE2>3.0.CO;2-M

Galef, B. G. (2012). Social learning and traditions in animals: Evidence, definitions, and relationship to human culture. *Wiley Interdisciplinary Reviews: Cognitive Science, 3*(6), 581–592. https://doi.org/10.1002/wcs.1196

Gallistel, C. R. (1990). *The organization of learning*. MIT Press.

Gallup, G. G. (1970). Chimpanzees: Self-recognition. *Science, 167*, 86–87.

Galpayage Dona, H. S., Solvi, C., Kowalewska, A., Mäkelä, K., MaBouDi, H. Di, & Chittka, L. (2022). Do bumble bees play? *Animal Behaviour, 194*, 239–251. https://doi.org/10.1016/j.anbehav.2022.08.013

Getty, T. (2006). Sexually selected signals are not similar to sports handicaps. *Trends in Ecology and Evolution, 21*(2), 83–88. https://doi.org/10.1016/j.tree.2005.10.016

Giraldeau, L.-A., & Dubois, F. (2008). Social foraging and the study of exploitative behavior. In H. J. Brockmann, T. J. Roper, M. Naguib, K. E. Wynne-Edwards, C. Barnard, & J. C. Mitani (Eds.), *Advances in the study of behavior* (Vol. 38, pp. 59–104). Academic Press. https://doi.org/10.1016/S0065-3454(08)00002-8

Goldstone, R. L. (1998). Perceptual learning. *Annual Review of Psychology, 49*, 585–612. https://doi.org/10.1146/annurev.psych.49.1.585

Goodenough, J., McGuire, B., & Jakob, E. (2009). *Perspectives on animal behavior* (3rd ed.). John Wiley & Sons.

Gould, S. J., & Vrba, E. S. (1982). Exaptation - A missing term in the science of form. *Paleobiology, 8*(1), 4–15. https://doi.org/10.1017/S0094837300004310

Green, B., Bardunias, P., Turner, J. S., Nagpal, R., & Werfel, J. (2017). Excavation and aggregation as organizing factors in de novo construction by mound-building termites. *Proceedings of the Royal Society B: Biological Sciences, 284*, 20162730. https://doi.org/10.1098/rspb.2016.2730

Griffin, D. R. (1981). *The question of animal awareness: Evolutionary continuity of mental experience*. The Rockefeller University Press.

Gul, F., Rahiman, W., & Nazli Alhady, S. S. (2019). A comprehensive study for robot navigation techniques. *Cogent Engineering, 6*(1), 1632046. https://doi.org/10.1080/23311916.2019.1632046

Häfker, N. S., & Tessmar-Raible, K. (2020). Rhythms of behavior: Are the times changin'? *Current Opinion in Neurobiology, 60*, 55–66. https://doi.org/10.1016/j.conb.2019.10.005

Halfwerk, W., & Slabbekoorn, H. (2009). A behavioural mechanism explaining noise-dependent frequency use in urban birdsong. *Animal Behaviour, 78*(6), 1301–1307. https://doi.org/10.1016/j.anbehav.2009.09.015

Harcourt, A. H., & de Waal, F. B. M. (1992). *Coalitions and alliances in humans and other animals*. Oxford University Press.

Hassenstein, B. (1965). *Biologische kybernetik: eine elementare Einführung*. Heidelberg Quelle & Meyer.

Hauser, M. D. (1996). *The evolution of communication*. MIT Press.

Hauser, M. D., Chomsky, N., & Fitch, W. T. (2002). The faculty of language: What is it, who has it, and how did it evolve? *Science, 298*(5598), 1569–1579. https://doi.org/10.1126/science.298.5598.1569

Held, S. D. E., & Špinka, M. (2011). Animal play and animal welfare. *Animal Behaviour, 81*(5), 891–899. https://doi.org/10.1016/j.anbehav.2011.01.007

Herbert-Read, J. E. (2016). Understanding how animal groups achieve coordinated movement. *Journal of Experimental Biology, 219*(19), 2971–2983. https://doi.org/10.1242/jeb.129411

Herbranson, W. T. (2017). Selective and divided attention in comparative psychology. In J. Call, G. M. Burghardt, I. M. Pepperberg, C. T. Snowdon, & T. R. Zentall (Eds.), *APA Handbook of comparative psychology: Perception, learning, and cognition* (Vol. 2, pp. 183–201). American Psychological Association. https://doi.org/10.1037/0000012-009

Heyes, C. M. (1994). Social learning in animals: Categories and mechanisms. *Biological Reviews, 69*(2), 207–231. https://doi.org/10.1111/j.1469-185x.1994.tb01506.x

Heyes, C. M. (1995). Self-recognition in primates: Further reflections create a hall of mirrors. *Animal Behaviour, 50*(6), 1533–1542. https://doi.org/10.1016/0003-3472(95)80009-3

Hirata, S. (2003). Cooperation in chimpanzees. *Hattatsu, 95*, 103–111.

Höfer, S., Raisch, J., Toussaint, M., & Brock, O. (2018). No free lunch in ball catching: A comparison of Cartesian and angular representations for control. *PLoS One, 13*(6), e0197803. https://doi.org/10.1371/journal.pone.0197803

Hoffmann, M. L. (2000). *Empathy and moral development: Implications for caring and justice*. Cambridge University Press.

Hollis, K. L., Dumas, M. J., Singh, P., & Fackelman, P. (1995). Pavlovian conditioning of aggressive behavior in blue gourami fish (*Trichogaster trichopterus*): Winners become winners and losers stay losers. *Journal of Comparative Psychology, 109*(2), 123–133. https://doi.org/10.1037/0735-7036.109.2.123

Hunt, G. R. (1996). Manufacture and use of hook-tools by New Caledonian crows. *Nature, 379*(6562), 249–251. https://doi.org/10.1038/379249a0

Hunt, G. R. (2021). New Caledonian crows' basic tool procurement is guided by heuristics, not matching or tracking probe site characteristics. *Animal Cognition, 24*(1), 177–191. https://doi.org/10.1007/s10071-020-01427-7

Jacobs, I., & Osvath, M. (2022). Tool use and tooling in ravens (*Corvus corax*): A review and novel observations. *Ethology, 129*(3), 169–181. https://doi.org/10.1111/eth.13352

Jordan, K. E., & Brannon, E. M. (2006). Weber's Law influences numerical representations in rhesus macaques (*Macaca mulatta*). *Animal Cognition, 9*(3), 159–172. https://doi.org/10.1007/s10071-006-0017-8

Kamil, A. C. (1998). On the proper definition of cognitive ethology. In R. P. Balda, I. M. Pepperberg, & A. C. Kamil (Eds.), *Animal cognition in nature: The convergence of psychology and biology in laboratory and field* (pp. 1–28). Academic Press. https://doi.org/10.1016/b978-012077030-4/50053-2

Kamil, A. C., & Cheng, K. (2001). Way-finding and landmarks: The multiple bearings hypothesis. *The Journal of Experimental Biology, 2043*, 103–113. https://doi.org/10.1242/jeb.204.1.103

Kamil, A. C., & Jones, J. E. (2000). Geometric rule learning by Clark's nutcrackers (*Nucifraga columbiana*). *Journal of Experimental Psychology: Animal Behavior Processes, 26*(4), 439–453. https://doi.org/10.1037/0097-7403.26.4.439

Kendal, R. L., Boogert, N. J., Rendell, L., Laland, K. N., Webster, M., & Jones, P. L. (2018). Social learning strategies: Bridge-building between fields. *Trends in Cognitive Sciences, 22*(7), 651–665. https://doi.org/10.1016/j.tics.2018.04.003

Kerepesi, A., Jonsson, G. K., Miklósi, Á., Topál, J., Csányi, V., & Magnusson, M. S. (2005). Detection of temporal patterns in dog-human interaction. *Behavioural Processes, 70*(1), 69–79. https://doi.org/10.1016/j.beproc.2005.04.006

Kingsley, H. L., & Garry, R. (1957). *The nature and conditions of learning* (2nd ed.). Prentice-Hall.

Kline, M. A. (2015). How to learn about teaching: An evolutionary framework for the study of teaching behavior in humans and other animals. *Behavioral and Brain Sciences, 38*, e31. https://doi.org/10.1017/S0140525X14000090

Knudsen, E. I. (2004). Sensitive periods in the development of the brain and behavior. *Journal of Cognitive Neuroscience, 16*(8), 1412–1425. https://doi.org/10.1162/0898929042304796

Laland, K. N. (2004). Social learning strategies. *Learning & Behavior, 32*(1), 4–14. https://doi.org/10.3758/BF03196002

Laland, K. N., Toyokawa, W., & Oudman, T. (2020). Animal learning as a source of developmental bias. *Evolution and Development, 22*(1–2), 126–142. https://doi.org/10.1111/ede.12311

Langley, C. M. (1996). Search images: Selective attention to specific visual features of prey. *Journal of Experimental Psychology: Animal Behavior Processes, 22*(2), 152–163. https://doi.org/10.1037/0097-7403.22.2.152

Leavens, D. A., Russell, J. L., & Hopkins, W. D. (2005). Intentionality as measured in the persistence and elaboration of communication by chimpanzees (*Pan troglodytes*). *Child Development, 76*(1), 291–306. https://doi.org/10.1111/j.1467-8624.2005.00845.x

Lehner, P. N. (1998). *Handbook of ethological methods* (2nd ed.). Cambridge University Press.

Lenkei, R., Faragó, T., Kovács, D., Zsilák, B., & Pongrácz, P. (2020). That dog won't fit: Body size awareness in dogs. *Animal Cognition, 23*(2), 337–350. https://doi.org/10.1007/s10071-019-01337-3

Lenkei, R., Faragó, T., Zsilák, B., & Pongrácz, P. (2021). Dogs (*Canis familiaris*) recognize their own body as a physical obstacle. *Scientific Reports, 11*(1), 2761. https://doi.org/10.1038/s41598-021-82309-x

Levitis, D. A., Lidicker, W. Z., & Freund, G. (2009). Behavioural biologists do not agree on what constitutes behaviour. *Animal Behaviour*, *78*(1), 103–110. https://doi.org/10.1016/j.anbehav.2009.03.018

Liker, A., & Bókony, V. (2009). Larger groups are more successful in innovative problem solving in house sparrows. *Proceedings of the National Academy of Sciences of the USA*, *106*(19), 7893–7898. https://doi.org/10.1073/pnas.090004210

Lorenz, K. (1981). *Foundations of ethology*. Springer-Verlag.

Magnusson, M. S. (2020). T-pattern detection and analysis (TPA) with THEME™: A mixed methods approach. *Frontiers in Psychology*, *10*, 2663. https://doi.org/10.3389/fpsyg.2019.02663

Marler, P., & Sherman, V. (1983). Song structure without auditory feedback: Emendations of the auditory template hypothesis. *The Journal of Neuroscience*, *3*(3), 517–531.

Marshall-Pescini, S., Basin, C., & Range, F. (2018). A task-experienced partner does not help dogs be as successful as wolves in a cooperative string-pulling task. *Scientific Reports*, *8*(1), 16049. https://doi.org/10.1038/s41598-018-33771-7

McCabe, B. J. (2019). Visual imprinting in birds: Behavior, models, and neural mechanisms. *Frontiers in Physiology*, *10*, 658. https://doi.org/10.3389/fphys.2019.00658

McGrew, W. C. (2013). Is primate tool use special? Chimpanzee and new caledonian crow compared. *Philosophical Transactions of the Royal Society B: Biological Sciences*, *368*(1630), 20120422. https://doi.org/10.1098/rstb.2012.0422

Mech, L. D. (1970). *The wolf: The ecology and behavior of an endangered species*. The Natural History Press.

Melis, A. P., Hare, B., & Tomasello, M. (2006). Chimpanzees recruit the best collaborators. *Science*, *311*(5765), 1297–1300. https://doi.org/10.1126/science.1123007

Merola, I., Prato-Previde, E., & Marshall-Pescini, S. (2012). Dogs' social referencing towards owners and strangers. *PLoS One*, *7*(10), e47653. https://doi.org/10.1371/journal.pone.0047653

Meyer, S., Nowotny, T., Graham, P., Dewar, A., & Philippides, A. (2020). Snapshot navigation in the wavelet domain. In V. Vouloutsi, A. Mura, F. Tauber, T. Speck, T. J. Prescott, & P. F. M. J. Verschure (Eds.), *Biomimetic and biohybrid systems. Living machines 2020. Lecture notes in computer science*, (Vol. 12413, pp. 245–256). Springer. https://doi.org/10.1007/978-3-030-64313-3_24

Miklósi, Á. (2007). *Dog behaviour, evolution, and cognition* (1st ed.). Oxford University Press.

Miklósi, Á., & Gácsi, M. (2012). On the utilization of social animals as a model for social robotics. *Frontiers in Psychology*, *3*, 75. https://doi.org/10.3389/fpsyg.2012.00075

Milinski, M. (1987). TIT FOR TAT in sticklebacks and the evolution of cooperation. *Nature*, *325*, 433–435. https://doi.org/10.1038/325433a0

Mineka, S., Davidson, M., Cook, M., & Keir, R. (1984). Observational conditioning of snake fear in rhesus monkeys. *Journal of Abnormal Psychology*, *93*(4), 355–372. https://doi.org/10.1037/0021-843X.93.4.355

Mitani, J. C., & Watts, D. P. (2001). Why do chimpanzees hunt and share meat? *Animal Behaviour*, *61*(5), 915–924. https://doi.org/10.1006/anbe.2000.1681

Morton, E. S. (1977). On the occurrence and significance of motivation-structural rules in some bird and mammal sounds. *The American Naturalist*, *111*(981), 855–869. https://doi.org/10.1086/283219

Nadkarni, P. M., Ohno-Machado, L., & Chapman, W. W. (2011). Natural language processing: An introduction. *Journal of the American Medical Informatics Association*, *18*(5), 544–551. https://doi.org/10.1136/amiajnl-2011-000464

Norris, K. J. (1990). Behavioral ecology and sociobiology female choice and the quality of parental care in the great tit *Parus major*. *Behavioral Ecology and Sociobiology*, *27*, 275–281. https://doi.org/10.1007/BF00164900

Oberauer, K. (2019). Working memory and attention - A conceptual analysis and review. *Journal of Cognition*, *2*(1), 36. https://doi.org/10.5334/joc.58

Oberliessen, L., & Kalenscher, T. (2019). Social and non-social mechanisms of inequity aversion in non-human animals. *Frontiers in Behavioral Neuroscience*, *13*, 133. https://doi.org/10.3389/fnbeh.2019.00133

Osthaus, B., Lea, S. E. G., & Slater, A. M. (2005). Dogs (*Canis lupus familiaris*) fail to show understanding of means-end connections in a string-pulling task. *Animal Cognition*, *8*(1), 37–47. https://doi.org/10.1007/s10071-004-0230-2

Peake, T. M. (2005). Eavesdropping in communication networks. In P. K. McGregor (Ed.), *Animal communication networks* (pp. 13–37). Cambridge University Press.

Peer, M., Brunec, I. K., Newcombe, N. S., & Epstein, R. A. (2021). Structuring knowledge with cognitive maps and cognitive graphs. *Trends in Cognitive Sciences*, *25*(1), 37–54. https://doi.org/10.1016/j.tics.2020.10.004

Pepperberg, I. M. (2006). Grey parrot numerical competence: A review. *Animal Cognition*, *9*(4), 377–391. https://doi.org/10.1007/s10071-006-0034-7

Piaget, J. (1936/1963). *The origins of intelligence in children* (M. trans. Cook, Ed.; 2nd ed.). International Universities Press.

Pinker, S., & Bloom, P. (1990). Natural language and natural selection. *Behavioral and Brain Sciences*, *13*(4), 707–784. https://doi.org/10.1017/S0140525X00081061

Pongrácz, P., Miklósi, Á., Kubinyi, E., Gurobi, K., Topál, J., & Csányi, V. (2001). Social learning in dogs: The effect of a human demonstrator on the performance of dogs in a detour task. *Animal Behaviour*, *62*(6), 1109–1117. https://doi.org/10.1006/anbe.2001.1866

Pongrácz, P., Miklósi, Á., Kubinyi, E., Topal, J., & Csányi, V. (2003). Interaction between individual experience and social learning in dogs. *Animal Behaviour*, *65*(3), 595–603. https://doi.org/10.1006/anbe.2003.2079

Pongrácz, P., Rieger, G., & Vékony, K. (2021). Grumpy dogs are smart learners - the association between dog-owner relationship and dogs' performance in a social learning task. *Animals*, *11*(4), 961. https://doi.org/10.3390/ani11040961

Preston, S. D., & de Waal, F. B. M. (2002). Empathy: Its ultimate and proximate bases. *Behavioral and Brain Sciences*, *25*(1), 1–20. https://doi.org/10.1017/S0140525X02000018

Price, J. J., Friedman, N. R., & Omland, K. E. (2007). Song and plumage evolution in the New World orioles (Icterus) show similar lability and convergence in patterns. *Evolution*, *61*(4), 850–863. https://doi.org/10.1111/j.1558-5646.2007.00082.x

Range, F., Horn, L., Viranyi, Z., & Huber, L. (2009). The absence of reward induces inequity aversion in dogs. *Proceedings of the National Academy of Sciences of the United States of America*, *106*(1), 340–345. https://doi.org/10.1073/pnas.0810957105

Range, F., Marshall-Pescini, S., Kratz, C., & Virányi, Z. (2019). Wolves lead and dogs follow, but they both cooperate with humans. *Scientific Reports*, *9*, 3796. https://doi.org/10.1038/s41598-019-40468-y

Rankin, C. H., Abrams, T., Barry, R. J., Bhatnagar, S., Clayton, D. F., Colombo, J., Coppola, G., Geyer, M. A., Glanzman, D. L., Marsland, S., McSweeney, F. K., Wilson, D. A., Wu, C. F., & Thompson, R. F. (2009). Habituation revisited: An updated and revised description of the behavioral characteristics of habituation. *Neurobiology of Learning and Memory*, *92*(2), 135–138. https://doi.org/10.1016/j.nlm.2008.09.012

Reid, P. J., & Shettleworth, S. J. (1992). Detection of cryptic prey: Search image or search rate? *Journal of Experimental Psychology: Animal Behavior Processes*, *18*(3), 273–286. https://doi.org/10.1037/0097-7403.18.3.273

Rendall, D., Owren, M. J., & Ryan, M. J. (2009). What do animal signals mean? *Animal Behaviour*, *78*(2), 233–240. https://doi.org/10.1016/j.anbehav.2009.06.007

Rodd, F. H., Hughes, K. A., Grether, G. F., & Baril, C. T. (2002). A possible non-sexual origin of mate preference: Are male guppies mimicking fruit? *Proceedings of the Royal Society B: Biological Sciences*, *269*(1490), 475–481. https://doi.org/10.1098/rspb.2001.1891

Rodgers, A., & Sferruzzi-Perri, A. N. (2021). Developmental programming of offspring adipose tissue biology and obesity risk. *International Journal of Obesity*, *45*(6), 1170–1192. https://doi.org/10.1038/s41366-021-00790-w

Rybak, I. A., Shevtsova, N. A., Lafreniere-Roula, M., & McCrea, D. A. (2006). Modelling spinal circuitry involved in locomotor pattern generation: Insights from deletions during fictive locomotion. *Journal of Physiology*, *577*(2), 617–639. https://doi.org/10.1113/jphysiol.2006.118703

Schafer, M., & Schiller, D. (2018). Navigating social space. *Neuron*, *100*(2), 476–489. https://doi.org/10.1016/j.neuron.2018.10.006

Schuster, S., Wöhl, S., Griebsch, M., & Klostermeier, I. (2006). Animal cognition: How archer fish learn to down rapidly moving targets. *Current Biology*, *16*(4), 378–383. https://doi.org/10.1016/j.cub.2005.12.037

Shettleworth, S. J. (2010). *Cognition, evolution, and behavior* (2nd ed.). Oxford University Press.

Silk, J. B. (2002). Using the 'F'-word in primatology. *Behaviour*, *139*(2–3), 421–446. https://doi.org/10.1163/156853902760102735

Silk, J. B. (2007). The strategic dynamics of cooperation in primate groups. *Advances in the Study of Behavior*, *37*, 1–41. https://doi.org/10.1016/S0065-3454(07)37001-0

Špinka, M., Newberry, R. C., & Bekoff, M. (2001). Mammalian play: Training for the unexpected. *The Quarterly Review of Biology*, *76*(2), 141–168. https://doi.org/10.1086/393866

Srinivasan, M. V. (2015). Where paths meet and cross: Navigation by path integration in the desert ant and the honeybee. *Journal of Comparative Physiology A: Neuroethology, Sensory, Neural, and Behavioral Physiology*, *201*(6), 533–546. https://doi.org/10.1007/s00359-015-1000-0

Számadó, S., Zachar, I., Czégel, D., & Penn, D. J. (2023). Honesty in signalling games is maintained by trade-offs rather than costs. *BMC Biology*, *21*(1), 4. https://doi.org/10.1186/s12915-022-01496-9

Tebbich, S., Taborsky, M., & Winkler, H. (1996). Social manipulation causes cooperation in keas. *Animal Behaviour*, *52*(1), 1–10. https://doi.org/10.1006/anbe.1996.0147

Tecwyn, E. C., & Buchsbaum, D. (2018). Hood's gravity rules. In J. Vonk & T. Shackelford (Eds.), *Encyclopedia of animal cognition and behavior* (pp. 1–9). Springer, Cham. https://doi.org/10.1007/978-3-319-47829-6_1535-1

Thinus-Blanc, C. (1988). Animal spatial cognition. In L. Weiskrantz (Ed.), *Thought without language* (pp. 371–395). Clarendon Press.

Thornton, A., & Raihani, N. J. (2008). The evolution of teaching. *Animal Behaviour*, *75*(6), 1823–1836. https://doi.org/10.1016/j.anbehav.2007.12.014

Tinbergen, N. (1951). *The study of instinct*. Oxford University Press.

Tinbergen, N. (1963). On aims and methods of ethology. *Zeitschrift Für Tierpsychologie*, *20*, 410–433.

Tolman, E. C. (1949). There is more than one kind of learning. *Psychological Review*, *56*(3), 144–155. https://doi.org/10.1037/h0055304

Tomasello, M. (2008). *Origins of human communication*. MIT Press.

Tomasello, M., Davis-Dasilva, M., Camak, L., & Bard, K. (1987). Observational learning of tool-use by young chimpanzees. *Human Evolution*, *2*(2), 175–183. https://doi.org/10.1007/BF02436405

Topál, J., Gergely, G., Erdőhegyi, Á., Csibra, G., & Miklósi, Á. (2009). Differential sensitivity to human communication in dogs, wolves, and human infants. *Science*, *325*(5945), 1269–1272. https://doi.org/10.1126/science.1176960

Townsend, S. W., Koski, S. E., Byrne, R. W., Slocombe, K. E., Bickel, B., Boeckle, M., Braga Goncalves, I., Burkart, J. M., Flower, T., Gaunet, F., Glock, H. J., Gruber, T., Jansen, D. A. W. A. M., Liebal, K., Linke, A., Miklósi, Á., Moore, R., van Schaik, C. P., Stoll, S., Vail, A., Waller, B. M., Wild, M., Zuberbühler, K., & Manser, M. B. (2017). Exorcising Grice's ghost: An empirical approach to studying intentional communication in animals. *Biological Reviews*, *92*(3), 1427–1433. https://doi.org/10.1111/brv.12289

Ujfalussy, D. J., Miklósi, Á., & Bugnyar, T. (2013). Ontogeny of object permanence in a non-storing corvid species, the jackdaw (*Corvus monedula*). *Animal Cognition*, *16*(3), 405–416. https://doi.org/10.1007/s10071-012-0581-z

Valletta, J. J., Torney, C., Kings, M., Thornton, A., & Madden, J. (2017). Applications of machine learning in animal behaviour studies. *Animal Behaviour*, *124*, 203–220. https://doi.org/10.1016/j.anbehav.2016.12.005
Vasconcelos, M. (2008). Transitive inference in non-human animals: An empirical and theoretical analysis. *Behavioural Processes*, *78*(3), 313–334. https://doi.org/10.1016/j.beproc.2008.02.017
Voelkl, B., & Huber, L. (2007). Imitation as faithful copying of a novel technique in marmoset monkeys. *PLoS One*, *2*(7), e611. https://doi.org/10.1371/journal.pone.0000611
Völter, C. J., Karl, S., & Huber, L. (2020). Dogs accurately track a moving object on a screen and anticipate its destination. *Scientific Reports*, *10*, 19832. https://doi.org/10.1038/s41598-020-72506-5
Waddington, C. H. (1966). *Principles of development and differentiation*. Macmillan.
Wagenaar, D. A., Hamilton, M. S., Huang, T., Kristan, W. B., & French, K. A. (2010). A hormone-activated central pattern generator for courtship. *Current Biology*, *20*(6), 487–495. https://doi.org/10.1016/j.cub.2010.02.027
Ward, A. J. W., Webster, M. M., Magurran, A. E., Currie, S., & Krause, J. (2009). Species and population differences in social recognition between fishes: A role for ecology? *Behavioral Ecology*, *20*(3), 511–516. https://doi.org/10.1093/beheco/arp025
Ward, C., & Smuts, B. B. (2007). Quantity-based judgments in the domestic dog (*Canis lupus familiaris*). *Animal Cognition*, *10*(1), 71–80. https://doi.org/10.1007/s10071-006-0042-7
Weldon, K. B., Fanson, K. V, & Smith, C. L. (2016). Effects of isolation on stress responses to novel stimuli in subadult chickens (*Gallus gallus*). *Ethology*, *122*(10), 818–827. https://doi.org/10.1111/eth.12529
West-Eberhard, M. J. (2003). *Developmental plasticity and evolution*. Oxford University Press.
Wheeler, B. C. (2009). Monkeys crying wolf? Tufted capuchin monkeys use anti-predator calls to usurp resources from conspecifics. *Proceedings of the Royal Society B: Biological Sciences*, *276*(1669), 3013–3018. https://doi.org/10.1098/rspb.2009.0544
Whiten, A. (2005). The second inheritance system of chimpanzees and humans. *Nature*, *437*(7055), 52–55. https://doi.org/10.1038/nature04023
Whiten, A., & Ham, R. (1992). On the nature and evolution of imitation in the animal kingdom: Reappraisal of century of research. In P. J. B. Slater, J. S. Rosenblatt, C. Beer, & M. Milinski (Eds.), *Advances in the Study of Behavior* (Vol. 21, pp. 239–283). Academic Press.
Wilkinson, G. S. (1984). Reciprocal food sharing in the vampire bat. *Nature*, *308*, 181–184. https://doi.org/10.1038/308181a0
Wilson, A. D., & Golonka, S. (2013). Embodied cognition is not what you think it is. *Frontiers in Psychology*, *4*, 58. https://doi.org/10.3389/fpsyg.2013.00058
Wynne, C. D. L. (2007). What are animals? Why anthropomorphism is still not a scientific approach to behavior. *Comparative Cognition & Behavior Reviews*, *2*, 125–135. https://doi.org/10.3819/ccbr.2008.20008
Zador, A. M. (2019). A critique of pure learning and what artificial neural networks can learn from animal brains. *Nature Communications*, *10*(1), 3770. https://doi.org/10.1038/s41467-019-11786-6
Zeiträg, C., Jensen, T. R., & Osvath, M. (2022). Gaze following: A socio-cognitive skill rooted in deep time. *Frontiers in Psychology*, *13*, 950935. https://doi.org/10.3389/fpsyg.2022.950935
Zhang, Y., Zhou, L., Zuo, J., Wang, S., & Meng, W. (2023). Analogies of human speech and bird song: From vocal learning behavior to its neural basis. *Frontiers in Psychology*, *14*, 1100969. https://doi.org/10.3389/fpsyg.2023.1100969
Zuberbühler, K., Noë, R., & Seyfarth, R. M. (1997). Diana monkey long-distance calls: Messages for conspecifics and predators. *Animal Behaviour*, *53*(3), 589–604. https://doi.org/10.1006/anbe.1996.0334

2

ETHOROBOTICS

Ethological approach to interactive, social robots

Judit Abdai and Ádám Miklósi

2.1 Introduction

Although the label of 'robot' was not coined until later, machines that we recognise as robots today have been in existence for a long time. These early machines were typically mechanical constructions, but their partially autonomous functioning provided significant advantages to humans. For instance, they could accurately measure time, as seen in clocks and watches, something that was difficult for humans to do. The integration of electric or petrol motors with complex mechanical structures led to the first 'robot-like' creatures (for definitions see below), such as cars, which allowed people to move through space with minimal physical effort.

The term 'robot' (derived from 'robota', meaning labourer in Czech) was first introduced by Karl and Joseph Capek in their literary works at the beginning of the 20th century. They reflected on the growing impact of industrialisation on workers' lives and the future of civilisation. The need for increased production of goods and competition among entrepreneurs led to the replacement of hand-made manufacturing activities with machines, a process that continues today. These machines, or robots, are increasingly taking over tasks previously performed by workers, whose roles are shifting from heavy and routine physical labour to overseeing, controlling, and monitoring the production process. The installation of the first robots into the car assembly line by General Motors (1961) is considered a significant milestone in this story.

Some researchers independently aimed to develop machines that could move around in simplified environments. The first electro-mechanical robot was built by Hammond and Miessner in 1912, well before the advent of computers. This robot, nicknamed the 'electric dog', could also be considered as the first bio-inspired robot because its functioning was based on the orienting mechanism of insects described by Loebl. However, Holland (2003) argued that the two turtle robots designed by Walter in 1949 were the first bio-inspired robots. Walter's (1950) aim was to come up

 DOI: 10.4324/9781003182931-3

with a new way to understand the mechanisms of the nervous system and provided a description of the robotic control in both biological and cybernetic terms.

The robot, named *Machina speculatrix* (an analogy to the taxonomic Latin labelling of species), was capable of various behaviours, including obstacle avoidance, avoiding bright lights, and recharging. Holland (2003) provides a careful reconstruction of the abilities of this robot. Several similar attempts were made during the first half of the 20th century, but they never gained much research interest. Even early ethologists interested in cybernetics, such as Hassenstein (1965), failed to take notice of these technical developments, leading to a missed opportunity for an early revolution in bio-inspired robotics.

During the following 50–60 years, robotics made significant strides in industry, taking over the production of a wide range of products from human workers. The next important advance came in the second half of the 1980s. For this period, Goodrich and Schulz (2007) identified two important innovations in robotics (see also Arkin, 1998; Brooks, 1986). First, newer behaviour-based robots utilised distributed sense-act loops to produce appropriate responses to external stimuli. Second, robotic architectures combined efficient low-level responsive behaviours with more complex cognitive structures, resulting in more flexible robotic behaviour and making them more interactive.

Although these early social robots showed some functional qualities, they were strongly limited by the level of technology available at the time. However, advancements in several areas, including mechanics, mechatronics and computer engineering, telecommunication science, and miniaturisation of computers and sensors, have made the efforts of a new wave of social robots more realistic in recent times.

2.1.1 Possible definitions for robots

There are many definitions on robots but none of them can capture the whole concept because of the wide operating range and difference in structure. Any definition is influenced by the nature of the agent utilised (e.g., industrial robot or socially interactive robot), or the actual research question studied. Definitions can also be categorised based on their focus, whether they prefer to define the type of the agent, the way of its inner control or would like to provide the basis for a mathematical and physical description of the relationship between the agent and its environment.

Often, these definitions also show some parallel to the descriptions applied to living organisms; however, it should be clear that recent (and for many years to come, also future) robots are many lightyears away from any biological agent because they lack important features, e.g., development, reproduction, regeneration, that is, true autonomy. Lastly, robots are also changing and improving. Inventions involving new materials, new mechatronic solutions, more rapid telecommunication tools and novel devices for communication, and the massive involvement of artificial intelligence make it impossible to come up with a future-proof definition.

According to Schweitzer (2003), '*A robot is a freely and re-programmable, multifunctional manipulator with at least three independent axes, for moving materials, parts, tools, or special devices on programmed, variable trajectories in order to fulfil a multitude of various tasks*'. This very general definition was practically adopted by the international ISO-Standard for industrial robots (see International Federation of Robotics – IFR, 2023). Similarly, the Robot Institute of America (RIA, 2019) describes a robot as a '*reprogrammable, multifunctional manipulator designed to move material, parts, tools, or specialized devices through various programmed motions for the performance of a variety of tasks*'.

Neither of these definitions seems to meet the expectations and needs of future development, they provide no reference to autonomy, mobility, or even specific problem-solving skills of robots.

For a broader perspective, here are some other recent definitions, which suggest that (a) robot(s) is/are:

> …an electromechanical device with multiple degrees-of-freedom (DoF) that is programmable to accomplish a variety of tasks.
>
> *(Williams, 2016)*

> …physical machines that embed elements of computational intelligence that enable them to behave autonomously (Bekey et al., 2008), often with the ability to operate for long periods of time without direct human control or supervision
>
> *(Prescott & Robillard, 2021)*

> …a machine that is able to interact physically with its environment and perform some sequence of behaviours, either autonomously or by remote control.
>
> *(Krause et al., 2011)*

> Robots are machines built to perform tasks using actions that are based on, or reminiscent of, humans or other animals.
>
> *(Webb, 2000)*

These definitions are quite heterogeneous and are sometimes in conflict. For example, robots may or may not be 'autonomous', and their mechanisms have to be electromechanical or could be based on any structure.

Alternative approach is offered by not referring to any embodiment but focusing on the general function of the robotic system as an adaptive organisation. According to Holland (1992) and Kaisler and Madey (2009), such systems change in the face of perturbations (e.g., changes in the environment) so as to maintain some kind of invariant state (e.g., survival) or achieve goals (e.g., reproduction) by altering their properties (e.g., behaviour, structure) or modifying their environment. Operationally speaking, such systems maintain some kind of invariant state or achieve goals by responding to perturbations.

Other definitions take the robots' relation to humans into account. Christaller et al. (2001) state that '*robots are sensomotoric machines for the extension of human mobility. They consist of mechatronic components, sensors and computer-based control functions. The complexity of a robot differs clearly from other machines by the larger number of degrees-of-freedom and the variety and the scope of its behaviours*'.

There have been attempts to categorise robots as either 'industrial' or 'social', with the former designed to operate in the absence of humans or at least separated physically, and the latter meant to interact with humans. While this distinction can be helpful, it does not fully address the issue of defining what a robot is. Although industrial and social robots may have different physical features, the underlying technology that enables their functionality is quite similar. Perhaps a more straightforward definition of a robot would be an electric or electromechanical device that has a physical body and extends human capabilities beyond its biological capabilities. This definition would encompass all types of robots, including industrial machines, mobile phones, and socially interactive robots.

2.1.1.1 The robotic mental architecture

The robot embodiment can be understood as an analogy to the body of living organisms. Just as animals have nervous systems that support mental processes ('thinking'), robots have dedicated hardware components that play a role in controlling their behaviour which is operated by a program or a software. However, both terms are too general and neutral, and thus a more precise term than 'software' is needed to refer to the controlling system of a robot.

Vernon (2019; Vernon et al., 2007), drawing from cognitive science traditions, proposed that the robot's mind should be conceptualised as a cognitive architecture. In contrast to emergent (connectivist) models, which differ fundamentally in structure (see Section 2.5), a cognitive system has a triadic structure of perception, processing, and action, as well as specific mental functions such as attention, reasoning, and anticipation (see also Section 2.5).

Cognitive architectures are often considered more complex yet flexible compared to non-cognitive architectures, which are typically simpler and more robust in design. However, it remains to be seen which type of architecture captures the complexity of the animal and human mind more accurately. Both approaches draw inspiration from the minds of living creatures. Similar discussions also occur in comparative studies of the animal mind, where often different types of models of the mind are contrasted (Heyes, 2000).

In the long run, it is probable that a hybrid architecture may offer a solution (see also Kotseruba & Tsotsos, 2020), even though these mental models may have some incompatible functions (Vernon, 2019). Therefore, for the purpose of this book, we refer to 'robotic mental architectures' (whether cognitive or non-cognitive) to draw parallels between the animal mind and the organised system of inner processes in robots.

Functionally, the animal mind has also been described as an inner model of the environment or other inner processes. In other words, there are mental units that are isomorphic with specific aspects of the inner and outer world (Gallistel, 1990; Webb, 2006). These units are called representations, which may be grounded by evolutionary processes (in animals) or created by the engineer and/or emerge as a result of learning. However, they are clearly separable from their local environment, and the agent's behaviour is the result of its mental architecture's manipulation or operation on these representations. The central part of robotic mental architectures mainly processes these representations. Although this concept may have some anthropomorphic connotations, and representations can have different structures, here, we follow the cognitive tradition (Vernon et al., 2007).

2.1.2 A challenge to robotics

The literature on social robotics and popular media often suggests that the integration of robots into human life is both inevitable and imminent. However, it is important not to take these assumptions too seriously. As we discussed, current social robots are still unable to perform most of the tasks that humans expect of them. Operating within the anthropogenic environment presents a highly complex challenge for any artificial system (Riener et al., 2023) (Box 2.1).

Box 2.1 The fall and the rise of embodied virtual agents?

The Computers are Social Actors (CASA) framework describes that humans relate to technological devices (originally computers) as to other humans, applying similar social rules when interacting with them (Gambino et al., 2020; Nass et al., 1994). The framework may be used to improve the experience and understanding of users by relying on already existing mental models (e.g., human–human communication).

One way to enhance this effect and help users guiding through more complex software is by providing the helper with animal/human-like features building on anthropomorphism. However, these do not always work as planned. One of the most (in) famous virtual agents is Clippit (commonly referred to as Clippy), the Office Assistant by Microsoft first introduced in 1996. Although in theory Clippy should have helped users and contributed to a positive experience, many people hated the agent (Xiao et al., 2003). Clippy was optimised for first-time use, and it did not adapt to the user's experiences and needs. This probably led to many users describing it as annoying, ignoring users' feedback, and overall rude and impolite. Thus, Clippy's human-like behaviour displayed inappropriately caused the agent's demise (at least as a helper).

Recently, especially with the growing number of digital assistants (e.g., Apple's Siri, Amazon's Alexa), some suggested that embodiment would improve their social presence, using non-verbal cues and trust in them. Augmented reality can provide visual embodiment for these agents interacting with the surrounding environment, improving user experience. For example, Kim et al. (2018) found that when an agent was visible, and it used gestures and moved around in the environment, participants had better sense of privacy (although still were not more likely to share private data), felt more comfortable, and rated the agent's reliability, helpfulness and social presence higher compared to encountering an agent without a body or lack of locomotion.

Whether visual embodiment of these agents is indeed useful from a user perspective (and not only for the companies), and the types of embodiment that would work best are still open questions. Based on the original CASA notion, humans 'mindlessly' apply the social scripts when interacting with a computer, but novel directions suggest that the extensive experience with technological devices in the past three decades has shaped our attitude towards these machines, including developing human–media interaction scripts (for a review see Gambino et al., 2020).

Let us suppose that social robots eventually gain such capacities. Would they still function as self-sustaining, embodied, and interactive agents, or are there alternative possibilities for emerging technologies? There are many complex tasks that previously required specific human involvement, but today's technology solves the same problems in completely different ways. For example, in the Middle Ages, monks copied texts written on paper, but Gutenberg's invention of printing completely revolutionised the process. Sending letters via traditional post has also been replaced by significantly different technologies, such as telegrams, faxes, and ultimately email. Nobody argued that monks or postmen should be replaced by socially interactive robots. Therefore, it is not necessarily true that all tasks performed by humans must also be accomplished in the same way by any kind of (social) robots, as we currently imagine them.

The extension of human capabilities by machines can follow four conceptually different, perhaps partially overlapping trajectories (see also Dautenhahn, 2007; Prescott & Robillard, 2021).

Disappearing technology: in his famous essay on the future of computers, Weiser (1991) noted 'the most profound technologies are those that disappear. They weave themselves into the fabric of everyday life until they are indistinguishable from it'.

Advancements in technology may lead to the transformation of many human-based tasks into 'smart' tools, reducing the need for social robots to solve them. This may imply that humans do not have to interact with as many robotic agents as the tasks currently performed by humans imply.

Machines: there are already many such equipment in use, like dishwashers or cars. In these cases, the key innovation was finding ways to enable machines to perform tasks that were previously done by humans. Often, this involved modifying the environment in some way, such as building good roads for cars to travel on. Conceptually, machines lack the features that make them appear 'alive' from a human perspective.

In certain situations, it may be possible to simplify social robots into conventional machines, despite our original intentions of replacing human workers with them. Although some people may still assign social characteristics to these machines, this would be a subjective observation rather than an inherent feature of the machine itself.

Human-like agents: currently, the one prevailing trend is to create human-like robots that, to varying degrees depending on the technology, resemble actual humans. While there have been some impressive achievements, this approach has several drawbacks. Despite extensive efforts, human-like robots only possess a superficial resemblance to real humans in terms of appearance, internal workings, and capabilities. Even the most advanced human-like robot can only perform a small fraction of the general skills that humans possess, and they are incapable of basic functions of living, such as self-maintenance. There is no reason to build a human-like social robot other than for people to ascribe emotion, intent, and the like to it. Misleading or inaccurate portrayals and manifestation of robotic technology can create unrealistic expectations or even dangerous assumptions on the part of humans.

Animal (non-human-like) robots: while there have been some attempts to develop animal-like robots that mimic the appearance and behaviour of real animals, such as dogs, these endeavours face similar challenges as those encountered in creating human-like robots. Therefore, an alternative approach advocated by ethorobotics is to design robots from scratch, without mimicking the embodiment of any specific living creature (see Section 2.4). This does not preclude the possibility that some similarities may exist between such robots and humans or animals, much like how dogs and cats share some traits, but any resemblances would arise from functional considerations rather than intentional design choices by the creators.

2.1.3 The 'Santa Claus' phenomenon

Christian culture introduced the tradition of a mythical figure, Santa Claus, presenting children with gifts around Christmas time (in some countries this takes place on the 6th of December or 7th of January). This tradition traces back to the legend of St. Nicholas, a bishop from the 3rd century (270–343 CE), who clandestinely supported impoverished children. Thanks to movies, media, and marketing, the figure of Santa Claus has become globally recognised.

While the process of purchasing and gifting presents is organised by parents, many prefer their young children to believe that Santa Claus is the true giver. In reality, it is not just the parents; the entire society actively participates in promoting this charade. The phenomenon of Santa Claus arises when community members collectively strive to maintain the illusion of something non-existent.

Children typically discover the nonexistence of Santa Claus around the age of seven, either on their own or through the revelation by others (Anderson & Prentice, 1994). However, the behaviour of adults in perpetuating this deception is even more intriguing. Why do they continue to foster this illusion and feel a sense of sadness when children learn the truth?

A parallel situation is unfolding today in the realm of robots. Those involved in creating robots often wish to convey the idea that robots are, or will soon be, remarkably similar to humans or animals, garnering unsolicited support from the media. Currently, we are uncertain about the benefits or potential harms of this behaviour. Sharkey and Sharkey (2021) and many others have highlighted that constructing human-like robots can mislead users by creating an illusion of similarity to humans, raising ethical questions.

Children between the ages of 4 and 6, a period when most still believe in Santa Claus, demonstrate the ability to differentiate between an animal-like robot activated by an experimenter and one manipulated by remote control (Cameron et al., 2017).

This suggests that older children may be aware that Santa Claus does not exist, yet they willingly participate in the adult-orchestrated game.

Extending this argument, adults may also conform to the rules of this collective charade. The issue with the Santa Claus phenomenon lies not in community members deceiving themselves, but in the case of robotics, it hinders the recognition of actual problems and challenges, especially in technology development.

Beyond the Santa Claus phenomenon, the way humans perceive and interact with robots, including machines, is not solely determined by their appearance, behaviour, or advanced interactive capabilities. It is shaped by various factors such as past experiences, cultural background, and technological literacy levels. Consequently, numerous experimental questions need exploration, and the answers to these questions may evolve over time as new generations become accustomed to interacting with robots.

2.1.4 Conclusions, prospects, and questions

The central premise of social robotics (and ethorobotics, as discussed below) is the idea that robots can be designed to function as socially interactive agents, rather than passive tools, with the potential to integrate into human society. While this concept may appear simple and offer promising predictions for the future, the history of this field of social robotics has been fraught with challenges and negative experiences. Despite interest and financial support, largely from academic science, most of the goals of social robotics have yet to be fully realised, and the field is only beginning to grapple with the real challenges of building truly effective and socially integrated robots.

Social robots, like any other machines, are inherently dependent on their environment for their existence and functionality. Therefore, it is not appropriate to create social robots for their own sake, without careful consideration of their intended purpose, including the extent to which they should be designed for social interaction. Instead of categorically distinguishing between social agents that are compatible with humans and those that are not, it is more useful to conceptualise the nature of these agents along a continuum, with varying degrees of social ability and function.

The social ability of robots can significantly impact how people perceive and interact with them, and this relationship is further influenced by individuals' general level of technological knowledge. Therefore, social robotics should not be viewed as creating these agents in isolation but rather in a constantly evolving and unpredictable societal context. As such, it is critical to consider the broader societal impact and implications of social robots and to design them in a way that reflects the changing needs and values of society.

2.2 The robots among us

Modern 2D, virtual agents can easily deceive viewers as sophisticated video images or by acting in movies, but humans can quickly detect any improper feature in a human-like 3D robot with little effort. Some researchers view this as a significant disadvantage that needs to be addressed in the coming decades, while others see it as an opportunity.

Encounters with real functioning robots are rare in our daily lives. Most social robots exist in scientific laboratories at universities or companies, with many impressive videos of them available on the internet. One of the most widely known and popular robots is the 'Big Dog', developed by Boston Dynamics a few years ago as a robotic carrier for the US Army. Today, smaller versions of the robot can be purchased for a few thousand dollars. According to media reports, one of these types of robots, nicknamed 'Spot', has the potential to assist civilians and professionals in various tasks. Here are two statements from managers describing their expectations (from the companies' home page, downloaded in August 2022, https://www.bostondynamics.com/products/spot):

- 'Every 12 hours, we follow the same path through the plant, and we're assessing situational awareness. It is important to communicate across shift changes and make sure you're handing off critical information. Spot is going to capture things we may not notice or that we're not always around to see.' Christopher Phillips, Production Technician, Woodside
- 'There are thousands of pounds of pressurized combustible material out there. High-pressure oil and gas can create risks for people working in close proximity. If we could have a robot with the proper sensors out there, we'd much rather do that.' Adam Ballard, Facilities Technology Manager, British Petrol.

While Spot exhibits many sophisticated features, it may not fulfil all of the managers' expectations when compared to other existing robots. Currently, Spot has limited sensory abilities and can only function for a limited duration of approximately 30 minutes. Therefore, what follows is more about the potential future of social robots in the next five to ten years, rather than a description of the present situation.

2.2.1 *Robots in the society*

While certain human functions may not be replaceable by robots, let us consider the possibility of interactive social robots being integrated into human social groups. In theory, there are numerous types of tasks in which robots could be potentially beneficial.

Social robots can be categorised as a subset of robots that possess the ability to move around within a limited space and exhibit at least one feature of interacting with humans or animals. Although this description may seem minimalist, it is realistic since many existing social robots are immobile and have limited interaction capabilities with humans (Bekey et al., 2008; Prescott & Robillard, 2021). However, this definition falls short of the expectations that most researchers hold for social robots (as discussed in Section 2.2.2). It also reinforces the perspective that social robots should not necessarily be considered a distinct category, but rather there are robots that may possess varying degrees of social traits based on their intended function. Based on both theoretical and experimental work, social robots could provide potential benefits at least in the following functions where the actual social relationship between the people and the robot is also different (see also Section 2.6.1). But in time new scenarios may also emerge in which social robots could be helpful.

2.2.1.1 Manual labour and service

Benefits: social robots have the potential to take over tasks that involve demanding and potentially dangerous physical activities or conditions, as well as those that put human health at risk (Lee, 2021). Examples of such tasks could include working in mines, rescuing people in hazardous situations, and other similar scenarios. Additionally, social robots are being developed to work alongside humans in situations that require close collaboration and coordination.

Currently, there are limited instances of social robots being deployed in real-world scenarios. Waiter robots can be considered as possible examples (Figure 2.1, see also Section 4.5) (Garcia-Haro et al., 2021). The creators of these robots must decide whether they intend to replace human waiters or provide assistance to them. The former can be challenging because waiters have to perform several tasks (often in parallel), including ensuring guests feel comfortable, discussing menu options, and providing attentive service. Robots may be better suited to replacing the task of carrying dishes and drinks between tables and the kitchen, which is typically carried out by 'food runners'. As such, robots could potentially take over the food runners' job in the catering industry (Box 2.2).

If working efficiently, these robots could provide a huge impact on the job market because their employment could be much cheaper (no need to pay tax and social security fees) and would enable quick and dynamic adjustment to the market needs for specific services. In the long term, such a trend could potentially lead to the

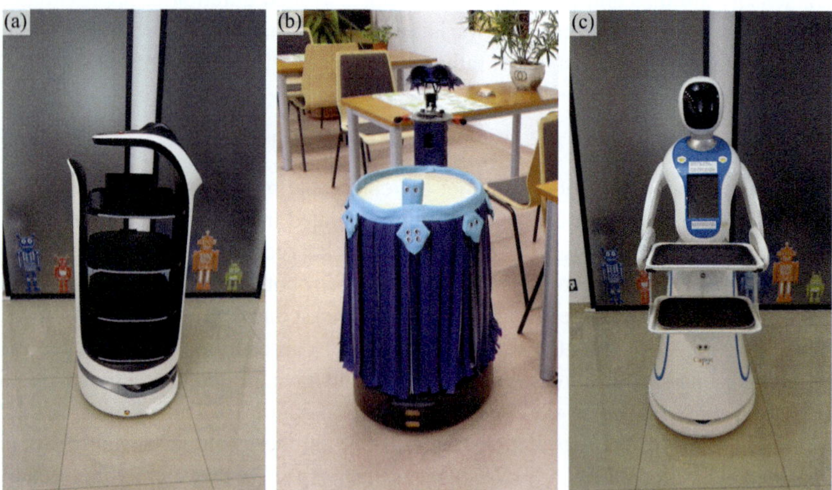

Figure 2.1 Waiter robots have different embodiments from resembling simple moving trays (although often having a screen with face-like touch screen), e.g., (a) BellaBot by Pudu Robotics; having more complex features such as a moveable head and/or extremities, e.g., (b) Biscee robot developed by the HUN-REN - ELTE Comparative Ethology Research Group, Hungary; or having humanoid form, e.g., (c) Amy service robot by Suzhou Pangolin Robot.

replacement of human workers, particularly in industries where the use of robots could significantly reduce labour costs.

Risks: the necessity for food runner robots to be social and interactive is less apparent. In situations where minimal human interaction is required, such robots may evolve into machines or partly disappearing technology. In many cases, the potential social features of the robot may lose their significance. Fink et al. (2013) observed that users began to view Roomba, a vacuum cleaner robot, as a tool rather than a social robot after a prolonged period of experience with it.

Box 2.2 Biscee, the waiter assistant robot

Biscee is a waiter assistant robot developed by the HUN-REN – ELTE Comparative Ethology Research Group in Budapest, Hungary. Although the robot can be controlled remotely and semi-autonomous motion is an option, autonomous behaviour is also available. Based on a map of the actual location, Biscee can navigate in space (including moving in large crowds) safely, and it actively avoids collisions with humans, dogs, or stationary objects. Biscee has been successfully used to disseminate candies at large events with about a hundred people for several hours without accidents (Figure 2.2).

Biscee is designed to interact with humans on a regular basis using simple behaviours (e.g., looking at faces) that can be spontaneously recognised by humans even without prior experience with the robot. Biscee's skills are supported by a few specific sensors (see Table 2.1).

Table 2.1 Sensors of Biscee robot

Sensor	*Function*
Lidar (HOKUYO UST-10L) in robot base	Localisation and navigation, object avoidance, leg detection (human detection)
2 video cameras (Basler ace acA1300–60gc Camera sensors, Edmund Optics 4.5 mm C Series lens) on robot head	Face detection (human detection), interactive functions (e.g., gaze alternation, orienting at faces) (currently one camera is in use for these functions)
Ultrasound sensors on robot torso	Around the robot's tray (6 sensors): close approach to stationary objects (e.g., tables), collision avoidance; sensor on top of tray (1 sensor): interactive function (e.g., sending the robot home)
RGBD camera in robot base	Currently not in use; possible future uses: navigation, object recognition

Figure 2.2 Biscee disseminates candies at an exhibition event and delivers food in a mock restaurant for guests.

2.2.1.2 Guiding, informing, and orienting

Benefits: although there are situations where humans help others find their way around, escort them to specific locations, or provide information, social robots have been developed to potentially replace humans in these tasks. Researchers have conducted field trials to investigate people's reactions to them (e.g., Chen et al., 2015; Kanda et al., 2009; Vijay et al., 2012). Such social robots have the advantage of providing the human partner with a feeling of intersubjectivity through their embodiment and ability to express themselves through gesturing, and they can assist by leading the way to the customer's destination.

Risks: despite significant effort, robots have not taken over this role in most places. There are many alternative solutions that fall in the category of disappearing technology. For example, using loudspeakers fixed on simple stands combined with touchscreens in shopping centres or providing headsets for visitors in a museum could be much cheaper and efficient solutions, particularly when the technology is able to respond to spoken words and is supported by conversional artificial intelligence.

2.2.1.3 Education and teaching

Benefits: social robots have the potential to serve as teachers or information providers in a wide range of settings, including both educational and non-educational environments (see also Section 4.2). They can effectively demonstrate and explain complex knowledge making it easier to transmit information in a repetitive manner. This capability proves exceptionally valuable when exhaustive interaction is necessary to attain the level of exposure essential for optimal learning. This is particularly evident in situations such as working with children who have disabilities (Alt et al., 2012).

One of the earliest examples of a social robot used for educational purposes was the 'museum tour guide' robot (RHINO) (Burgard et al., 1998). This robot was cylindrical in shape and featured two small video cameras mounted on its top. Although its social interaction capabilities were limited to object avoidance, indicating direction of movement through camera orientation, and emitting a sound when faced with an unexpected object, it was an early demonstration of social robots' potential to be used in educational settings. Tanaka et al. (2007) allowed 18–24-month-old toddlers the opportunity to interact with a human-like robot (QIRO) in a nursery. The researchers found that the quality of interaction between the toddlers and the robot improved over time, and that the children began to treat the robot more like a social partner than a mere toy. Based on these findings, the researchers suggested that social robots have the potential to be beneficial in various communities of children.

Since these early studies, child–robot interaction has received increasing interest. More sophisticated robots displaying better interactive skills proved to enhance performance of both typical children (e.g., Kory-Westlund & Breazeal, 2019) and children with handicaps (e.g., Scassellati et al., 2018; Wood et al., 2021). In a review, Rohlfing et al. (2022) have identified a wide range of skills (e.g., language learning, meta-cognition, error-detection) that can be supported and facilitated by robots with specific design and abilities. It seems that both short- and long-term such social robots may contribute to the development of social competence in children (see Table 2.2).

Risks: early experiments involving social robots as educational tools or teaching aids sparked intense discussions about their usefulness and ethical aspects (Sharkey & Sharkey, 2010, 2021). While these robots may initially generate interest among both adults and children, this attractiveness may decline as the novelty wears off. Critics have raised concerns that excessive interaction with human-like robots may interfere with the development of species-specific human behaviour (Kubinyi et al., 2010) and reduce motivation for interacting with humans (Turkle, 2011). Moreover, the greater social and emotional investment required to interact with humans compared to robots (Bryson, 2018) may cause some children, particularly those with insecure attachment to their parents or anti-social attitudes, to avoid human interactions (Konok et al 2017). Furthermore, humans can form a unique attachment bond with a particular social robot, leading to a sense of loss in the event that the robot is removed or malfunctions.

2.2.1.4 Personal service

Benefits: social robots can serve as social partners, similar to animal companions or ladies' companions, and provide social feedback to their human partners. Their primary function is to offer opportunities for shared activities, and their presence can alleviate loneliness by providing physical and emotional comfort (Taylor et al., 2023). Some social robots are capable of delivering these services, engaging in complex conversations (functioning as chatbots) and providing visual and auditory entertainment, such as displaying dancing actions or playing music, for their human companions. It is not necessary for these robots to resemble humans because animal companions, such as dogs and cats, have already proven successful in this role, and there is a long list of animal-like 'robotic pets' available (Melson et al., 2009).

Table 2.2 The required abilities of robots fulfilling a specific role (based on Rohlfing et al., 2022). Perceptual abilities of the robot include skills required for the perception of specific communicative signals; cognitive abilities refer to the processing of the information in various ways; and dialogical abilities enable the robot to interact with the social partner

Role	*Definition of the role*	*Required abilities of the robot to fulfil the role*			*Studies on language learning and literacy (examples)*
		Perceptual	*Cognitive*	*Dialogical*	
Socioemotionally supporting	Socioemotional support during learning	Perceiving whether the child is engaged in the learning activity and thus respond appropriately	Modelling the optimal frequency of feedback and animacy during learning, to avoid the robot becoming a distraction	Initiating dialogue to present the robots' sentience and engagement, and to offer encouragement when needed	Chen et al. (2020)
Assisting	Socioemotional assistance to the learner or taking over the tasks of teachers	Perceiving and recording learner behaviour	Interpreting and assessing behaviour in terms of task performance to provide appropriate input	Initiating and/or maintaining dialogue to offer encouragement and provide feedback	Engwall and Lopes (2022)
Prompting	Inviting the learner to use language expressively	Understanding what the learner is saying	Interpreting the learners' speech	Initiating and/or maintaining dialogue	Lin et al. (2022)

Role playing	Acting out a certain role	Recognising nonverbal social signals (e.g., facial expressions)	Training to produce proactive and socially appropriate behaviour	Responding adequately	Ali et al. (2019)
Displaying incorrect behaviour	Error-prone tutee for the human tutor	Understanding the instructions, explanations and corrections given by the human tutor	Improving the performance without surpassing the tutor's performance	Establishing a cooperation cycle with the tutor	Tanaka and Matsuzoe (2012)
Encouraging metatalk	Initiating an interaction/dialogue	Perceiving the child's utterances or deliberately initiating certain peculiarities in its perception that can be reflected upon	Modelling child's actions and inferring knowledge related to the task	Initiating/eliciting a specific type of dialogue and engaging in an exchange about the communicative situation	Hsiao et al. (2015)

Risks: people have diverse preferences for forming partnerships, which can be influenced by their living circumstances. In the past, watching television was a popular pastime, and more recently, social media platforms like Facebook, Instagram, and video games have become popular means of (virtual) social interaction for many lonely individuals. Therefore, it remains uncertain whether social robots can effectively compete with these alternative forms of social contact. Furthermore, the introduction of social companion robots may hinder people's ability to connect with others, potentially contributing to their isolation rather than mitigating it (Primack et al., 2017). Similar trends may emerge by the use of robotics sex partners that may also lead to changes in inter-personal (sexual) relationships (Danaher, 2017).

2.2.1.5 Curing and assisting

Benefits: social robots are increasingly being recognised as valuable tools for various interventions, with two primary scenarios being commonly cited as examples. One is their potential use as caregivers for the elderly, either in nursing homes or private homes. The other is their role as therapeutic tools, for example, for children diagnosed with autism spectrum disorder (ASD), where they can supplement human tutors in special education (see also above). There is also the possibility that using robots in such assisting situations could free up time (quality time) for human professionals to engage in higher-quality interactions. Kouroupa et al. (2022) reviewed several studies that aimed to improve social skills in children living with autism, using various types of robots, including humanoids and zoomorph agents. These studies targeted a range of social competence skills, including communication, engagement in joint play, imitation, emotional behaviour regulation, joint attention, and eye contact. The findings showed that over 72% of the studies reported some improvement in social skills as a result of the interventions.

In a review of several studies on socially assistive robots in elderly care, Abdi et al. (2018) found that such robots can serve a range of roles, including providing companionship and aiding in mental and physical training. Some robots have also been designed to provide physical assistance, such as helping with walking or staying in touch with family (e.g., KOMPAÏ, see details in Asgharian et al., 2022). While there is potential for this technology to ease the workload of nurses, it is unlikely to replace their role in the near future.

Risks: providing therapy for children living with disabilities or assisting the elderly with restricted physical or mental skills can be a more challenging task than interacting with typical individuals. Therefore, constructing a social robot for curing and caring may be more difficult than creating one for typical interaction. Children with ASD may be more receptive to robots showing more machine-like behaviour due to their specific conditions (e.g., Moore et al., 2018), while elderly people may be less accustomed to such technology (Di Giacomo et al., 2019). However, excessive use of social robots may deprive the elderly of opportunities to form and maintain social relationships with their human companions. Therefore, social robots should only be included as tools in any care for people with restricted skills, and all humans should keep their right to remain part of society.

2.2.2 Human–robot interaction (HRI)

The potential, challenges, and risks of human–robot interaction (HRI) have been topics of discussion in philosophy, psychology, literature (particularly science fiction), and movies for many years, long before the development of robust computers. Interactive robots have been developed slowly over time, and with the beginning of the 21st century, the study of HRI has become a truly scientific discipline, requiring the collaborative work of researchers from various fields, including engineering (mechatronics, informatics), behaviour (ethology, psychology), cognitive science, and social science. As the likelihood of integrating social robots into human groups increases, specialists dealing with ethical and legal issues are also becoming involved in the work. Although interdisciplinary collaboration can offer significant advantages for HRI, the need to converge scattered ideas can slow down the pace of progress.

In considering the broad field of human–machine interaction, research on human–computer interaction (HCI) can be seen as a precursor to HRI (Rogers, 2005). While many reviews on HRI emphasise the differences between these two fields, much can still be learned from the research and outcomes of HCI. Over the past 40 years, there has been significant progress in computer technology, and the successes or failures of HCI have had economic and financial implications for companies based on direct feedback from consumers. This is not typically the case for most HRI research.

The fundamental concept behind HCI was to design computers and software to cater to human needs, making it as user-friendly as possible. Although the physical hardware of computers has remained relatively unchanged, the computing capacity, including computation speed, memory, parallel processing, portability, networking, and so on, has experienced exponential growth. This development has facilitated the introduction of more accessible and user-friendly software, making interactions with computers easier (but see also Box 2.1). Consequently, for newer generations of users, computer interaction will have become an everyday aspect of life. Even children as young as three years old interact with devices such as mobile phones, often learning to use them independently (Konok et al., 2016; Mitra & Crawley, 2014).

2.2.2.1 Perspective on HRI

Dautenhahn (2007) proposed three distinct perspectives for HRI research, with the aim of comprehending the interaction between humans and robots (Goodrich & Schultz, 2007). These viewpoints align with the visions and objectives of the researchers involved.

Robot-centred HRI: this approach focuses on designing mechatronic agents capable of maintaining a certain level of autonomy (see below) during interactions with humans. This involves enabling the agent to support its own proper functioning (e.g., navigation) while simultaneously collaborating with the human partner to achieve a shared goal. Much of this research requires ensuring the reliable operation of both hardware and software components necessary for executing actions.

Robot cognition-centred HRI: a social robot is also considered a cognitive agent, with its own internal states that guide decision making about which actions to execute at any given time. It is important to note that cognitive processes are organised hierarchically, so local decisions do not interfere with the overall goal of solving a specific problem. The effectiveness of these decisions relies on the available data and the structure of the decision-making system. Researchers with backgrounds in artificial intelligence and cognitive science may adopt this perspective in their approach to studying HRI.

Human-centred HRI: the contribution of humans is essential for a team to achieve its goal in HRI. Humans should be involved in the interaction in a natural way, with only the minimum knowledge required about the robot. The robot's actions should be executed in a manner that is acceptable and comfortable for the human partner. This perspective of research focuses on studying the effects of social robots on humans using various tools to observe human behaviour and collect information through questionnaires and interviews. Psychologists and ethologists are particularly suited to work within this perspective of HRI.

A further perspective within HRI could be *task-centred*, which aims to evaluate the most efficient way of solving a problem with a specific team or facilitate the construction of robots that can form a better team with humans. This approach may also involve devising innovative solutions to problems that require a new type of robot, for which the human partner may require additional training (see Section 2.4.1).

2.2.2.2 Social robots in HRI – a minimalist approach

Research in HRI encompasses a diverse range of robots with varying capabilities. Some robots resemble remote-controlled laptops mounted on moving platforms, while others possess more complex embodiment with moveable arms, hands, and heads. Therefore, in this discussion, it is crucial to establish a clear definition (including minimal capabilities) for social robots. Although many attempts have been made to define social robots, we suggest a minimalist approach that considers any robot with the following autonomous (programmed) features to be a social robot (see also Box 2.2):

- *Mobility*: social robots should be capable of moving in the same space as humans and other robots autonomously, without requiring continuous human supervision;
- *Object Detection and Avoidance*: to navigate shared spaces safely, social robots must be equipped with sensors and a mental architecture to detect and avoid stationary and moving obstacles;
- *Human Interaction*: social robots should be able to interact with humans in a way that is simple but natural (e.g., texting should be excluded). The exact nature of the interaction depends on the application and context;
- *Task-oriented*: the interaction between the robot and the human should have a clear, measurable objective that can be evaluated in terms of performance.

While this description is consistent with some views, there are notable differences as well (Scholtz, 2003). Some researchers emphasise the importance of interaction without necessarily aiming to achieve a specific task or common goal. However, this approach overlooks the broader objectives of HRI, which is to develop robots for specific tasks (Section 2.2.1). Social interaction should always serve a purpose, and the human–robot team should have a goal to be achieved. Moreover, the tasks need not be highly complex, as some robots may have limited capabilities.

Social robots require specific hardware and software solutions to navigate safely in their environment. Importantly, some of the most widely used social robots, for example, NAO, Kaspar, and Leo are not mobile (for details about NAO and Kaspar, see Chapter 4). Thus, one may differentiate between 'static' and 'mobile' social robots for now. Nevertheless, navigation and object avoidance are essential features for any social robot that aims to collaborate and have a partnership with humans.

Although some social robots may lack certain features, they can still be employed in HRI experiments, albeit as simplified models. It is important to note that such scenarios do not diminish the significance of the above requirements for social robots that aim to interact with humans in real-world environments (Table 2.3).

Table 2.3 Roles and proximity patterns of robots used in different areas (taken from Goodrich & Schultz, 2007)

Application area	*Remote/proximate*	*Roles*	*Example*
Search and rescue	Remote	Human is supervisor or operator	Remotely operated search robots
	Proximate	Human and robot are peers	Robot supports unstable structures
Assistive robotics	Proximate	Human and robot are peers, or robot is tool	Assistance for the blind, and therapy for the elderly
	Proximate	Robot is mentor	Social interaction for children with autism spectrum disorder
Military and police	Remote	Human is supervisor	Reconnaissance, de-mining
	Remote or proximate	Human and robot are peers	Patrol support
	Remote	Human is information consumer	Commander using reconnaissance information
Edutainment	Proximate	Robot is mentor	Robotic classroom assistance
	Proximate	Robot is mentor	Robotic museum guide
	Proximate	Robot is peer	Social companion
Space	Remote	Human is supervisor or operator	Remote science and exploration
	Proximate	Human and robot are peers	Robotic astronaut assistant
Home and industry	Proximate	Human and robot are peers	Robotic companion
	Proximate	Human is supervisor	Robotic vacuum
	Remote	Human is supervisor	Robot construction

2.2.2.3 Social robots in HRI – a maximalist but at present unrealistic approach

Fong et al. (2003), Breazeal (2003b), and Dautenhahn (2007) present a comprehensive perspective on social robots that can function as equal partners with humans. While several social skills are critical to maintaining a productive collaboration between robots and humans, even current social robots lack most of these (after more than 15 years of the original reviews mentioned above). Nonetheless, numerous efforts have been made to realise these skills in specific social robots built to engage in realistic social interaction in the field. Below, we outline the five most significant dimensions based on Fong et al.'s (2003) work:

- *Attractiveness*: an agent's morphological and/or behavioural characteristics should make it appealing to its human partner. Some of these cues may increase the human tendency to anthropomorphise the robot. These cues may bear resemblance to their human counterparts (e.g., human-like crying), but more generalised cues may also be effective (e.g., high-pitched sounds). An attractive robot reduces the chance of humans fearing it (Section 2.3.3);
- *Sociability*: the agent should display an inclination to be a part of a social group and exhibit decreased inter-individual distance (increased proximity) in certain situations (e.g., attack by enemy) (Section 2.6.2);
- *Situatedness*: the agent has to recognise itself in the network of group companions. Through individual recognition, it acquires knowledge about the social structure of its group and determines the ways of interaction (Section 2.6.1);
- *Competence*: this feature encompasses an agent's ability to act and minimise conflict or conflicting situations within the group. Social competence entails various complex social skills, such as empathy, rule-following, sharing, and pedagogy. It relies on extensive experience, some of which is specific to the actual group, and is based on representing the inner states of others (Section 2.6.1);
- *Interactivity*: robots prefer social interaction and mainly rely on socially acquired information through communication or social learning (e.g., emulation, imitation). They should also possess skills for transmitting knowledge to humans and agents by teaching and may communicate with humans through language or other non-linguistic signals (Sections 2.6.3 and 2.6.4).

It is apparent that current social robots are far from exhibiting these social skills in conjunction and at an acceptable level of performance. Additionally, specific robots may succeed in possessing only a subset of these skills due to the nature of their relationship with the human partner. For certain types of collaboration, simpler social skills may suffice. As a general principle of efficacy, social robots should demonstrate a minimal set of social features (the lowest possible while still fulfilling their intended function) to prevent human individuals from becoming too engaged or developing unrealistic expectations of them.

2.2.3 Animal–robot interaction

While ethorobotics is primarily formulated within the context of HRI, it is essential to note that HRI constitutes a specific subcategory within the broader domain of animal–robot interaction (ARI) (Figure 2.3), focusing on the human species. Although the general aims of ARI and HRI are somewhat different, there are several similarities in the two fields. Research in ARI has the following objectives:

1. Identifying the embodiment and behavioural elements is crucial for a specific species to accept an artificial agent as a social partner.
2. Verifying the potential of the artificial agent to become a social partner comparable to living conspecifics.
3. Employing the artificial agent as a social partner to enhance the reproducibility and replicability of experiments, providing greater control over the partner's behaviour and physical features.
4. Evaluating the real-life applications of artificial agents (see below).

Figure 2.3 Examples of animal-robot interactions. (a) A cat is approaching a UMO (Sphero Ollie) that previously displayed animate motion cues (Photo by Judit Abdai; Abdai, Uccheddu, et al., 2022); (b) a dog is waiting for the UMO to retrieve food for him (Photo by Judit Abdai; Abdai, Bartus, et al., 2022); (c) a rat is interacting with a robotic iRat (Photo by Laleh Quinn; Quinn et al., 2018); (d) a small fish shoal displays cohesive tracking towards the robot fish (Photo by Maurizio Porfiri; Aureli et al., 2012) (figure from Abdai & Miklósi, 2023).

Just like in case of HRI, the aim is to design robots that are capable of engaging in one-to-one social interactions and/or to be integrated into a group of individuals (Romano et al., 2019). Fish-like robots were successfully integrated into a group of different fish species, for example, they acted as leaders, recruiting a fish from refuge and changing the direction of motion in a group of ten fish (Faria et al., 2010). Furthermore, Halloy et al. (2007) also showed that a robot can influence the collective decision making of cockroaches (*Periplaneta americana*) when choosing shelter, and Quinn et al. (2018) found that rats (*Rattus norvegicus*) release a rat-like robot, similarly as they do with conspecific, especially if the robot displayed helpful behaviour before.

Although most often ARI research focuses on the application of robot partners to study the importance of specific aspects of behaviour in social interactions. In addition, recent findings of Polverino et al. (2022) show that using robotic fish mimicking the native predator of eastern mosquitofish (*Gambusia holbrooki*), a highly invasive species, leads to long-term negative influences on the life history and reproduction of the invasive species, but do not elicit antipredator response in tadpoles (*Litoria moorei*) (native species negatively impacted by mosquitofish). Although research was carried out in the laboratory, the results implicate the potential utilisation in natural conditions.

Probably the highest similarity between HRI and more conventional ARI (at the moment) is research on dog–robot–human interactions. Studies show that dogs engage in communicative interactions with human-like robots (PeopleBot or NAO) (e.g., Kasuga & Ikeda, 2020; Lakatos, Janiak, et al., 2014; Morovitz et al., 2017), following pointing to a specific location or obeying to a verbal command. Dogs also interact with a dog-like robot (AIBO), but here researchers found that the reaction depends on the age of the dog, the experimental context, and the physical features of the AIBO (having fur or not) (Kubinyi et al., 2004). This research raises the possibility that robots may be involved in the life of companion dogs as 'assistants', either providing social company while being alone, or caring for them (e.g., feeding), just like social robots are developed to help humans in various contexts.

Importantly, human-like (or even dog-like) appearance is not necessary to successfully establish dog–robot interactions. Dogs interact with self-propelled agents without organism-shaped bodies (Unidentified Moving Objects, UMOs) in various social situations. Results show that animate motion cues (Abdai et al., 2017; Abdai, Uccheddu, et al., 2022), reactivity to the partner's behaviour and helping behaviour (e.g., Abdai et al., 2015; Capitain et al., 2023; Gergely et al., 2015) is sufficient for dogs to interact with these agents. During these interactions, dogs display similar behaviours towards the UMO that are used in communicative interactions with humans, even after a few minutes of direct experience. Furthermore, dogs remembered the behaviour of the UMO for at least a month (Abdai, Bartus, et al., 2022) indicating the potential long-term use of the agent.

Considering the intraspecies variability in dogs and that one of their main social companions is a heterospecific partner (humans) it may not be surprising that the partner's behaviour, independently of its embodiment, is important for getting engaged in various interactions. Thus, similarly as in the case of robots developed to interact with humans, these agents should be developed for a specific purpose (see also Section

2.3.5) and should be considered as a new species (see also Section 2.3.4). Although there is no information about whether the 'uncanny valley' (Section 2.3.3) emerges in dogs, dog- or human-like appearance and behavioural skill set can introduce serious, unnecessary limitations in the design of robots for interacting with dogs.

2.2.4 How to measure HRI performance

The purpose of deploying robots is to achieve specific tasks, and socially interactive robots aim to take on some or all of the tasks typically performed by humans. To evaluate the performance of social robots, researchers should establish a baseline using the performance of humans or human teams. However, there is limited exact data on human performance in natural conditions, making it difficult to determine what constitutes acceptable performance for social robots. This lack of data presents a challenge for researchers in identifying or hypothesising a suitable performance standard for social robots.

To provide a simple example, what is the typical frequency of interpersonal collisions in an office environment? It may be assumed that such incidents either never occur or are quite common. However, the frequency of collisions depends on a variety of internal and external factors, and one's subjective perception of an 'acceptable' frequency may vary. We may attribute collisions to factors such as clumsiness, sensory problems, or fatigue once they occur too frequently. Measuring the rate of human error is especially important in situations where errors can result in significant

Table 2.4 Human error rates based on data from the industry (selected items taken from Smith (2005)) and compared to mistakes made during everyday actions at home. Similar data are not available for social interactions neither from humans nor from social robots functioning under field conditions

Action/behaviour	*Error rate*	*Error rate*	*Error rate*
	Read/reason	*Physical operation*	*Everyday behaviour*
Overfill bath			0.00001
Fail to notice cross road			0.0005
Leave light on			0.003
Let milk boil over			0.01
Read single alphanumeric wrongly	0.0002		
Read five-letter word with good resolution wrongly	0.0003		
Read five-letter word with poor resolution wrongly	0.03		
Do simple algebra wrongly	0.02		
Set switch (multi-position) wrongly		0.001	
Type or punch character wrongly		0.01	

damage, such as operating a nuclear plant, or in activities where flawless performance can provide a significant advantage (Table 2.4). Contrastingly, the shared propensity for making mistakes renders both humans and robots quite similar. Paradoxically, this characteristic could be leveraged to enhance the acceptability of robots among their human companions (Box 2.3).

Box 2.3 To err is human. To err is robot?

Depending on the specific task of the robot, different levels of error may be forgiven. In industrial robots, the robot has to be extremely precise at all times, in fact, the most important features of these robots are their positional accuracy and repeatability. In their cases, one single erroneous act within a series of million actions can be a serious issue.

In the case of social (and etho)robots, one may envisage a larger tolerance depending on the specific environment. Although today's robots are flawed, investigating the influence of these errors in HRI can reveal the type of error and/or context when it may be acceptable to make mistakes. This can further contribute to develop strategies how robots can recognise, communicate, and successfully recover from a mistake in social interactions.

Although some studies found that humans prefer error-free robots over erroneous ones, recent research suggests that acceptance depends on different factors, and mistakes committed by the robot may also increase their likeability. It has been suggested that the pratfall effect may also extend to social robots, that is, making a mistake makes a human more likeable (Aronson et al., 1966). Errors committed by a partner or social agent that otherwise expected to function perfectly, may improve its acceptance. In support of this idea, some studies found that humans tend to like more a robot that commits mistakes than a robot that functions flawlessly (e.g., Mirnig et al., 2017; Ragni et al., 2016). Hoffmann et al. (2020) also showed that negative ratings of likeability, trust, and acceptance of an erroneous robot disappeared when the robot used a human-like way of speaking compared to a machine-like phrasing. Furthermore, if the robot can detect that it made a mistake and react to it, it can contribute to its likeability. Lee et al. (2010) found that a service robot receives negative rating when committing errors, but when it apologised or offered a compensation it increased its likeability. Although regarding errors, researchers mainly relied on errors in speech (social norm violation) or combination in speech and action (technical error), similar questions should be explored in robots that do not use speech (see also Box 2.7).

To be able to recognise when it makes a mistake, researchers suggested that robots should process feedback from humans. Studies found that humans react differently when the robot makes an error than they behave during an error-free interaction (Cahya et al., 2019). People seem to display more head movement (e.g., head tilt), and more and longer gaze shifts (between the robot and the task) and facial expressions during erroneous situations. Thus, detection of human behaviour may provide information to robots whether they committed an error, and thus be able to rapidly react to it.

Although carrying out specific tasks and functioning without errors would be the best, it is highly unlikely to be realistic. Hecker and Moses (2013) created a multi-agent simulation of physical robot experiments based on the behaviour of foraging ants. They found that agents evolved specifically in an erroneous world (sensor errors: positional and resource detection errors) performed better under these circumstances than those evolved in an error-free world. Preparing robots for a world in which errors occur can contribute to their success in that specific environment.

2.2.4.1 Forms of validity

While there have been numerous discussions on the use of benchmarks (a reference point for measurements) in social robotics, there has been comparatively little attention paid to the validity of the measures employed. For example, consider a social robot designed to assist nurses in escorting elderly individuals from their private rooms to a communal dining area in a nursing home. Since it is not feasible to test the robot in a nursing home during development, experiments are conducted in a laboratory setting that is typical of HRI research. To ensure that the laboratory results accurately predict the robot's performance in real-world situations, researchers typically investigate four facets of validity.

Face validity pertains to the extent to which laboratory observations resemble real-world scenarios in an elderly care facility. In other words, does the test accurately represent the actual situation? As the name suggests, face validity is a subjective assessment made by researchers, and thus it is the researcher's task to decide whether the actual testing situation is similar to the real situation in which the social robot will be working.

Content validity concerns whether the test measures the expected performance accurately. Researchers must determine the critical parameters of the job and whether the test can measure these features in a reliable way. For instance, in the case of an escorting social robot, key measures may include the number of people escorted per hour or the smoothness of the escorting process, avoiding any sudden movements. The number of measures required varies depending on the complexity of the task, and the challenge lies in reproducing them under laboratory conditions.

Construct validity is used to demonstrate that the test provides a specific measure for the phenomenon in question. Social robots designed for elderly care must behave differently from those that interact with younger adults. Therefore, using young adults as test subjects may limit our ability to predict how the robot will perform with elderly people. Very often the human participants are not 'blind' with regard to the hypotheses of the research. Many reports on testing social robots involve students or researchers working in the laboratory or at the department.

Criterion validity refers to the potential to generalise the results of a particular test to other conceptually similar situations. Thus, the situation used to measure the escorting skills of the social robot should yield similar results to other existing tests or should have the potential to predict the robot's behaviour in other situations that may arise in elderly care.

2.2.4.2 The 'Clever Hans effect' in social robotics

In social situations, the subject's performance may be influenced by the experimenter's or bystanders' behaviour. This effect was named after a horse that was believed to have superior mental capabilities, which were later revealed to be the result of human influence. The 'Wizard of Oz' (WoZ) method was introduced in social robotics, which involves a well-trained person operating the robot remotely, unbeknownst to the participants. This method enables social robots to demonstrate skills that they do not possess (Riek, 2012).

Initially, the WoZ method was meant to be a steppingstone towards gradually replacing human supervision with real computing, but in practice, it has become the final method in itself and has not been followed by technological replacements. This method poses a significant challenge for measuring HRI performance because the human technical and social skills required for operating the robot cannot be separated from the robot itself. Even if operators strictly adhere to pre-established rules of action, their on-the-fly judgment of the situation, influenced by the subjects, may result in biased interactions. The use of WoZ technology for measuring human or robot performance violates many forms of validity, especially if one assumes that the observed performance can be related to real-life scenarios using the same autonomous social robot (see also Box 2.4).

Box 2.4 Types of robots used in human–robot interaction tests in the laboratory

Three types of artificial embodied agents are commonly employed to test human's (or animals') reaction to social robots:

1 *Remote-controlled robots*: these robots are operated by humans who remain hidden from the experimental subjects. Human operators navigate the robots and control their actions from a separate location. This type of control allows for precise manipulation and specific interactions between the robots and the subjects under study;
2 *Semi-autonomous robots*: this category encompasses robots that operate under a mixed control system, combining both automated processes and human intervention. Certain aspects of the robot's behaviour and actions are automated, guided by

pre-defined algorithms or rules. However, human control is also present, allowing for direct intervention or adjustments when necessary. This hybrid approach provides flexibility and enables a balance between autonomous operation and human oversight;

3 *Autonomous robots*: these robots operate independently and interact with the experimental subjects without any human control. They possess their own decision-making capabilities and exhibit autonomous behaviour based on pre-programmed instructions or learned algorithms. Autonomous robots are designed to carry out tasks and engage with subjects in a self-directed manner, enabling more natural and dynamic interactions (Figure 2.4).

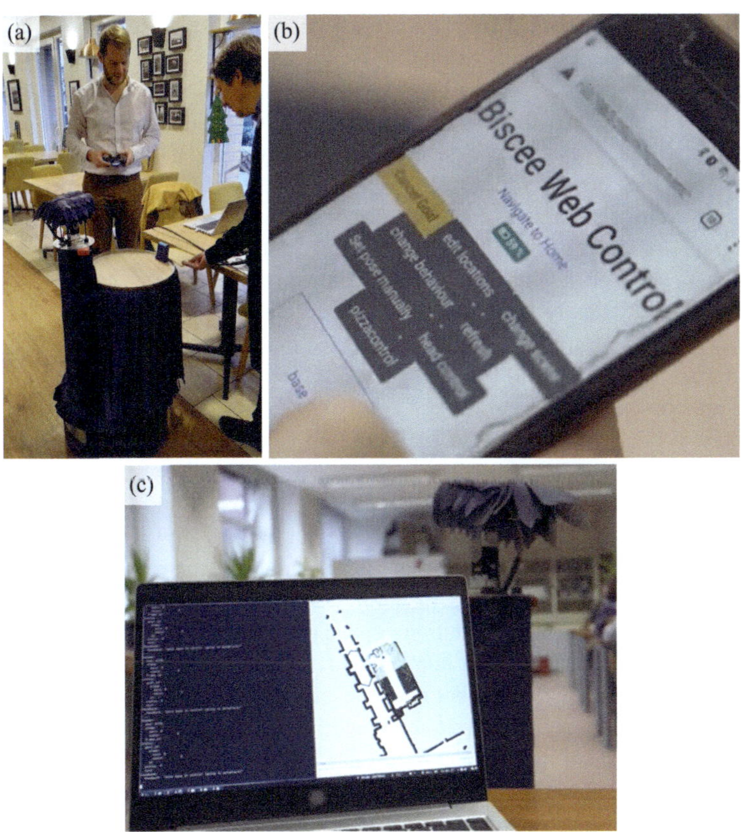

Figure 2.4 (a) Biscee is controlled by the human experimenter; (b) Biscee has a few pre-programmed actions that are displayed on command by the experimenter; (c) Biscee is moving autonomously in a room.

2.2.4.3 Measuring performance: task-centred benchmarking

Benchmarking is a complex and comparative process used to assess the performance of a system, to determine if it meets the expectations set by its creators. When developing a social robot, utilising effective benchmarks is crucial to ensure that it can perform well under natural conditions. However, evaluating a robot against benchmarks becomes more complicated as the agent's expected activities become more complex. Additionally, there are practical and financial limitations to how much effort can be invested in this process. Therefore, benchmarks should be broad enough to capture the robot's full functionality but few enough to make the evaluation process effective (e.g., Feil-Seifer et al., 2007; Steinfeld et al., 2006). Despite discussions on the topic, little practical progress has been made. To construct common benchmarks, a community effort is necessary, as most researchers and industrial developers design social robots for similar tasks (Wisspeintner et al., 2009).

Gao and Huang (2022) offer a comprehensive overview of how the navigation abilities of social robots can be objectively tested in environments populated by people. They propose several types of interactions (Figure 2.5) based on simple human interactions that may occur as individuals pass by. Importantly, these interactions have associated data on human performance, forming a foundation for comparisons between robots, humans, or different types of social agents. For instance, metrics such as average displacement error or final displacement error, measurable in terms of length, can be precisely determined. Therefore, conducting a sufficient number of tests and collecting data, including human control scenarios, could yield statistically validated performance measures. Furthermore, the results can be assessed within the context of expected distance values derived from existing literature, such as Hall (1966) (Table 2.5), but cultural and individual differences may be present.

Social robots are expected to perform complex tasks in a dynamically changing environment, which should be reflected in the benchmarks used to evaluate their performance. The RoboCup@Home competition offers a promising benchmarking method (Holz et al., 2013; Wisspeintner et al., 2009) (Box 2.5), utilising a 'bio-inspired' test battery consisting of various tasks for the robot to complete. The robot constructors do not have prior knowledge of the test arena or specific tests, but they may have some expectations about the possible tasks in an environment designed to simulate a person's home. The researcher gets one and a half days of preparation time to set up the robot once the test battery is revealed. This approach is crucial as it reflects a real-world situation, where social robots are expected to rapidly adapt to novel and unforeseen events.

Although some issues with the evaluation of social robots have yet to be resolved, the RoboCup@Home competition provides an opportunity to benchmark both task-level performance and specific within-task skills. To perform well in this test battery, which reflects environmental fluctuations, social robots must have task-oriented functionality, possess a wide range of skills (such as speech and face recognition), exhibit robust self-sustaining performance (including navigation and collision avoidance), and have stable decision-making systems. This complex challenge

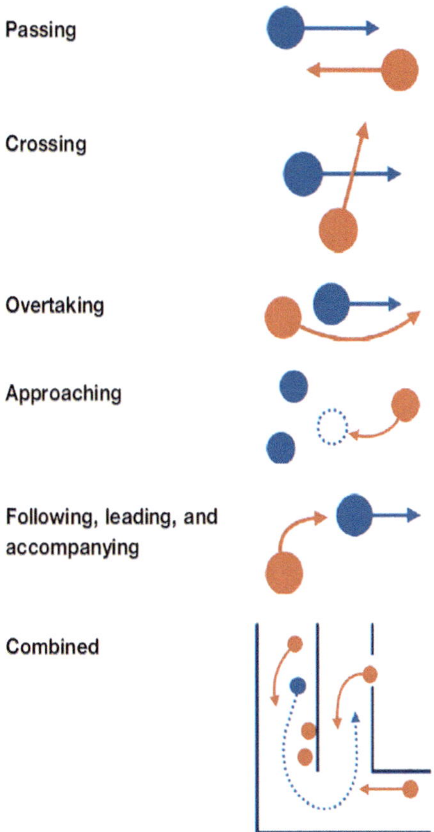

Figure 2.5 Common scenarios encountered in spaces with human interactions (from Gao & Huang, 2022). *Passing*: Two agents move in opposite directions, typically in confined spaces where they must alter their paths to navigate around each other. *Crossing*: Two agents intersect each other's paths while either or both are in motion. *Overtaking*: Two agents travel in the same direction, but one has a higher speed, resulting in the faster agent overtaking the slower one. *Approaching*: Two agents either move towards each other or converge upon a stationary agent while stopping at a socially acceptable distance. *Following, Leading, and Accompanying*: An agent takes the lead or follows (from behind or on the side) another agent while maintaining a socially acceptable distance. Additionally, two agents may move together, maintaining appropriate spacing.

Table 2.5 Inter-personal distances as reported by Hall (1966)

Function	*Description*	*Range (m)*
Intimate interactions	Intimate space (e.g., parent and child)	<0.45
Friendly interactions	Personal place (e.g., friends)	<0.45–1.2
Buffer zone for coexistence	Social space (e.g., on the bus with strangers)	1.2–3.6
Public interaction	Public space (e.g., on the street with strangers)	>3.6

motivates constructors to develop general mental architectures rather than localised solutions for different problems. The RoboCup@Home methodology also allows for the incorporation of new tests into the battery as technology and researchers' knowledge continue to improve.

Burghart et al. (2007) provide a distinct perspective on benchmarking for social robots that take into consideration different levels of complexity. At Level 0, the robot demonstrates the ability to perform a known task based on prior learning, even under altered conditions. At Level 1, the robot follows instructions from humans to carry out actions related to Level 0 performance, while the possibility of errors arises, enabling the measurement of the robot's corrective behaviour. Level 2 involves performance that is deemed acceptable under natural conditions. To achieve success at

Box 2.5 RoboCup@Home competitions

Competitions provide one important way to compare and also improve skills of social robots and at the same time they provide a more realistic scenario testing actual level of performance than tests in the laboratories. The RoboCup@Home competition provided such an opportunity (Wisspeintner et al., 2009). The actual tests designed for these robots more than 10 years ago would still provide a serious challenge for many present-day robots.

The functional abilities that have been considered in the competitions are the following (from Holz et al., 2013; Table 2.6): (1) Navigation, the ability of path-planning and safely navigating to a specific target position in the environment, avoiding (dynamic) obstacles. (2) Mapping, the ability of autonomously building a representation of a partially known or unknown environment online. (3) Person Recognition, the ability of detecting and recognising a person. (4) Person Tracking, the ability of tracking the position of a person over time. (5) Object Recognition, the ability of detecting and recognising (known or unknown) objects in the environment. (6) Object Manipulation, the ability of grasping, moving or placing an object. (7) Speech Recognition, the ability of recognising and interpreting spoken user commands (speaker dependent and speaker independent). (8) Gesture Recognition, the ability of recognising and interpreting human gestures. (9) Cognition, the ability of understanding and reasoning about the world.

The test scenarios used in the RoboCup@Home (for the extended version, see Wisspeintner et al., 2009; Figure 2.6):

Simple tests

Introduce test: a robot has to autonomously enter the scenario and move to a position in front of the audience. It has to introduce itself and the team using speech, gestures, slides, or multimedia.

Table 2.6 The overall performance (scores) of the teams during the first five competitions. The first value is the maximum score from a team, the second is the average score of the finalist teams (taken from Holz et al., 2013)

Ability	*2008 (%)*	*2009 (%)*	*2010 (%)*	*2011 (%)*	*2012 (%)*
Navigation	40/25	47/40	33/20	61/26	52/23
Mapping	100/44	100/92	21/10	33/10	10/4
Person recognition	32/15	69/37	57/23	48/16	62/15
Person tracking	100/81	100/69	100/72	100/76	62/33
Object recognition	29/8	39/23	6/1	25/10	56/20
Object manipulation	3/1	48/23	29/8	49/21	73/27
Speech recognition	87/37	89/71	50/38	76/59	90/56
Gesture recognition	0/0	0/0	62/26	100/49	88/37
Cognition	-	-	17/3	68/24	32/8
Average	**41/21**	**61.5/44.4**	**41.6/22.4**	**63.3/32.5**	**58.2/24.8**

Figure 2.6 RoboCup@Home competition scenarios from 2007 (a) and 2008 (b) (figure from Wisspeintner et al., 2009).

Fast Follow: a robot's task is to follow a person from one entrance of the environment to the other.

Fetch and Carry: the robot has to find and retrieve a certain object which it then needs to return to the human instructor.

Who is Who?: the robot has to detect and recognise three persons (two of them unknown to the robot) spread out around an area near to the entrance of the scenario.

Lost and Found: the robot has to find and identify as many out of three objects as possible. The referees pick the objects randomly from the set of objects and distribute them randomly throughout the scenario just before the test starts.

Complex tests

PartyBot: the robot's task is to find, recognise, and/or remember multiple unknown people randomly distributed throughout the entire environment (standing and sitting) and to tell them apart later on when serving a drink.

Supermarket: the robot needs to retrieve certain household objects from a shelf for a person randomly chosen from the audience who does not know how to operate the robot.

Walk and Talk: a human leader guides the robot through the scenario that was completely rearranged beforehand (and is therefore unknown to the robot) and has to teach specific locations only using natural language. The robot then has to prove that it has correctly learned those locations by having to navigate to certain places in random order after a speech command is given.

Cleaning Up: the robot needs to collect a set of five unknown objects (i.e., not from the set of known objects) dispersed throughout the scenario. Objects can be anything that can be expected to lie around in a household.

this level, the robot must exhibit basic social skills, including maintaining attentive contact, regulating distance, and communicating non-verbally.

2.2.4.4 Measuring performance: human-centred benchmarking

The impact of interactive social robots on team partners or bystanders must be evaluated, as these robots are designed to work in human environments. Human partners are typically well-trained to work in teams with other humans, so it is essential to determine how teaming with a robot either facilitates or delays problem-solving. These measures can be obtained directly from interaction participants or from observers who evaluate the test from a third-person perspective (such as by watching videos of the interaction).

Several questionnaires have been developed to measure the user's experience of the interaction. Some questionnaires focus on the user's evaluation of the robot as an agent, while others aim to measure the user's perception of the interaction.

Bartneck et al. (2008) propose measuring the user's perspective using a Likert-scale and specifically targeted questions along five dimensions:

- *Anthropomorphism*: how similar is the robot to a human? (e.g., fake–natural; machine-like–humanlike; unconscious–conscious; artificial–lifelike; moving rigidly–moving elegantly);
- *Animacy*: how lifelike is the robot? (e.g., dead–alive; stagnant–lively; mechanical–organic; artificial–lifelike; inert–interactive; apathetic–responsive);
- *Likeability*: how attractive is the robot as a teammate? (e.g., like–dislike; unfriendly–friendly; unkind–kind; unpleasant–pleasant; awful–nice);
- *Intelligence:* how competent is the robot socially and in its environment? (e.g., incompetent–competent; ignorant–knowledgeable; irresponsible–responsible; unintelligent–intelligent; foolish–sensible);
- *Safety:* how safe is it to interact with the robot? (e.g., anxious–relaxed; agitated–calm; quiescent–surprised).

Steinfeld et al. (2006) propose measures to estimate the quality of the interaction between the social robot and the human partner based on the user's experience. The authors do not specify a particular method for collecting this data. However, answering the following questions could provide a good estimate of the user's experience:

- *Interaction Quality*: to what extent does the interaction with the robot resemble a human–human interaction? Does the robot achieve an acceptable level of synchronicity during the task?
- *Persuasiveness*: to what extent does the robot influence the behaviour and emotions of the human partner? Does the robot give the impression of being in control of its part of the task?
- *Trust*: to what extent does the human partner rely on the robot? Does the partner feel that the collaboration is effective and not wasting their time?
- *Engagement*: does the robot give the impression that it is interested in the user's behaviour, actions, and emotions? Does it attend to the interaction and focus on the task at hand?
- *Compliance*: how well does the robot adapt to the social situation in general? Does it follow rules? Is it able to receive verbal instructions? Does its behaviour synchronise with that of humans?

2.2.4.5 Primary and secondary benchmarks

In applied research, it is important to distinguish between primary and secondary benchmarks (endpoints) in order to clearly define the main objective of the project. The variables defined as primary benchmarks measure the key achievements (or their failure) that are expected from the project. These benchmarks must be defined before the start of the investigation, and the results have a significant impact on the future of the project. If the system fails to meet the assumptions, that is, the measurement of the primary benchmarks does not support the expected outcomes, the positive results of the secondary benchmarks are of little significance.

When the human and the robot work together, they have a clear goal or task to achieve, and the result of their interaction should be defined as a primary variable. For example, if the task is for the robot to follow a human and move to a specific location, then the robot's performance in completing this task should be the primary measure. In contrast, the user's likes or dislikes of the robot or their attribution of human-like skills to it should be considered as secondary benchmarks, as they are less important than the successful completion of the task.

In addition to measuring success or failure as a binary variable, it may be useful to establish additional variables for measuring performance related to subgoals, which can provide continuous data on performance. For example, the time taken to complete the task could be included as a primary benchmark, providing continuous data on efficiency, while time for reaching specific subgoals could also be taken into account. For the testing, repeated measurements, by using preferentially more than one agent, are necessary to collect enough data for proper statistical analysis, as performance may vary over time or due to changes in the experimental conditions.

2.2.5 Conclusions, prospects, and questions

Social robotics is founded on the idea that, in certain situations, human partners can be replaced by highly advanced artificial agents. However, it quickly became apparent that there is a substantial disparity between the existing technology and the tasks that these robots were expected to perform. A variety of solutions have been implemented to address this issue in laboratory settings, including remote control by humans or humans serving as surrogates for the robots. Furthermore, challenges arise due to the difficulty of measuring the performance of these robots, particularly because they must collaborate with humans, and both team members may make mistakes. To benchmark their performance, it is also necessary to demonstrate that replacing humans with robots or including a robot with a specific task does not diminish team performance. However, data on human teams are lacking in this regard.

While it remains possible that social robots could eventually assume human roles in various contexts, recent trends in their application are more conservative. These involve breaking down complex human tasks into simpler subtasks that can be accomplished using current technology. Rather than aiming to create waiter robots, present-day robots are better suited to replacing runners in restaurants, which have fewer complex tasks to perform. The overly optimistic expectations of early social robotics may be replaced by more practical objectives. Employing a bottom-up approach supported by ethorobotics could establish a stronger foundation and potentially avert past failures, ensuring a more focused trajectory.

2.3 What is ethorobotics?

'Why do these animals behave as they do?' – this is a *post hoc* question from an observer's point of view (Section 1.1). The formulation of this question for ethorobotics should take a different form: *why should these robots behave as they do?* The answer to this question should precede the construction of these agents.

2.3.1 Definition

Ethorobots are agents that operate based on ethologically sound rules encompassing both the morphological ('static') and behavioural ('dynamic') characteristics of the phenotype. Ethorobotics draws heavily from biology, with a special focus on evolution theory. To create these agents, ethorobotics relies on existing and future technology, using the same building blocks (both hardware and software) as other fields of robotics. A key goal of recent robotics is to build agents that can function in human environments, making social characteristics a primary concern for ethorobotics. While ethorobotics does not represent a fundamentally new approach to social robotics, it takes a distinct perspective and aims to establish a biologically supported theory for robot development (see also Nazir et al., 2023). Therefore, many of the insights and ideas that have emerged over the past 20 years of social robotics research are repurposed or reformulated to provide a broader evolutionary perspective, freeing social robotics from its human-centred (anthropocentric) views.

Ethorobotics aims to achieve several objectives (Miklósi et al., 2017), which will be explained in detail in the following chapters. First, ethorobots should complement humans as cooperative partners. Second, their main purpose should be related to their primary function. Third, they should have adequate autonomy to perform their job. Fourth, they should possess appropriate social competence to form a successful team with people. Thus, this group of robots represent special classes of agents, analogous to the species concept in biology (e.g., Duffy, 2006) (see Section 2.3.4).

Ethorobots must not (1) destroy natural human relationships and (2) compete with humans. However, they should (3) develop a social partnership with humans that matches the required level of cooperation, and they (4) should be acceptable for integration into our communities.

Although robotics cannot and does not follow the rules of natural evolution, some analogies are worth investigating. Thus, the four questions of ethology (see Section 1.2) are also applicable to robot technology, especially if there is an expectation that (some) robots will coexist with humans (and other animals) in the anthropogenic environment (Miklósi et al., 2017; Rahwan et al., 2019). So, let's consider those four questions in turn:

Ethorobots not only have a function but must have a *function*. In ethology, the function of behaviour is directly related to survival. Similarly, a specific type of robot only survives if it proves useful for humans, that is, its existence helps humans to perform better. Tools in human history tell a similar story. Importantly, the function is always strongly related to the environment, which is also subject to change, especially if it is under human control. Ethorobots that are effective in one environment may not perform well in another.

In biology, *evolution* represents an uninterrupted autonomous chain of events concerning populations. The adaptation to a changing environment takes place slowly, constrained by earlier manifested adaptations (in a different environment), often metaphorically described as 'moving between local maxima'. This issue may not be as serious for robots because they can be created de novo, without any constraints from previous existence. However, there are several examples where technical development was constrained by previous states (e.g., compare the structure of the stagecoach and the construction of railway carriages in England). The exaggerated trend to copy humans in social robots could also be seen as a kind of self-imposed constraint by the constructing community. But just as evolution is also 're-using' adaptations as its fate unfolds, ethorobotics should also be open to varying and relying on previous solutions evolved in living beings (e.g., legs) or in artificial systems (e.g., wheels) (see also Section 2.3.6).

Biological systems are very conservative in their basic *mechanisms*, while technology can change more rapidly and can combine different mechanisms in a more efficient way. Therefore, it is relatively easy to test the efficiency of different mechanisms, find the best ways of operation, and develop hybrid systems. Although there is the possibility of copying biological mechanisms for robot technology (e.g., neural networks in computing), the differences between the 'living substrate' (organic material) and the rigid 'ore-based' technology provide many limitations.

A wide range of biological organisms undergo *development*, which plays an important role in local adjustment to the actual environment. Development itself has

Table 2.7 Comparison of the behaviour of living entities and ethorobots based on Tinbergen's four questions (based on Rahwan et al., 2019)

	Living beings	*Ethorobots*	*Living beings*	*Ethorobots*
Ultimate	Function		Evolution	
	How does the specific behaviour contribute to the survival and fitness of the animal?	Only robots that are useful for humans can survive; needs to work well in the specific environment	How does the phylogenetic history influence the behaviour and what kind of selective pressures shaped the behaviour?	Can be designed de novo, but previous solutions in living entities and artificial systems can be ‚reused'
Proximate	Mechanism		Development	
	How does the behaviour work to fulfil its function based on genetic, physiological and cognitive mechanisms; basic mechanisms are conservative?	Due to rapid technological changes and less constraints, the most efficient mechanism can be selected	How does the behaviour changes during different life stages and due to different experiences?	Learning abilities should be involved, including plasticity to be more effective in the specific environment and learning preferences of the user

a specific evolutionary history, resulting in physical or mental structures with increasing complexity. However, despite being an obvious feature of living beings, the role of development has been the least utilised in artificial agents, aside from adopting learning mechanisms in some social robots. Modularity, both for physical components and the robotic mental architecture of robots, could be utilised as a parallel to development in biological agents.

Considering the aspects of function, evolution, mechanism, and development can be beneficial for robot constructors to approach the design process with an open mind (Table 2.7; based on Rahwan et al., 2019). Note that these perspectives should be considered in this order when applied to building a novel generation of robots. Function holds a central position due to its association with survival, and the notion of evolution should remind us of the vast possibilities for variability. Mechanisms that support function can originate from various sources, and development, which involves plasticity in adjusting to the environment, is crucial for robots to reach their full potential. By considering these four aspects, robots can directly interact with a dynamic physical and social environment.

Box 2.6 Why human-like robots may not be the future?

Although (the attempt to) creating human-like robots is widespread in the field of social robotics, designing a believable human is extremely difficult and when getting close, we may roll down into the uncanny valley (see Box 2.8). For a humanoid robot to act credibly around humans, even small details should function properly. For example, facial expressions play an important role in social interactions, and some are processed without awareness (e.g., fearful expression elicit fearful response). However, studies showed that not all emotional facial expressions are recognised correctly in robots. In general, studies found that adults have issues recognising disgust and fear in the facial expressions of a robot. Furthermore, even if participants can identify the emotional displays correctly, researchers found a decreased amygdala activity upon observing the facial expressions of robots, compared to that of humans (Hortensius et al., 2018).

In social interactions, many non-verbal cues and signals are applied that are decoded by the receiver spontaneously. However, depending on the specific context and characteristics of the expression, they may convey different information. The eyebrow flash is displayed the same way across cultures (see also Figure 2.7 from Eibl-Eibesfeldt, 1989): (a) starts with a fast onset, variable length of apex, and a slow offset, (b) to which if other facial muscles (action units, AU, see also Box 3.17) join, they always start and stop together, and (c) while the AU99 (defined as the contraction of AU1 and AU2; Grammer et al., 1988) is shown, no new AU is added. The rapid eyebrow flash is present during greeting but also occurs during interactions upon expressing agreement, sympathy, friendly joking, and it can be an ostensive signal and may also contribute to indicate trustworthiness (Frith, 2009). However, it can also express indignation in case of which the brow raise takes more time and is usually accompanied with other movements such as turning the head away. Thus, configuration of the eyebrow flash, accompanying movements (or lack of them at specific moments), and the context should be matched perfectly otherwise the robot may elicit eeriness.

Figure 2.7 Illustration of the eyebrow flash (Eibl-Eibesfeldt, 1989).

With specific research on small details of all facial and bodily expressions, and error-free application in robots we may design a behaviour that is spontaneously recognised as if a human would interact with us. However, an important question is whether all the work required to achieve this goal is indeed worth it, if the aim is 'simply' to create robots capable of meaningful social interactions with humans.

2.3.2 Ethorobotics vs social robotics

It may be debateable why there is a need to introduce a novel term: ethorobotics. The main reason is to establish a close connection between biology and robotics (and label of biorobotics has been invented in a different meaning, see Dario, 2005). Social robotics can provide a broader category, partly because all recent efforts to build robots, which are able to interact with humans, can be amassed under this title. Ethorobots can be envisaged because there are a few conditions they have to fulfil (see Table 2.8); however, ethorobots can be also much more variable because there is no expectation of being human-like that is a main trend in social robotics (see also Box 2.6). Ethorobotics is based on evolutionary theory, so it has its own interpretation about the future of robots in the human society, as part of evolution. Ethorobots represent one form of humans' extended phenotype (Dawkins, 1982).

Take the example of the android AMECA robot that falls clearly into the category of social robots. Will Jackson, the developer of the robot said that 'the goal of the company

Table 2.8 Most important conditions to be fulfilled by ethorobots compared to the existing trends manifested in the development of social robots

	Social robotics	*Ethorobotics*
Function	Focus on social functioning	All robots have to be useful in at least one specific task; both social functions and non-social functions are important
Embodiment	Preference for human-like embodiment	Robot should not look like humans
Communication	Preference for human-like modes of communication	Robots should use all kinds of ways to communicate, relying on the general laws of animal communication. They may be endowed with a unique way of communication (e.g., own voice). Robots should be able to communicate with each other
Autonomy	Very variable, no specific requirement has been proposed	Robots should always display some autonomy (including mobility) that is in line with their functioning, WoZ method is not acceptable
Testability	There are existing benchmarks but there has been no systematic application	The functioning of robots should be validated by quantifying their performance in benchmarking test batteries compared to the performance of humans

[is]... to develop robots that are able to interact in human-like ways, with humans' (Figure 4.10). This approach represents a huge restriction on the development of social robots, and so AMECA can never become an ethorobot. Despite the constructor's wish AMECA will remain always a mechanic agent however close it can mimic humans. This can be easily admitted if someone reads the reasoning on the four questions above.

What could serve as a genuine example for ethorobotics, inspiring its main principles? An illustrative case is drawn from the realm of dogs (Box 2.7). Dogs have frequently been referred to in social robotics as specific examples when envisioning robots for specialised tasks. For instance, there have been endeavours to create robots modelled after dogs to assist the visually impaired (Walia et al., 2010), and similarly, specific robots have been developed to undertake tasks analogous to trained dogs involved in search and rescue missions (Delmerico et al., 2019). While these instances highlight dogs as specific biological models, this species can offer broader inspiration for constructing a variety of non-human-like robots capable of interacting with people.

Humans readily engage with dogs in diverse social scenarios, considering them as family members and utilising them for various roles. This acceptance and interaction with non-human entities could extend to robots with non-human-like embodiments but attuned social behaviour. In essence, if humans are open to incorporating dogs into their social communities, treating them as family, and employing them for various tasks, then robots with non-human-like forms but similar socially tuned behaviours could seamlessly integrate into human social environments.

Box 2.7 Dog as a model for ethorobots

Dogs can serve as a model species to design the social behaviour of ethorobots capable of functioning well in the human environment (Gácsi et al., 2013). Dogs have coexisted with people for 16,000–32,000 years, a period during which their social behaviour evolved the potential to facilitate close interactions with humans. Despite inherent limitations in their physical traits and behaviours, dogs have gained a sophisticated repertoire of social skills comparable to humans, enabling effective cross-species interactions in diverse situations. While dogs exhibit sociocognitive skills akin to humans, their mental mechanisms may differ.

Engaging with humans on a daily basis, dogs go beyond companionship, collaborating in various tasks such as hunting, herding, or serving as assistance dogs. Observing dogs during these activities allows for a detailed understanding of their behaviour, unveiling effective time patterns in complex interactions and inferring behavioural rules (Miklósi, 2007). Translating these observations into mathematical and physical rules, and programming robots based on such rules, holds the potential for fostering similar interactions between humans and robots.

Humans may accept mistakes made by robots, especially when accompanied by apologies or compensation (see Box 2.3). Similarly, dogs, despite making mistakes, have collaborated with humans for millennia without relying on verbal apologies.

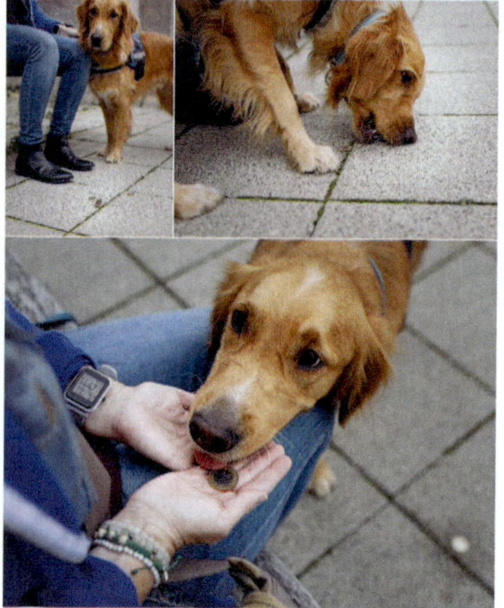

Figure 2.8 An assistant dog helps a human in various tasks (Photos by Melinda Egyed (upper), Artemis photo (bottom left) and Gábor Szabad (bottom right).

Studying how dogs respond to errors, recognise their mistakes, and attempt to solve problems could serve as a valuable model for developing similar strategies in ethorobots (Gácsi et al., 2013) (Figure 2.8).

2.3.3 The uncanny valley hypothesis

The initial and continuing trend of many robot developers to create human-like or human-identical robots has raised an important theoretical question that echoes old Freudian traditions about the fear from strangeness (Jentsch, 1997). Mori's (1970) well-known 'uncanny valley' metaphor has also resonated with many researchers because they have all likely experienced situations where encountering something human-like (e.g., a dead person) has caused a sense of eeriness rather than familiarity.

Mori's original hypothesis suggested that the more similar a creature is to humans, the more uneasy and uncomfortable humans feel towards them. This idea, known as the 'Uncanny Valley Hypothesis' (UVH), has been widely discussed in theoretical and experimental research. As there is extensive literature on this matter, this issue will only be briefly introduced here (e.g., Miklósi et al., 2017; Wang et al., 2015; Zhang et al., 2020). The UVH has been popularised due to its ability to provide both a good question for theoretical discussions and testable predictions for experimental research.

Box 2.8 Ethorobotic concept to avoid the uncanny valley

The uncanny valley hypothesis was envisaged by Mori (1970) who proposed that robots looking and acting more and more like humans would increase their likeability, but at a certain level of high similarity, they elicit avoidance due to their eeriness (Figure 2.9a). Since then, several studies found support for the uncanny valley hypothesis, emerging

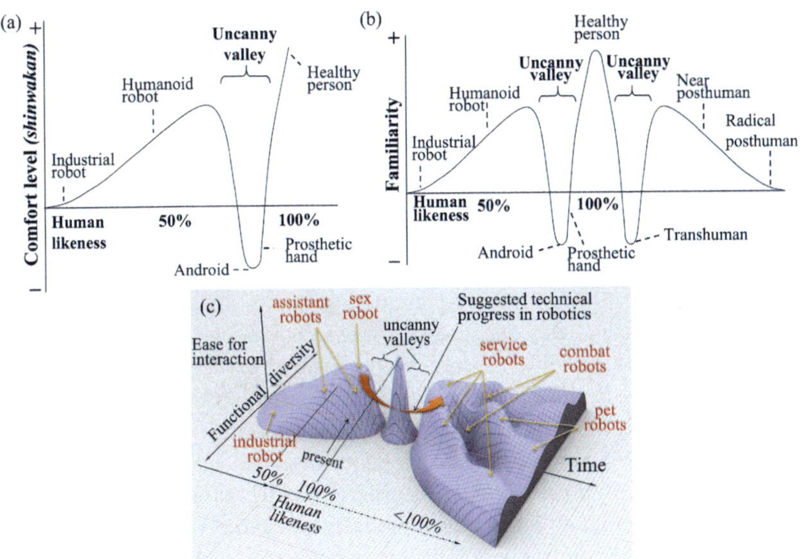

Figure 2.9 (a) A modified reproduction of Mori's (1970) original 'uncanny valley' hypothesis by showing only the reaction to moving agents. (b) An extended version of Mori's idea by Jamais Cascio (from http://www.openthefuture.com/2007/10/the_second_uncanny_valley.html). (c) The three-dimensional space depicts human-likeness, functionality and ease of interaction. After the first uncanny valley (which we are progressing into), we may reach a peak in human likeness, but further improvement in robots can result in a second uncanny valley before moving past the perfect human likeness. Ethorobotics approach suggests a jump around the valleys by focusing on the functionality rather than the embodiment. Figures are taken from Miklósi et al. (2017).

at around 12 months of age in humans (Lewkowicz & Ghazanfar, 2012), and also in macaques (Steckenfinger & Ghazanfar, 2009). However, in the past two decades, the number of findings contradicting the existence of uncanny valley has grown. Studies show that when faces are presented, only displaying abnormal features elicit the uncanny valley effect and close but not perfect features are not sufficient (Seyama & Nagayama, 2007), and it may also depend on other features of the images displayed (e.g., colour vs grey-scale) (Carp et al., 2022). Presentation of the stimuli may also be crucial, for example, Mori (1970) suggested that dynamic stimuli would elicit more robust responses than static ones.

Jamais Cascio (unpubl.) raised that in the future, social robotics may create trans- and posthuman robots (Figure 2.9b). Instead of staying in a perfect human-like state (having a maximum familiarity), improved social robots acquire superior abilities compared to a human. For example, human-like robots have 360° visual field or wings for flying. However, depending on the level of difference compared to a human, these robots would initially fall in a new uncanny valley before having radical posthuman robots overcoming the issue of close but not perfect familiarity.

Ethorobotics approach on the other hand suggests that being trapped in a double uncanny valley can be avoided by introducing robots that occupy a different niche (Figure 2.9c). Focusing on the functionality when designing not only the behaviour but the embodiment as well can help to avoid the issue of human likeness while keeping their capacity for social interaction with humans.

One way to study the UVH is by presenting a variety of stimuli that range from abstract or animal-like to very human-like, using techniques such as morphing for 2D images. Researchers typically measure the response of participants through subjective ratings of eeriness, human likeness, and familiarity, or through objective measures such as looking time at the presented stimuli. According to Mori's hypothesis, there is a non-linear relationship between humanness and eeriness. Specifically, as stimuli become more human-like, there is generally an increase in interest and attraction, but there is a point at which attraction decreases and eeriness increases before the stimuli reach 100% humanness. The interpretation of results from studies on the uncanny valley hypothesis is quite diverse, likely due to the use of different measures in these studies. Without delving into the specifics (see also Box 2.8), the main findings can be summarised as follows:

Support for the existence of UV: several studies have investigated the relationship between the level of human likeness in stimuli and the degree of eeriness reported by participants. MacDorman and Ishiguro (2006) found that morphing images with mechanical features into human-like pictures resulted in the highest levels of eeriness at the morphed stimuli. While this effect supported the UVH, the peak level of eeriness was expected to occur at images with more human-like features. As partial support for UVH, Ferrey et al. (2015) demonstrated that by morphing pictures of two animal species or robot/animal to human pictures, the most negative effect was elicited at the

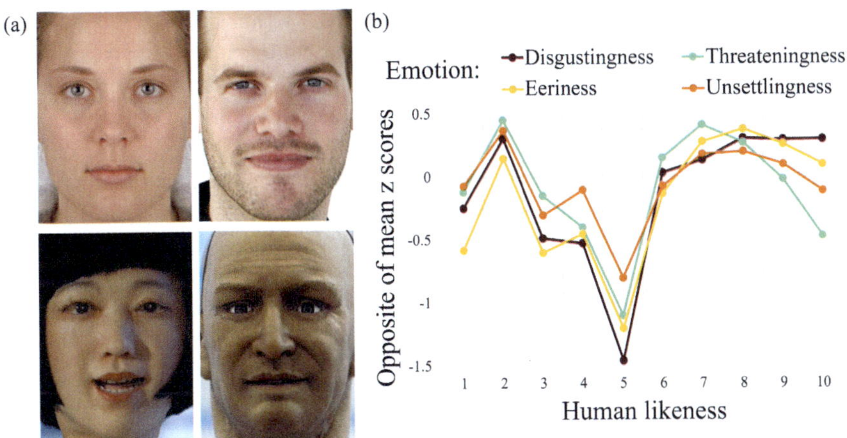

Figure 2.10 Wang and Rochat (2017) tested human subjects (N=62) with 89 typical and uncanny faces (a) (photos were provided by Shensheng Wang). Subjects' responses to questions on four negative emotions (disgustingness, eeriness, threateningness, and unsettlingness) as a function of human likeness from 1 to 10) are shown (b) as the opposites of mean z scores by 10 bins. The authors argued that the results support the presence of an uncanny effect, although the 'valley' was expected to be closer to the maximum value of human likeness.

midpoint of the morphing. In addition, Mathur and Reichling (2016) found that when people rated 80 real-world android faces then likeness was markedly reduced towards androids that aimed to closely mimic humans. Finally, a study by Seyama and Nagayama (2007, Study 2) revealed that morphing eyes and heads unrealistically in a series of pictures ranging from a drawing of a cartoon character to a picture of a human elicited eeriness. In sum, despite supportive trends, no clear results indicated that the eeriness only emerges at close similarity with the human (see Figure 2.10).

Lack of support for the existence of UV: several studies have failed to provide support for the existence of the UVH. For example, Seyama and Nagayama (2007, Study 1) found no support for the UVH when they morphed coloured drawings of humans into real human pictures, even though participants could differentiate between a geminoid (a robot that mimics a living human) and a human individual. Similarly, Bartneck et al. (2009) and Hanson (2005) did not observe an effect when they used stimuli such as morphing cartoon portraits into real human pictures.

The role of experience: Złotowski et al. (2015) found that the level of eeriness experienced by individuals interacting with a geminoid or humanoid robot decreased after repeated interactions. Chen et al. (2010) showed that exposure to images with enlarged eyes can lead to a shift in preferences towards that direction. Additionally, research by Lewkowicz and Ghazanfar (2012) has demonstrated that infants begin to display a preference for real human faces after the age of 12 months, suggesting that developmental experience may also play a role in shaping our responses to human-like stimuli. This suggests that early encountering potentially eery robots may lead to decreased avoidance later in life.

Evolutionary origin: macaque monkeys (*Macaca mulatta*), show less attention towards realistically morphed images compared to both unrealistic and real monkey faces (Steckenfinger & Ghazanfar, 2009). This finding suggests that the UVH may not be specific to humans but could also apply to non-human primates.

In their review, Wang et al. (2015) noted that the UVH is supported more often by studies using morphed images; however, the explanation of the elusive nature of this phenomenon is more complex. It is important to stress that there are no replication attempts, and each study is using a different methodology, including the fact that UVH may emerge more robustly if real behaving (moving) 3D stimuli would have been used. The stimuli used, the way of presenting the questions or measuring the behaviour, the subjects' experience and many other factors can significantly alter the results (Zhang et al., 2020). Often various hypotheses are contrasted against each other while they refer to different types of causality (Diel et al., 2022; Wang et al., 2015), and thus they are not exclusive (see also Miklósi et al., 2017).

The biological phenomenon is probably rooted in the need to recognise and often also make a series of hierarchically structured choices among conspecifics and heterospecifics, in addition to avoid them, if they show deviant behaviour. An individual should show a general preference for conspecifics for mating, but it should also recognise the opposite sex, and should be able to choose based on 'quality' (to increase the fitness of the offspring), including genetic relatedness (to avoid inbreeding).

Choices take place in various contexts, e.g., when one is looking for collaborative partner or a friend or alliance among group mates. It is very likely that animal species have been selected for good perceptual and recognition systems to be able to take the best choices. Importantly, humans are often also not aware of their very fine-tuned abilities in this regard. For example, human subjects base their judgements of 'liveliness' on the subtle differences in the musculature in and around the eyes (Looser & Wheatley, 2010). Other animals may rely on other features for making a judgement, but more importantly, these skills seem to be very widely distributed in the animal kingdom. The preference and the perceptual basis of the choice are context dependent. Individuals could be more selective when their decision only concerns a short period of time than when they have to make a choice for their life.

Most experiments carried out so far did not capture an important aspect of the interaction. In the real world, a choice takes place over a longer time period, that is, individuals rarely base a choice on 'first impression'. First experience is very important, but it is usually followed by some evaluation when one gains further inputs. This temporal process (may take only a few seconds) may give rise to the 'uncanny' experience when the very first impression is contradicted by further experience. This phenomenon is often referred to as 'violation of expectation' which at the behavioural level can lead to avoidance, controlled by emotional states such as frustration or disappointment or surprise (Burgoon, 1992).

Humans' sophisticated ability to perceive minute differences in appearance and behaviour makes it very difficult (and very expensive) to build robots that are 'identical' to humans. Thus, one may ask, if humans were always able to differentiate between a human and a robot then why should any robot look like a human?

2.3.4 Ethorobots as species

Although robotics has been around for about 100 years, it is still in its early stages of development. Due to the unpredictable nature of technological advancements, it may be wise to keep all possibilities open for the future. Rather than focusing on creating robots that are 100% human-like, it is more practical to construct robots that can enhance human efficiency. A growing number of researchers advocate for considering robots as a distinct non-human species (Duffy, 2006), referred to as 'ethorobots' in this book. The primary goal of ethorobotics is to create robots that can benefit and improve human life and not that they mimic a perfect human.

Human-ethorobot interaction rests on the following foundations (Miklósi et al., 2017; see also Section 2.6):

1. Humans have much experience to interact with beings other than humans. Sharing life with many species of domesticated animals provides a good example;
2. Many aspects of sociality (e.g., attachment, hierarchy, collaboration) rely on general rules among animals. People can thus spontaneously learn how to interact with ethorobots;
3. Ethorobots present no concern for the UVH, they do not 'fall into the valley' (Duffy, 2006); rather, they elicit interest from humans as being different but displaying key features of social behaviour;
4. Communicative signals (often indicating a specific emotional state) in animals follow many general rules (Morton, 1977), and thus humans can learn to rely on and produce those signals rapidly. This is also supported by comparing the neural responsiveness to various non-human communicative signals (e.g., Andics et al., 2016; Dubal et al., 2011).

Practical advantages of ethorobots (based on Miklósi et al., 2017) are as follows:

1. There is no need to assume competition between humans and ethorobots even if the latter gains better skills in some domains (see also below);
2. Although at present, existing animal species often provide some exemplars for robots but ethorobots can be mosaics of any animal species or involve other features that are absent from biological agents (Duffy, 2006; Rachwan et al. 2019);
3. If ethorobots are endowed with social behaviours that are shared by many social species, then humans react to them also, but they are still aware that these are machines (Nass & Moon, 2000);
4. Human disappointment in ethorobots can be minimised because the non-human appearance and behaviour may not elicit unrealistic expectations (Haring et al., 2018; Prescott & Robillard, 2021);
5. Much less ethical issues have to be solved if the ethorobots do not share or share only a few human characteristics (See Section 2.7);
6. There is no need for assigning 'sex' or 'gender' for these robots. Sex is an evolved feature for reproduction and is not needed for ethorobots.

There are many practical arguments that ethorobots should be distinctive from humans and represent a different category of agents supported by technological development. Populations of a specific ethorobot should be regarded as being analogous to a species, just like in the case of living organisms.

2.3.5 The niche concept

The concept of the niche is fundamental in ecology as it explains why there are numerous, often seemingly similar species, and provides a framework to understand the relationship between species and their environment. Although there is an ongoing debate about the concept in general, a summary of the main idea should be sufficient (Trappes et al., 2022).

The environment can be envisioned as a multidimensional virtual space, where each dimension represents specific resources (e.g., type of food) or constraints (e.g., existence of certain predators). It follows that any two niches should not share at least one dimension. A species is considered successful if it outcompetes other species in a specific niche.

This ecological insight can be applied to ethorobots by determining their expected niche. This involves identifying the features of the environment, including the possible ways of interaction with humans, and providing a detailed description of the task(s) with the expected performance targets (benchmarks, see Section 2.2.4) to be carried out by the robot.

Consequently, each species of ethorobot would be developed for a specific niche, and its performance would determine its survival. Ethorobots would not compete with people because their niches would be constructed in a way that does not overlap with those of humans.

2.3.6 The robot phenotype: morphology, behaviour, and performance

In biology, the phenotype is generally defined as the sum of all observable and measurable traits of an organism. It is also the substrate of selection, as individuals' survival (and their fitness) depends on their phenotypes. Although individuals belonging to a species share the same phenotypic traits, they vary in the expression of these traits, which provides the basis for selection.

Developing appropriate measures for the phenotype is a crucial task for biologists because, in principle, anything can be regarded as a trait and measured by some means. Although there are no exact rules for this task, subjective judgments and the experience of many generations have led to some common knowledge. For example, many biologists consider the conspicuous colour of birds to be a useful trait because it helps differentiate between species, and such traits often play a role in mate choice, thus influencing survival. Since the early days of zoology and botany, taking measurements and testing and comparing their usefulness for differentiating species or estimating fitness has been an essential task.

In this respect, ethorobotics is still in its infancy because each type of robot is usually represented by a single exemplar (or two), so no variation in the traits is expected. However, starting from the concept of benchmarking (Section 2.2.4), one could construct a list of potentially important phenotypic traits for ethorobots. Depending on their proposed function, some traits would be common for all, while others could be very specific.

While measurability is a key in the case of any phenotypical trait, the definition of the trait itself and how exactly sampling should be carried out, is often problematic. General categories for phenotypic categories may include (1) design, appearance, look; (2) embodiment including appendages; (3) inner hardware components; (4) software features; (5) behavioural skills and capabilities; and (6) performance.

2.3.7 Robot taxonomy

Currently, the number of functions and working of interactive robots, as well as any other social robot type, is quite low. Most robots are given fancy names (such as NAO or AIBO), while others are identified by a label consisting of letters and numbers, likely reflecting some preferences of the constructor. However, the names of the latter robots can be difficult to remember. As the number of robot types increases, there will be a need for a more objective way of labelling them. This process should begin by defining categories and variables upon which these robots can be distinguished.

The process of establishing categories and relationships among them has historical roots in both chemistry and biology. In chemistry, knowing the rules for creating molecule labels enables us to determine the basic composition of a chemical and its relationship to other similar compounds, as well as estimate some of its most important chemical properties. In biology, the field of taxonomy aims to define fundamental categories, such as species, and is working to establish possible relationships among them. This process is complicated due to the hierarchical nature of the relationships and the historical aspect of the 'story of life', which must be taken into consideration. Determining phenotypic variables is always a critical aspect of establishing category boundaries and constructing a useful framework for biological organisms, and it can often be a matter of discussion.

Social robots have been categorised both in terms of shape and form as well as their proposed function. Here is list of categories based on the appearance and design of the social robot (e.g., Walters et al., 2008; see also Figure 2.11):

1 *Mechanoid robots:* these robots are designed to have a purely mechanical appearance and do not aim to resemble living organisms. This design choice may be due to cost-effective engineering solutions or a deliberate decision by the constructors. Some research laboratories may create hybrid robots that have both mechanic and zoonoid or humanoid features. An example: Kismet;
2 *Zoonoids*: these robots are designed to resemble animals, either intentionally by the constructors or inadvertently due to their design. Zoonoids may also be based on fictional creatures or cartoons. An example: AIBO;
3 *Humanoids*: these robots share many physical features and capabilities with humans, but they are still distinguishable from humans due to visible mechanical parts. An example: NAO;

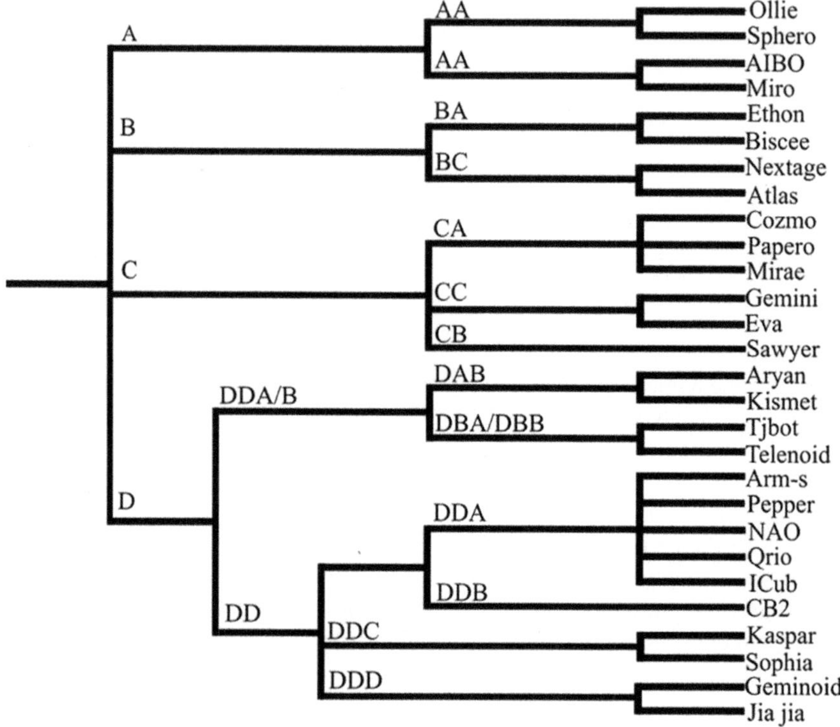

Figure 2.11 Evolutionary tree-like illustration of different robot types; based on human-likeness of the ABOT Database (https://www.abotdatabase.info/) (Phillips et al., 2018). The tree was created based on three different features, corresponding to the three letters indicated on the branches: similarities in facial feature, body manipulators, and surface look to humans (respectively). ABCD letters indicate the level of similarity (scores were retrieved from the ABOT Database except for Biscee which is not presented in ABOT; assignment to the four categories is arbitrary). A: no to very little similarity (ABOT score <0.3); B: little similarity (ABOT score <0.6); C: medium similarity (ABOT score <0.8); D: high similarity (ABOT score ≤ 1). The tree was created based on the categories (ABCD) of the three features using iTOL 6.8.1 (Letunic & Bork, 2021).

4 *Androids*: these robots are designed to look and behave as closely as possible to humans, with the aim of convincing human partners that they are interacting with real people. An example: Geminoid;
5 *Genoids*: this is a specific type of android that aims to replicate a specific person, like a monozygotic twin. The constructors may use advanced technology to create a lifelike appearance and behaviour of the person. An example is the robot mimicking the roboticist Hiroshi Ishiguro.

A different kind of social robot categories can be established on the basis on their suggested functions. Goodrich and Schultz (2007) proposed the following categories

(1) Search and rescue; (2) Assistive robots; (3) Military and police; (4) Edutainment; (6); Space; (7) Home; and (8) Industry (see also Table 2.3).

Categorising social robots based on subjective insights can be problematic since different arrangements cannot be easily transformed into one another, and some robots may not fit into existing frameworks. To address this issue, the goal for social robotics should be to develop a categorisation method that is applicable to a broad range of robot designs and functions.

One approach could be to focus on phenotypic variables, which describe and characterise social robots and use statistical procedures to establish a systematic framework for categorisation. This method is similar to the efforts of biologists to create an overarching system for classifying all organisms (c.f., numerical taxonomy). By using this approach, social robotics could avoid a priori categorisations that are based on limited insights and subjective opinions. Instead, it would create a more objective and comprehensive system that could accommodate future robots that do not fit into existing categories.

2.3.8 Technological evolution

The evolution of human activities, which are essential for our existence, is often compared to the evolution of living organisms. While this analogy may hold up well for human culture, it is less clear whether the material constructions of human activity also follow the rules of evolution (Basalla, 1988). While human culture evolves through the accumulation of knowledge and ideas, material constructions of human activity may be subject to different processes. For example, technological advancements may result from deliberate design and engineering rather than natural selection. Therefore, it is necessary to examine whether the rules of evolution apply to the material constructions of human activity and how they may differ from those governing living organisms.

Duffy (2006), Ziemke and Sharkey (2001), and others noted that living systems are autopoietic, meaning they have the ability to sustain and reproduce themselves. In contrast, extant artificial agents are described as being allopoietic, that is, they are devoid of any reproductive capacity, but nevertheless they can accommodate to their actual environment, they can achieve goals by executing some operations in and on the environment, and they can also produce agents other than themselves (see Table 2.10).

This distinction between living systems and machines is likely to persist in the future, preventing artificial agents from being subject to 'true' evolution. It is important to recognise the contrasting situations of biological evolution and technological development:

1 Biological evolution operates through natural selection, allowing for the emergence of new species and adaptations over long time. In contrast, technological development relies on deliberate design and engineering, with the aim of creating new tools and machines to achieve specific goals as rapidly as possible.
2 While artificial agents may continue to advance in sophistication and capability, they will likely remain fundamentally different from living systems, and their development will be guided by human intentions and goals rather than natural selection.

3 The robustness of living systems emerges gradually over time through a process of testing and benchmarking that takes millions of years. For example, it took 500 million years for cellular organisms to emerge, highlighting the lengthy and deliberate process of biological evolution.
4 Although both evolution and technological development can lead to the emergence of sophisticated and robust systems, the processes underlying each are fundamentally different. Biological evolution generally (there are many exceptions also) progresses from less complex to more complex agents and operates on a timescale that is orders of magnitude slower than technological development. The ontological continuum of biological systems also contrasts with the independent emergence of machines. Basic clades such as 'fishes' or 'apes' have a relatively long evolutionary history, reflecting the slow and gradual nature of biological evolution. The lifespan of any given type of machine is comparatively very short, with technologies such as steam engines being replaced by fundamentally newer (ontologically discontinuous) inventions (e.g., electric engines) within less than hundred years.
5 Sophisticated sensory and perceptual processing skills typically evolved earlier in evolutionary history than complex mental mechanisms for processing sensory information. Similarly, the development of robust motor functioning and control was also well established early on. For example, the vertebrate eye is estimated to be approximately 500 million years old, while the emergence of the tetrapod leg dates back to around 390 million years ago. However, in the context of technological development, introducing novel cognitive architectures often precedes the availability of high-quality sensors and perceptual skills, as well as the development of robust locomotory systems, and thus perceptual and motor capacities may not always provide the necessary level of robustness to support complex mental functioning.

Evolution has to be conservative and new adaptations are constrained by the history of the organisms. No such limitations exist in technology development, agents can be constructed from scratch. This may be an advantage but also involves risks, because of the lack of testing for robustness. Interestingly, evolutionary robotics may rely on selecting better mental architectures or embodiments based on evolutionary rules (Floreano et al., 2008).

For the evolution, the main target of selection is the individual who also possesses a reproductive system. As noted above, artificial agents are not self-replicating, so not the individual but the researchers and specifically the human society decides about their fate. While for more traditional objects, tools, and equipment many exemplars exist displaying large variations, in current robotics typically only a few robots represent their class.

The use of evolutionary modelling is gaining importance in the analysis of product families. This approach not only provides insights into the significance of specific features (traits) of product development, but it can also help identify future avenues and possibilities. Biswas et al. (2022) applied evolutionary modelling to iPhones, using a range of phenotypic variables such as product design specifications (e.g., display size, memory capacity, and camera resolution) and the presence or absence of specific design features (e.g., camera, GPS, and outer casting). Through this analysis, they ranked the iPhones and identified the most important features of this product family, such as the operating system, processor, and touchscreen, which also play a key role in future development. Interestingly, the performance of newer iPhone models does not always advance over time and may even show worse performance than previous models (Box 2.9).

Box 2.9 Evolution of cell phones

There are several parallels in the emergence of living entities and artificial agents, some of which are mirrored by the market of cell phones (see Biswas et al., 2022; Faragó & Miklósi, 2012). We can find the initial population of previous models where the specifications are influenced based on their functions, changing based on the limiting environment (customer requirements). The different models need to adapt to similar environment, and thus constant innovation, and new or improved functions are needed to be able to spread better than the competitors. Biswas et al. (2022) described that the three major drive (c.f. selective) forces are the market conditions, customer requirements, and technology requirements.

Regarding living entities, new traits can only evolve from earlier ones and similar agents are more closely related. Drawing phylogenetic trees can help to see how the different agents diverged from the earlier versions. Faragó and Miklósi (2012) created the

Table 2.9 Functional similarities between evolutionary phenomena and trends in the changes of cell phone characteristics

Characteristics	*Trend*	*Description*	*Evolutionary term*
Dimensions	Miniaturisation	Selective pressure of mobility	Selective pressure
Screen resolution	Higher	Improvement in quality; as an ‚evolutionary correlation', cameras were built-in to models having high-resolution screens	Evolutionary correlation
Energy	Trade-off between battery capacity and energy consumption of the device	Improvement and occurrence of additional functions increase energy consumption in newer/high-end models; thus, they require batteries with higher capacities. This could drive the industry to the development of more efficient batteries.	Trade-off/ selective pressure

(*Continued*)

Table 2.9 (Continued)

Characteristics	*Trend*	*Description*	*Evolutionary term*
Communication	Appearance of new forms (SMS, MMS, internet)	The introduction of the digital mobile network allowed the appearance of new ways of communication, thus entering and adapting to a novel environment becoming a 'new species'	Radial adaptation
Appearance	From candybar, through clamshell to large touchscreen	The initial models were designed based on wired phone receivers, and following forms were designed step-by-step from these. Reducing the size (see above) led to clamshell form and later touchscreens were introduced. Appearance is probably also influenced by fashion trends.	Evolutionary convergence; non-related species develop similar characteristics due to similar selective pressures
Antenna and physical keyboard	Disappeared	By improving the technology the antenna and physical keyboard has become unnecessary and disappeared in newer models	
Signalling	Improvement in playback, vibration is stable		Selective mutation occurred in case of vibration

phylogenetic tree of cell phones by treating them as biological systems, using the oldest Nokia model as an outgroup for the following models. Their tree grouped together the phones belonging to the same series, and older models were closer to the root of the tree. Their results strengthened that principles of evolution can be applied to an artificial system (Table 2.9).

The analysis of product family evolution and forecasting relies on several aspects, including a history of development, market testing and evaluation in terms of financial return on investment, and variations in individual product phenotype. While present-day social robots are not yet at this stage, it is worth preparing for similar possibilities. Vacuum cleaner robots, which have already reached the everyday market, could provide a good starting point for investigating the evolution of product families. Previous studies on the Roomba vacuum cleaner have also explored its potential for social interaction (Fink et al., 2013).

In contrast, evolutionary studies may be premature for ethorobots, which mainly exist in research laboratories. However, given the complexity of these future products, developers and researchers should start thinking about potential solutions to provide a unifying framework for such robots, which could facilitate their categorisation and the discovery of functional connections.

2.3.9 Conclusions, prospects, and questions

Robots that are designed to be human-like or possess human qualities could pose significant dangers, as they have a greater potential to influence people than agents belonging to another species. This is especially concerning in today's world, where deception is rampant and opportunities to influence people's opinions through the media are abundant. The negative effects on individual autonomy can be significant, which is why it is essential to prevent agents from being 'humanised' in any way. The more human-like a robot becomes, the more ethical issues may arise.

Instead, it is important to ensure that humans perceive robots as 'others'. This can be achieved by defining the function of the robot, which in turn defines its niche. By occupying a specific niche, we can minimise the risk of ethorobots becoming competitors for us. Additionally, the intensity and intimacy of social interactions with robots should be kept at a minimum level necessary for successful collaboration.

Ethorobotics provides a fresh approach to elucidating essential aspects of problem-solving behaviour and cognition (see also Nazir et al., 2023 for a similar argument). By focusing on the behavioural construction of human–robot or animal–robot interaction, researchers have the opportunity to liberate this predominantly engineering process from anthropocentric and anthropomorphic mental projections. Through assessing the efficacy of behavioural functions such as attachment, empathy, imitation, and more, there is reduced reliance on elusive psychological theories designed to simulate these phenomena.

2.4 Planning of ethorobots

Let us assume that an ethologist is interested in studying the prey-catching behaviour of cats. By systematically observing cats when they hunt for mice, the ethologist can develop a detailed description of the hunting behaviour, including the types of actions displayed and their sequences. Cats may employ various tactics, switching among them based on the situation, and these tactics may differ depending on the individual cat's experience, among other factors. The ethologist can focus on the specific actions that ultimately lead to catching prey and develop a mental model of cat hunting behaviour. After many hundreds of hours spent with observation, would the ethologist be able to design a robotic mental architecture for a hunting cat? Has he collected all the data needed for this task? How should he know whether he has all the data that is needed?

While the hunting behaviour of cats can be quite variable, it is unlikely to reach the complexity of human behaviour when considering a specific function. There is an intensifying discussion about the potential of social robots to replace or assist nurses in elderly homes, but there is limited data on the behaviour of nurses that could be used to develop robotic mental architectures. Typically, we rely on verbal accounts of nursing activities, such as job descriptions or subjective reports by nurses about their daily activities. However, these do not provide large amounts of behavioural data about the activities of nurses and their patients that would describe real situations, and about how specific decisions were made.

Ethorobots that operate in anthropogenic environments must be designed as collaborative agents. However, it is not clear which are those behavioural and mental traits that are essential for collaborative interactions (Section 2.6.5). In addition, complex emergent features may become also important, such as reliability, responsibility, reflectiveness, persuasiveness, and believability, all of which are vital components of a trustworthy social partner (Section 2.6). The exact social skills that a robot is expected to possess depend on its intended relationship with people, which, in turn, is determined by its specific function. For example, a robot working in an office setting would require different social skills than a robot operating as a team member in a rescue mission.

In the following discussion, we explore the essential aspects required to plan an ethorobot's physical and mental skills. Instead of relying on subjective assumptions and incomplete knowledge of specific functions, we propose a bottom-up approach that aims to gain a deeper understanding of embodiment, animacy, mental architecture, etc., drawing on parallels from animals. These concepts may assist the reader in developing more effective ethorobots for specific functions.

2.4.1 What are ethorobots for? Solving problems

Problem-solving behaviour is central to animal survival, and so it is the key feature of robotic function. The definition introduced in Section 1.5 also applies to ethorobots. In the field of industrial robots, it is crucial to provide a detailed description of specific problems. These machines exhibit very rigid, highly precise, and error-free performance, unlike humans. Social robotics has often formulated problems and tasks in a vague manner, leading to implicit assumptions that ultimately hinder the successful performance of these robots. Therefore, in ethorobotics, it is essential to define problems and tasks clearly before implementing any solutions. Since some ethorobots are intended to replace humans, planning their actions can be based on analysing human problem-solving behaviour. However, in many cases, the robot can solve tasks in different ways through appropriate modifications to the environment: the difference of washing up dishes by a human and a dishwasher provides a good analogy.

Ethorobots are able to solve problems if they have the necessary basic skills (perception, mental architecture, action), and they are able to display an organised behaviour. Here we concentrate on the latter aspect, while some details of the technology are discussed in Chapter 3.

The process of determining organised behaviour to solve specific problems exhibits many similarities to the analytical approach of ethology. Ethologists not only define the components of behaviour but also investigate how an animal executes tasks and achieves its goals. This approach has resulted in the development of a hierarchical model of behaviour (Section 1.4, Box 1.1), so there is a close parallel when analysing the nest-building activity of a blackbird and the work of a human bricklayer (see also in Stanton, 2006). The primary distinction lies in the assumption that the ethologist can infer that the bird's activity has evolved to be generally beneficial ('on average efficient') in its specific environment and that the bird possesses all the necessary skills to complete its tasks. As a result, the problem-solving model of behaviour described by ethologists does not require a high level of precision in all details ('backward engineering'). Similar more intuitive and applied approaches also occur in dog training when the trainer has to plan the target behaviour of the dog that is often also achieved by working 'backward' from the target. In this case, the last action is trained first, followed by the action before the last and so on (see backward chaining, Blumberg, 2002).

In contrast, the task analysis of the bricklayer must thoroughly examine every aspect of the problem in detail because the resulting model construct of the problem-solving behaviour is intended to be tested in a robot ('forward engineering'). In this case, all minute skills have to be revealed so that the robot is endowed with the necessary embodiment and mental abilities. Such a comparative analysis may conclude that the robot cannot display a specific skill (e.g., it would be very expensive), then one has to find out how the task and/or the environment needs to be re-structured for the robot to succeed.

Box 2.10 How to carry out hierarchical task analysis?

Hierarchical task analysis (HTA) describes an activity in terms of goals and its sub-goals linked by plans. It has been suggested that instead of a rigidly prescribed technique, HTA is based on broad principles, and it can have multiple representations due to the flexibility of the method. Stanton (2006) described the following typical steps to prepare HTA:

1 Define the purpose of the analysis;
2 Define the boundaries of the system description;
3 Access a variety of sources of information about the system to be analysed;
4 Describe the system goals, sub-goals, and the relationships among them;
5 The number of immediate sub-goals under any super-ordinate goal should be minimised (between 3 and 10);
6 Link goals to sub-goals and describe conditions under which sub-goals are triggered;
7 Stop re-describing the sub-goals when you judge the analysis is fit-for-purpose;

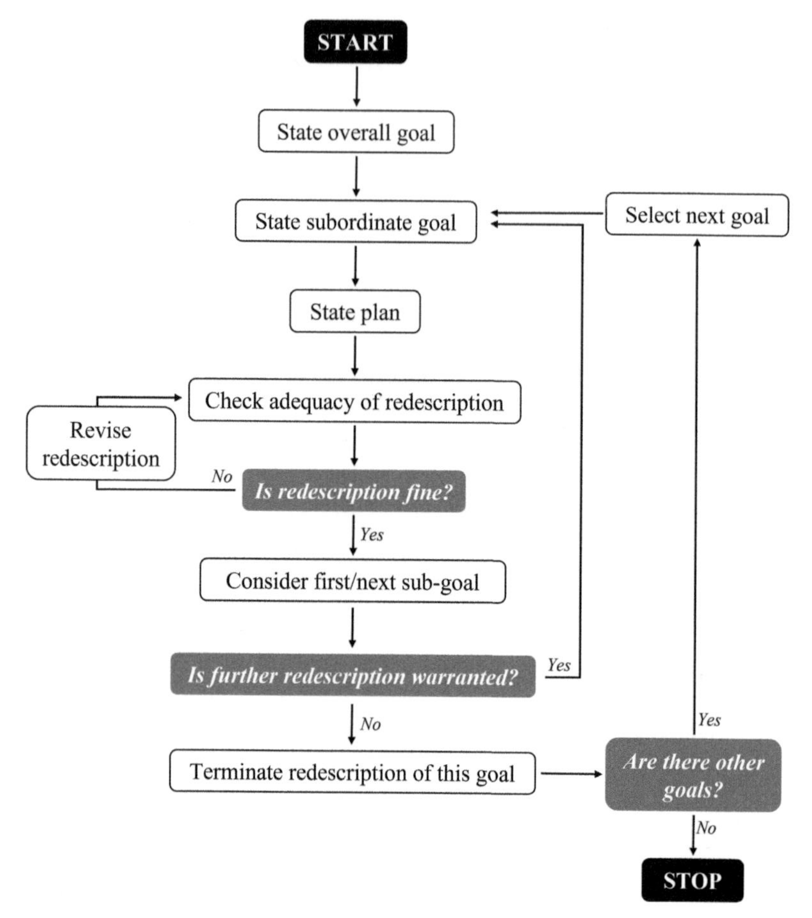

Figure 2.12 Depiction of the sub-goal hierarchy of steps 4–7 (based on Stanton, 2006).

8 Verify the analysis results with subject-matter experts;
9 Revise the analysis if necessary.

It should be noted that task analysis is only successful if there are clear definitions of actions/behaviours (see Section 1.3) (Figure 2.12).

HTA offers a useful method (Annett & Stanton, 2000; Stanton, 2006) for dealing with problem-solving behaviour in ethorobots. The core idea is rather straightforward: human behaviour, as well as the behaviour of an agent, is perceived as goal oriented, and the ongoing sequence of actions is decomposed into hierarchically organised goals and subgoals. This hierarchical structure is considered as a plan designed to achieve the primary goal, incorporating (if necessary) action cycles, feedback loops (positive or negative), decision points, and other elements where the fulfilment of the conditions is determined in advance. Stanton (2006) identified three relevant principles to be followed (see also Box 2.10):

Goal identification: there has to be a generally agreed goal that can be measured objectively. In the context of ethological investigations on problem-solving, an exact determination of the goal may not always be necessary. For instance, in the case of nest building by a bird, there is a certain point at which the male bird perceives the nest as ready and switches to other activities. There are actually no (or very few) data on how the bird decides that the nest is complete. However, when it comes to tasks assigned to a 'nest-building robot', it becomes imperative to define an objective and measurable value for 'nest readiness'. In the scenario of a social robot assisting the elderly in reaching a common room, it is important to precisely determine when the task is considered complete. For example, a specific criterion could be met when the elderly person is comfortably seated in one of the chairs in the room after being escorted by the robot from their own room.

Structure determination: in order to effectively achieve the overall goal, the behaviour involved should be broken down into suboperations and subgoals, utilising the same criteria as mentioned earlier. In this regard, the task description in ethology shares similarities with the approach in robotics, as the completion of subtasks also signifies the end of a specific action. For instance, in the case of a bird collecting twigs to build a nest, the subtask ends when the bird has a detached tree branch in its beak and is ready to fly back to the nest. Similarly, in the task of escorting the elderly, there may be subtasks such as leaving the room, walking to the end of the corridor, and so on. The fulfilment of each subgoal can be objectively defined by specific parameters, such as the person's spatial location.

Operational relationships: identifying the nature of hierarchical relationships between the main goal and subgoals can be a challenging aspect of the process. While devising an optimal plan may seem relatively straightforward, it is important to ensure that the structure of problem-solving behaviour remains robust and adaptable to environmental changes, as well as capable of handling errors. For example, the bird needs to recognise the best type of branch for its nest and start collection again if the branch accidentally falls from its beak. Similarly, the robot should continuously monitor whether the elderly person is following, and it should intervene in a specific way if it appears that the person has become disoriented or veered off the intended path.

Goals and subgoals are considered the main focal points of problem-solving behaviour, representing the primary endpoints (see Figure 2.13). However, it is also possible to identify secondary endpoints that hold specific significance or offer additional benefits. For instance, when an ethorobot interacts with an elderly person, the primary endpoint would be reaching the main goal, such as escorting the person to a specific location. In parallel or as an additional aspect, a secondary endpoint could be to establish communicative exchanges between the human and the agent to facilitate the interaction. Achieving the secondary endpoint, in this case, is optional and may not directly impact the fulfilment of the primary goal. In different scenarios,

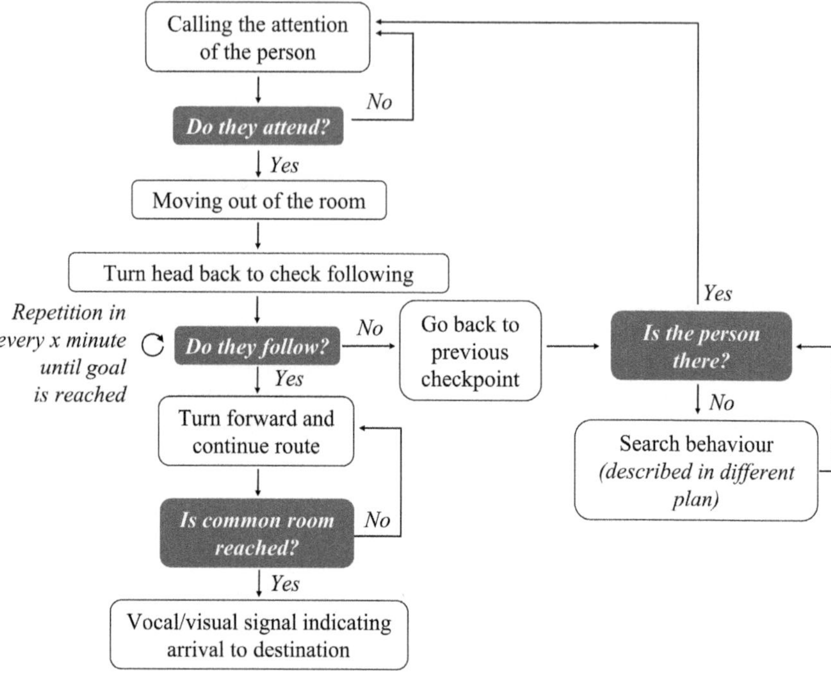

Figure 2.13 Hierarchical diagram for leading an elderly person from their own apartment to the common room by the robot based on HTA. This approach could be extended by indicating at each step what kind of information does the robot need to perform the next step. For example, 'Do they attend?' goal should be described as what is attention (e.g., looking at the robot), by what means is this verified (face recognition software), how robust is this recognition (e.g., viewing angles, the robots should look actively by head turning to search for a looking face), how long should the face look at the robots before the next action is started, etc.

problem-solving plans can be redesigned, causing initially secondary endpoints to become primary ones based on specific requirements or changing circumstances.

While the HTA approach provides a third person view on problem-solving, models of situation awareness (SA) prefer a first-person view. According to Endsley (1995, 2015), 'situation awareness is the perception of elements of the environment within a volume of time and space, the comprehension of their meaning, and the projection of their status in the near future'. Note that this perspective is heavily centred around human experience, making it challenging to directly apply to ethorobots. When using the SA approach, one must attempt to place himself in the 'shoes' of the robot. Nonetheless, there have been efforts to incorporate HRI into theatrical performances. For instance, Jochum et al. (2016) organised a performance in which human actors interacted with real robots to simulate scenarios related to elderly care. The researchers then gathered the audience's opinions on the observed events and their perception of the human and robot actors. Alternatively, a different approach could involve humans assuming both the role of the robot and the elderly person. The actor playing the robot's role could provide insights into their perceived SA, offering valuable input for the design of problem-solving abilities in future ethorobots. When enacting the role of the social robot the actor should be able to recall the three levels of SA (Scholtz, 2003): (1) Perception of the essential elements in the environment, (2) comprehension of the on-going situation by establishing a holistic view, and (3) finally, projection of future status based on knowledge and the perception at level one. This analysis should occur at each step of the problem-solving behaviour.

Integrating SA with HTA has the potential to enhance task descriptions and establish a stronger foundation for problem-solving. To achieve this, the algorithm outlining the problem-solving behaviour should be augmented at each stage with the relevant situational parameters that the robot needs to be aware of. By incorporating SA and HTA together, a more comprehensive understanding of the task context can be achieved, enabling the robot to be more effective by making more informed decisions.

Despite significant advancements in the methodology of task description and problem-solving behaviour, the HRI can still present challenges, particularly due to the wide range of possible reactions displayed by the human partner. The limitations of the robot's sensory and mechanical capabilities necessitate careful planning to avoid situations where the robot lacks the necessary skills and tools to respond appropriately. In social situations, alternative solutions may also be available. For instance, an ethorobot could request assistance or help from its human partner, similarly to how humans interact with each other. While such situations should be rare (otherwise the robot is not considered useful), it is essential for the agent to possess the required behavioural skills to employ this tactic if necessary.

One possibility in social robotics is to aim to solve simple problems first based on task analysis, and if successful, then the agent can be exposed to more complex problems. In other words, the levels of the hierarchical constructions should be minimised at the start. Alternatively, the complex problems could be divided into smaller units which should be solved independently from each other before these are combined in real life.

2.4.2 The concept of embodiment

The use of tools has always been a part of human natural history. For a significant period of this evolution, these tools were passive physical objects, and their primary function was largely determined by their physical shape. Early tool manufacturing focused on identifying the most suitable materials and optimal shapes to enhance their performance.

However, the notion of a tool underwent a transformation with the advent of software and computers. This shift created the impression that certain tools, specifically software, could exist without a tangible body. Unlike traditional tools, the functionality of software was no longer strongly tied to the physical hardware. A single computer could execute a wide range of diverse programs, allowing for a multitude of tools to be accessed and utilised.

This may explain why the concept of embodiment, having a physical body, had to be re-examined by cognitive science in the 1980s. The emergence of robot technology reversed the previous trend, as similar programs were used to control different types of embodied agents. The notion of embodiment became significant when comparing the performance of robots and humans. It is also considered a major paradigm shift in cognitive science (Ziemke, 2003). Earlier attempts to model the mind primarily relied on computational methods, disregarding the fact that natural minds are inseparable from their physical bodies. The concept of embodiment highlights that minds influence the environment through the body's actions, and in turn, the body's actions impact the mind. There is an ongoing interaction between mind and body, shaped by a long evolutionary and developmental history. Embodiment has posed significant challenges for the development of artificial minds in cognitive science. For the advancement of ethorobotics, it becomes crucial to understand how the mind and body must work together, rather than solely focusing on enhancing abstract cognitive capacities in agents. As observed in chess-playing, artificial agents can surpass world chess champions, yet they may still encounter difficulties in physically moving the chess pieces on the board.

The concept of embodiment encompasses various aspects. It can refer to the requirement of agents operating in the real world to possess a physical body, the types of bodies capable of harbouring cognitive capacities, or how agents should be constructed to achieve specific goals. These different perspectives on embodiment often arise when comparing organic (living) agents with artificial agents (human-made machines) (Deng et al., 2019; Maturana & Varela, 1987; von Uexküll, 1982; Ziemke, 2003) (see Box 2.11). Ethorobots exist solely in a three-dimensional space, always possessing a physical body, and are bound by the fundamental laws of Newtonian physics. Consequently, in these systems, the body and cognitive capacities are inherently tightly coupled (Dautenhahn, 2002; Pfeifer and Scheier, 1999). As mentioned earlier, having a body presents specific challenges for the mind, as it necessitates a particular set of computations to maintain the spatiotemporal integrity of the body. Interestingly, these challenges often manifest as sensory or perceptual issues in object recognition (e.g., Falck et al., 2020). However, any mind must solve similar problems concerning its own body and its relationship to the surrounding environment.

Box 2.11 Comparison of living organisms and artificial machines

Von Uexküll (1928) was among the first, followed by many other researchers, to aim at pinpointing the main differences between living and artificial machines. The topic was further explored by Maturana and Varela (1987) and Ziemke and Sharkey (2001), a period coinciding with the emergence of computers and the introduction of AI-based mental architectures, bringing about new levels of complexity in machines. A central question arises: does the increasing, albeit still limited, similarity between living and artificial machines imply that they can achieve the same level of functionality, or do the fundamental differences (see Table 2.10) present an insurmountable obstacle? One may argue that structural differences do not necessarily preclude closely similar performance (e.g., consider legged versus wheeled robots). Alternatively, the vastly different construction of these agents, especially the fact that living organisms require reproduction 'to survive', may delineate these two distinct embodiments. Currently, it is challenging to envision how artificial machines could closely resemble biological organisms unless they replicate all aspects of their existence. Listing (some of) the distinctive features of living organisms may inspire engineers to use these concepts in the construction of the new generation of social robots, e.g., increase hierarchy in the organisation of embodiment.

Table 2.10 Some of the key characteristics that distinguish autopoietic from allopoietic systems were introduced by Maturana and Varela (1987). This table extends these features that also strictly connected to embodied agents. Autopoietic systems, typical of living beings, are characterised by their ability to self-produce and maintain their own components. In contrast, machines are recognised as allopoietic systems, generating outputs or products external to themselves

	Autopoietic social agents (e.g., dogs)	*Allopoietic social agents (e.g., AIBOs)*
Homeostasis – stable inner state (s)	Yes	Partially
Reproduction of own parts	Yes	No
Repair of damage	Partially	No
Self-supply of energy	Yes	No/partially
Systems connecting parts of the hierarchical structure	Transport system ('blood'), nervous system, protecting (immune-) system	'Nervous system' (electric cables)
Hierarchical organisation	Strong (all higher structures (e.g., organs) consists of similarly organised units, cells)	Weak (higher structures are not composed of similarly organised units)
Rules of operation	Rules are generated within the system (autonomous)	Rules are provided from outside (heteronomous)
Material interaction with the environment	Exchange (input and output) of materials (e.g., food) that are obtained or discarded	No exchange
Complexity and development	Multicellular organisms increase their complexity during development	No change
Reproduction (production of similar entities)	Yes	No

When ethorobots move in physical space, there must be a mechanism in place to prevent the violation of the continuity and solidity principles (Section 1.6.3). Naturally, there are various fundamentally different approaches to achieving this, each with its own advantages and disadvantages. These solutions may bear similarities or differences to those found in living agents. Regardless, these challenges contribute to the mental complexity of embodied agents when compared to their non-embodied counterparts.

It is equally important to recognise the intricate relationship between any embodied agent and its environment, a subject that falls within the realm of ecology. This insight is echoed, for instance, by Dautenhahn et al. (2002):

> A system S is embodied in an environment E if perturbatory channels exist between the two. That is, S is embodied in E if for every time t at which both S and E exist, some subset of E's possible states with respect to S have the capacity to perturb S's state, and some subset of S's possible states with respect to E have the capacity to perturb E's state.

This definition is significant as it not only highlights the importance of body-environment interaction but also provides a means of quantifying the complexity of this interaction.

In biological systems, the presence of these perturbatory channels is intricately linked to survival and arises as a result of evolutionary processes. These channels also define the niche occupied by the agent (species). Additionally, in principle, the number of such channels can serve as a basis for quantitative comparisons among agents, reflecting their complexity. For instance, comparing a rigid tube-like agent with a similar agent possessing a movable head (both existing within their respective environments/niches), it is likely that the latter has more channels and states for interacting with its surroundings. The complexity of interaction with the environment is sometimes associated with the degree of freedom (DoF) in movements (see Section 3.5). However, it is important to note that there are numerous other ways to exert influence on the environment that do not solely rely on motion.

The above definition outlines a necessary but not sufficient condition for body-environment coupling, as it does not specify the outcome of state changes and their impact on the agent. In biological systems, it is ensured that these mutual perturbations and state changes, on average or in the long run, contribute to the agent's survival. If a robot is deemed useless for any reason, it is likely that the latter rule has been violated. This means that despite its ability to perform complex perturbations and state changes, it fails to 'survive'. The embodiment of AIBO robot did not guarantee survival. This robot was designed to resemble a dog, which is capable of engaging in intricate interactions with its environment but did not become a social companion for humans or dogs (Kubinyi et al., 2004).

In biological agents, there is a certain correspondence between sensory and motor abilities. Part of the reason for this is that sensors are not only necessary for gathering distant information but also for monitoring the agent's body and the consequences of its actions. In the case of robots, it is not sufficient to construct an arm capable of picking up a bottle; it should also be equipped with a sensory and perceptual system that can monitor performance under various conditions. Robots are destined to fail if there is an incongruity between their motor and sensory capabilities. For instance, they may be able to speak but lack the ability to understand speech or have limited capabilities in that regard. Similarly, the availability and use of conversational agents (CAs; e.g., ChatGPT) often lead to similar misunderstandings or unfounded expectations. While these tools may be intelligent as conversational partners on a laptop or cell phone, they can only be helpful for embodied agents if there is a close correspondence between their perceived mental abilities based on 'chatting' and the physical capabilities of the robot.

It is crucial to recognise that embodiment entails increased costs for both biological and embodied artificial agents compared to virtual entities. Thus, 'cheating' (giving the impression of a non-existing functional capacity) in embodied agents should be rarer because, according to the evolutionary theory, these costs must be compensated for through energy investment (Searcy & Nowicki, 2010). This not only leads to higher competition in the traditional Darwinian sense but also limits the opportunity for cheating, wherein the same level of performance is achieved without the necessary energy investment. This means that embodied agents are more 'honest' in their behaviour, they have less room for cheating.

Embodiment also implies a sense of presence from the perspective of human partners. If an ethorobot presents itself as a socially interactive embodied agent, users naturally expect efficient and much richer social interaction. If the ethorobot fails to meet these expectations, it reduces the likelihood of acceptance by people, regardless of its perceived mental abilities.

2.4.3 The structure of embodiment

The process of providing robots with a physical body raises questions about its practical implementation. In many cases, researchers or designers aim to mimic an exemplar, often the human body. However, they frequently encounter various constraints during this process, resulting in a product that significantly differs from a real human body. While a robot may resemble a human superficially, the notion of 'similarity' becomes vague in this context. Describing a robot as 'humanoid' is based solely on the fact that both entities have a head, torso, two legs, and two arms, but in reality, they may resemble any monkey or bear. Similarly, AIBO robots are considered dog-like solely due to their body proportions and head shape resembling those of a dog, while otherwise being similar to many other four-legged mammals.

In contrast to this somewhat arbitrary approach, evolution has spent millions of years re-constructing and refining body structures in response to environmental challenges and constraints. Thus, typical multicellular organisms share a common body plan with specific modifications that adapt them to their past or present environment. This evolutionary history, intimately connected to survival, ensures that body parts represent functional capacities. Unfortunately, engineers often overlook this aspect when adding body parts to robots, which may appear similar to real organisms but lack any functional purpose. For instance, human-like robots are often equipped with arms that have limited or no functionality in practice.

In the case of ethorobots, the body should be assembled based on functional requirements, without the need to adhere to the shape or structure of any specific natural being. Engineers should strive to keep the body as simple as possible and only incorporate additional features if they serve a specific function. Here, we summarise some insights based on a very general bilateral body plan (see also Fukuhara et al., 2022 and Section 3.5 for details).

Trunk (or torso): the trunk serves as the central part of an organism, housing vital organs that are essential for its proper functioning. In some cases, organisms may have a body that also incorporates necessary sensors. If appendages are present, the trunk needs to incorporate the necessary connection mechanisms.

Head: it is a specific unit that is typically separated from the trunk, situated at its tip. It defines the forward direction of movement of the body and is often equipped with numerous sensors, supported by significant processing and overall controlling capacity exhibited by the brain.

Neck: a neck primarily functions as a flexible and elongated connection between the trunk and another unit, typically the head.

Appendage: an appendage refers to a projecting part of the trunk that possesses a distinct appearance or serves a specific function. Typical appendages are the head and limbs. While there are no strict rules, appendages are typically differentiated based on their function rather than their mechanism. For example, ‘legs’ are used for body movement, while ‘hands’ are employed for object manipulation. Most appendages exhibit symmetry, with one or more on each side of the body, but there are also asymmetric appendages such as tails.

The challenge in defining these denominations partly stems from the evolutionary history of living organisms. It is common for appendages to change their function over millions of years or acquire multiple functions. For instance, the foreleg in mammals evolved into an arm and hand in monkeys, and in the case of many mammals, the tail, which initially served as a balancing tool, now primarily functions in communication as a signalling mechanism.

It is interesting to note that despite the structural and operational differences, we use the term ‘legs’ to describe most appendages that facilitate movement in insects and mammals on the ground. On the other hand, we differentiate the trunk of an elephant from the arm and hand of a monkey, even though they serve similar functions.

It should be added that bilateral bodies appear to be symmetric, but this is not the case at all levels of functioning. For example, there are two legs on each side of the body, but they may differ in small details and may be used asymmetrically, including their mental and neural control. For example, adult humans display a strong right-hand preference for manipulation (Rogers & Andrew, 2002).

This brief overview can assist in planning the actual embodiment of ethorobots. The important lesson to be learned is that the body plan of living organisms is primarily based on function rather than a specific appearance or design (Fukuhara et al., 2022). By utilising the typical building blocks mentioned above, a wide range of body plans have emerged, enabling organisms to thrive in diverse ecological conditions. This logic should be applied to ethorobots as well. The body plan should reflect the crucial challenges posed by the robot's environment, with every component of the embodiment serving a clear function, while unnecessary parts designed merely for superficial similarity should be eliminated.

2.4.4 The nature of embodiment

The properties of systems are typically determined by the materials they are constructed from and their internal structure. While this understanding is straightforward in industrial applications, it raises a more philosophical question in the context of artificial creatures. Can these entities, built from materials of fundamentally different nature, truly emulate the characteristics of living beings? One may ponder whether current technology allows for the creation of engineered agents that are indistinguishable from or resemble closely humans or animals. Taking a realistic perspective and assuming that technological advancements will remain relatively incremental over the next 20–25 years, with robotics primarily relying on inorganic compounds, the likelihood of achieving humanoid (or animal-like) robots is limited. Here is an incomplete list outlining some of the significant distinctions:

Flexibility and rigidity: animal bodies exhibit a remarkable level of flexibility, allowing them to assume various configurations by adjusting the orientations of their body parts at the joints. Additionally, they can alter their overall size and shape, becoming smaller, slimmer, or larger and more extended as required. This flexibility is further enhanced by the presence of soft surfaces, including skin and muscles, which can be moulded to some degree. In certain scenarios, even slight changes in size, on the scale of a few centimetres, can prove crucial (e.g., rescue robots designed for navigating in caves).

Scaffolding: in animals, the body's scaffolding can exist internally (e.g., endoskeleton in vertebrates) or externally (e.g., exoskeleton in insects). Both structures contribute to the construction of highly efficient walking, flying, and swimming organisms. However, when it comes to exoskeletons, there are limitations regarding scaling up, as they do not facilitate the design of larger bodies. The type of skeletal structure also impacts the overall flexibility of the body and the range of motion provided by the joints.

Locomotion: for advancing on the ground there are two major possibilities. The invention of the 'wheel' allowed a relatively simple and cheap solution for this problem. At the functional level, this technology can outperform any other solutions. However, it is an artificial solution and needs a very different support for propulsion and also for stopping. In addition, wheeled vehicles are also facing problems on more complex terrain, so for them to be efficient the environment has to be changed (e.g., floors, roads, rails). Considering the body-mind issue (see above) wheeled agents may never become similar to natural ones. In contrast, legged robots have gained popularity, and their design often mimics the structure of animal limbs (Ritzmann et al., 2004). Although they serve a similar function, the inner mechanical structure of robotic legs can vary significantly depending on the preferred technology utilised (e.g., Fukuhara et al., 2022; Section 3.5.3). Nevertheless, despite its complex technology, legged robots have gained impressive skills in the last few years.

Brain or brains: natural organisms may have a single centralised nervous system (e.g., vertebrates) or a more distributed one (e.g., insects). At present most robots follow the constructions of vertebrates but a distributed controlling system can have several advantages, especially in hazardous conditions.

The construction of the body (and behaviour) presents an important scenario for trade-offs. It is likely that decisions on locomotor abilities, size, sensors used, or sensory capabilities are also determined by financial and technological constraints that the designer should be aware of. In turn, these aspects may also influence the reliability and endurance of physical aspects of functioning.

2.4.5 The environment and embodiment

While the concept of embodiment has received significant theoretical attention (e.g., Dautenhahn et al., 2002) in the field of social robotics, comparatively less focus has been placed on the environment in which they function. There is often an implicit assumption that robots will operate in environments where humans are present. However, these environments can often be unsuitable for typical robots due to factors such as narrow spaces, slippery surfaces, uneven terrain, darkness, excessive light, and more.

The human body quickly accommodates to such irregularities, even if they occur infrequently or at a low rate. In contrast, similar unexpected situations can render robots ineffective. For wheeled robots, even a small doorstep can present an insurmountable obstacle, not to mention the presence of stairs. While it is possible to design robot-friendly environments, doing so may restrict the overall usefulness of the robot and incur additional costs. Thus, a trade-off exists between the cost of the robot itself and the expenses associated with providing specific surroundings for the robot.

From an evolutionary standpoint, it appears that investing in the capabilities of the agent (i.e., the robot) is more common than investing in agents that modify their environment. It may be worthwhile to allocate greater efforts towards the advancement of legged robots capable of navigating diverse environments effectively. There seems to be a prevailing trend favouring the development of legged robots with a wider range of utility (Section 3.5.3).

Alternatively, one may consider the robot's constraints as an advantage because they foster to keep the robot in its niche. A robot that can operate only on smooth surfaces remains within the buildings and there is no chance of it to escape.

2.4.6 Modularity and segmentation

Modularity is a significant biological concept used to describe the structure of biological agents. Even a casual observation of familiar animals reveals that they are composed of distinct but interconnected parts, known as modules. Head, legs, and tail are common examples of these modules found in many animals (Section

2.4.3). However, modularity is not limited to gross morphological structures alone; it emerges at all levels of biological organisation. As a result, modules are typically organised in hierarchical fashion and play crucial roles in both evolution and development (Eble, 2005; Klingenberg, 2014).

In general, modules are defined as units in which the integration among the constituent parts is higher than the integration between other units. If we conceptualise the organism as a network, modules consist of hubs with strong connections within the module, while connector hubs facilitate interaction among different modules (Klingenberg, 2014).

Segmentation is a closely related concept to modularity, prominently observed in the animal body when it is divided into similar repeating units (ten Tusscher, 2013). This characteristic not only enhances the organism's flexibility but also serves as a means to evolve new body parts (Chipman & Edgecombe, 2019). The emergence of a new segment can contribute to the functionality of the organism's body while offering an avenue for further changes without significantly impacting earlier.

Modules of organisms are often organised in a nested hierarchical way (Dyke, 1988) that has a specific significance in the developmental processes. For example, in case of heterochronic development local changes in one part of the body are not interfering with ongoing functions at other units. The hierarchy of the nervous system, which aligns with the structure of the body, facilitates parallel and distributed control. This organisation enables the organism to efficiently coordinate its various components. However, for the organism to function effectively, it also requires mechanisms that ensure the integration and coordination of these partially independent units when necessary. This inter-connection is maintained by transport systems (e.g., blood) and the nervous system. Need for harmonisation includes all kinds of behaviours from simple locomotion (Chapter 3) to the display of emotions (Section 2.5.5).

Currently, robotics has not fully embraced the principles of modularity and segmentation found in biological organisms. Unlike living beings, robots, as embodied agents, do not exhibit the processes of evolution or development in the true biological sense. Instead, they are created from scratch by engineers (see also allopoietic systems, Box 2.11). Components such as heads, legs, and arms are often by-products of the planning process. While this approach may be understandable, there are potential advantages to plan exchangeable units and subunits which may also provide some additional benefits. For instance, the length of a segmented torso could be easily modified by adding or removing units, without the need to extensively alter other parts of the robot. The use of joints to connect these segments can enhance the robot's overall flexibility, enabling finer adjustments in length or height, akin to stretching movements in biological organisms.

2.4.7 Autonomy

Autonomy is commonly defined as the capacity of a system to operate independently and exhibit self-organisation. However, this seemingly straightforward concept encounters significant challenges when confronted with the wide array of biological and artificial agents in existence. Thus, the four questions of ethology (Section 1.2) allow for a holistic description that transcends specific definitions. One key distinction between living beings and robots lies in their respective autonomy. Living beings must demonstrate some degree of autonomy across various dimensions, encompassing cognitive, behavioural, and physiological aspects. In contrast, robots can also exhibit autonomy but are often quite limited in this capacity (Moreno et al., 2008; Ziemke, 2008, see also Box 2.11).

Function: biological agents are required to exhibit autonomy in all functions necessary for their survival, unlike embodied artificial systems such as social robots, which often rely on human assistance for various tasks. These artificial systems are typically limited in terms of autonomy and are usually independent only in relation to one or a few specific functions.

Evolution: living systems acquire their autonomy and achieve functional integrity through the process of evolution, without the need for external intervention. In contrast, artificial systems gain autonomy solely through human action. Thus, these two types of systems differ fundamentally in this respect.

Mechanisms: it is valuable to differentiate between constitutive processes, which contribute to the system's maintenance, including homeostasis and structural integrity, and interactive processes, which involve the system's interaction with the environment. Both living and artificial agents exhibit features of autonomic mechanisms, but interestingly, in artificial agents, there is often a stronger emphasis on autonomous behaviour during interaction, with constitutive processes being supported by human assistance. In biological agents, robust constitutive processes form the foundation for the emergence of interactive processes. The ability to sustain deformations is a notable aspect in this regard.

Development: in present-day embodied artificial social agents, autonomy plays a minimal role in ensuring emerging stability and integrity. Instead, human inference and decision making largely determine the represented autonomy of such systems, marking a stark contrast to biological organisms. From a developmental perspective, autonomy is a characteristic that an agent acquires through interactions with the environment. It implies that no two agents can be exactly the same, much like the situation with two monogamous twins.

When comparing biological organisms, such as bacteria and mammals, there is a subjective perception that mammals possess a higher degree of autonomy. This observation can be partially quantified by attempting to define levels of autonomy. As discussed earlier, the distinctions between the autonomy of living beings and artificial embodied agents highlight significant differences (Moreno et al., 2008; Ziemke, 2008).

While comparing biological and artificial agents may not yield profound insights, this comparative approach can aid in the development of autonomy measures for artificial systems. One possible strategy is to focus on behaviours that are under autonomous control. For instance, in the case of a waiter assistance robot, it can be considered autonomous if it can consistently transport dirty plates from tables to the dishwasher without human intervention. The robot's level of autonomy increases if it is also capable of autonomously recharging itself to maintain its primary function, and so on. Goodrich and Schultz (2007) proposed a similar idea, defining autonomy as the duration of time during which the robot can operate without requiring human attention (Boden, 1996).

Insights into the degree or levels of autonomy can also be obtained by contrasting it with automation (Rodseth & Vagia, 2020). However, it is important to note that there is no clear-cut distinction between the two concepts. Automation involves engineering a machine (agent) that can replicate a specific human task within a narrowly defined environment. The notion of autonomy arises when considering the complexity of the task, the complexity and uncertainty of the environment, and the presence or anticipation of human assistance. From a logical standpoint, the autonomy of an agent is greater when it demonstrates more complex performance in intricate environments with minimal human assistance.

From a third-person perspective, autonomy holds significant importance, particularly in the context of social robots. The level of autonomy exhibited by robotic agents can influence how others perceive and interact with them. Agents that display limited autonomic features and fall short in the eyes of observers may be disregarded or dismissed. Thus, autonomy plays a vital role not only in the functioning and survival of a system but also in shaping the dynamics of social interactions. When a robot exhibits autonomous behaviour, it tends to capture attention, generate engagement, and elicit positive attitudes from human counterparts. The perception of autonomy in robotic agents often fosters a sense of authenticity and agency, enhancing the overall HRI experience (Bartneck et al., 2007).

Conversely, a lack of perceived autonomy may lead to indifference or rejection. When robots appear to lack autonomy, they may be viewed as mere tools or passive objects rather than active agents capable of independent action. This perception can diminish the sense of engagement and connection between humans and robots, limiting the potential for meaningful interaction.

One might think about the analogy between the state of being 'dead' in biological systems and being 'switched off' in the case of artificial agents (Bartneck et al., 2007; Spatola, 2019). In both scenarios, the systems lose the capacity to function autonomously. When living organisms lose their autonomy, they are regarded as deceased, representing an irreversible transformation in their state. In the case of biological agents, autonomic functions may temporarily cease for brief periods, such as during sleep or lethargy or hibernation (during periods of cold or drought). However, the agent can autonomously reactivate itself once again. When artificial agents possess the skill to autonomously reinitiate their operations in response to subtle alterations in their surroundings, it can be seen as a manifestation of heightened autonomy. This capability allows them to adapt and resume functioning independently, akin to the resilience observed in living organisms. Consequently, the presence of such reactivation skills can be considered an indicator of increased autonomy in artificial agents.

2.4.8 Animacy and restlessness

When seeing another person or a non-human animal, it is clear that they are *animate*, but what properties make something animate or inanimate? From the viewpoint of observers, an object can be perceived as animate based on different static or motion cues. Importantly, the perception is present without experience in diverse animal species, and it is considered '…to be fairly fast, automatic, irresistible and highly stimulus driven…' (Scholl & Tremoulet, 2000). By incorporating certain characteristics into a robot's behaviour, it is possible to evoke the perception of the robot as an animate entity, even without intricate features. This, in turn, can facilitate the robot's acceptance as a social agent. Researchers often employ such cues intuitively, such as designing robots with eyes, to enhance the perception of animacy. However, the ongoing and rapidly expanding research on the specific cues that influence animacy perception offers further possibilities in robot development.

Regarding static cues, face-like configuration has been described as the most powerful cue. Having two blobs placed horizontally on the top and one below at the middle is enough to elicit preference in a wide range of species, including humans. Simple motion cues such as those indicating self-propelledness are also important features to animacy. These motions indicate the ability of the object to generate and cease movement without (visible) external cause (Scholl & Tremoulet, 2000). Bilateral body plan also constrains several aspects of the motion, most notably forward-facing motion. Alignment between the main axis of an object and the direction of its motion is also important to elicit animacy perception. When multiple moving objects are presented, spatiotemporal contingencies in their motion can provide specific information. Bassili (1976) found that temporal contingency is crucial to perceive the interaction, while spatial configuration has a role in identifying the nature of the interaction.

These cues have been found to evoke visual preference in the early months of human life, potentially guiding the attention of newborns towards animate stimuli and aiding their understanding of behaviour (Di Giorgio et al., 2021). Importantly, within the first year of life, humans not only exhibit a greater focus on animate objects but also develop distinct expectations towards them. When constructing an ethorobot, incorporating these simple cues can facilitate the spontaneous recognition of the robot as a social partner, leading to the development of basic expectations regarding its behaviour. While these expectations may be not as complex as those associated with humanoid robots, they contribute to the 'believability' (Section 2.5.4) of the ethorobot and facilitate the learning process of its capabilities. Given that these cues are commonly observed in animate agents, integrating them into ethorobots can enhance their social interaction potential and aid in the acquisition of relevant skills.

In studies on HRI, animacy is often represented by more complex features, such as dog-like or human-like embodiments. For instance, Kahn et al. (2006) discovered that preschool children were more likely to attribute animacy to an AIBO robot than to a stuffed dog, even referencing mental states in their verbal reports. Furthermore, Okita and Schwartz (2006) observed that children from different age groups were more inclined to perceive animacy in an AIBO robot that was 'dancing' compared to one that remained stationary. Notably, the differences observed primarily pertained to perceptions of intelligence rather than the biology or agency of the robot.

An essential characteristic of living systems is their inherent 'activeness or restnessless', which can be observed by others through various bodily movements, including full-body motion, appendage gestures, and subtle actions like breathing or balancing. In contrast, robots tend to remain completely motionless when stationary between actions, which can diminish their perception as animate agents. As mentioned earlier, biological agents take shorter breaks in active motion, and even during these periods, they typically exhibit subtle but noticeable movements (e.g., during sleep). For autonomous robotic agents, incorporating small movements, even when stationary, can be advantageous in maintaining the appearance of an animate entity. By including these subtle motions, robots can create a more dynamic and lifelike presence, which contributes to their perceived animacy. Terzioğlu et al. (2020) equipped an industrial robotic manipulator with a breathing-like behaviour during manipulation tasks alongside humans. By continuously moving the robot's joints while keeping the end-effector's position fixed, the robot's body spans expanded and shrank, simulating breathing. This behaviour alone increased the likeability, animacy, social presence, intelligence, and adaptability of the manipulator, as observed in the study. The findings highlight the positive impact of incorporating naturalistic behaviours, such as breathing-like motions, in the design of autonomous robots for enhanced human interaction.

2.4.9 Readiness

Rapid reaction to (unexpected) environmental events is vital for both animals and in ethorobots. Being prepared for a wide range of actions is crucial for survival in nature (Wong et al., 2017). Continuous readiness allows for swift responses and seamless execution of action sequences, while omittance or failures in sensory monitoring can lead to interruptions, slowing reaction time and requiring complex feedback mechanisms for correction (Box 2.12).

Box 2.12 Personality, temperament, and reaction times

There has been a trend to differentiate temperament and personality. In contrast to the general definition of personality (see the main text) for adult individuals, temperament has been referred to as developmentally early features of behavioural reactions or tendencies which at time of emergence have not been influenced by environmental factors. The convergent behaviour tendencies of young offspring (e.g., chicks, puppies, and children) are better described as manifestations of temperament.

In line with this, temperament can be also envisaged as the form of reaction according to three main features (see Figure 2.14). Note that some categories of personality are also characterised by different features of reaction times. Choleric people (see Figure 2.16) are said to display reactions with short latency in contrast to phlegmatic ones.

- *Reaction time (latency)*: it indicates the time of showing any behavioural reaction toward a stimulus or to an event after its onset;
- *Intensity*: it refers to the engagement of the subject displayed to the stimulus or event (amplitude/degree/gauge/force). The measurement of intensity depends on the

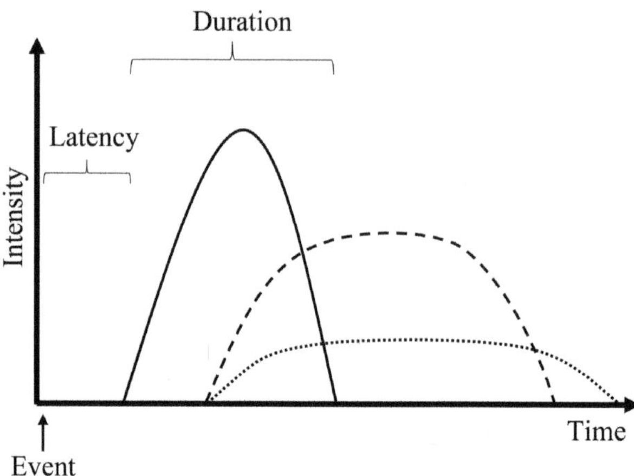

Figure 2.14 Examples of different modes of reacting to specific events. Full line indicates quick response with high intensity and short duration; dashed line indicates late response with high intensity and medium duration; and dotted line depicts slow response with low intensity for an extended period.

nature of the encounter, and may involve different aspects, such as attentional focus or measures of physical action (strength of hitting, loudness of voice);

- *Endurance (duration)*: the duration of interest or engagement shown towards a stimulus or an event.

For example, upon perceiving a person in the way, the human/robot may start to avoid the person 3 m in advance (quick response), but only changes its route for a few cm (low intensity) and returns quickly to the original route (low endurance); or rather starts to avoid last minute (slow response), but changes route completely (high intensity) and keeps the changed route for a few metres after (high endurance).

A similar approach was implemented by Kim et al. (2008) who used speed, velocity, and frequency to produce eight gesture types. The gestures were displayed by a speaking robot and a listening robot. The participants were asked to evaluate the personality of the robot based on their gesturing.

From a functional perspective, we can identify three distinct processes connected to readiness: monitoring, exploration, and habituation. *Monitoring* involves continuously sampling the environment through various sensors, such as visual observation, auditory perception, and olfactory sensing. The collected information is stored in short-term memory until a more recent monitoring activity refreshes the current knowledge of the environment.

When placed in a novel situation, individuals typically engage in exploration behaviour. This behaviour allows agents to gather environmental information effectively (e.g., Gibson, 1988). During *exploration*, the movement of animal or robot is organised to cover the entire area while travelling the shortest possible routes. This enables the collection of information about the surrounding state in the shortest time. Furthermore, exploration naturally decreases when no new information is obtained, a phenomenon often referred to as *habituation* (Section 1.11.7).

An important component for readiness is the mental preparation for an action. In functional cognitive terminology, such preparation is often referred to as expectancy, meaning that there are preparatory mental processes that support swifter reaction in typical situations (Burgoon, 2016). This phenomenon is often revealed by observing the subjects' reaction to a novel event which follows a series of habitual events. In the case of sensory events, this process can be shown using a habituation/dishabituation paradigm. For example, after a longer initial viewing duration, dogs decrease watching a human who is interacting with the same object for a few times (habituation response) in the presence of another object. However, watching time increases again when the same human changes his preference for the other object (Marshall-Pescini et al., 2014). This dishabituation is said to occur because the dog's expectation (prediction) about the human action was violated by the change in preference. Analogue processes can be detected regarding preparation for some actions (Flanagan & Beltzner, 2000). The force needed for counteracting gravity when an object is handled is estimated in advance by visual sensory input and previous experience. If we trick an observer by significantly increasing the weight of a familiar object (such as an apple), they exhibit compensatory lifting hand actions after the initial grasping moment. This compensatory action is in response to the unexpected heaviness of the object, which caused their arm to descend when handing over the apple.

2.4.10 Personality: consistent tendencies in behavioural differences among individuals

The inclusion of a personality is often considered essential for social robots to enhance their human-like perception and facilitate human partners in predicting their behaviour (Fong et al., 2003; Mou et al., 2020). Moreover, robots with diverse personalities may exhibit improved functionality in specific situations and can complement each other in collaborative tasks. This notion finds support in studies examining human personalities. For instance, research has shown that individuals with more extroverted personalities tend to have a lesser preference for remote work (Gavoille & Hazans, 2022), and programmers with varying personalities have achieved better performance when working in pairs compared to those with similar personalities (Sfetsos et al., 2006).

A recent review by Mou et al. (2020) examined 40 studies on social robots and identified various operationalisations of robot personality. These included visual appearance, language, vocal features, movement, countenance, haptics, interaction, and proxemics. This growing interest in endowing robots with personality highlights the significance attributed to this concept in social robotics. However, the application of the biological personality concept in social robotics also raises some problematic issues that need to be addressed.

Box 2.13 What is the difference between a personality trait and an emotion?

In the HRI literature, there is often some confusion about these two concepts (e.g., Kishi et al., 2013) or even used interchangeably. Indeed, one could say that an animal is fearful, or it shows fear. There is certainly some relationship between the two concepts, but one should keep them distinct. Characterising an individual as fearful means that there have been many situations when that particular animal behaved more fearful than the population average. This statement holds also if the animal does not behave fearfully in the future because it never experiences in any negative stimulation. This kind of fearfulness is a personality trait that is stable across situations and time. In contrast, animals may display the emotional state of fear regardless, whether they have a fearful personality if shocked by some specific events. Fearfulness as a personality trait may, however, indicate that that specific individual may show a stronger fear reaction than another animal having a bolder personality. Similar logic does not work so well with disgust that is also a common emotional state, but it has a very weak association with any personality trait.

Accordingly, unlike emotional states, which have a transient nature, personality traits are unique long-term patterns of behaviour that manifest themselves as consistent tendencies in terms of internal motivations, emotions, and decision making. Emotional states and personality models should be separate entities within the

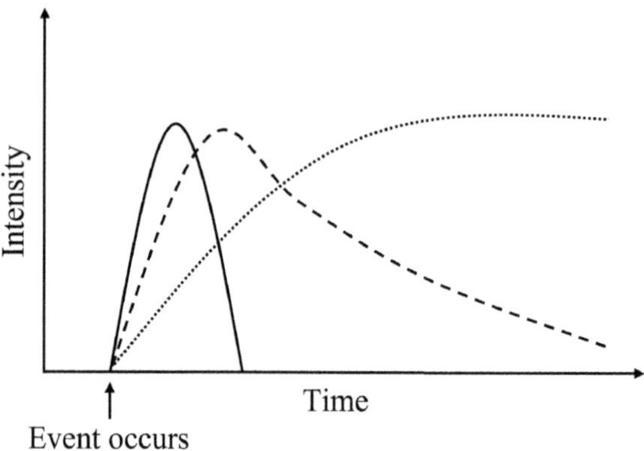

Figure 2.15 Typical emotional states are short-lived, lasting from seconds to minutes (continuous line). However, certain emotional states can have longer-lasting effects, ranging from days to weeks, on behaviour. These prolonged effects are termed moods (dashed line). If a mood consistently influences an individual's behavioural tendencies with high intensity, it should be considered pathological (dotted line). This occurs, for example, when the mood of 'sadness and fearfulness' transforms into a state of depression. Note that this does not necessarily indicate an organismic (causal) relationship between sadness and depression.

architecture, while specific personality states may influence the displayed emotion. Fearfulness, as a personality trait, should affect the characteristics (e.g., latency, intensity, duration) of the displayed emotion when sensing fear-eliciting stimuli (Figure 2.15).

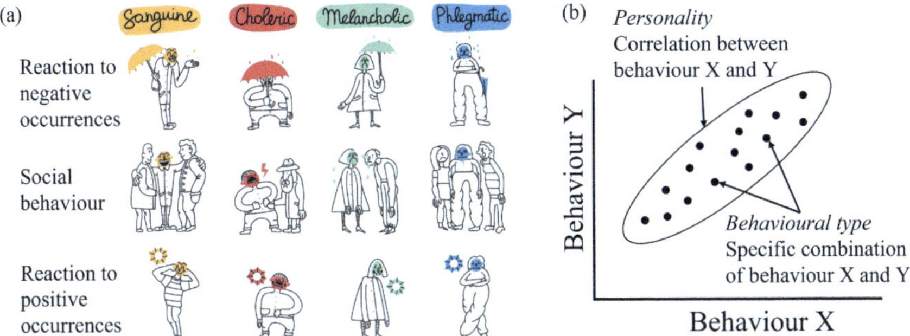

Figure 2.16 Galenus (129–199), the Greek medical writer and scientist was the first who introduced a categorical model of human personality. Categories as choleric, phlegmatic, melancholic, sanguine are still very popular (a) (Picture by Freepik, Flaticon). In the case of (b) trait models, several behavioural traits measured on continuous scales determine the personality. Here behaviours X and Y suggest some correlation, and this indicates the presence of a common factor that may be described as personality trait. Personality traits considered to be more robust and fundamental when more behavioural variables show associations (based on Bell, 2007).

2.4.10.1 A general definition

The concepts of temperament, personality, character, and behavioural syndrome aim to capture the stability of an individual's behaviour over time and across different contexts (Bell, 2007) (see also Box 2.13). In humans, these concepts also encompass cognitive processes, such as how individuals perceive and think about objects, events, and social partners in their environment. While the underlying statistical models for these concepts are similar, the term 'personality' is commonly used due to historical and practical reasons. Contemporary personality models define a multidimensional behavioural space based on personality traits, which are represented statistically by correlating sets of behaviours consistently displayed by individuals across different contexts and over time (Bell, 2007) (Figure 2.16; and see also Box 2.12).

2.4.10.2 Two types of personality models

Categorical models, also known as 'typologies', assume that there are distinct and well-separated aspects of behaviour. These models categorise individuals into specific personality types based on non-continuous traits. An example of a categorical model is the Myers-Briggs Type Indicator, which sorts people into 16 personality categories by using extreme ends of four complex human traits: introversion/extraversion, sensing/intuition, thinking/feeling, and judging/perceiving (Kim & Shim, 2007).

Trait models rely on various numbers of measurable behavioural traits and establish secondary personality traits based on the highest observed variance through statistical modelling, such as principal component analysis. In trait models, personality traits are treated as independent and continuous variables when determining an individual's personality. The number of measured behavioural traits can scale up or down the complexity of the trait model. For instance, a small number of behavioural traits (or questionnaire items) may result in one or two primary personality traits.

The structure of personality models is determined by the number and relationships among categories and personality traits. A smaller number of categories or traits results in a simpler description of an individual's personality, while a greater number allows for a more comprehensive understanding of the 'whole' personality. The key distinction between categorical models and trait models lies in their approach to characterising individuals. In categorical models, each category automatically encompasses all variables, providing a comprehensive characterisation. In contrast, trait models allow for describing an individual by several independent personality traits. The success of personality models is ultimately evaluated based on their utility: can they effectively predict an individual's behaviour in both familiar and novel situations? This predictive capability is a crucial factor in assessing the effectiveness and practicality of personality models in determining how specific individuals may be advantaged or disadvantaged when exposed to various environments or tasks (e.g., Penney et al., 2011).

2.4.10.3 Biological relevance

Individual differences are observed in living organisms, and recent research suggests that the concept of personality applies to a wide range of species. If an organism displays measurable behaviour traits that remain stable over time and in different contexts, a model of personality can be developed. Although systematic studies are lacking, comparative analyses indicate that similar personality traits can be found in phylogenetically distant clades, such as insects and mammals (Gosling & John, 1999; Koski, 2014).

Personality traits, defined by behaviours displayed in different contexts over time, have the potential to reveal homologies and divergences among animals, including humans. One commonly studied trait is boldness (or approach avoidance), which reflects an individual's tendency to explore novel environments, objects, or engage in risky situations. The presence of such behavioural tendencies has been documented in a variety of species (Wilson et al., 1994).

Behavioural complexity is influenced by the degree of freedom of movement in an organism's body (see Section 3.5). Generally, organisms with greater freedom of movement have the capacity to exhibit more complex behaviours. This expanded repertoire of behaviours can provide a robust measure for assessing personality traits and also enable the discovery of novel traits. Organisms with limited freedom of movement may exhibit a restricted set of behaviours, potentially constraining the complexity of their personality traits.

Additionally, the complexity of interactions with the environment can impact an organism's personality traits. For instance, personality traits related to social tendencies may manifest differently in solitary and social species. Social species, engaging in a wider variety of social interactions, are likely to exhibit more sophisticated dimensions of personality compared to those living alone. As an example, one might hypothesise that both group-living lions and solitary tigers possess the boldness personality trait, but they may differ in the structure of a personality trait related to social tendencies, such as agreeableness.

2.4.10.4 Physical features and personality

Humans readily attribute personality traits to objects, such as cars (Windhager et al., 2008) or even virtual ones, like circles or triangles (Heider & Simmel, 1944). Car fronts, for example, can be perceived as arrogant, angry, friendly, or assertive/withdrawn. Interestingly, these attributions are observed even in cultures with limited car experience (Windhager et al., 2012). Evolutionary arguments suggest that predicting others' personalities can enhance fitness (Little & Roberts, 2012). Furthermore, certain physical attributes, like size, are associated with specific personality tendencies. Larger individuals, for instance, often display traits like boldness and curiosity due to their lower predation risk and increased chances of winning fights (Dingemanse & Wolf, 2010). This may explain why larger conspecifics may be perceived as bolder or more aggressive, especially at first encounter by others. Many similar findings highlight the evolutionary basis of attributing personality traits to objects and the link between physical attributes and specific traits.

These tendencies have significant implications for robot design as first impressions can be crucial. Consequently, larger or excessively large robots are more likely to be perceived as bold, obtrusive, or aggressive compared to smaller ones. However, it is important to note that the association between physical size and behavioural traits is not necessarily strong. Therefore, these perceptions can be altered by endowing the robot with behaviours that indicate opposite trends in personality traits. For instance, exhibiting smooth and slow behaviour, as well as avoiding direct approaches, may help mitigate any potential negative effect of large robots on people's perceptions.

2.4.10.5 Testing robot personality with humans

Most experimental approaches assume that the human partner should be able to assess the personality of the robot, or at least some of its personality traits, during the initial encounter. To achieve this, human subjects are asked for their opinions, or their behaviour is observed when interacting with robots that exhibit two or three 'extreme' personalities. In a study by Celiktutan and Gunes (2015), the NAO robot was endowed with two distinct personalities: an 'extroverted' robot that displayed hand gestures, frequent posture shifts, and spoke quickly with a higher voice pitch, and an 'introverted' robot that maintained an almost static posture, spoke slowly with a lower voice pitch. The NAO robot, which was controlled by an operator using the Wizard-of-Oz (WoZ) method (Section 2.2.4), engaged in a turn-taking dialogue by asking a few questions. A subsequent questionnaire revealed that human participants with different personality traits were able to differentiate between the 'introverted' and 'extroverted' robots. For instance, more extroverted individuals enjoyed the interaction with the similar robot, while no such association was found in the case of the 'introverted' NAO.

While numerous studies have reported similar findings, it is important to acknowledge that these experiments deviate from natural interactions in several ways (see Box 2.4):

1 The use of the WoZ method introduces methodological challenges because the operator's influence on the robot's behaviour can be significant;
2 The operator, in addition to executing the predetermined character, may also be influenced by the subjects' reactions and unconsciously modify the robot's behaviour;
3 These experiments do not provide a technological solution for robot personality, as the autonomous behaviour of the robot is replaced by the WoZ method. Therefore, strictly speaking, the robot does not possess an internalised personality function as it has not been implemented autonomously.

In these experiments, the assumption is made that participants are able to recognise differences among personality types or between extreme values of personality traits. However, this assumption is not realistic because natural individuals rarely exhibit extreme personalities. Judgments about someone's personality typically emerge after observing or interacting with him in various contexts and situations over a longer period of time, in line with the fundamental definition of personality.

Personality represents a significant tendency in behaviour rather than being a fixed characteristic. Individuals also accommodate their behaviour to different situations. Personalities differ in their capacity for adjustment. For example, an extroverted person may not always speak loudly, but they statistically tend to be louder in most contexts compared to an introvert. Similarly, a robot with high levels of boldness would not be expected to display indiscriminate boldness in all situations. Instead, it should generally exhibit more boldness more often compared to less bold agents. This flexibility is an important aspect of any personality, and endowing robots with static personalities may not make them appear less machine-like.

2.4.10.6 The application of the personality concept to ethorobots

In the context of ethorobots, the trait model of personality offers a broader perspective. Ethorobots can have various embodiments and be deployed in different roles. However, regardless of their specific design or function, the use of corresponding ethograms to collect behavioural data can facilitate the identification and comparison of personality traits across different robots (see also Andriella et al. (2021) for an experimental approach).

1 The embodiment of ethorobots plays a crucial role in shaping their (perceived) personality, including factors such as flexibility in locomotion and movement. The presence or absence of certain physical features, such as a robot with a head compared to one without, can significantly impact the complexity of their behaviour towards novel objects. Unlike a robot without a head, which can only approach or withdraw from an object by adjusting the distance between its body and the object, a robot with a head can engage in more sophisticated interactions. For instance, it can simultaneously withdraw its body while moving the neck forward to interact independently with the object. These conflicting actions between the body and head may convey a sense of 'ambivalence' in the robot's behaviour, a trait that may not be evident in a robot without a head.
2 Personality is a dynamic phenomenon influenced by both the physical and social environment through ongoing interactions. In the case of ethorobots, it is important to enable them to accommodate and modify their personality traits based on their experiences over time. This process can also compensate for any limitations imposed by their physical features.
3 It is beneficial for ethorobots to have the capability to recognise certain personality traits in their human partners. When personalities align, collaboration and teamwork tend to be more effective. Therefore, ethorobots should possess the ability to display a compatible personality that complements the human partner they are interacting with.
4 As a social partner, ethorobots could become role models. In specific scenarios, such as dangerous situations, robots can exhibit personality traits that serve as a model for humans present. For instance, they should demonstrate awareness of the danger while maintaining control of the situation making the people less worried which increases the chances of escape.
5 Additionally, ethorobots could adjust their own personalities based on the personality of the specific person they are interacting with. This flexibility allows for a more personalised and successful interaction, enhancing the overall human-robot relationship.

2.4.11 Conclusions, prospects, and questions

This chapter has focused on the importance of planning the embodiment and behaviour of ethorobots. Living organisms offer a rich source for inspirations to build functional ethorobots, and there are several possible insights from evolutionary biology and ecology that can help in the planning process. For example, thinking in terms of building blocks for designing ethorobots for specific functions, somewhat in the line with the Lego-robot, has still not captured the imagination of engineers. This could be achieved by increased collaboration between companies producing robots and also more standardisation so that different components can be used for assembling robots with diverse functional capacities.

Similar standardisation is needed to benchmark robots' performance before unrealistic claims are made. This could have the form of competitions (e.g., Robocup or similar challenges) or widely accepted test procedures implemented in all laboratories could increase trust in this technology.

There are also several behavioural features that could make the ethorobot more vigilant and similar to living beings. Increasing animacy and restlessness or providing the impression of greater autonomy could also make the social interaction more intimate.

Ethorobots should be also individuals by differing from each other in behavioural terms. As described having a personality has nothing to do with anthropomorphism but is a typical feature of animals with more complex organisation. As some personality traits offer an advantage for specific jobs, ethorobots should be endowed with the most efficient traits. Personality reflects only tendencies, so small variations are not only acceptable but could make the robot individually distinctive from their artificial companions.

2.5 Ethorobotic perspectives for building robotic mental architectures

Multicellular organisms have evolved neurons, specialised cells that organise and control animal behaviour. Neurons form networks with specific nodes, or local 'hubs' (e.g., ganglions, brain nuclei), and in both invertebrates and vertebrates, a hierarchical structure has emerged with the evolution of a central hub, the brain, which provides the highest level of neural control. However, vertebrates exhibit a greater concentration of neural control in the brain compared to invertebrates. It is widely accepted that these neural structures are essential for the emergence and proper functioning of the mind.

In the field of robotics, a key question is how closely the brain-mind structure should be replicated to achieve animal/human-like functioning. Currently, the following considerations are relevant (for similar arguments, see also Jonas & Kording, 2017; Miłkowski, 2018; Shagrir, 2010, 2018):

1 The nature of the mind is believed to be rooted in the underlying organic neural structure and substance. Therefore, constructing an artificial mind (mental architecture) using non-organic materials is unlikely to replicate the animal mind as we know it.
2 Artificial architectures may achieve equal or superior performance compared to animal minds. Thus, it may not be necessary to replicate the exact functioning of the brain and mind because of the specific functions of robots.
3 Organic brains and minds have undergone extensive evolutionary history, and their functioning is optimised for specific ecological niches rather than human-centric high-tech performance. Some experts assume that despite intensive research, our knowledge about the brain is very limited and only 5%–10% of the brain mechanisms have been discovered. This may provide a limitation for those who want to build an artificial mind based on contemporary anatomical and physiological insights.

4 Animal brains and minds have evolved in conjunction with a physical body. Therefore, any mental architecture aiming to mimic an organic counterpart must have full control over the respective embodiment.
5 Unlike artificial computational systems, brain functioning must be able to sustain its own existence and adapt to variations in energy supply or temporary/long-term loss of functions.
6 Using the human brain/mind and body as a reference for replication may not be practical due to various idiosyncrasies in human functioning, including cultural and historical influences. Most psychological research has been conducted on individuals from Western, Educated, Industrialised, Rich, and Democratic (WEIRD) societies, which may introduce biases when applied to artificial minds (Henrich et al., 2010).

2.5.1 Architectures and their function

Both living brains with minds and artificial architectures, consisting of hardware and software, exhibit system-level networks. These architectures are comprehensive networks that sustain the fundamental functionality of an agent, encompassing cognition, motivation, emotion, and enabling autonomous existence.

Both natural minds and artificial mental architectures rely on their own knowledge base, supported by adjustment and learning processes. These processes involve short- and long-term behaviour changes that enhance performance in response to changing conditions in the physical and social environment. Natural architectures are characterised by an open 'information flow', constantly receiving or discovering environmental input to update network parameters (Vernon et al., 2007).

Conceptually, architectures can be divided into three main parts, following a traditional input-processing-output paradigm, which is also analogous to animal minds (Section 1.4) (Figure 2.17). Perception is preprocessing of data from sensors, central mechanism (cognition) deals with handling both internal (memory, motivation, emotion) and external (perception systems) information to arrive at final decisions, and the action system executes and oversees the outcomes in a coordinated manner. Importantly, this separation is purely practical and does not imply that the three main units should be considered distinct either physically (in terms of hardware) or in terms of actual functioning. Perceptual and motor systems also involve processes that are recognised as cognitive.

In their review, Nocentini et al. (2019) examined a limited set of research studies (N = 56) involving social robots that demonstrated certain functionalities employing a mental architecture and/or behavioural model. The objective of these studies was to replicate specific human mental skills to achieve success in specific contexts, utilising diverse embodiments and mental architectures. This approach is akin to modelling particular functions observed in animal behaviour, such as territorial behaviour, attachment, exploration, navigation in social hierarchies, and more. Although these mental architectures share a similar structure, the integration of these networks into a fully operational social agent remains an area that requires further investigation.

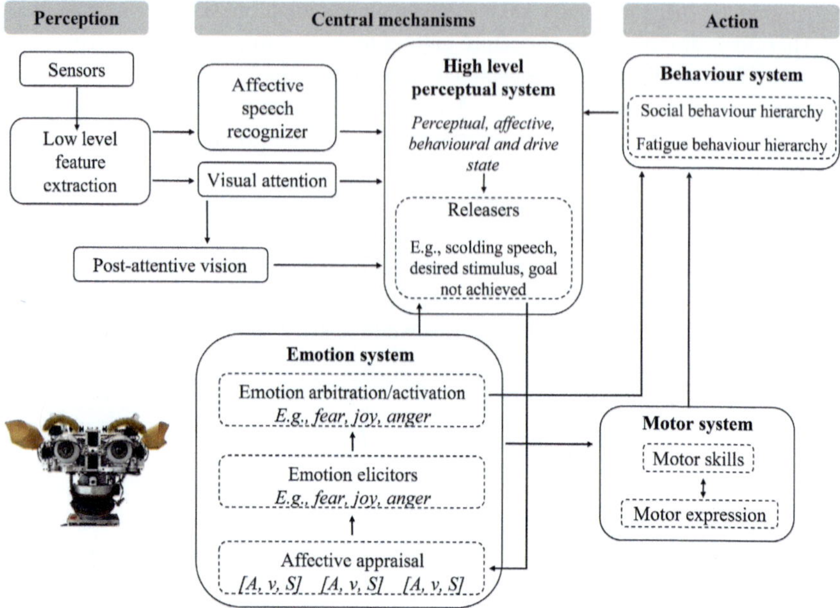

Figure 2.17 The cognitive architecture of the Kismet robot (a) rearranged based on the perception-central mechanism-action system (re-drawn and simplified from Breazeal, 2003a; Vernon et al., 2007). This model is very similar to that emerging from ethological studies. It is focusing on the behavioural output but also supports complex internal computations. The framework includes the following features and components that are common for many cognitive architectures: working memory, semantic memory (frames), episodic memory (implicit), procedural memory (implicit), and attention control (via expectation generation and matching). Using similar (standardised) structure in representing robotic architectures would improve their comparability.

2.5.2 Paradigms for artificial robotic architectures

Extensive reviews have identified two primary paradigms in the design of artificial minds; however, in practice, hybrid solutions that aim to combine the strengths of both paradigms are the most prevalent (for specific details see Section 3.6). Determining which of these representations, if any, accurately reflects human cognitive processes remains an open question and has been a subject of debate for the past three decades. Works by researchers such as Vernon et al. (2007), Samsonovich (2010), Nocentini et al. (2019), and Kotseruba and Tsotsos (2020) can shed further light on this topic (see also Table 2.11, Figure 2.18).

1 Symbolic (cognitivist) systems are rooted in the intuitive understanding of the human mind, as their units (symbols) reflect how we perceive our mind conceptualising knowledge about the environment and ourselves. These symbols possess an isomorphism with elements of the environment, to some extent. These

Table 2.11 Comparison of cognitivist and emergent paradigms, based on most important characteristics (based on Vernon et al., 2007)

Characteristics	*Cognitivist*	*Emergent*
Inter-agent epistemology	Agent-dependent	Agent-independent
Embodiment	Not implied	Cognition implies embodiment
Relevance of autonomy	Not necessarily implied	Cognition implies autonomy
Semantic ground	Percept-symbol association	Skill construction
Perception	Abstract symbolic representation	Response to perturbation
Action	Causal consequence of symbol manipulation	Perturbation of the environment by the system
Motivation	Resolve impasse	Increase space of interaction
Anticipation	Procedural or probabilistic reasoning typically using a priori models	Self-effected traverse of perception-action state space
Adaptation	Learn new knowledge	Develop new dynamics
Temporal constraints	Not entrained	Synchronous real-time entrainment
Computational operation	Syntactic manipulation	Concurrent self-organisation of a network
Representational framework	Patterns of symbol tokens	Global system states

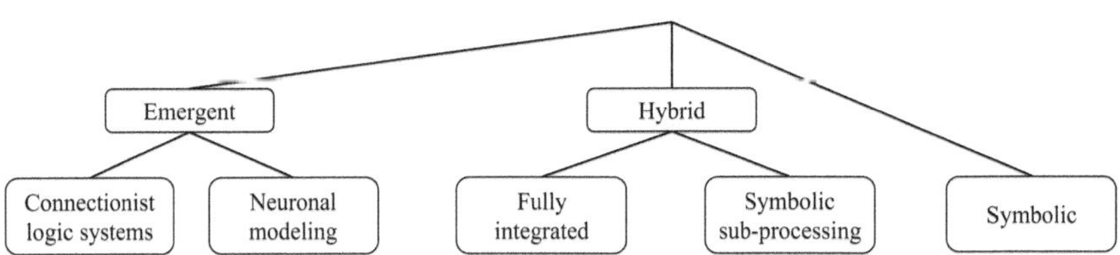

Figure 2.18 A taxonomy of cognitive architectures based on the nature of representations and processing mechanisms (modified from Kotseruba & Tsotsos, 2020). Functionally similar debate in animals is 'associationism' versus 'cognitivism' (e.g., Barrett, 2012; Shettleworth, 2010) (see Section 1.5). The combination of cognitivist and emergent systems represents the most popular in practice (see also Section 3.6).

systems operate by manipulating these symbols according to sets of rules. Symbolic systems excel in planning and reasoning tasks but exhibit less flexibility and robustness. Their functioning is more transparent, as the symbolic content or 'knowledge' can be physically located.

2 Emergent (connectionist) systems rely on diverse network-like architectures composed of nodes. Parallel processing occurs across significant parts of the network,

and isomorphic elements of the environment are distributed throughout the entire system. The efficient operation of emergent systems is based on learning, which enables them to adapt better to dynamic environments. In emergent systems, knowledge assumes a distributed nature, becoming a characteristic of the entire network.

It appears that neither approach, on its own, is capable of exhibiting capacities that can rival those of living systems. Consequently, from this standpoint, it becomes less significant which of the two approaches more closely resembles the mechanisms at work in the brain. As mentioned, roboticists adopt an opportunistic approach by combining elements from both types of systems (hybrid systems) to maximise the functionality of their creations. For instance, a symbolic system can be programmed to represent its own body, while an emergent system needs to learn about its body through interactions. Natural architectures may be predisposed (preprogrammed) regarding body structure, but they also possess the ability and mechanisms to discover specific aspects of their embodiment.

2.5.3 Levels of co-determination: evolution and development

In biological organisms, there exists a significant interdependence between the mental architecture and the environment. This relationship operates on two distinct levels that correspond to different time scales. Over the course of evolution, mental architectures gradually adapt to the environment in which they function. Constrained by their evolutionary history, these mental architectures perform optimally to support the survival of the species (Shettleworth, 2010). Failure to adapt can result in extinction.

Natural architectures primarily develop through a process wherein continuous exploration of environmental information and experiences shapes emerging systems. The ongoing interaction between the architecture and the environment ensures that these mental systems exhibit robustness and flexibility, essential for successfully adapting to new challenges. According to Vernon et al. (2007), this developmental interaction leads to 'co-determination', meaning 'that the agent constructs its reality (its world) as a result of its operation in that world'. Therefore, autonomous development (see also autopoietic systems) critically relies on processes involving (1) networks of competing and cooperating distributed multifunctional subsystems, (2) architectures capable of accommodation and self-modification, (3) anticipatory and prospective capabilities, and (4) a body that establishes a coupling between the architecture and the environment (Vernon et al., 2007).

Developmental processes play a crucial role in expanding the functional potential of the mental architecture, enabling it to handle novel and more complex situations. Based on these principles, the Interaction theory suggests (Section 2.5.4) that a significant portion of the system's knowledge originates from active engagement with the environment (e.g., Trevarthen, 2011). This assumption holds particular relevance in the realm of social interactions, as the emergence of complex social representations in the mind remains a topic of inquiry. Gallagher (2001) proposes that direct behavioural interactions with others form the basis for constructing mental models of the minds of others. This theory of primary intersubjectivity (also discussed by Moll et al., 2021) proposes that early, intense, and continuously enriching interactions between infants and their social partners (such as parents and siblings) eventually give rise to specific types of mental representations. These representations enable the mind to anticipate and predict the behaviours of others, contributing to social problem-solving abilities as well. It is important to emphasise that this interactivity occurs as a gradual process and goes in tandem with the development of the body, including the neural system. As time progresses, interactive processes extend from simpler interactions to more intricate and sophisticated ones.

2.5.4 Inner states: motivation

To consider the significant role of motivation, one has to take seriously the issue of what makes an agent behave or move in general. Energetically costly actions and movements are needed to achieve any kind of goals, but it is also based on a spatial and temporal coordinated events that have to be resilient to environmental perturbations in the short- and long-term. Thus, motivation had been introduced as concept of inner mental states that control behaviour to ensure that the organism achieves its goals, and it is assumed that artificial agents also need motivational systems to act autonomously and to be able to take most appropriate decisions (Konidaris & Barto, 2006).

2.5.4.1 Key components of models on motivation

Despite many very different ways to conceptualise motivation, there have been some important recurring insights that may provide input for a useful motivational model for ethorobots (Section 1.4). It should be noted that for the time being ethorobots do not need to be equipped with a complex human-like motivational system. A much simpler approach could be adequate here, based on simplified versions of motivational theory. Only a few key concepts are highlighted below (also based on Berridge, 2004; Pool et al., 2016).

Homeostasis: it is widely believed that organisms require specific internal states to operate optimally. These states are typically characterised by a specific range, often referred to as the 'optimal range'. The motivational system plays a crucial role in maintaining the systems' parameters within this range. Although there are likely numerous such states, the challenge lies in keeping a majority, if not all, of them within their respective optimal ranges for a given period. Importantly, the optimal ranges can vary depending on the specific challenges confronted by the organism.

Push and pull: the initiation of an action arises from the interplay between intrinsic ('push') and extrinsic ('pull') processes, although theories often attempt to explain motivation from only one perspective. When the inner states of an organism fall below optimal values, behaviour can be activated. For instance, a decrease in blood sugar level, indicating a depletion of available energy resources in the body, violates homeostasis. This violation prompts the animal to restore the inner state through a behavioural action, such as eating (as described by the drive reduction model, Atkinson, 1964). While the consumption of food is crucial for restoring the inner state to normal, it appears that the display of eating behaviour itself is also under motivational control (as discussed below). Additionally, spontaneous behaviour exhibited in response to novel stimuli encountered for the first time can also be seen as a case of intrinsically motivated action. For example, when newly hatched chicks follow a moving object upon perceiving it. Stereotype behaviour emerges often because there is no possibility to execute the action in its natural form. Individuals of species typically possessing large territories show much walking in captivity (that is within the small cage) probably elicited by the active state of a motivational system (e.g., Mohapatra et al., 2014).

The process of learning about goals exerts a certain 'pulling' effect. Inexperienced animals typically possess somewhat vague mental representations of potential goals. Through interaction with goals and supported by learning mechanisms, animals are better equipped to make appropriate decisions in the future. When an animal senses a goal that previously had a positive impact on its inner state, there is a higher likelihood that it will influence its behaviour. The development of the ability to discriminate and recognise goals from non-goals is a vital aspect for any organism. For instance, herbivores must be able to identify and differentiate between various types of 'good' foods and potentially harmful ones.

Appetitive and consummatory behaviour: given the multitude of possibilities for interacting with the environment, motivation plays a critical role in regulating the behavioural output of the animal. Goal-directed behaviour typically consists of two distinct phases. The appetitive phase is characterised by actively searching for the goal, which involves not only exploration but also the ability to differentiate between goals and non-goals (e.g., distinguishing grains from gravel in chicks). During this phase, the animal must also make decisions regarding whether to persist in the current search or switch to another activity (see also search image, Section 1.11.6). Consummatory behaviour represents the direct interaction with the goal once it has been reached. While the appetitive phase is closely tied to the intrinsic aspects of motivation, the consummatory behaviour carries significant consequences in terms of environmental feedback and learning.

Attractiveness: although different types of goals can elicit the same consummatory behaviour, animals may exhibit a preference for one goal over another. This ability to be selective holds biological significance, as certain goals (such as specific types of food) may offer greater advantages, such as being richer in calories. The variation in preference arises from three key factors (Berridge, 2018). First, researchers distinguish the individual's effort to obtain a particular goal, referred to as 'wanting', which can be observed through the strength of the behavioural response towards the goal. Second, goals acquire another attribute known as 'liking', which represents the positive or hedonic effect of the goal on the animal. Lastly, the memories associated with the goal introduce a cognitive component to the final decision-making process. The interplay of these three inputs – wanting, liking, and cognitive factors – accounts for both the common hierarchical structure observed in motivated behaviour and the differences observed among individuals (see incentive salience hypothesis, Berridge & Kringelbach, 2015).

Motivational states: while many examples of motivation focus on the regulation of eating or drinking, it is important to recognise that potential goals in animals are much more diverse. This raises the question of whether each goal has its own distinct motivational state, potentially requiring a multitude of different states to account for motivated behaviour. Although a definitive answer to this question remains elusive, animals appear to possess the ability to compare the motivational value of specific goals at a given moment, often referred to as a 'common currency' (Cabanac, 1992). This calculation becomes more feasible if we consider the idea that the mind constructs a multi-dimensional motivation space, defined by shared variables that apply to all motivational behaviours. This approach aligns well with the notion that motivational states develop over time, with the increasing number of dimensions reflecting the animal's growing experience and interactions with the environment. Furthermore, the motivational space may undergo changes in adults. Actions yielding novel yet positive outcomes can lead to expansions in the motivational space, accommodating new goals and incorporating them into the overall motivational framework (Steel et al., 2021, Box 2.14).

Box 2.14 A simplified model of a mammalian nervous system integrating the hedonic and homeostatic components of food intake regulation (adapted from Campos et al., 2022)

During the pre-ingestive phase, various senses such as vision and smell perceive food stimuli. After food ingestion, sensory information from taste, smell, consistency, and temperature constitutes a second stimulus within the oral cavity (post-ingestive stimuli). This input is relayed to cognitive areas, but at this stage, the energetic and macronutrient content (metabolic content) of food has not been sensed. As food moves through the gut, it activates specific hormones that send signals to the hypothalamus and the brainstem. Information about the mechanical distention (fullness) of the gut is conveyed to the brainstem. The metabolic and mechanical feedback, along with reward feedback, is integrated within brain areas associated with aversion, cognition, reward, motivation, memory, and decision making (both cortical and subcortical areas). From these regions, information projects to higher cognitive brain centres to ultimately regulate eating behaviour.

Even this simplified (the involved brain structures and molecular pathways were omitted for clarity) model shows the complexity of one of the ancient motivational systems. The decision for looking for food and stopping consumption involves almost all regions of the brain in addition to the sensory and motor areas (not shown here). Multiple feedback loops through perception and involvement of experience are also important components of the feeding behaviour (Figure 2.19).

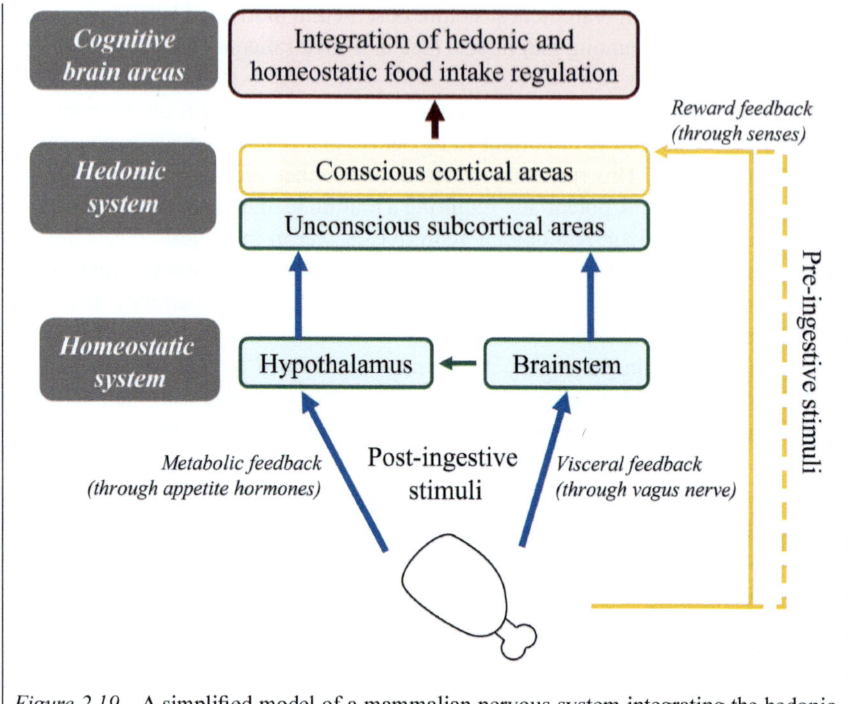

Figure 2.19 A simplified model of a mammalian nervous system integrating the hedonic and homeostatic components of food intake regulation (adapted from Campos et al., 2022).

2.5.4.2 Circadian rhythms and motivation

Natural organisms are exposed to various rhythmic features in their environment, and these changes often have implications for the fitness of the species. In response, animals have evolved adaptive mechanisms that allow for adjustments that also include motivated behaviour (Antle & Silver, 2016). Rhythmicity in the environment also provides reference points for determining the optimal timing of certain activities based on resource availability (Section 1.6.2). Since only one behaviour can be performed at a time, temporal effects can help bias one motivated tendency over another, reducing competition among motivational states. There are three ways in which circadian rhythms influence motivated behaviour (Antle & Silver, 2016). Circadian rhythms may provide the temporal structure that allocates different goals to different parts of the day. They coordinate physiology and metabolism to ensure that the body can most effectively fulfil the underlying needs associated with the goals. Finally, the circadian system supports the prediction and anticipation of the availability of temporally restricted goals. An example of applying this approach can be seen in the work of Stoytchev and Arkin (2004), who incorporated a motivational system into office robots. They found that the use of motivational variables helped the robot achieve its goals more effectively and made it more resistant to interruptions. By leveraging the principles of circadian rhythms and integrating them into the robot's motivational framework, the researchers were able to enhance its performance and adaptability.

2.5.4.3 Motivational systems for ethorobots

The integration of in-built motivation into ethorobots presents intriguing possibilities, although its specific function and the necessary rules for behavioural control remain unclear. Bio-inspired approaches have been widely employed in this area (e.g., Cañamero, 1997), with researchers drawing parallels between brain substrates involved in motivation control and their artificial implementations. Khan et al. (2021) developed a neuron network model that simulated the glucose-insulin system responsible for regulating blood sugar levels, thereby influencing feeding, drinking, and resting behaviours. While this work is appealing, its direct application to typical robots may pose challenges, as the rules governing energy use and regulation differ between biological organisms and artificial systems. When discussing the potential of their findings, these authors also alluded to a parallel between the model's 'hunger state' and the fuel level in a car (or robot). While this analogy holds some validity, the focus should be on implementing the components of a generalised biological motivational system and understanding their interactions, rather than attempting to replicate specific biological mechanisms.

Similar to living systems, the motivational states of ethorobots should align with their specific functional challenges. While the basic need for energy is evident, engineers must identify additional functions that ensure the robot's survival and effectiveness in its intended role. Consequently, artificial motivational states may vary depending on the specific functions of the robots. For instance, a waiter assistant robot should not only excel at collecting dirty plates and cutlery but also be motivated to behave in an accommodating and friendly manner towards guests and waiters. Thus, motivational states such as 'resting', 'charging', 'clear table', and 'social acceptance' could effectively control such a robot. This parsimonious design recognises that present-day robots are simpler entities compared to many animal species with more intricate behaviour patterns. Therefore, there is no need to overly complicate their motivational systems. However, it remains a question whether incorporating more complex biological features of motivation, such as wanting or liking (as mentioned earlier), would provide additional advantages for artificial systems.

The classic example used to illustrate motivational states is hunger. The actions that emerge when an animal experiences hunger, are often straightforward and intuitive. However, laboratory experiments and research on humans have revealed that motivation for eating is actually a highly complex phenomenon in living beings (Box 2.14).

The nature of hunger states can vary among animals depending on their specific energy requirements. Some animals need a continuous intake of energy and can only survive for short periods without food, often just a few hours. In contrast, other animals may eat infrequently, perhaps only once a week or even less frequently, and can withstand weeks of starvation without significant risk. Animals that inhabit environments with periodic food scarcity may exhibit different motivational states when it comes to collecting and storing food. In these cases, storing or caching food may indicate the activation of a distinct motivational state.

The complexity of behavioural patterns that can arise from the 'hunger state' becomes evident when considering a few key insights.

- *Alliesthesia*: the positive value or reward associated with goals is enhanced when experienced in a state of higher motivation. For example, food can be perceived as more rewarding when consumed after a prolonged period of starvation (Cabanac, 1971).
- *Incentive contrast*: a single experience with better quality reward (goal) leads to a more pronounced behavioural response in subsequent encounters. For instance, rats exhibit faster running speeds in a maze when they encounter a positive change in food quality, while the opposite occurs when there is a negative shift (Flaherty, 1982).
- *Action motivation*: motivational states can directly influence actions, wherein animals may engage in specific behaviours even in the absence of a specific goal. For instance, rats may exhibit chewing movements even when they are intravenously fed, a behaviour known as vacuum activity (Dawkins, 1988).

In summary, animals possess intricate motivational systems, but implementing the full complexity of these systems in robots may currently be unnecessary.

2.5.5 Inner states: emotion

Undoubtedly, emotions and emotional behaviour have a significant impact on human communication. They influence how individuals perceive and judge one another, their ability to cooperate, and the decisions they make. However, it is important to avoid a solely anthropocentric perspective that distorts the understanding of the primary role of emotional states in living organisms.

One way to conceptualise the mechanistic function of emotions is to recognise that embodied creatures require a system that enables the entire body to react in a coordinated manner to different situations (Al-Shawaf et al., 2016). For instance, fear is a common emotional state shared by many organisms. When faced with danger, an animal may choose to escape. However, for this escape manoeuvre to be successful, all parts of the body must react rapidly and in synchrony. This coordinated reaction is facilitated by the rapid activation of a specific emotional state. To achieve this reaction, emotional states encompass important physiological aspects, including abrupt and significant changes in neural and hormonal functioning (Mauss et al., 2005) to execute an effect on the body.

Another perspective highlights the function of emotions as a means to gauge the significance of events, whether they have a positive ('reward') or negative ('punishment') impact on the individual. In this view, emotional states serve as a link between actual experiences and the subsequent reactions of the animal to similar events in the future (Rolls, 1999).

Modern ethology does not typically question whether animals experience emotional states, but rather focuses on understanding the complexity of these states and how they can be differentiated (Anderson & Adolphs, 2014; de Waal, 2011; Mendl et al., 2022; Panksepp, 2010; Paul & Mendl, 2018).

2.5.5.1 Traditional approach to emotions

The anthropocentric approach focuses on identifying and categorising human emotions and understanding their role in human social interactions. One way to differentiate emotions is by distinguishing between primary and secondary emotions. Primary emotions are considered universal among humans, recognised across cultures, and characterised by specific facial expressions. For instance, Ekman and Davison (1994) proposed six primary emotions in humans: sadness, happiness, anger, surprise, fear, and disgust. Secondary emotions are believed to be more complex and influenced by social and cultural factors.

In humans, primary emotional states are often associated with observable facial and bodily signals (see Section 3.5.8). By watching someone's behaviour or actions, an observer can make inferences about his or her underlying emotional state. For example, when seeing a person with a 'big smile', the observer can reasonably assume that the person is experiencing happiness. Especially in humans, facial expressions serve as signals that facilitate others to gauge the emotional states of individuals (Box 3.17).

2.5.5.2 Ritualisation of emotional behaviour

Emotional behaviour, as a form of communication, originates from a long evolutionary process known as ritualisation (see Section 1.8). The behaviour exhibited by an animal in response to environmental events initially served a primary physical function. For instance, a strong fearful emotional state would trigger physical reactions such as body contraction, slowed or halted movement, and protective retraction of appendages. These rapid and coordinated actions enhance the prey's survival by reducing visibility and making it harder to be captured.

In social interactions, there can also be potential dangers for both parties involved. As fights can be risky for both individuals, avoiding unnecessary conflicts becomes advantageous. In such cases, the behavioural responses elicited by the attacking individual can evolve into communicative signals. If potentially weaker individuals display fearful behaviour and avoid confrontation before an attack, it can help prevent fights from occurring. Consequently, elements of fearful behaviour gradually acquire a signalling function over evolutionary time. A similar process can be observed with anger, where behavioural actions associated with successful attacks also become communicative signals to opponents (LeDoux, 2012).

Emotional behaviour in animals, particularly those living in complex social systems, can also involve learned components. These modifications of emotional signals are often passed down from one generation to the next and can be specific to particular populations or cultures (Section 1.8). The incorporation of learned features into emotional signalling occurs during an individual's lifetime and is referred to as developmental ritualisation (Tomasello, 2008).

2.5.5.3 A functional approach to emotions

In line with the above discussion, Plutchik (2001) proposed a functional system that offers an ethological explanation for a broad spectrum of emotional states (Box 2.15). According to this model, emotional states are primarily associated with specific behavioural functions, such as territoriality and social interaction. These emotional states emerge as responses to particular events, such as threats, encounters with unpalatable objects, or unexpected occurrences. Their main purpose is to trigger coordinated actions, such as attacking or escaping, ultimately leading the animal to achieve a specific state, such as safety or acquiring resources.

Box 2.15 A functional approach to emotional behaviour

Plutchik (1980) introduced the idea that emotional states should be connected to specific functions that are critical in the survival of animals (see Table 2.12). Emotional states are evoked by specific events or stimuli that typically occur under some conditions and the evaluation of the situation leads to the emergence of a specific emotional state that organises and controls an adequate behaviour response. Although Plutchik and many followers of this approach tried to use different wordings for the functions, emotions, and behaviours, there are linguistic constraints for separating these features, and ethological studies also show that there could be an overlap between the action repertoire displayed in different contexts. This tabular arrangement also shows the limitation of the categorical approach for modelling emotions.

Dimensional models of emotions are probably more adequate to become part of the mental architecture controlling behaviour. Russel (1980) developed a two-dimensional model based on multidimensional scaling and also human evaluation of emotional terms (Figure 2.20a). This model has two main axes: arousal or intensity (low–high) and pleasure or valence (negative–positive). From an ethological aspect, changes in emotional states have important behavioural consequences that such models should have at least a simple component reflecting this. Russel's intensity and valence model should be extended to include an axis for approach-avoidance (Stance) (Breazeal & Brooks, 2005) tendencies (Figure 2.20b). The change in distance does not need to involve the whole body. Reaching towards an object can be considered as an approach ('showing interest'), while spitting out a bad food indicates avoidance.

Table 2.12 The table follows Plutchik's approach but was modified at several points. The 'Function' of the behaviour was moved to the most left side, and the concept or examples in the other columns were changed, amended (taken over from Panksepp, 2005; TenHouten, 2017)

Function/ outcome	*Stimulus/event*	*Cognitive component (appraisal)*	*Emotional state (less-more intense)*	*Overt behaviour*
Protection, safety	Threat, challenge	Danger	Fear – panic	Escape
Destruction, competition	Competitor, obstacle	Enemy	Anger – rage	Attack
Gain resources, reproduction	Gain of valued object (e.g., food)	Possession	Joy – happiness	Retain or repeat
Reintegration/ Reattach to lost object	Loss of valued object (e.g., companion)	Abandonment	Sadness – grief	Withdrawal, crying
Attachment, cooperation	Group member	Friend	Acceptance – trust	Proximity seeking, grooming, play
Avoid or eject poison and danger, rejection	Unpalatable object	Poison	Disgust	Remove, distancing, vomit
Exploration, monitoring	Home range	Examine	Expectation	Moving around, sensory investigation
Familiarisation, orientation, learning	Unexpected event	Exploration	Surprise	Stop sensory investigation

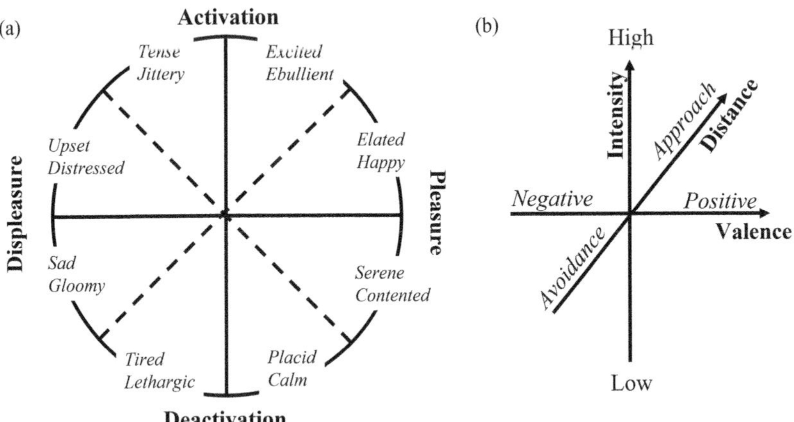

Figure 2.20 Dimensional models of emotional states: (a) simplified depiction of Russel's two-dimensional model (from Plass & Kalyuga, 2019) and (b) tendencies to move or behave towards (or away) from a physical or living stimulus/object is reflected in the ethologically inspired model.

2.5.5.4 Emotional categories or emotional space

The categorical model of emotions, characterised by specific linguistic labels and context-specificity, has received considerable attention in research. Efforts have been made to identify the precise neural structures associated with particular emotional states, such as fear or happiness. Affective neuroscience has made significant progress in uncovering brain systems that could be linked to specific emotional states, including seeking, fear, rage, lust, care, panic, and play (e.g., Panksepp, 2011). However, it is important to note that there is ongoing debate and lack of consensus in the literature regarding the definition and categorisation of emotional states, especially with regard to animal-human comparison. Different authors propose different numbers and labels for mental states considered as emotions, leading to a lack of harmonisation in the fields of cognitive psychology and affective neuroscience.

An alternative approach to modelling emotions is to envision a multi-dimensional mental space for emotions without strict boundaries. In this case, the key challenge lies in defining the dimensions of such a spatial model. Cochrane (2009), drawing on the work of Russell (2003) and Scherer et al. (2006), suggests eight dimensions for the mental–emotional space: (1) attracted–repulsed, (2) powerful–weak, (3) free–constrained, (4) certain–uncertain, (5) generalised–focused, (6) future directed–past directed, (7) enduring–sudden, and (8) socially connected–disconnected. Similar dimensional models have been proposed by other researchers as well (Kim et al., 2020) (Box 2.16).

Regardless of the specific conceptualisation, dimensional models of emotions offer several advantages, especially if utilised in robots (see below).

1. By adopting such models, concerns regarding anthropomorphism can be mitigated. These models can effectively encompass emotional states in animals without necessitating a direct comparison between an animal emotional state (e.g., happiness) and its human counterpart.
2. There is no need to focus on the use of labels for a specific emotional state, but a good model should pinpoint where a specific emotion may reside inside the multidimensional structure.
3. They provide a more flexible framework that allows for individual differences and the potential for blending or variation between more common emotional states. These intermittent emotional states can arise spontaneously and provide advantages in specific situations.
4. Dimensional models can capture the complex and nuanced nature of emotional experiences, taking into account the interplay of multiple dimensions rather than rigid categories.

Box 2.16 Guilty dogs and guilty robots

Behaviour inspired by dogs could be employed to prompt individuals to attribute emotional states to a non-humanoid robot. Some dog owners assert that dogs exhibit guilty behaviour (Figure 2.21a) if they have done something wrong, but thus far, no evidence has surfaced to support the existence of such an underlying emotional state (e.g., Hecht et al., 2012). Nevertheless, dogs exhibit various behavioural features, likely reflecting slight fear, that may give the impression of guilt. Owners were less inclined to punish their dogs in any manner if the canine companion displayed guilty behaviour following a misdeed.

Lakatos, Gácsi, et al. (2014) investigated whether human participants tended to attribute guilty behaviour to a robot (Figure 2.21b) in a relevant context. This was examined by assessing whether human participants could detect if the robot transgressed a predetermined rule based on the robot's greeting behaviour. In the study, participants had to teach the robot a specific behaviour (spin) while simultaneously teaching a dog not to knock over bottles. The bottles were positioned on the floor behind a tall and opaque barrier, rendering them invisible from a distance (Figure 2.21c). Following the teaching phase, the robot demonstrated that it had learned to avoid the bottles when moving across the room.

After a brief interval during which participants left the room, they returned and were asked to determine, based on the robot's behaviour, whether the robot had knocked over bottles in their absence. Participants were more likely to report that the bottles were

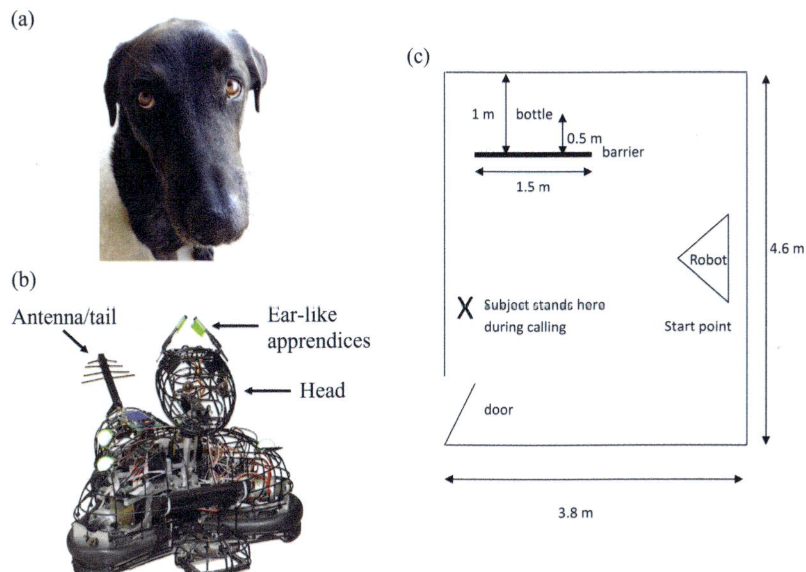

Figure 2.21 (a) 'Guilty' behaviour shown upon greeting by many dogs if they did something wrong in the absence of the owner (photo by Lab Look, Flickr). (b) The holonomic zoomorph robot ('Mogi Robi') used in the guilt-attribution experiment. It was remotely controlled by a hidden experimenter. The head, antenna, and ear-like appendages could be moved at different intensities according to a protocol. (c) Top view of the experimental room and the position of the subjects, the robot, and the hidden bottles (from Lakatos, Gácsi, et al., 2014).

knocked over when the robot displayed 'guilty' behaviour upon their re-entry (For the control group the robot displayed typical greeting behaviour). The 'guilty behaviour' was defined using a dog ethogram. It started by stopping in front of the subject, lowering the head, ear-like appendices, and antenna, and was followed by reversing half a meter. The robot then wagged its antenna low and looked at the subject by raising its head and turning it in the appropriate direction.

These results further support the idea that observing facial emotional expressions is not an essential requirement for humans to accurately assess emotional states. Providing ethorobots with such capacity could enhance their acceptance, especially when they make mistakes.

2.5.5.5 Emotional behaviour in ethorobots

Numerous arguments in the literature (e.g., Fong et al., 2003) support the notion that exhibiting emotional behaviour can be advantageous for social robots engaged in interactions with humans. Extensive research has been conducted to fulfil this premise, drawing primarily from human affective psychology, and diverse approaches have been employed to equip robots with the ability to display emotional states (Cavallo et al., 2018; Fellous & Arbib, 2005; Hieida & Nagai, 2022).

Contrary to the mechanistic definition and the role of emotional states in organising behaviour, emotional behaviour in robots often lacks a fundamental impact on their behaviour's underlying structure. It does not emerge as an intrinsic feature, as seen in animals. Instead, robots display positive emotional signals primarily to enhance their acceptance by human partners (Hudlicka, 2004). Frequently, emotional signals are constrained to head/faces only. While human observers may be able to interpret these signals when prompted, it is crucial to note that these solutions do not replicate natural circumstances. In contrast, when an animal experiences a fearful emotional state, it affects its entire body, not just its face. Moreover, the emotional state influences its overall behavioural tendencies. Employing exaggerated human-like emotions on an animal-inspired artificial agent (e.g., iCat, Kismet) further adds to the peculiarity of the interaction.

A more ethologically valid approach involves utilising the entire embodiment of the interactive agent to convey emotional signals. This can encompass changes in movement speed or trajectory along the approach-avoidance axis. Visual signalling using various appendages, combined with vocalisations, provides a technologically feasible means of displaying emotional behaviour (see Section 1.8). The combination of visual and acoustic features can create a more impactful emotional signalling system than relying solely on one means of communication.

In the study conducted by Lakatos, Gácsi, et al. (2014), participants were instructed to interact with a wheeled animal-like robot. This robot displayed either fear or happiness when confronted with a ball. Depending on the situation, the robot responded in one of two ways. When the subject propelled the ball towards the robot, it exhibited either of these behaviours. The 'happy' robot would raise and wag its antenna and ear-like extensions, approach the ball consistently, and return it to the human participant. Conversely, the 'fearful' robot, upon detecting the ball, ceased movement of its antenna, drew nearer to the ball, oriented its head towards it, and then abruptly lowered its antenna and ear-like extensions while attempting to maintain distance from the ball. Human subjects demonstrated an ability to differentiate the robot's internal states based on these behaviours.

Taking a natural approach to designing emotional behaviour for robots is particularly important for emotions that lack specific facial expressions. For example, implementing the emotion of jealousy to a social partner can be significant in social robots. Behavioural patterns associated with this emotion may include attention-seeking behaviours, actions aimed at separating intruding partners, or even display challenging behaviour towards them (e.g., Abdai et al., 2018).

The implementation of specific emotional behaviours not only enhances HRI but may also promote the acceptance of the artificial agent. Ethorobots, like any system, may make mistakes, cause unintended harm, or fail to execute a task. In such situations, displaying an emotional state that corresponds to human feelings of guilt can inform the human partner that the robot is aware of the problem. This type of signalling can be highly advantageous in robots that are prone to errors due to incomplete design or are required to operate in complex situations where errors may occur more frequently (Box 2.16).

2.5.6 Physical problem solving

Research on interactive robots has predominantly focused on the social aspects of behaviour rather than on the physical challenges they face. This top-down approach is understandable from an anthropocentric perspective but contradicts the evolutionary history of biological agents. In nature, organisms initially evolved robust motor skills to cope with various environmental perturbations, and social behaviours were gradually added, particularly in the case of vertebrates. In robotics, however, this reverse approach has led to some agents capable of displaying emotional behaviour without possessing adequate mobility skills. Consequently, it may be time to partially shift the agenda and prioritise efforts towards enhancing the physical problem-solving abilities of socially interactive robots.

2.5.6.1 How to move?

In principle, ethorobots designed for human interaction can utilise various modes of locomotion (see Section 3.5), including wheels, legs, and potentially even flight. While there are no biological examples of organisms employing wheels, there are instances of organisms utilising legs or flight. Ethorobots interacting with people typically operate on terrestrial surfaces, and thus among the available options, using legs or wheels is the most feasible choice. Both enable freedom of movement in two-dimensional spaces, but wheeled robots have certain limitations, and their effectiveness relies heavily on the terrain's quality (Section 2.4.4). Consequently, the choice of locomotion may dictate the environments in which these robots can effectively perform their duties.

2.5.6.2 Development and plasticity

Mental architectures in living beings are developing over time and possess inherent plasticity. The first step is to separate parts of the body from that of the environment. It has been suggested that human infants possess a 'contingency detection module'. Perfect correlation between motor commands and spatial changes marks off own body from all other objects in space (Gergely & Watson, 1999). Typically, mental systems for body control need to mature and get integrated into one overarching architecture. In precocial species, at least some behaviour systems have to be functional shortly after hatching or birth, while in altricial animals this process is more delayed. Development has to deal with possible changes in the body (e.g., growth) and is paralleled by increased integration of different control systems.

Short-term or long-term plasticity of these mental architectures comes into play with qualitative or quantitative changes in the performed actions. Getting more skilful during lifetime by using body parts differently, more often, or with higher precision is paralleled by long-term changes in mental control. Short-term changes come about when, for example, a tool in the hand is integrated into body schema for the duration of use (Vaesen, 2012).

2.5.6.3 Monitoring and predicting

The environment is changing at some pace either as a result of further external forces or because the animal is moving from one location to another. Maintaining rapid reactivity necessitates continuous monitoring and predicting. By comparing the current mental state with new information, the animal can decide whether there is any need to change its behaviour (see also Test-Operate-Test-Exit principle, Pezzulo et al., 2006) (Section 3.6.3). This kind of monitoring is typically based on many senses and should be accomplished without disturbing other ongoing processes. The animal may execute the next move based on the assumption ('prediction') that there is no change, but it is advantageous if the mental functioning allows for interruption if the monitoring activity discovers alterations. Obstacle avoidance provides a good example for the operation of such processes. Animals start to alter their path at some distance when they aim to by-pass and object and they are also able to shape this trajectory continuously, if, for example, the object moves towards them.

Monitoring is a costly activity, and therefore its frequency depends on the extent to which changes in the environment are expected. Individuals also differ in their monitoring behaviour which can be particularly important when the animal travels to a (relatively) new location. Better knowledge about the environment is advantageous for the individual to locate more resources but absence of caution may expose it also to risks (e.g., predation).

2.5.6.4 Physical problem solving in terms of goals

In the course of solving a physical problem animals move from A to B or change their physical relationship with objects (e.g., the crow takes a nut in its beak). From the observer's point of view, it seems as if they have executed goal-directed actions (see also Box 2.17). While this may be true functionally, there are two questions to be answered. First, was this action preceded by a mental decision to achieve the goal, and second, whether the goal achievement has some specific mental representation? Although the former question can be reformulated as an enquiry about intentionality, which is outside of the present discussion, one could raise the possibility that intention could be regarded as a measure of the strength/resistance (or determination) of goal-oriented behaviour. McFarland (1989) describes three fundamentally different ways for agents, like ethorobots, to arrive at their goals.

1. A *goal achieving system* includes a mechanism that changes (stops) the behaviour once the goal is achieved but the displayed behaviour is rather under environmental control. This mechanism can be useful in unpredictable environments, for example, prey aiming to get away from repeated predatory attacks may find an escape route in this way by rapid random movements.
2. A *goal seeking system* reaches its goal by following a set of different rules without necessary having a specific representation of the goal. Spiders do not have a mental representation of their web they weave; rather, they follow a set of rules which ensure the final form of this structure. For such rules to work, the environment has to be relatively stable because there is little possibility for correction of errors.
3. A *goal-directed system* involves a representation of the goal-to-be-reached, which has a major role in guiding the behaviour. Goal-directed systems can show high variability in complexity, especially with regard to correction mechanisms for attaining the target. For example, nest building is a goal-directed behaviour because the actual status of the nest informs the bird about the next move.

Box 2.17 Solving problems by reaching goals

The human observer tends to attribute goal-directed behaviour to any organism. For instance, when observing a fish larva swimming rapidly in a small dish, one might assume it 'wants' to escape. However, there are reasons to believe that the architecture controlling goal-oriented behaviour differs in complexity in agents, particularly concerning the mental representation of the goal.

In the figure depicting a top view of an agent in a small arena, the agent begins to search for food, although there is none available (Figure 2.22). The aim of leaving the arena can be achieved in different ways. A goal-achieving system activates random swimming behaviour, eventually leading to the localisation of the opening, allowing the agent to extend its search. A goal-seeking system employs a set of rules (go forward, go along the obstacle, stop at gap, turn towards gap). It is important to note that these

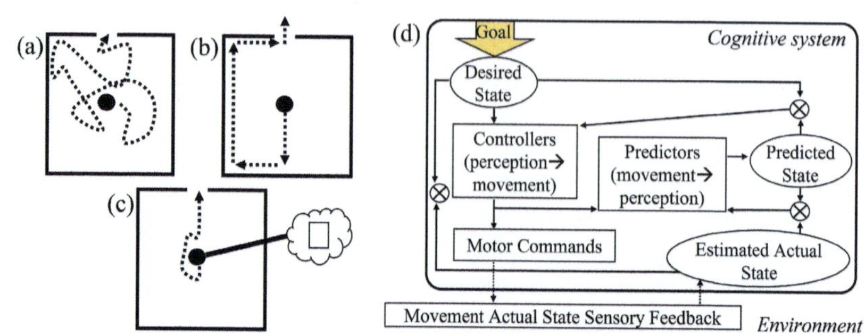

Figure 2.22 (a) Goal-achieving, (b) goal-seeking, and (c) goal-directed agents finding their way out of an arena (the small cloud stands for some kind of mental representation of the 'opening'). (d) Hommel (2022) describes goal-directed behaviour in terms of a comparator model. The goal defines the desired state, which controllers translate into motor commands eventually executed, and predictions of expected outcomes. The re-afferent information is compared with the predictions, resulting in an estimate of the discrepancy (the error), ranging from 0 in the case of perfect performance to higher values depending on the degree of failure (based on Hommel, 2022).

agents lack a representation of the opening. In contrast, a goal-directed system relies on a mental representation of an 'opening'. Such an agent first visually searches for an opening and then approaches it directly. A simplified overview of such goal-directed systems is described by Hommel (2022).

2.5.6.5 *The concept of emergent behaviour*

As mentioned above, purposeful behaviour can manifest in various goal-oriented systems. The behaviour of an agent is regarded as emergent when an observer (or a constructor) has evidence that this characteristic could not be the result of a specific mental algorithm or a particular decision-making process. In general, such phenomena emerge as a consequence of animals or artificial agents following a set of rules in specific situations.

For instance, consider a group of wolves chasing prey, forming a complex spatial arrangement lead by the leader wolf. This intricate pattern, however, can be explained by the wolves adhering to two simple rules: (1) closely follow the prey to increase the chances of obtaining meat if the hunt is successful, and (2) maintain the maximum distance from other wolves (Muro et al., 2011).

Similarly, in an experiment by Maris and te Boekhorst (1996), small robots equipped with sensors were programmed to divert their path upon detecting an obstacle. During one specific set of trials, the front sensors were deactivated, while those on the side and behind remained active. These robots pushed the blocks they encountered forward until they passed along other obstacles making them turn away. When multiple such modified robots were allowed to navigate freely, the blocks were being pushed to a specific location, giving the appearance of tidying up the area. Notably, this seemingly 'novel' and goal-oriented behaviour was not explicitly programmed into the robots' control structure. Instead, it emerged as a result of the interaction among the robots, their behavioural rules, and the objects in the environment (blocks and other robots). Crucially, the notion of goal-oriented behaviour, specifically tidying up, arose solely in the mind of the observer and not within the architecture of the robots.

Ethologists, such as Tinbergen (1952), have also documented emergent behaviours, often termed as displacement or 'derived' activities. For instance, birds engaged in agonistic contexts may exhibit seemingly irrelevant behaviours, like suddenly cleaning their bills or feathers. Although these behaviours are goal-oriented (self-cleaning), their emergence is attributed to the activation of conflicting motivational states (e.g., fight or flight).

The phenomenon of emergent behaviours holds several implications. First, it is anticipated that the frequency of emergent behaviours may rise with the complexity of the mental architecture. Second, emergent behaviours could serve as substrates for learning mechanisms if they yield positive consequences for the agent. Stereotypic behaviours may emerge this way. Last, over a more extended time frame, emergent behaviours undergo selection and integrate into the typical behavioural repertoire, acquiring their own distinct functions.

2.5.6.6 Mental architectures for physical problem solving

Below, we present a partial list of physical challenges that different animal species may encounter in their natural environments. The cognitive and neural solutions employed by these species can vary, exhibiting varying degrees of complexity and flexibility. Some of these solutions could be the result of adaptations and predispositions. Ethorobots, acting in the anthropogenic environment, face mostly similar challenges as humans or other animals of comparable size, and thus examples for solving physical problems are taken predominantly from studies on mammals. Over the years, engineers have introduced many bio-inspired or artificial solutions to such problems. Nevertheless, the focus of the following section is to provide a short summary on the insights on physical problem solving (see also Wiener et al., 2011) that could be useful for ethorobotics. Although concepts are often treated separately, a tighter interaction is expected among these systems under natural conditions.

2.5.6.7 Mental architectures for the body

Researchers and engineers have yet to establish the necessary mental architectures required to achieve full functionality of the human body. Furthermore, there are disagreements in the nomenclature surrounding this topic. To provide a concise overview, we refer to the review by Hoffmann (2021). Although valuable insights have been gained from experiments in human cognitive psychology and advancements in neurobiology, our understanding of the similarities between humans and non-human species remains limited. While the human body exhibits remarkable morphological complexity and a high number of degrees of freedom, animal bodies encompass a wide range of structural and mechanistic solutions. Without comprehensive comparative data, it becomes challenging to determine how the structure of the body impacts mental architectures, if at all. While certain basic processes and rules may be shared among various living organisms, the evolution of specific body parts, such as the versatile primate arm and hand or the elephant trunk, may fundamentally influence the mental architectures associated with the body.

Based on mostly human-centred knowledge, Hoffman (2021, and references therein) suggests three aspects ('taxonomies') of architectures which together or in a kind of co-action seem to be able to deal with all aspects of body control (de Vignemont, 2020; Pitron & de Vignemont, 2017). Examples refer only to typical applications, while it is widely accepted that body-centred actions are often supported by the co-action of many sub-systems (a to b/c):

1 *Taxonomy based on the nature of the main input(s):*
   - a Sensorimotor body control (e.g., architecture needed for executing actions on the environment);
   - b Visuo-spatial body control (e.g., architecture needed for moving in space);
   - c Conceptual body representation (e.g., linguistic representation of the body).
2 *Taxonomy based on function:*
   - a Body schema (e.g., sensorimotor control of the body used for planning and executing actions);
   - b Body image (inner representation of own body as perceptual and cognitive, etc. state, e.g., emotional behaviour, body image distortions).
3 *Temporal taxonomy:*
   - a Long-term or 'offline' body representation (e.g., size and shape of torso and appendages);
   - b Short-term or 'online' representations (e.g., current position and posture).

2.5.6.8 Architectures for navigation

In recent years, comparative research has focused on the central processes of animal navigation, leading to the identification of a few general mental mechanisms that are at work in most cases (e.g., Åkesson et al., 2014; Shettleworth, 2010; Section 1.6.1).

Wiener et al. (2011) provided a systematic overview on the 'mental tools' animals use for navigation. Despite the fact that especially in the case of navigation engineers have alternative artificial (non-natural) tools to develop navigation skills for ethorobots, an overview of the necessary features for this behaviour function could be profitable. The 'navigation toolbox' consists of four levels (see also Box 2.18; Chapter 3).

Sensorimotor system: in principle navigation can rely on any senses depending on the source of cues provided by the environment and the goal to be achieved. Different senses also have different affordances in terms of mental effort, and they offer different kinds of effect range, flexibility, error correction, etc. Vision seems to be the most typical sensory input for navigation, but animals can also rely on sounds (ultrasounds) or olfactory cues for finding their way around. At this level sensor input is tightly connected to simple motor responses such as approach/avoidance or template matching.

Spatial primitives: this level utilises input from the sensorimotor system to construct more systematised outputs. This is achieved by computations comparing internal and external information such as aiming to keep optic flow equal on both sides, or to use a beacon for orientation. This level may also combine information from different sensorimotor systems (i.e., sensor-fusion) to compensate for lack of information. Decisions on heading, movement velocity, and acceleration may be predominantly determined at this level.

Spatial constructs: information obtained at lower levels are used to calculate absolute or relative distances that may allow for planning routes in the most optimal way (e.g., shortest, fastest). Such constructs may also provide the agent the ability to locate itself and also to make calculations for navigation based on vectors or multiple bearings. This level may also supply computations for mental constructs often described as 'cognitive maps' (Section 1.6.1).

Spatial signals (Non-spatial representations): this level includes other non-navigational representations of spatial knowledge. Human language or graphic representations are typically forms of this level, but one may consider that bee dance, indicating direction and distance of a target object belongs here, similarly to a situation when animals may learn an acoustic cue associated with a specific location.

Box 2.18 Spatial representation in animals

Animals often display remarkable navigation skills, such as homing and shortcuts; however, the precise mental mechanisms underlying these abilities remain uncertain. Hypotheses have been proposed, suggesting that mammals (including humans) and possibly bees and ants could employ a Euclidean cognitive map – an accurate mental representation of spatial relationships. The ability to make novel detours and shortcuts has been considered crucial for validating this concept (see Section 1.6.1). Despite intensive experimentation, hypotheses on cognitive maps have not gained substantial support. Warren (2019) introduced an alternative hypothesis, proposing that spatial representation is better modelled by a labelled graph, where paths between places are enriched with local metric information. While this model does not constitute a true cognitive map, it still enables individuals to successfully navigate novel detours and shortcuts (Figure 2.23).

a *Route knowledge*: a line graph where nodes represent places and edges correspond to actions. Routes are represented as a single chain of places and actions, allowing travel only along familiar paths.

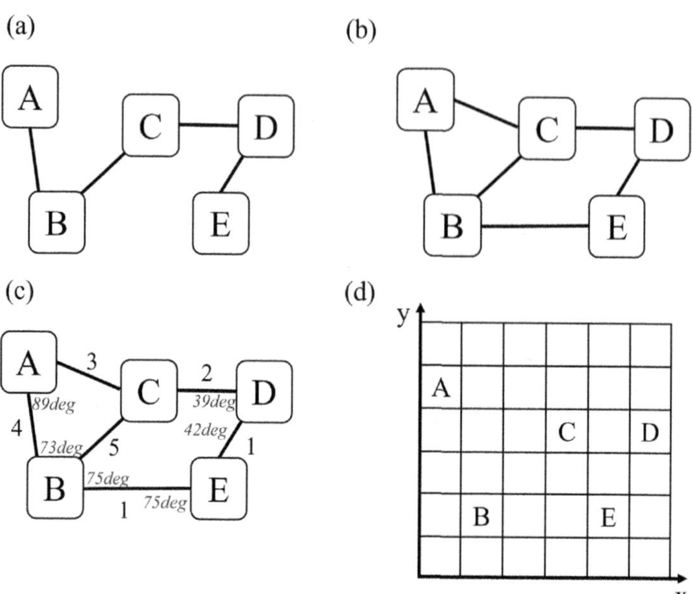

Figure 2.23 Warren (2019) distinguishes four alternative hypotheses to explain spatial representations in animals. While this distinction is conceptually very important, it does not exclude the possibility that these models are not exclusive, that is, minds may harbour more than one type of such spatial representation (based on Warren, 2019).

b *Topological graph*: a place graph where nodes denote places and edges represent paths between them. Notably, this 'place graph' captures only the connectivity between places without embedding them in a coordinate system. While a topological graph enables novel routes and detours by recombining edges in new sequences, it lacks metric distance and angle information, hindering the ability to plan the shortest detour or a novel shortcut.

c *Labelled graph*: consisting of edge weights denoting approximate path lengths and node labels denoting approximate intersection angles. This graph is supplemented by local metric information (e.g., views, landmarks). Despite its lack of precision and geometric inconsistency, a labelled graph can support making shortcuts.

d *Euclidean map*: places correspond to coordinates in a metric coordinate system.

2.5.6.9 Distal and proximate interactions with objects

Besides navigation different types of interaction with objects make up the other part of physical problem solving. There is considerable overlap between the architectures involved in navigation and interaction with objects. Thus, to review the mental architectures for interaction with objects, the framework by Wiener et al. (2011) offers the useful tool (Section 1.6.3).

Sensorimotor system: different sensory systems pick up different features of objects that also determine the range of interactions. Vision is the most widespread sensor to be used but objects can be also detected by acoustic, olfactory, or tactile inputs. Visual information may be also dominant in supporting sensorimotor transformations which is needed for future interaction. A dog watching a ball on the ground may extract two different kinds of information. Vision helps to recognise the object as a ball by its shape, colour, or material but in parallel the same information is utilised to prepare the body for an interaction with it which includes approach and grabbing by the mouth. Such parallel processing followed by interaction of these subsystems is a general pattern of neural networks.

Object primitives: objects possess several physical features that have to be evaluated for efficient interaction with them. At this level mental architectures should be able to deal with tracking of object movement and plan trajectories for approach (to target head on) or avoidance (to evade collision). Goal-seeking behaviour is combined with spatial information to solve detour problems in which the animal has to move away from the goal before eventually reaching it. Object primitives are also engaged in following moving objects and distance regulation. Among other features, differentiation among large and small, recognition of objects being partially occluded, or telling apart animate and inanimate objects is also associated with this level.

Object constructs: constancy, solidity, connectivity, weight, and largeness have to be sensed or inferred by the sensory and the perceptual systems. The nature of such constructs in animals may vary to a large extent because sensors may constrain the information to be gained. Lack or limited seeing ability certainly narrows the possibility to establish complex representations about objects. Multi-modal representations of objects also belong to this level, as well as ability to develop any principles with respect to numerosity.

It is still debated to what extent non-human animals possess abstract representations of these features or their problem solving is based on local rules, which are established after habitual experience collected during interaction with objects.

Object signs: interaction with objects may lead to mental representations (signs) that have a dyadic relationship with their target (Pierce, 1998). Icons show close a physical resemblance to the object to be represented (e.g., picture of a ball), indices indicate some specific connection or relationship with the object (e.g., footprint in the mud), and finally, symbols refer to an arbitrary association, such as word labels of objects. Note that this is a simplification of relationships between the signifier and signified. For example, human spoken words as symbols have a more complex mental function as an artificial sound that may be associated with an object. Nevertheless, there is evidence that the mental architecture in some non-human animals can deal with simpler forms of object signs.

2.5.7 Social problem solving

Ecology describes species interaction (living together) as a symbiosis based on positive, negative, or neutral effects on the partners. For species engaged in a mutualistic

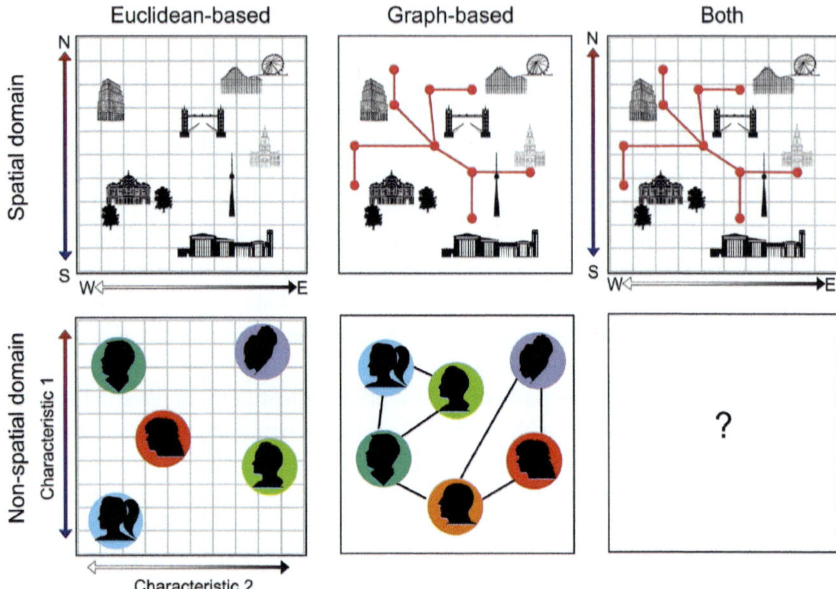

Figure 2.24 Potential similarity between map-based and graph-based mental representations of physical and social space is explored by Peer et al. (2021). In this model, all topographic and labelled graphs (see Box 2.17), as proposed by Warren (2019), are not distinguished. In the spatial domain, knowledge can be purely map-based, with locations coded in terms of Euclidean coordinates (e.g., latitude and longitude), or purely graph-based, where locations are nodes and paths between locations are links. Simultaneously, map- and graph-based representations can coexist, allowing for flexible switching between the two. In non-spatial (social) domains, knowledge is map-based when information is encoded in continuous dimensions, and graph-based when encoded in terms of distinct links between items. For example, individuals in a social group might be represented in a space of personality traits (map-based) (Section 2.4.10) or in terms of social connections (e.g., group hierarchy) within the group (graph-based). Currently, it remains unclear whether a flexible combination of graph- and map-like representations exists in non-spatial domains. (Figure from Peer et al., 2021)

relationship, the interaction has only positive effects. While in the case of commensalism, only one species benefits in the absence of any negative effect on the other. Ethorobots robots (as representatives of another species; Section 2.3.4) are expected to develop a mutualistic relationship with humans. Industrial robots typically share a commensalistic relationship, while military robots could be regarded as predators or parasites from an ecological aspect.

Interestingly, there is a trend to look for commonalities between solving physical and social problems, both at the level of behaviour and architecture. Such overlaps may indicate also the conservative nature of evolution by reusing mental skills of physical problem solving to overcome more complex social problems (Figure 2.24).

2.5.7.1 Ethorobots as social agents

Social robots are often described as agents that are able to verbal and/or non-verbal communication, having animal/human-like appearance, and are equipped with displays or actuators supporting interaction (Prescott & Robillard, 2021), agents that are sociable, evocative, situated and have human-like cognition (Dautenhahn, 2007), or agents that capable for interaction with humans.

There are two main problems with such approaches. First, the term agent is used very loosely and most often is conceptualised from a human point of view (e.g., Jackson & Williams, 2021) and second, definitions are often based on a set of behavioural descriptors lacking a holistic explication of the idea. Our concept of agency is closely connected to the function of an agent, that is, providing a physical solution to a social problem. Social agency should not be regarded as an anthropocentric concept (Jackson & Williams, 2021), but rather it should have general validity for biological entities, and therefore agency should be defined from a biological perspective.

The relationship and contrast between physical and social agency may serve as a good starting point.

1 Robots can rely on various, including unnatural (e.g., using laser) ways to solve physical problems, in contrast the sensory capacities, social skills, and behaviours of a social robot must be in line with and match specific expectations of the other social agent;
2 A socially interactive agent has to capitalise on its physical problem-solving skills, but it deals with a more complex situation due to the autonomous nature of the partner in contrast to non-autonomous physical entities in the environment;
3 A social partner (another agent) can be of any embodiment and nature, and thus its actions are generally less predictable in comparison to physical entities, because it has its own inner states and decision-making systems that rely on different inputs from those of the subject;
4 A certain level and complexity of physical skills are necessary to solve social problems, that is, physical problem-solving behaviour may provide limitations to any social ability displayed;
5 Solutions to social problems critically rest on co-construction among the agents. This requirement adds a new dimension to physical problem solving because the agents have to coordinate their behaviour both physically (e.g., collision avoidance) and 'non-physically' (e.g., communication) to achieve good performance;
6 Some limited linguistic skills are often added to social robots to boost interactive possibilities, but these chatbots are not organic parts of the robot' social power just as if they would run on a computer;
7 When constructing an agent to solve physical problems, the engineer needs to adopt a single perspective, that of the agent. However, when developing agents to tackle social problems, the constructor must consider two perspectives: that of the agent (subject) and that of the partner involved.

Thus, it may be useful to borrow a concept, which was originally applied to the notion of distributed cognition (Johnson, 2001) to define social agency. Accordingly, social agents have the necessary behavioural skills and goals (inner states) to participate in interactions with each other, they regard each other as if they could interact (mutual recognition of being autonomous), and as a result of (some) co-construction, they achieve physical changes in the world that would not be possible when acting alone (Section 1.10). This definition may be perceived as restrictive because 'exchanging' communicative signals on their own are not acknowledged as manifestations of social agency, only if the interaction leads to physical changes, e.g., the robot goes to the door and opens it for a disabled person to get through. However, this view is fully in line with the ethorobotic approach placing the emphasis on measurable behavioural performance of the social agent.

Social agents typically face three types of general problems for which different, but overlapping skills are needed (Section 1.7):

1. Sociability involves various forms of collective behaviour in which the agents establish and maintain group cohesion that provides specific advantages for the individuals, e.g., saving energy when flying in a flock (see also Section 2.6).
2. It is the interest of the social agent to make the other doing an action either immediately or sometime in the future. Communication is the typical behaviour for the former situation (Section 2.6.3), while teaching refers to the latter case (Section 2.6.4).
3. During cooperation, social agents act together to achieve a common goal which would not be possible by themselves (Section 2.6.5).

2.5.7.2 Mental aspects of social actions

It is very important to consider that as social agents act independently form each other, their inner states are different. Thus, when one agent initiates and interacts with another one, it has to put in much effort to achieve a change in the ongoing behaviour of the partner. For example, if a weightlifter aims to lift a heavy weight over his head, he has to be concerned only with its action in relation to the object on the ground. However, if an ice skater lifts his partner, then several additional social aspects may come into play, such as reliability, responsibility, trustfulness, etc. Taken over from human psychology, these attitudes are also considered in social robots (Lazzeri et al., 2013; Liu et al., 2022; Pollini, 2009; Sweeney, 2023). But just as social agency is not restricted to humans, these features of social agents should also be defined in general terms. Importantly, there are many nuances in these linguistic labels that have to be ignored, also because there is also some overlap among them indicating individual or cultural variations in human thinking and reflection. Importantly, each above feature of (human) interactive behaviour contributes to the success of the social interaction and thus to higher level of joint performance. These features may be characteristic of the social agent per se or may be restricted to some aspects of its functioning (context specific). For example, the meaning of being reliable or reflective is very different when used in sport or regarding work in the household.

Here, we propose an operational approach to these concepts, by emphasising one main aspect, not excluding alternative propositions.

Reliability reflects stability in the way of actions carried out despite perturbations from inner states (e.g., tiredness) and states of the environment (e.g., type of partner), including passage of time. Thus, reliable social agents typically execute highly predictable actions in the same manner.

Reflective agents are concerned with the re-evaluation of own or other's actions. The emphasis is on using experience to improve future behaviour and performance, that is, monitoring the difference between actual output and desired output, and making this deviation ('error signal') smaller by relying on feedback mechanisms.

In contrast, *responsive* social agents rely on a kind of feedforward control system, that is, they prefer to predict (possible) outcome of the action and change the systems' actual behaviour to maximise the performance (or minimise the damage) in a preventive way. The recognition of potential consequences and modelling various outcomes *a priori* is a key feature of responsive systems.

A social agent is considered *believable* if (in the long term) the partners making gains (increase their fitness) as a consequence of responding in the expected manner to the agent's behaviour or action. Thus, outputs of a believable agent carry a high level of 'truth' in relation to their inner state and/or outer world, in terms of communication believability is closely connected to being honest.

Persuasiveness of a social agent means that it has wide-ranging skills to influence the other's behaviour by using different, complementary behavioural strategies. This reflects a complex way of acting, including high sensitivity to the partners' possible inner states as well as to the environmental context.

These above skills in concert increase social agency but they may also emerge independently from each other. For example, something can be reliable without being believable (e.g., a consistently flawed source of information), and something can be believable without being reliable (e.g., a one-time convincing but unverifiable claim). Nevertheless, group success crucially depends on both increased reliability and believability. In addition, features including reliability and believability are important for engagement on communicative and collaborative interactions, often described as 'trust', and measured by the joint performance of the participants (Salem et al., 2015b).

2.5.7.3 Developmental aspects of social agency

The developmental trajectory has important effects on social behaviour that have been rarely taken into account when developing the social behaviour capacities of interactive robots. Social behaviours develop gradually from simpler to more complex forms, and there is a close temporal relationship between performing a social action (e.g., pointing with hand and finger) and being able to display and adequate reaction to similar behaviour of the other (e.g., following the direction of the pointing gesture in space) (Gallagher, 2001). Neither aspect is present in social robots because their fully fledged skills are provided by the engineers at once, at the start of their operation.

At least for humans, social behaviour develops in a world ruled by societal constructs such as law, technology, art, etc., all being part of culture in the broader scene.

This dependency on social environmental feedback is growing with the complexity of social abilities and has a huge effect on the manifestation of behaviour. It is an open question of how this aspect of the social environment should concern engineers of social robots. Again, a simpler solution could be to program a robot in a certain way, but this may not reflect the natural situation when the constituent elements of the culture and their functioning is discovered step by step by the individual.

2.5.7.4 Social theories about the mind

Psychology aiming to understand the developmental mechanisms of human social problem-solving skills has come up with several, often non-exclusive models. Anthropocentric approaches, such as mind reading, focus on the mental representations of the other and the self as it seems to be a general consensus that the attribution of mental states to the other is a central feature of human cognitive machinery to tackle social problems. At the moment there are no mindreading robots that parallel human skills, nevertheless a short overview of the present concepts may provide useful insights.

There are many different accounts of how the human mind may operate to predict the behaviour of others. Here, we highlight two main mentalistic concepts in a simplified manner (see e.g., Goldman, 2012, in Merritt, 2021), and contrast them with a more behaviour-oriented view which could be more fruitful for an ethorobotic approach. It should be noted that all such theories are formulated in a child developmental context. Although this is understandable from a human psychological perspective, we have to avoid this anthropocentric approach to producing general models of mind reading.

The *Theory of Mind* (ToM, Leslie et al., 2004) model assumes that the mind has an inbuilt capacity to represent the other's mental state in terms of intentions, desires, beliefs, etc. (Wellman, 2018). This skill is deployed automatically when the subject encounters a social agent by making specific predictions about the other's state of mind. This prediction is the function of the specific behaviour of the other and/or the actual or past context. The main idea is that ToM mechanisms continuously generate assumptions about the partner's behaviour that help the subject to prepare and execute adequate responses. The consequences are evaluated, and the system is updated whether the prediction was proved to be true or false.

Simulation theory (ST; e.g., Goldman, 1989) proposes that subjects use their own knowledge base to predict the other's mental states. After observing the partner behaving somehow in certain situations, the mind runs a simulation based on its own experience as input to arrive at a specific prediction of the state of mind, and then the result of this prediction is attributed to the partner.

Here is a non-exhaustive list of considerations that should make one careful about mentalistic theories:

1 The ToM and ST are often contrasted although both heavily rely on similar folk psychological approaches when explained from a philosophical or psychological perspective. Most textbooks immediately tell stories for the reader to make the

approach feasible. ToM is described as making hypotheses (like a scientist) about the other's mind, while ST puts the subject in the 'partners' shoes'.

2 There are no strong arguments that ToM and ST are mutually exclusive theories, probably a combined mental strategy would bring the best results in predicting the other's mental state.
3 Both ToM and ST must rely on massive experience gained in interaction with the environment, especially with other agents. There has been much effort to push the emergence of ToM to early months of infancy (Slaughter, 2015), but these manifestations also revealed much less complex cognitive power.
4 Given the very restricted possibilities to experiment with human infants, the experience-based developmental contribution to ToM or ST remains obscured. For example, it is difficult to estimate what kind of knowledge is gained under what sort of conditions during the first four years of life when infants start to understand false beliefs (Slaughter, 2015). Very likely any mentalistic skill is based on massive learning experience, starting with many specific default behavioural preferences (Choi & Luo, 2023).
5 Despite immense amount of research little evidence of ToM or ST was found in non-human animals (Krupenye & Call, 2019), despite the fact that, for example, mirror neurons (fundamental features for the ST argument, see Gallese & Goldman, 1998) were first detected in rhesus monkeys.
6 Non-human social animals are engaged in many complex social interactions comparable to that of humans. This suggests that there is a need for a theory that makes the evolutionary transition from non-mentalistic models to the mentalistic ones.

Not denying the importance of mentalistic theories of the social mind, there is at least one alternative possibility of how a social agent can acquire adequate responsiveness to others. Various forms of interaction theories (e.g., De Jaegher et al., 2010; Gallagher, 2001; Trevarthen & Aitken, 2001) propose that mental constructs about the other emerge as a consequence of different kinds of second-person behavioural interactions. Thus, (young) social agents may not excel by having sophisticated theorising or simulating potential; rather, they are built to engage in very complex behavioural interactions based on some preferences in responding, sensitivity to and interest in the other's action (social exploration), and ability to measure the efficacy of the interaction in various scenarios. According to Trevarthen and Aitken (2001), agents perform three types of behaviours: (1) actions regulating body functions; (2) actions with physical objects aiming to take full control; and (3) actions with living objects for co-constructing specific events. The third type of actions/interactions provides the scaffolding for the bottom-up emergence of mental states that may also refer to the other (see also Figure 2.25).

In the initial 18 months of life, significant changes occur in the infant's awareness of other individuals and their motivations for communication, all without language. Various crucial transitions in self and other awareness can be noted at specific ages. These transitions guide the child towards a cooperative interest in actions and objects, as well as cultural learning. The development of behaviour listed in Trevarthen and Aitken (2001): (A) Imitation of expressions. Smiles to voice. (B) Fixates eyes with

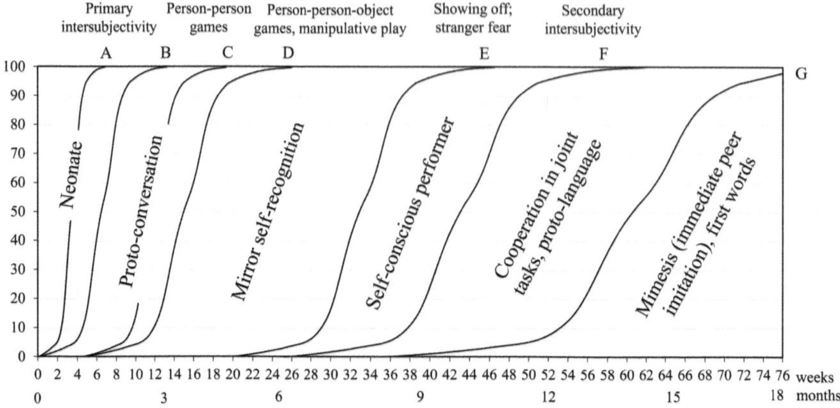

Figure 2.25 Developmental trajectories for human infants based on a summary by Trevarthen and Aitken (2001). Social agency develops in fundamentally different ways in human infants (and possibly in other social animals) compared to social robots. In this period, the rules governing behaviour gradually internalise, as seen in typical autopoietic systems. Note that most developing skills are based on experience gained during interactions with others. In contrast, in robots, social rules are provided by the human engineer in the absence of any social interactions. This disparity in the developmental process likely leads to very distinct forms of social agency.

smiling. Protoconversations. Mouth and tongue imitations give way to vocal and gestural imitations. Distressed by 'still face' test. (C) 'Person-Person' games, mirror recognition. Smooth visual tracking, strong head support. Reaching and catching. (D) Imitation of clapping and pointing. 'Person-Person-Object' games. Accurate reach and grasp. Binocular stereopsis. Manipulative play with objects. Interest in surroundings increases. (E) Playful, self-aware imitating. Showing off. Stranger fear. Persistent manipulation. Babbling and rhythmic banging of objects. Crawling and sitting, pulling up to stand. (F) Cooperation in tasks; follows pointing. Declarations with joint attention. Proto-language. Clowning. Combines objects, executive thinking. Categorises experiences. Walking. (G) Self-feeding with hand. Beginning of mimesis of purposeful actions, uses of tools, and cultural learning. May use first words.

The interaction theory has the advantage that it provides a common platform for human and non-human social animals (Merritt, 2021), in addition it can also deal with the problem of how social agents of different species may establish a common mentalistic framework. Such situations occur between owners and their animal companions (e.g., dogs and cats) or between humans and social robots of various embodiments and behavioural skills.

2.5.7.5 *A note on linguistic skills of ethorobots*

In modern societies, language is the main form of communication because much of human interaction is about exchanging 'information'. Over the last 50 years, technology has developed a range of solutions to produce artificial agents that can either use language to provide information (e.g., automatised public address systems) or act as conversational partners for humans (e.g., Alexa, ChatGPT). Although such systems are very useful and help people in various ways, there are a few reasons why their integration into ethorobots can be problematic (see also Prescott & Robillard, 2021).

1. CAs do not require any specific embodiment, mobile phones, laptops, or simple 'loudspeakers' can be equipped with such functions;
2. CA's main role is to provide a kind of 'inquiry service', helping the human to get the needed information rapidly and in a natural way (e.g., instead of searching manually on the Internet);
3. By only listening, a well-constructed CA can be mistaken for a human that confuses people. Thus, producers often have to ensure that the user is aware that he speaks to a machine;
4. CAs are better in production than in perception, and thus communication with humans is asymmetric. This often requires much effort by the human that makes this way of interaction often quite challenging;
5. From an ethological point of view, language is species-specific human behaviour, and only humans possess this communicative tool. Thus, adding this feature to ethorobots with various embodiments can confuse the human partner;
6. CAs can provide information using huge databases. However, their contextual awareness is very poor, that is, their verbal knowledge is not integrated with their knowledge about themselves, their actual embodiment, and environment/context.

Thus, any specific embodiment provides limited advantages for a CA at present. Although it is likely that social robots will be endowed with linguistic skills, in the not-too-distant future, it is less clear how actually, this will happen.

2.5.8 Conclusions, prospects, and questions

Ethological and psychological research can offer valuable insights for ethorobotics, particularly when these principles are de-anthropomorphised. This process can be achieved through three key approaches. First, it is essential to establish precise operational definitions that are not overly reliant on psychological, cultural, or philosophical interpretations of the concepts. For instance, even within the realm of social robotics literature, there often exists confusion regarding the distinction between 'emotion' and 'personality', or interchangeable terms are employed when discussing social cognitive tendencies such as trust and believability. Similar to how definitions are established for the physical components of robots, there is a need to have equally clear operational description of units or processes of the mental architecture.

Second, interaction with the environment provides a realistic scenario for living or, non-living agents to develop beneficial behavioural strategies. Therefore, researchers should dedicate attention to a more intricate measurement of these phenomena, especially how living agents perform outside the laboratory, in real life. Contemporary sensory technology complemented with AI-based data processing provides enhanced tools for this purpose; however, researchers have to anticipate manifold increase in the scale of acquired data. This appears inevitable for those who want to go beyond theorising about mental functioning based on small data sets. The development of newer social robots should be community effort among researchers with different background and people who are the end users of the product. There is a need for co-design or expert-based co-design based on an iterative and interacting process including problem identification, technical implementation deployment, and testing (Winkle et al., 2021).

The quantitative measurement of robot behaviour should become a standardised procedure. However, this is useful only in the case of autonomous robots. Instead of relying solely on a simple description of the robots' behaviour by the human partner or their responses to questionnaires, the obtained data should offer both an objective third-person view of the robot's actions (behavioural coding) and data acquired directly from the robot's sensors.

Third, we have to take seriously that humans are not anomalies but one evolved species, among many millions. This underscores the fact that theories or models exclusive to humans are destined to fall short in comprehending the essence of the human mind. For understanding the human mind, we need common, unifying models encompassing both human and non-human cognitive processes to become essential. These are also needed to build successful ethorobots 'from scratch'.

2.6 Social behaviour functions in robots

In psychology, there is some ambiguity regarding how social relationships are defined (Reis et al., 2000). From the ethorobotic perspective, the foundation of any relationship lies in repeated physical ('face to face') encounters in which agents recognise each other as individuals, and they modify each other's behaviour influenced also by their previous encounters. Relationships cease if specific gaps occur in the temporal pattern of contact.

In addition, the individualised aspect of the relationship is significant. This has been taken for granted in psychological definitions, but this is not an evident feature in non-human agents. If a social robot lacks this capacity, then it may not be able to engage in true relationships with people, despite its capacity for social interaction.

Following Reis et al. (2000), it is also important to note that social agents are by definition part of at least one kind of relationship within their group, but typically social relationships have a hierarchical and also nested structure. This means that one agent can be engaged in many, different types of relationships in parallel, but non-human animals share typically relationships only within their actual group. Humans are exceptional because they can be members of several groups at the same time (family, workplace, club, etc.). It may be interesting to note that companion animals (e.g., dogs and horses) commonly belong to two distinct groups: (1) the human family providing care for them and (2) a group of conspecifics with whom they also spend time (e.g., dog training schools and parks; horse herds). Ethorobots may also belong to this category as they could develop individual relationships with both each other and human partners.

2.6.1 Robots in the human social network

Numerous theories are dedicated to projecting the nature and quality of human–robot relationships (e.g., Prescott & Robillard, 2021). An ethological or psychological examination would be meaningful only if human–robot relationship were much more widespread, providing comprehensive data, including variations. Until such time, it remains feasible to outline a theoretical framework concerning the potential configuration of HRI on a societal level.

Ethorobots have the potential to integrate into diverse human social groups like families, schools, and workplaces. Yet, any bond formed with humans is determined by the fact that they represent another species. Consequently, these relationships inherently bear an asymmetric nature, wherein the role of leader predominantly falls upon humans (Section 2.6.5), while ethorobots might occasionally assume responsibilities of leading the course of actions.

This scenario does not preclude the possibility of humans and robots developing intricate interactions that mirror those transpiring among humans (Section 1.7.1, Box 1.4). This implies that humans and robots could establish attachment bonds, form teacher–pupil dyads, or engage in various collaborative actions (Section 2.6). Their social relationships might encompass a spectrum ranging from despotic to egalitarian; however, given the overarching role of ethorobots, a predisposition towards more mutually beneficial connections should prevail (Danaher, 2020).

Navigating the intricate social network of humans remains a formidable challenge for social robots. While these scenarios may not materialise in the immediate future, it is worthwhile to contemplate the potential issues that might arise when, for example, a social robot with teaching duties has to establish and maintain relationships within a school setting. The standing and reputation of such a teaching-oriented social robot could hinge significantly upon its rapport with various community members, including human teachers, pupils, parents, and support staff. These robots necessitate a flexible and adaptable repertoire of social skills. They should have clear objectives for cultivating distinct relationships with members of diverse subgroups, while also demonstrating the capacity to gain insights from their interaction experiences.

Behaviours employed for attachment, communication, and cooperation can be capitalised upon to establish and maintain a social network of relationships with humans provided they are controlled by an appropriately structured mental architecture (Haslam, 1994). The successful integration of dogs within human societies suggests that this task is not impossible for another species (Miklósi, 2015).

Social networks may exhibit common features with spatial networks (also conceptualised as network of routes or maps) (Peer et al 2021). The similarity in mental organisation might also find support in evolutionary theory, suggesting that the ability for complex spatial representations could have emerged earlier than the skills needed to navigate intricate social structures. By repurposing (exaptation) the capacity for spatial representation to map social relationships, there may have been a limited need to evolve novel neural structures and processes.

2.6.2 Attachment

Attachment is a core concept in human social relationships (Bowlby, 1969), but this was not realised until recently in social robotics (Rabb et al., 2022). While in the case of humans, the attachment has been conceptualised in the framework of mother–infant interaction, for ethorobotics, a more general approach is needed since in HRI there can be very different manifestations of attachment. Importantly for a general attachment theory, the provider (a general synonym for attachment figure/caretaker) and recipient (similar role to that of the infant/offspring) are used to refer to the roles of the two individuals. Note that there are many cases where the attachment is symmetric, that is, both partners are mutual providers and receivers (e.g. romantic relationship in adult humans)

Box 2.19 The Strange Situation Test

The test was designed to expose the recipient to mild stress by observing their behaviour in an unfamiliar place (an experimental room) and exposing them to a friendly yet unfamiliar person (the stranger). This potentially dangerous context is essential for activating the attachment system. The behaviour of both the provider and the stranger

Table 2.13 Description of the episodes of the Strange Situation Test

Episode	*Present in the room*	*Scenario description*
0	Provider and recipient	Provider and recipient are introduced to the experimental room
1	Provider and recipient	Provider and recipient are alone. The provider sits in a chair, next he carries objects from one place to another, the recipient is free to do anything
2	Provider and recipient and stranger	Stranger enters, interacts (talking) with the provider, then approaches recipient to interact (play/talking); Provider leaves conspicuously
3	Recipient and stranger	Stranger initiates interaction (play/talking) with the recipient, stops if subsequent attempts are not reciprocated; Stranger leaves conspicuously
4	Provider and recipient	Provider enters, interacts (talking) with the recipient; Provider leaves conspicuously
5	Recipient (alone)	The recipient is free to do anything
6	Provider and recipient and stranger	Stranger initiates interaction (play/talking) with the recipient, stops if subsequent attempts are not reciprocated; Stranger leaves conspicuously
7	Provider and recipient	Provider enters, interacts (talking) with the recipient; Provider and recipient leave

is semi-determined, but the recipient can freely interact with either persons or the environment. Both the provider and the stranger should exhibit comparable interactive behaviour during the test and remain passive after entering to observe the recipient's greeting. It is beneficial to have two separate doors for the provider and the stranger: this facilitates to determination of recipient's preference when attempting to leave the room (Table 2.13).

It is important to note that attachment between two partners is partly the result of their past interactions. Therefore, the attachment style is influenced by the provider's general behaviour. Recipients can develop different attachments to various individuals within the social group.

Attachment is a behavioural system that (1) regulates distance ('proximity seeking') between two specific individuals (provider and recipient), in addition to the (2) activation of positive communicative displays upon encountering. The attachment system (3) activates search behaviour in the case of the partner's incidental disappearance ('separation behaviour'), (4) maintains regular (physical) contact ('secure base') and (5) provides protection in danger ('safe haven') (Ainsworth et al., 1978) (Box 2.19).

The main *function* of attachment system is to ensure that social beings are not alone while they explore their environment. The physical proximity also supports that individuals can learn from each other and provide emotional support if needed. Naturally, attachment systems are especially important in the case of mother–infant relationships because the offspring is most vulnerable to predators and lacks experience about its physical and social environment. However, attachment also emerges between human adults in a romantic relationship, or between friends (Hazan & Shaver, 1987). Research has provided evidence for mutual attachment between dogs and their owners (Topál et al., 1998). This indicates that an attachment relationship may emerge and persist between living beings and artificial agents, such as ethorobots (Rabb et al., 2022) (see also Box 3.24).

The attachment system works differently between specific dyads because it depends on the actual needs of partners, their individual differences, including personality and the experience they gain during this relationship. This affects the balance between exploring the environment and being on their own and spending time in close social contact. In line with this, developmental psychology identified different types or styles of attachment between mother and infant (Box 2.20), which may have a long-term effect on the social behaviour of children as they grow up (e.g., Zimmermann et al., 2001).

The attachment system gets activated when two social agents (provider and recipient) start to develop an individualised relationship that has functional significance for both of them. In living organisms, specific morphological and behavioural signals and specific hormonal changes (e.g., when having an offspring, entering a monogamous relationship) play an important role. Depending on the species, attachment relationship may develop within a few hours (e.g., precocial species) or weeks/months (e.g., altricial species).

Box 2.20 The functioning of the attachment system in living and artificial agents

Attachment is a key attribute of many social beings, and ethorobots should possess this capability to function effectively as human companions. The attachment system's operation has been elucidated in various species, such as humans and dogs. From an ethorobotic standpoint, it is important to de-anthropomorphise this phenomenon, using terminology that reflects the agent's behaviour rather than presuming specific emotional states. Additionally, terms with potentially negative connotations should be avoided.

Table 2.14 summarises three primary attachment styles relevant to ethorobots. Originally introduced by Ainsworth et al. (1978) to categorise parent–infant relationships, this model needs some modification to make it more general. References to 'mother', 'caretaker', and 'owner' are replaced with 'provider', and 'child', 'partner', and 'dog'

Table 2.14 Summary of three attachment styles

	Secure – Teaming Balanced	*Unsecure – Avoidant Unbalanced*	*Unsecure – Ambivalent Unbalanced*
Distance regulation (Secure base)	Alternating movement between provider and environment depending on environmental input	Preference for exploration in the environment	Preference for provider over the environment
Proximity seeking	Time in physical contact with provider	Little time in physical contact with provider	Time in physical contact with provider
Choice of protection (Safe haven)	Proximity to provider in case of danger	No preference for the provider in case of danger	Proximity to provider in case of danger
Separation	Regulated activities to escape/reunite	Rare attempts to escape/reunite	Unregulated and forceful activities to escape/reunite
Reestablishment of social contact	Regulated activities and proximity seeking during greeting	No proximity seeking during greeting	Alternation between proximity seeking and avoidance during greeting

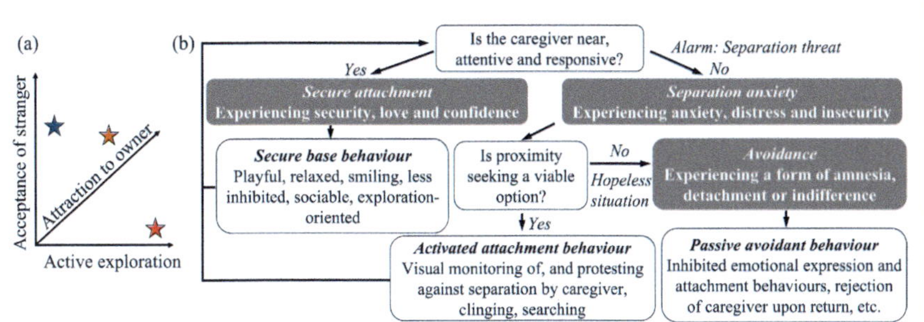

Figure 2.26 (a) A dimensional model of attachment based on three key behavioural factors: exploration of the environment, acceptance of stranger, attraction to owner ('attachment') (Topál et al., 1998). This model was developed by the means of a factor analysis based on behavioural observations and subsequent assessments. Stars in the 3D space represent individuals who achieved the relevant scores for the three behavioural variables in the Strange Situation Test (b) A control theory model of the attachment system, based on Bowlby (1969) and Ainsworth et al. (1978).

are substituted with the term 'recipient'. Operationally, it might be more practical to use 'balanced' and 'unbalanced' instead of 'secure' and 'insecure' concerning the time budget spent close or away from the provider. The left column outlines the main sub-functions and contexts of attachment contributing to the recipient's survival and the provider's increased fitness. The cells of Table 2.14 contain brief, simplified, and qualitative behavioural descriptions towards the provider specific to each attachment style. These aspects of the behaviour are predominantly scrutinised for categorisation of the attachment.

In practice, the balanced-secure (teaming) and unbalanced-avoidant styles could be effectively applied to ethorobots. The former aligns with a typical manifestation of attachment style, while the latter might be suitable for ethorobots with a more distant or official relationship with humans (e.g., office clerks, security guards, and cleaning staff) (Figure 2.26).

2.6.2.1 Ethorobots and attachment

Endowed with the corresponding skills, an ethorobot can be both provider and recipient in an attachment relationship with humans. This may depend on the function or aim of the interaction, for example, as providers ethorobots can offer safety for humans in need, but human can also teach unexperienced recipient ethorobots novel skills. Providers and recipients should also differ in their look and or modify their behaviour to elicit the expected behaviour from the partner. For example, providers should indicate that they are able to protect, if necessary, by being skilful, assertive, and determined, while recipients should indicate the need for help and support if being alone or facing novel and or frightening situations.

Rabb et al. (2022) have reviewed attachment theory to build ethorobots, which can act as an attachment provider. Similarly, to the above concept, they emphasise the importance of the secure base, safe haven, proximity seeking, and separation behaviour. Importantly, they also summarise by providing some examples, what kind of problems can such ethorobots solve when they act as providers for humans. At present, no social robot is able to fulfil all functions, but in principle, current technology would allow for combining all capabilities listed below (see Collins et al., 2013).

Secure base:

1 Physical needs: ethorobots can help people with physical problems, for example, assist walking of elderly (e.g., Broekens et al., 2009);
2 Intellectual needs: ethorobots can educate, teach their human partners for learning new skills (Belpaeme et al., 2018);
3 Connection to others: ethorobots could be topic of conversation among people, for example, in the case of robot-assisted therapies with Paro (Cruz-Sandoval et al., 2020);
4 Dependable availability: ethorobots could provide some security functions ('being present') for people living alone (Vulpe et al., 2021).

Safe haven:

1 Palliate physical distress: ethorobots may engage in touching ('hugging') interactions with humans indicating social closeness and physical support (Shiomi et al., 2017);
2 Palliate mental/emotional distress: ethorobots could display both empathy and sympathy towards their human users when the latter is in need to share negative affect (Kahn et al., 2012).

It is also important that any kind of attachment is based on mutual individual recognition. The provider and the recipient need this ability not only for distance regulation per se but also for differentiating their specific relationship from that formed towards other partners because both partners can be part of several attachment relationships in parallel. Importantly, to achieve such individualised relationships, the partners must go beyond recognising each other specific morphological and behaviour features. Mutual adjustment to each other's actions, shared experiences, and specific memories are all essential components of an attachment relationship that by the nature of these processes needs a certain amount of time to develop (Rabb et al., 2022).

Rabb et al. (2022) also introduce the concept of attachment spectrum, defining weak and strong versions of attachment. Strong attachment emerges when all four main components are functioning, like in the case of human attachment. Weak attachments are characterised by not meeting all criteria. This view may be a shorthand to differentiate attachment systems, but it also gives room for subjective interpretations and systems with different functions could be described by weak attachment (Konok et al., 2015) and could be confused with attachment styles (see Box 2.20). It is more useful to specify what of the four main functions are working in a specific social relationship, and if necessary, one may refer to incomplete versus complete attachment.

As noted above, ethorobots can also play the role of the recipient in an attachment relation when they lack some functionalities and may need to rely on (occasional) human help or are deficient of some knowledge that can be provided by humans (Hiolle et al., 2012; Oudeyer et al., 2005: playground experiment). Hiolle et al. (2014) provide an experimental approach to show the viability of this concept and argue that such kind of interactions not only widen the robots' acting range but also enhance the emotional experience of humans. Both in laboratory observations and experiments in a natural situation they could show that the social robots' performance improved when they acted as recipients in an interaction with humans. In a between-subject arrangement, humans were allowed to interact with a 'needy robot', which solicited regular attention (by looking at the human and vocalising) and spend more time close to the human or an 'independent robot', which showed these actions at a lower frequency and spend more time away from the human. Thus, the two robots differed in the balance of their exploratory and social tendencies.

The results showed that an optimal frequency of 'needy behaviour' helped the robot to learn during the interaction. Humans recognised the differences between the two robots by describing the 'needy robot' as one needing more help and being less independent, and also interacted with it by displaying more positivity and engagement. Thus, relatively small differences in the robots' behaviour were effective in changing human attitudes towards them. There are many situations when at the start robots may lack specific skills and human assistance and teaching may speed up the learning process in comparison to individual learning. Playing the recipient's role of an attachment relationship could be an important aspect for developmental robotics (Cangelosi & Schlesinger, 2015).

Ethorobots could also differ in their attachment styles towards their partners (see also Box 2.20). While the typical and most common pattern, labelled as secure attachment indicates a balance between exploration tendencies and a social proximity and also possible tendencies for cooperation ('teaming'), other possible styles, described as avoidant attachment is characterised by a strong imbalance towards heightened interest in the physical environment (Ainsworth et al., 1978). This later style could be seen as similar to the 'independent' robot suggested by Hiolle et al. (2014). Manipulating the ethorobot's attachment style could influence the effect on the human partner, for example, separation from an avoidant/independent robot would be less emotionally challenging. The attachment style of the robot may be also adjusted in relation to its actual function. Different attachment styles could be useful when interacting with children or playing the role of a receptionist.

2.6.2.2 Prosocial aspects of attachment

A diverse array of behaviours is employed by individuals as a means to activate the attachment system and maintain social contact during interactions. These affiliative or prosocial behaviours leverage various other mental mechanisms, including communication and imitation.

In humans, the dynamic use of *eye contact*, typically lasting no more than three to four seconds, signifies social preference. Individuals displaying such gazing patterns are perceived favourably by others (Binetti et al., 2016). Humans also establish eye contact with social robots, and there exists neurophysiological evidence suggesting that such social gaze could enhance people's preference for such agents (Kompatsiari et al., 2021).

Emotional contagion occurs when one partner's emotional state aligns with that of another. This synchronisation of emotions necessitates mental investment due to the differences in the partners' initial emotional states. Such skills in ethorobots could have a positive impact on HRI (Xu et al., 2015).

Mutual mimicry (chameleon effect) emerges when two partners display similar postures or facial expressions during interaction, effectively mirroring each other's actions. This synchronisation of behaviour typically signifies an increased affinity for one another (Lakin et al., 2003). Such mimicry also enhances the acceptance of social robots (Burns et al., 2018).

Matching preferences occur when partners share similar interests in objects, like food, or other aspects of the environment. Evidence from studies involving human infants suggests that individuals tend to favour agents who share their own preferences (Choi & Luo, 2023). Analogously, the manifestation of similar behaviour by ethorobots could yield comparable effects.

Many of these behaviours can be implemented in contemporary ethorobots, regardless of their physical embodiment. Whether in isolation or in concert, such abilities have the potential to enhance human-robot partnerships, particularly in situations where humans and robots interact promptly without an extensive prior acquaintance, such as in the context of service robots in retail (Chuah & Yu, 2021).

2.6.2.3 Challenges of attachment in HRI

One must be aware that humans may develop spontaneously strong bonds towards ethorobots showing all attachment criteria. This could be a problem if the robotic agent may not be able to play the role of a partner in long term. Any deployment of such ethorobots should be planned in advance carefully and with caution (Yamazaki et al., 2023).

Thus, it is important that ethorobots should not replace natural attachment relationships among humans. The end of an attachment relationship with a robot should be prepared by gradual increase of the time spent apart from the robot, and in parallel alternative (human) partnerships should be offered (if needed). This process could be more difficult with ethorobots showing complete attachment behaviours in comparison to robots with incomplete attachment systems.

2.6.3 Communication

Contemporary research in HRI predominantly centres around a combination of psychological (non-biological) and engineering interpretations of human communication. Regrettably, this approach tends to overlook a substantial body of biological research (see Kunold & Onnasch, 2022). Psychological approaches focusing on the role of human behaviour within communicative interactions, often assume that any action performed by humans constitutes an act of communication. This perspective is epitomised by the widely cited statement by Watzlawick (1967), 'One cannot not communicate', which has gained significant attention (e.g., Maniscalco et al., 2022). However, this assertion is demonstrably overstated and erroneous (see Section 1.8.1 for a formal communication definition). For instance, fleeing or concealing oneself from an attacking lion is not a signal but rather a survival-oriented action. On the contrary, running way (in a teasing way) during play could serve as a signal to encourage the partner's pursuit. The assumption made by an observer that the prey of the lion experiences fear is an instance of eavesdropping, involving the observation of an interaction between two individuals, rather than a case of communication. Therefore, a more precise yet less sensational interpretation would be that, at the very least in the context of humans, all actions possess the potential to acquire communicative functionality.

A theoretical challenge surfaces when certain researchers differentiate between 'intrinsic' and 'extrinsic' communication. They propose that the former occurs when an agent must draw an 'inference' from the actions of others, whereas in the latter scenario, there is 'direct' communication through speech or specific gestures (e.g., Gildert et al., 2018). According to this perspective, any action one does fall either into the 'intrinsic' or 'extrinsic' category of communication, following Watzlawick's logic. However, as argued earlier, 'intrinsic' communication should be excluded from communicative interactions based on the ethological definition of this behaviour.

On the contrary, the engineering perspective relies heavily on the sender–receiver paradigm, accentuating the concept of 'information exchange' (Kunold & Onnasch, 2022). However, this viewpoint falls short when it comes to elucidating the fundamental purpose of communication and frequently overlooks the intricate interplay between 'message' and 'meaning' within the specific context of communication (see Section 1.8). In order to facilitate biologically grounded communication, robots must possess situational awareness and a spectrum of capabilities. These capabilities are essential to ensure that the conveyed message effectively reaches the intended receiver, and that the receiver has an enhanced chance to decode the meaning of the signal (Box 2.21)

Unfortunately, these nuanced aspects of communication often go unnoticed in HRI. In many cases, communication is confined to methods like texting or basic verbal exchanges, such as those when interacting with a CA via a keyboard.

2.6.3.1 Robots as conversational agents

CAs allow the human user to interact with it using natural language. The widespread use of AI allows for various types of CAs that can provide substantial help to the human user. Such systems can be implemented on any kind of hardware, but they have not been integrated with other behavioural systems and awareness capabilities of robots. Eventually, ethorobots may develop into hybrid agents that have linguistic capabilities despite being non-humans. Thus, the user may imagine that there is a 'person in the machine', even though there is not in reality. To avoid this situation, ethorobots may develop a specific way of talking (robotic dialect) so that they can be always distinguished from real people. It will be an experimental and economic question to what extent and in what form people consider these technologies as useful. Given many uncertainties at these times, the issue of ethorobots as CAs seems to be outside the interest of this book.

2.6.3.2 Ethorobots as heterospecific agents

This chapter has introduced many arguments for regarding ethorobots as heterospecific agents, despite all efforts to make them similar to humans, as much as possible (Section 2.3). Importantly, this situation does not constrain communication; rather, it decreases competition between people and robots. This context also provides a better understanding of the challenges developing successful communication between humans and ethorobots when partners have to invest in the interaction both from a design and temporal point of view.

Humans have a natural way of communicating and they do not change it, at least not without specific reason, when interacting with other agents, including, for example, companion animals (Miklósi et al., 2017). Thus, for engaging in successful communication, ethorobots need to be endowed with a specific set of skills when playing the role of either the sender or the receiver (see Box 2.21).

Box 2.21 The communicative cycle model

Effective communication between agents often involves intricate mental processes aimed at maximising the efficiency of the system by aligning the 'message' with its intended 'meaning'. The communicative cycle model seeks to elucidate this dynamic. Figure 2.27 illustrates the interaction between the sender and the receiver, recognising that an agent may possess both systems, which can function collaboratively.

Consider a scenario where a dog (the sender) spots a piece of food on a shelf out of its reach. Having received help from its owner before, the dog decides to involve the owner, who is reading in the same room. Initiating the signal occurs because the Inner State T0 value differs ('is lower') from the Inner Goal State (obtaining the food). The 'Attention State Detector' determines if the owner is watching, while the 'Context State Detector' provides information about the surroundings (e.g., presence of others or food location). These detectors influence the signal's form and direction. The dog looks at the owner and barks (Signalling System). If the owner remains engrossed in the newspaper, the Action Detector (owner does not move) and Body State Detector (no food in the mouth) provide no input, maintaining Inner State T1's difference from the Goal State. The signalling cycle

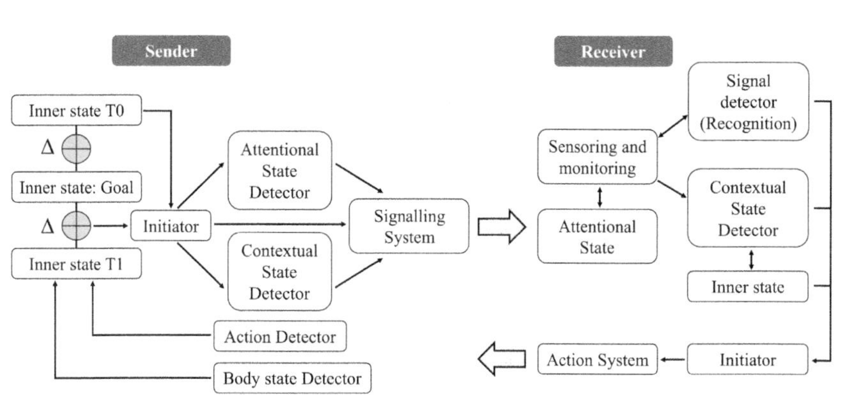

Figure 2.27 The communicative cycle model (see text for explanation).

is repeated based on inputs from the 'Attention State Detector' and 'Context State Detector'. If the owner eventually attends to the shelf, the Action Detector notes a change, increasing Inner State T1 based on past experience. However, the absence of input from the Body State Detector means signalling continues. When the owner provides the food, communication ceases as Inner State T1 aligns with the Goal state.

Despite the owner (receiver) reading the newspaper, she remains aware of her environment. Noticing the approaching dog, she shifts her attention and may recognise the signal. To decide on an action, the owner considers input from the Contextual State Detector (e.g., the dog's last meal) and Inner State (e.g., finishing the page). Eventually, the owner may act in response to the dog's signal.

While not all communicative interactions are as complex, social agents need to be prepared for such cases. Analogously, similar situations arise among clients in a network where data transmission fidelity is crucial. The Transmission Control Protocol (TCP), a primary Internet protocol, employs multiple mechanisms to mitigate errors. For instance, (1) the receiver sends acknowledgments (ACK) to confirm data receipt; if no ACK is received within a set time, the sender assumes unsuccessful delivery and retransmits the data. (2) TCP implements flow control to prevent data overload, using windowing mechanisms to regulate transfer rates and prevent congestion. (3) Additionally, TCP employs a checksum mechanism to detect errors; if a segment is corrupted, it is discarded, and the receiver may request retransmission.

2.6.3.3 *Ethorobots as senders*

Drawing from the ethological definition (Section 1.8.1), the formulation of communication design should inherently favour the sender. This bias towards the sender arises from the sender's imperative to get greater benefits by initiating the interaction. In addition to emitting the most contextually suitable signal, the sender must also guarantee that the receiver's behaviour is subsequently altered in the anticipated manner. Thus, communication rests on at least the following abilities:

1 Detecting the attentional state of the receiver and change it if necessary;
2 Choosing the most effective signals based on the message and the environmental conditions (increase signal/noise ratio);
3 Detecting the recipients' action;
4 Evaluating the outcome of the signalling (success/failure);
5 Executing corrective actions in the case of failure.

2.6.3.4 Ethorobots as receivers

In common scenarios, the receiver might not possess foreknowledge of an impending communication event, and there might not be any advantageous outcome for the receiver resulting from the interaction. Consequently, it becomes important to determine whether the receiver should respond and, if so, the appropriate manner in which to do so. The key skills on the receiver's side are the following:

1 Directing attention to the partner (c.f., selective attention);
2 Processing the signal and its meaning;
3 Checking the accuracy and honesty of the signal;
4 Deciding to send a signal as a response or not is based on the actual inner state including memory about previous interactions.

Although an extensive body of literature focuses on human–robot communication, the presence of the aforementioned skills is seldom examined. It is also widely accepted that no single social robot currently possesses the ability to exhibit all of these attributes (Kunold & Onnasch, 2022). Notably absent are comprehensive research studies delving into the success of intricate communicative interactions occurring in real-world scenarios between widely used socially interactive robots (such as NAO and Pepper) and individuals. A recent review on NAO (Amirova et al., 2021, see also Section 4.2) underscores numerous potential applications for the robot, despite its lack of most functionalities enumerated above. However, many interactions mentioned in the review do not truly qualify as instances of communication; for example, activities like playing football with children or engaging in storytelling.

2.6.3.5 The role of eye contact in communication

While there remains a scarcity of complex studies in HRI that integrate both verbal and non-verbal cues in natural communication, there exists a wealth of controlled observations focusing on the utilisation of eye contact between robots and humans. Capturing and sustaining human attention remains a significant challenge for ethorobots (Das et al., 2015). In social interactions, eye contact generally serves two primary functions. Analogous to many other mammals, extended and unwavering eye contact, often referred to as 'staring', is indicative of a potential threat in humans. Brief and recurrent glances typically signify interest and/or shared attention (as discussed above).

It is important to note, however, that the mere establishment of eye contact does not inherently constitute a communicative interaction, nor does the act of observing where a partner is looking. Eye contact becomes a signalling mechanism only when the sender employs it in a tactical manner to enhance the effectiveness of subsequent signals. In other words, eye contact becomes a signal when it is followed by another form of communication. The subsequent examples help illustrate this distinction:

Eye contact as a communicative action: in one study, the iCub robot directed its gaze towards the eyes of human participants before subsequently turning its head to the left or right. In a control condition, the iCub's gaze was directed downward towards a table, eliminating any eye contact. Notably, participants focused their attention on the robot for a longer duration during the eye contact condition. Moreover, they reacted more swiftly to the gaze shift to the left or right, but only when it was preceded by eye contact (Kompatsiari et al., 2019, for similar observations between humans and dogs, see Téglási et al., 2012). Adopting a communicative standpoint, the establishment of joint attention through eye contact increased the iCub's (as the sender) likelihood of having its gaze shift tracked by the human observer. Consequently, the robot was able to effectively guide the observer's attention to a specific location in space.

Eye contact as a non-communicative action: in game-like interactions, humans had to find out what object a NAO robot was 'thinking' about. While the robot could be queried about certain features of the objects, a distinct behaviour emerged during some of these interactions: the robot would direct a short gaze towards one of the target objects. This behaviour was termed 'leakage' (Mutlu et al., 2019). The researchers set out to investigate whether human participants could deduce the object more quickly when the NAO robot made eye contact with them. Overall, the robot's glancing action indeed expedited participants' accurate guesses. However, it is important to clarify that this interaction does not amount to communication; rather, it constitutes a case of social interaction where the observer benefits from interpreting the other's gazing behaviours. It is essential to note that signals are never 'leaked', as per their definition: they are conveyed by the sender to elicit a beneficial change in the receiver's behaviour.

Endowing robots with the ability to convey emotional signals (Section 2.5.5) often serves as a means to enhance their accessibility and acceptability among humans. Such capabilities also play a pivotal role in facilitating communication and fostering social connections (Breazeal & Brooks, 2005). Emotional signals effectively impart a heightened sense of liveliness to social robots. Nevertheless, the pertinent question revolves around the sender's benefit to disclose its inner states, and more fundamentally, whether a robot can have 'true' emotional states at all.

Consistent with the aforementioned, ethorobots should manifest emotional behaviours when they intend to prompt changes in human behaviour. The sender's benefits are often apparent in some scenarios. For instance, an ethorobot expressing anger might dissuade its partner, whereas signals of sadness could evoke comforting responses and forgiveness. But why would a robot experience happiness or display disgust? The former could contribute to fostering stronger relationships, particularly in collaborative settings, while the latter might serve the purpose of safeguarding others from adverse experiences.

2.6.3.6 Design of embodied signalling systems

As clarified above, not all characteristics are employed for communication, but both morphological ('static') and behavioural ('dynamic') traits can serve as signals based on their evolutionary history, that is, having the potential to be utilised in communication. This crucial understanding is frequently overlooked in the robotic design, as there is often a lack of awareness that the body or its components can be interpreted as signals. For example, placing a relatively large head on a small body is a typical signal for being a young mammal, and other morphological features of a head add to this effect (Hecht & Horowitz, 2015).

The ethorobotic approach also acknowledges the fact that numerous physical or behavioural signals are shared across different animal species due to evolutionary homologies or convergent processes (Section 1.8). While the use of high-pitched speech ('motherese') towards infants was initially observed in humans, recent research has unveiled that humans employ a similar form of signalling when interacting with dogs ('doggerel') (Gergely et al., 2017). Additionally, observations have shown that dolphin mothers emit higher-pitched calls while interacting with their calves (Sayigh et al., 2023).

The arguments presented by de Gelder (2009) can also serve as a foundation to underscore the significance of utilising the entire body as a signalling system in communicative interactions.

1 Faces represent a very narrow display for communicative signals;
2 The generality of bodily signals makes them less sensitive to cultural differences;
3 Including bodily signals can make signalling more graded, thus helping in displaying smaller differences in emotional states;
4 Signals including whole body could increase transmission efficiency by signal redundancy and by increased detection from larger distances;
5 Bodily signals could be also processed in the absence of the receiver's full attention;
6 Bodily signals may also help the receiver to note the context of the signalling.

Box 2.22 Conveying emotional signals

Breazeal (2002) created a zoomorphic robotic head named Kismet, designed to convey communicative signals. Breazeal's model garnered significant attention, however, despite its simplicity and expressiveness in signalling emotional states, it has found limited application in social robotics. The table illustrates the arousal, stance, and valence states (low, medium, high) displayed on the robot head (see also Figures 2.28 and 2.20). This concept can be extended to various embodiments, including the robot's body itself. Employing 3D modelling of the emotional space enables the representation of 'transient' or 'ambiguous' emotions with intermediate values, enhancing the agent's animacy. Introducing asymmetric displays can further enhance the naturalness of conveying emotional states.

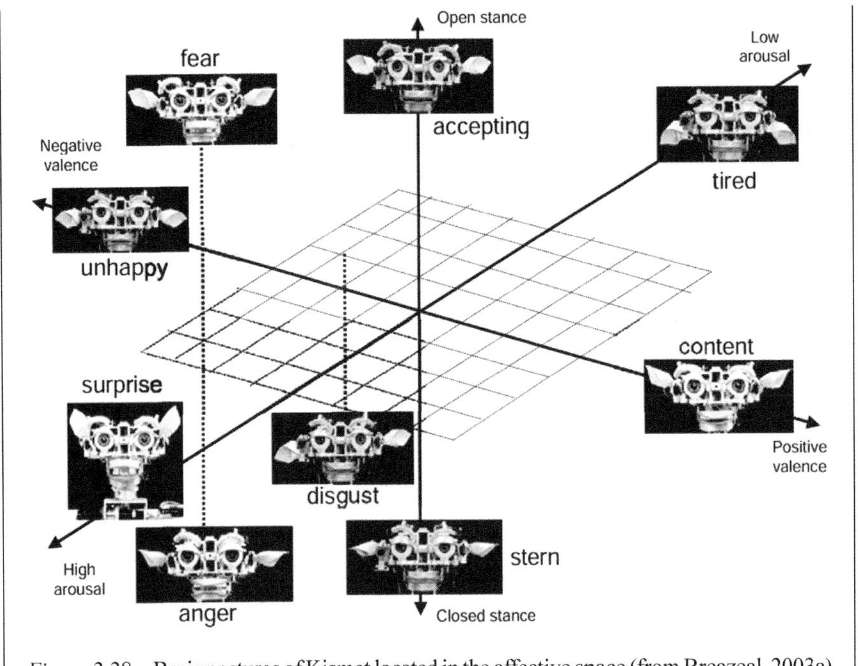

Figure 2.28 Basis postures of Kismet located in the affective space (from Breazeal, 2003a).

Thus, ethorobots should be developed by utilising a more general approach of signalling capabilities. Breazeal and Brooks (2005) suggest the use of a three-dimensional affect space for determining the signal structure of emotional states (see also Russell, 2003); Section 2.5.5, Boxes 2.15 and 2.22): (1) arousal/intensity (low/high), (2) valence (negative/positive), and (3) stance (approach/avoidance).

The prospect of formulating a comprehensive emotional state model opens the door to develop a more universal signalling system grounded in the specific embodiment of ethorobots. This approach stands in contrast to the human-centric method that aims to replicate primarily human facial emotional signals. To validate this more general principle, Korcsok et al. (2018) devised a virtual agent that exhibited abstract signals corresponding to five basic emotions (Table 2.15). The visual agent underwent alterations in motion, size, and colour to represent distinct emotional states. Interestingly, both Hungarian and Japanese participants demonstrated comparable proficiency in identifying the conveyed emotions. Among the emotions, anger was the most accurately recognised, while surprise proved to be the least successfully identified.

A parallel study was conducted in the auditory domain, involving variations in call length, fundamental frequency, and complexity (e.g., repetition) of simple sine wave sounds. These alterations were guided by biological rules delineating vocal features and emotional states (as discussed in Section 1.8.6, Morton, 1977). Notably, sounds characterised by lower fundamental frequency and shorter call lengths were rated

Table 2.15 Confusion matrix for the answers of the matched Japanese and Hungarian participants. The grey cells indicate the percentage at which each emotion received the biggest score. The correct emotions are indicated by dark borders (from Korcsok et al., 2018)

	Japanese						*Hungarian*					
	Displays (%)						*Displays (%)*					
Emotions receiving the biggest score	*Sadness*	*Surprise*	*Fear*	*Happiness*	*Anger*	*Neutral*	*Sadness*	*Surprise*	*Fear*	*Happiness*	*Anger*	*Neutral*
Sad	76	6	37	0	1	40	43	5	12	6	0	22
Surprised	5	44	9	15	8	11	7	36	6	7	0	6
Afraid	19	9	49	7	3	23	36	17	59	13	3	6
Happy	0	34	5	25	1	20	5	29	21	40	8	62
Angry	0	9	0	53	87	7	10	14	2	34	90	4

by human listeners as having a more positive valence. Conversely, calls featuring a higher fundamental frequency were consistently rated as more intense across all emotion categories by listeners (Korcsok et al., 2020).

2.6.3.7 The problem of grounding and developmental ritualisation

In natural situations, communication between agents is typically driven by specific objectives, and agent partnerships are not established arbitrarily. In customary HRI experiments, there is an anticipation for the robot and the human participant to promptly initiate communicative exchanges. This initiation is frequently facilitated by the experimenter, who outlines the scenario, introduces the robot to the human, and delineates the execution of the interactive task. Although such experimenter assistance is imperative for the success of the experiment, it can inadvertently alter the genuine communicative dynamics of the situation.

The issue of establishing a common ground is frequently overlooked in HRI research, despite its pivotal role in understanding how signal utilisation arises in social agents (Harnad, 1990). HRI research frequently makes the implicit assumption that all signals are universally understood by the participants. However, this assumption is unrealistic, as both senders and receivers need to acquire knowledge about the nuances of signalling, including when and how to do so. In biological agents, this learning process is particularly critical during early development, and its relevance extends into later stages also.

Hence, a more typical scenario in HRI involves learning about each other's signals over an extended period of interaction. For instance, the robot might employ an initial set of context-specific vocal signals in its regular communication. Initially, a human recipient might not immediately grasp the signal meanings, but comprehension could emerge after a series of spontaneous interactions. If the human fails to decipher the specific signal within a designated timeframe, the signal might be phased out, prompting the robot to introduce a novel signal. The learning process is expedited when the two agents employ similar signals. This phenomenon is referred to as developmental ritualisation (Section 1.8) and is foundational in establishing a strong alignment between signal meaning and message. This alignment leads to the creation of shared mental representations that underlie signalling behaviour. Consequently, the often-quoted Wittgensteinian assertion, 'If a lion could talk, we would not understand it', can be surmounted through the establishment of shared signalling systems and the potential for shared social development. Successful communication between agents is firmly rooted in a shared social space (Krämer et al., 2011).

2.6.3.8 Last note on linguistic communication

It is beyond question that verbal engagement with uncomplicated devices, such as mobile phones, offers substantial assistance in myriad situations that are not entirely attainable through non-verbal communication alone. Thus, ethorobots will inevitably engage in linguistic interactions with humans. While some of the existing solutions are remarkably promising, the prevailing approaches tend to regard language as an efficient tool for sharing explicit knowledge rather than as a skill for orchestrating interpersonal relationships. These applications often focus on the capability to decipher the semantic content of questions and answers through machine learning techniques and natural language modelling, employing vast databases. Nevertheless, despite the advancement of sophisticated technologies, achieving natural verbal communication with social robots still remains an evolving endeavour, influenced by several factors (see Chapter 3). For prospective planning, a few key points are worth consideration (Strohmann et al., 2023):

1. Implementing CAs on social robots may seem relatively straightforward, but this approach can lead to practical challenges due to technical limitations. For instance, the robot might encounter difficulties in accurately capturing human speech, particularly in noisy environments;
2. Many CAs excel in generating output rather than receiving input. People may get frustrated if the expectations formed based on the robot's output capabilities are not met by its capacity to comprehend them;
3. The robot's grasp of linguistic nuances could be restricted by its lack of awareness about its own embodiment and contextual information;
4. Robots should steer clear of attempting to mimic humans solely through linguistic abilities. As such, it could be beneficial if 'speaking' robots employed self-generated speech, distinct dialects, and their own unique phonemic production methods to differentiate themselves from humans;
5. When it comes to computers, people are accustomed to text-based or verbal interactions. However, to establish more natural communication, robots should also be capable of incorporating non-verbal signals in tandem with verbal interactions, aligning with the discourse's content.

Chemero (2023) expresses scepticism regarding the prospect of large language models (LLMs) achieving human-like qualities in the near future solely through extensive training on internet data. The key issue, according to Chemero, is not whether such a system can exhibit 'intelligence', but rather to what extent it can resemble human characteristics. He argues that LLMs, lacking a physical body, fundamentally differ from humans in experience. Without perception, action execution, and the essential needs for sustenance and reproduction, LLMs cannot truly emulate human-like qualities.

Chemero's argument aligns with the ethorobotic perspective, asserting that systems of different origins can only imitate specific functions of other systems and never capture their entirety. Despite these differences, LLMs may effectively replace specific language functions, such as dynamic information exchange about 'fact-like' linguistic representations, which were traditionally provided in a more passive way by sources like books, encyclopaedia, or Wikipedia.

2.6.4 Social partners as information sources

Just as social learning constitutes a fundamental skill for social animals, ethorobots should similarly harness this potential to efficiently acquire knowledge about their environment, decreasing the need for individual trial-and-error learning, which can be more laborious and riskier (Section 1.9). Although individual and social learning mechanisms are often juxtaposed, each holds distinct advantages contingent on the context. However, in species that live in groups or where parental care is prevalent, the perpetual presence of companions confers an advantage, facilitating reliance on various forms of social learning. Social animals have genetic predisposition for curiosity in the activities of their peers, enabling them to learn about others' actions and their outcomes.

The field of social robotics recognised the utility of social learning relatively early, allowing inexperienced robots to learn from experienced humans through natural interactions (Breazeal & Scassellati, 2002; Schaal, 1999). More recently, Fang et al. (2019) elucidated numerous benefits linked to incorporating social learning capabilities into a social robot. While their discussion centred on imitation learning, their perspective can be extrapolated to encompass a broad spectrum of social learning forms (Thomaz & Cakmak, 2009).

Accelerated learning: observing experienced individuals directs the observer's attention to specific environmental aspects, accentuated by the demonstrator's behaviour. This targeted focus guides the observer towards executing similar actions, as opposed to selecting from an extensive array of potential behaviours, as in the case of individual learning in the absence of a demonstrator. In social interactions, the acquisition of new behavioural aspects occurs faster. Thomaz and Cakmak (2009) directly compared the effectiveness of individual and social learning using the Leonardo robot. In an experimental setup, the robot was exposed to human demonstrations (with some guidance) to learn simple action-related tasks. Leonardo displayed swifter task acquisition in the social condition, as measured by various parameters (e.g., the number of actions executed before reaching the first goal), compared to individual learning.

Augmented learning capacities: social learning serves as a valuable supplement to individual learning mechanisms, functioning in tandem and complementing each other. Learning through observation engenders good approximation to act, while further refinement is achieved through individual learning.

Natural interaction: in numerous scenarios, the ability to learn through observation (or auditory perception) does not demand significant investment from the experienced demonstrator (although teaching provides an exception, Section 1.10.6). This effortless knowledge transfer facilitates the acquisition of experiences that observers might not have obtained on their own.

Group-level dissemination: given that all group members possess social learning skills, any novel behavioural pattern (innovation) introduced by an individual can swiftly propagate throughout the entire group, saving a lot of individual investment in learning.

Group and inter-individual cohesion: social learning, particularly imitative capacities, synchronises group activities and contributes to both inter-individual attraction and emotional contagion (Section 2.6.2).

2.6.4.1 Social robots learn from humans

In conventional scenarios, social robots acquire knowledge from humans. Humans possess the know-how and naturally convey this knowledge by demonstrating tasks to the observing robot. This type of interaction is referred to as socially guided learning, as delineated by Thomaz and Cakmak (2009). They also argue that social learning should not be limited solely to imitation (in the strict sense; see Section 1.9), but various other forms of social learning can be utilised. Breazeal and Scassellati (2002) similarly encompass a range of potential social learning modes feasible within HRIs.

Social robots not only learn about actions demonstrated by humans but also details regarding objects and objectives. By observing a human, a robot can learn the sequence of retrieving items from a table, even though it might adopt its distinctive approach when lifting and putting them down. This type of social learning (object facilitation and emulation) proves especially valuable when the robot's physical embodiment diverges from that of the human demonstrator, making a direct imitation of the action unfeasible. This instance underscores the necessity for social robots to possess a versatile learning system capable of deducing the most useful and specific information from the demonstrator's behaviour (see Box 2.23).

Box 2.23 What can thirsty robots learn from a thirsty human?

In realistic scenarios, social learning relies on various mental mechanisms to learn from the actions of others. While this capability has evolved in many social species, similar versatility is still absent in social robots.

Imagine a 'thirsty' wheeled social robot as the observer in the following scenario: a human enters the room and approaches a box on the floor. He opens it and retrieves a glass. After placing it on the table, he goes to a cupboard, unlocks it using a key in the door. Then he takes out a bottle of water, takes off the cap, and pours the water into the glass for about one to two seconds, puts the cap back, and then proceeds to shake the bottle up and down five times. Finally, he returns the bottle to the cupboard, closing the cupboard door using its legs before leaving the room. What could the 'thirsty' ethorobot learn from this observation and what should it do? (For definitions of terms in parentheses, refer to Section 1.9).

1 The robot has acquired knowledge about the locations of glasses and bottles within the room (local enhancement);
2 The robot is likely to copy the actions of taking out the glass and transporting it to the table (response facilitation);
3 The robot might attempt to open the cupboard door using the key, although its pulling action could differ from that of the human (emulation);
4 The robot realises how to pour water into the glass (selective imitation, emulation based on experience);
5 The robot may choose to replicate the (seemingly redundant) action of shaking the bottle five times (imitation/over-imitation) or may opt not to do so (selective imitation);

6. Given its lack of legs, the robot would utilise its arm to close the door (emulation).

This example illustrates that in a realistic scenario, various facets of social learning may come into play. The observed behaviour is evaluated by considering aspects like novelty, efficacy, and behavioural or morphological constraints.

Crucially, a distinction exists between novice learners (typically offspring in animals) and more experienced learners (adults) in terms of information acquisition through observation. The former may readily adopt newly learned behaviour, while the latter likely possess own experiential insights about the situation. In standard experimental settings, naïve and inexperienced social robots are directly exposed to a human demonstrator, facilitating learning through observation. This scenario proves advantageous for acquiring new skills, yet it is less congruent with more authentic interactions where the observer's decision to learn depends on its past experiences and contextual cues. The critical aspect of self-evaluating one's existing knowledge alongside information gained from observation is often overlooked in social robotics. Although any fresh experience might offer functional value, unchecked incorporation of novel information can also yield adverse outcomes, potentially proving less efficient than pre-existing behaviours or even outdated for the specific environment. This capacity, known as selective imitation, may safeguard the observer against the pitfalls of misinformation.

2.6.4.2 Humans learn from social robots

The other direction of learning by observation has been investigated mostly in infants and children, although it may have some validity for adult also. Social robots may prove to be useful when deployed in educational or training contexts (see also Section 2.6.5), especially when there is need for extensive interaction. There have been many attempts to use social robots for improving affective and motor skills of children living with autism (see Raptopoulou et al., 2021). Fitter et al. (2019) demonstrated that 6–8-month-old infants are able to imitate the leg movement of a NAO robot. This finding paves the way for using social robots in situations where there is a need for repeated exercises, for example in infants born with cerebral palsy.

Nevertheless, while mere exposure to a demonstrator might suffice for infants, a teaching context should be established for children or adults to ensure meaningful performance when learning about the behaviour of social robots (see Section 2.6.5).

2.6.4.3 Social learning as a form of exploration

Social animals inherently possess a strong propensity for learning through observation, even in the absence of external support. This trait could also be harnessed in ethorobots. Similar to social animals, ethorobots could also find success by leveraging socially acquired information. Notably, among non-human animals, the significance of social learning diminishes with age, as personal experience grows and the potential for novel knowledge acquisition from group members becomes restricted within a relatively stable environment. However, this paradigm does not apply to modern human societies characterised by swift technological and cultural shifts. Consequently, proficient ethorobots should demonstrate an ongoing capacity for continuous social learning or the effective transmission of such acquired knowledge.

2.6.5 Cooperation

The primary objective of integrating ethorobots into human societies is to leverage their capacity for assistance, making cooperative skills a fundamental aspect for social robots. These skills can serve a wide range of functions, from simply 'being present' to aiding in intricate tasks such as locating missing individuals or assembling objects. Mechanistically, cooperation represents the apex of HRI, necessitating the harmonious operation of an array of highly advanced hardware and software tools. While fundamental biological mechanisms of cooperation are shared across both non-human and human entities (Section 1.10), the latter excels particularly in collaborative endeavours involving object manipulation – a trait rarely observed among non-humans.

Drawing from Matheson et al. (2019), one can identify four distinct forms of HRI related to cooperation, differentiated by variations in temporal and spatial coordination. In natural scenarios, these forms of collaboration may intersect, unfolding concurrently and adapting in accordance with the ongoing task. Here are a few aspects for consideration of the context:

1 *Coexistence*: the human and robot share proximity without direct interaction;
2 *Chaining*: both the human and robot operate in close proximity, undertaking separate tasks within a chained action sequence. Actions of one partner follow those of the other in a synchronised temporal pattern, despite spatial separation. For example, a mobile robot regularly delivers objects for the human partner to work on;
3 *Synchronisation*: the human and robot concurrently engage in different aspects of the same task, working in parallel within a chained action sequence. This involves both spatial and temporal synchronisation. An example would be one partner creating holes while the other inserts screws into the same wooden plate;
4 *Complementation*: the human and robot alternately contribute to the same task aspects within a chained action sequence, maintaining both spatial and temporal synchronisation. For instance, one partner holds a screw while the other screws it in.

2.6.5.1 Task description and analysis

Collaborative interactions often exhibit a high level of complexity. Most joint tasks not only involve exchanges between partners but also follow a hierarchical organisation, encompassing numerous sub-tasks and simultaneous actions. Task analysis (Section 2.2.4) serves as a valuable tool for structuring the robot's behaviour in

Figure 2.29 Snapshots acquired during the collaborative assembly of the table by a naive human subject and Baxter robot. (1) Baxter provides the tool to the participant; (2) the robot supports the human by holding the tabletop while the human screws the leg in place; (3) the user has finished his task and observes the robot freeing the workspace from the box of linkages. Such experiments are important for finding out how to organise the physical and communicative interaction between the two participants, but it can also be seen that the actual level of performance of the team is very low compared to working alone or a team of two humans (from Mangin et al., 2022).

collaborative interactions. However, this approach assumes that the engineer possesses comprehensive knowledge of all anticipated interaction details, aiming to achieve the desired goal.

Mangin et al. (2022) introduced Hierarchical Task models as a framework for describing collaborative interactions. This model accommodates various scenarios, including alternative, parallel, and sequential actions, each requiring distinct levels of partner involvement. Through real-world experimental scenarios, they demonstrated the robot's ability to autonomously determine appropriate instances for assisting the human, selecting the most suitable supportive action from a range of available options (Figure 2.29).

2.6.5.2 Goals and roles

Cooperation is commonly defined as the interaction between agents through joint actions aimed at accomplishing a shared goal that cannot be attained individually. While this description is probably valid for collaboration in animals, humans can often reach the same goals alone (e.g., carpenter can make a table alone or with a partner but then the process takes usually longer).

The partners may share the goal of the activity (symmetric cooperation) or not, that is, a partner can also be helpful if it does not know about the final aim (asymmetric). Given the inferior abilities of cooperative robots, HRI is characterised mainly by asymmetric forms of interactions (Mangin et al., 2022). Roncone et al. (2017) also show that a cooperative robot can also be helpful if it attends only to the local task and has no representation about the final goal of the interaction.

In cooperative interactions that rely on synchronisation or complementation, one partner typically assumes the lead role while the other follows. In cases of asymmetric roles, only the leader possesses knowledge of the final objective. However, both symmetric and asymmetric cooperation scenarios may entail situations where either partner can assume a leadership role for certain action sequences. A suitable analogy can be found in guide dogs assisting the blind. Although the dog predominantly makes decisions due to its ability to perceive environmental obstacles, the owner holds knowledge of the ultimate goal (destination of their trip), often determining when to initiate or halt movement. This dynamic exemplifies asymmetric cooperation, yet leadership roles are exchanged periodically based on context (Naderi et al., 2001).

Pereira et al. (2002) effectively orchestrated an interaction between two cooperating robots tasked with jointly carrying a box without dropping it. Navigating a complex obstacle-laden environment, only the leading robot possessed knowledge of the intended destination of the box. However, the follower robot could assume the leader's role in instances where the team encountered obstacles and needed to change direction to proceed. This role interchange is particularly crucial in complementing cooperation, where the partners offer diverse expertise.

2.6.5.3 Coordination, action perception, and execution

Coordinated actions establish a causal link between the behaviours of the two partners. To achieve this, the partners must possess the capability to anticipate each other's actions and choose the most suitable response from their range of options based on the current context (Gildert et al., 2018). Joint action can be considered a unique scenario in which two agents either operate in close synchrony or simultaneously. Coordinated actions can be realised through a variety of approaches, which are not mutually exclusive:

1 Making predictions about the other's behaviour through observation and learning or relying on one's own knowledge. This process is grounded in the structured occurrence of action sequences, and prediction accuracy improves over time. Partners begin by observing each other's action sequences and adjust their predictions as necessary. Wang et al. (2013) illustrate this method by programming a robot to engage in table tennis with a human partner.
2 Targets are typically brought under visual control 1–2 s before executing an action on them (e.g., Mennie et al., 2005). In essence, a partner directs his gaze towards objects before initiating any action (anticipatory look). This behavioural tendency can be exploited by a partner to forecast the subject's intended action when the partner follows the gaze movement (gaze following, e.g., Silverstein et al., 2019).
3 Contextual observation further refines predictions regarding potential actions or goals. For instance, encountering a hammer and nail positioned on a desk in front of a partner might heighten expectations for a hammering action.
4 Action synchronisation can be facilitated by external rhythmic stimuli dictating the tempo, analogous to the role of a conductor.
5 Obtaining sensory measurements of physical forces exchanged between interacting partners, either directly or transmitted through a solid medium, as exemplified by Pereira et al. (2002), e.g., jointly carrying a cupboard.

As outlined by Mangin et al. (2022), the perceptual and motor limitations inherent to collaborative robots constrain the scope of tasks in which they can effectively engage with humans. Generally, robots tend to primarily carry out explicit human commands, rather than actively participating in spontaneous cooperation. Today's robots typically exhibit deficiencies in one or more of the subsequent skills (also referenced in Lorenzini et al., 2023):

1. *Insufficiency in partner's action recognition*: there are multiple solutions, such as video imagery or wearable sensors, for tracking motion. However, even sophisticated methods can be susceptible to environmental factors, such as alterations in illumination.
2. *Lack of environment/context recognition*: most collaborative robots lack the capability to utilise contextual cues for anticipating potential actions by their partners, except narrow representations that are needed for the specific task.
3. *Constraints in action execution speed and precision*: frequently, the mechanics controlling movements are inadequately swift to execute the intended action, and the precision required for delicate, object-specific actions remains challenging – particularly in real-world settings. For numerous tasks involving objects, achieving optimal movement precision on the order of millimetres is essential, rather than centimetres. Robots also lack feedback loops to check whether tasks or subtasks were executed successfully or failed.
4. *Restricted range of action types*: even robots equipped with advanced arms and hands exhibit a limited repertoire of actions, often displaying clumsiness that diminishes interaction enjoyment and efficiency.

2.6.5.4 Communication in cooperation

Communication that runs concurrently with cooperative activities can enhance performance. Signalling (not necessarily in verbal form) can serve to indicate either the current or future actions of partners ('I will grasp this end'), their intended objectives ('Let's transport it to the window') or synchronise their activities ('Let's elevate the table now'). Gesture-based cues, such as alternating gazes or pointing, can also function as directional indicators.

When testing the non-verbal interactive Leonardo robot, Breazeal et al. (2005) observed that collaboration was notably improved when the robot employed a diverse array of signals and engaged in more intricate activities. Human attention to Leonardo's behaviour, when it exhibited a comprehensive repertoire of actions and signals, led to a decrease in error rates and an increase in task satisfaction.

2.6.5.4.1 CHOOSING PARTNERS

Achieving successful task completion is the goal of any collaborative endeavour. In common scenarios, partners assess each other's performance and subsequently opt to continue collaborating with their previous partner or, when feasible, seek an alternative. Various attributes of the partner can influence these preferences, with some factors being contingent on the socio-economic and cultural contexts of the involved individuals (Mou et al., 2020).

Proficiency: a robot frequently committing errors or impeding task execution could deter future collaboration. There have been several instances when collaborative robots, serving as receptionists, waiter assistants, or carriers, were abandoned by human personnel due to work delays and escalated issues. Interestingly, in experimental settings, individuals found robots more appealing when the robots sought assistance or displayed the motivation to rectify errors or retry a task (Honig & Oron-Gilad, 2018).

Task disposition: a robot's behaviour, or attitude, can either closely align or deviate from the specific task at hand. In a particular study, participants interacted with a robot assigned to guide them through exercises (Goetz et al., 2003). The participants exhibited greater compliance when the robot maintained a serious demeanour, as opposed to a more playful one.

Personality: certain personality traits prove more suited to specific types of work. Individuals displaying a higher degree of 'Openness to Experience' (a dimension within the Big Five Personality model, McCrae & John, 1992) tend to excel in home office settings, whereas extroverted individuals show less inclination towards working from home. Alternatively, particular personality traits may render an individual more appealing as a collaborative partner (Gavoille & Hazans, 2022). Following a collaborative interaction, extroverted partners tended to anthropomorphise the robot to a greater extent and reported a heightened sense of closeness (Salem et al., 2015b).

Matching: commonly, it is presumed that partners with congruent personality traits perform better in collaborative tasks due to the potential for enhanced synchrony. However, an alternative perspective suggests that diverse personalities that complement each other might hold an advantage in collaborative tasks when their roles differ significantly. Neither human studies nor HRI investigations have definitively resolved this matter; the phenomenon appears complex and could be influenced by an array of factors (Robert Jr. et al., 2020).

2.6.5.4.2 TEACHING

Teaching encompasses a multifaceted phenomenon (Section 1.10.5), intertwining mechanisms of social learning, communication, and cooperation. In a teaching interaction (Sage & Baldwin, 2010), a knowledgeable individual, the teacher, guides and directs the accumulation of experiences in the learner until the skill transfer is achieved. This process unfolds through a sequence of interactions when both individuals share the same physical space (Box 2.24).

Teaching relies on distinct behavioural mechanisms that foster attunement between the teacher and the learner. Knowledge transfer can occur through both non-verbal and verbal behaviours. At the level of overt behaviour, these mechanisms encompass ostensive signalling, joint attention, gaze, and point tracking, as well as pointing itself. The unique strength of social robots, in contrast to other technologies (such as online teaching), lies in their physical presence, which bolsters the efficacy of instruction, particularly among younger pupils. Additionally, social robots are favoured in scenarios where the educational process incorporates acquiring knowledge about physical interactions with objects (Belpaeme et al., 2018).

A meta-analysis (Belpaeme et al., 2018) found that teaching studies predominantly involved children aged 5–6 years and above. While an overarching positive impact on both affective and cognitive outcomes was observed, it is worth noting that the results were frequently contrasted with non-teaching conditions, rather than being directly compared to human teachers' achievements. As anticipated, the robots' embodiment and behaviour consistently exhibited a strongly social demeanour, with the NAO robot emerging as the most prevalent choice for these experiments. Interestingly, no distinct advantage associated with any specific behaviour was found, as the presence of attention-getting behaviours, joint attention, or empathetic responses yielded similar improvements.

Box 2.24 How to train your AIBO

AIBO robot learns how to display an action on human command (Kaplan et al., 2002). Implementing a method inspired by dog training to teach robots is a promising approach for instructing agents in performing specific actions upon receiving verbal commands. Kaplan et al. (2002) developed a relatively straightforward technique for teaching AIBO commands, incorporating both classical and operant mechanisms for effective learning (called also 'Clicker training'; Fugazza & Miklósi, 2014). AIBO has a distinct topology of actions (see pictures) that dictates the likelihood of one action following another (see Figure 2.30). Notably, certain transitions are impossible; for instance, AIBO cannot walk after lying down without first standing up. It is the programmer's task to determine the probabilities of possible transitions. This model mirrors partly the behaviour of a real dog, and a dog trainer naturally possesses implicit knowledge regarding these transitions.

As a demonstration, the aim of the trainer is that AIBO displays digging on command. Digging behaviour occurs when the robot is sitting and uses its left front paw to scratch the ground. Without going into the details, the training consists of the following steps (see Kaplan et al., 2002).

1 AIBO learns about primary reinforcement, where touching its head at a pressure-activated sensor is akin to petting a dog, and the command 'Bravo' is associated with it;
2 The system looks for events in memory within the last five seconds whenever a primary reinforcer is detected. These events become potential secondary reinforcers;
3 After an event is detected more than 30 times, it becomes a secondary reinforcer. A happy signal, like wagging the tail, indicates recognition. In this case, the utterance 'Good' is used;

4 If AIBO hears the command 'Good', it checks the last behaviour produced. If no action occurred in the last ten seconds, nothing happens. Otherwise, the action is selected and becomes the basis for future variations;
5 AIBO displays further actions for a minute, attempting to show behaviours similar to the selected one. Every 20 seconds, a new action is displayed, with the chance determined by topology and transition probabilities;
6 The trainer rewards ('Good') actions similar/closer to the desired one (e.g., AIBO is using its leg to 'say' Hello). Upon displaying 'digging', the dog is rewarded with the primary reinforcer, 'Bravo';
7 The command 'Dig' is emitted, associating the word with all marked phases in the route towards the command during training;
8 AIBO learns that the 'Dig' command specifies only the last action. As AIBO repeats the action sequence, making it shorter, it receives the primary reward only if digging is displayed upon the 'Dig' command (Figure 2.30).

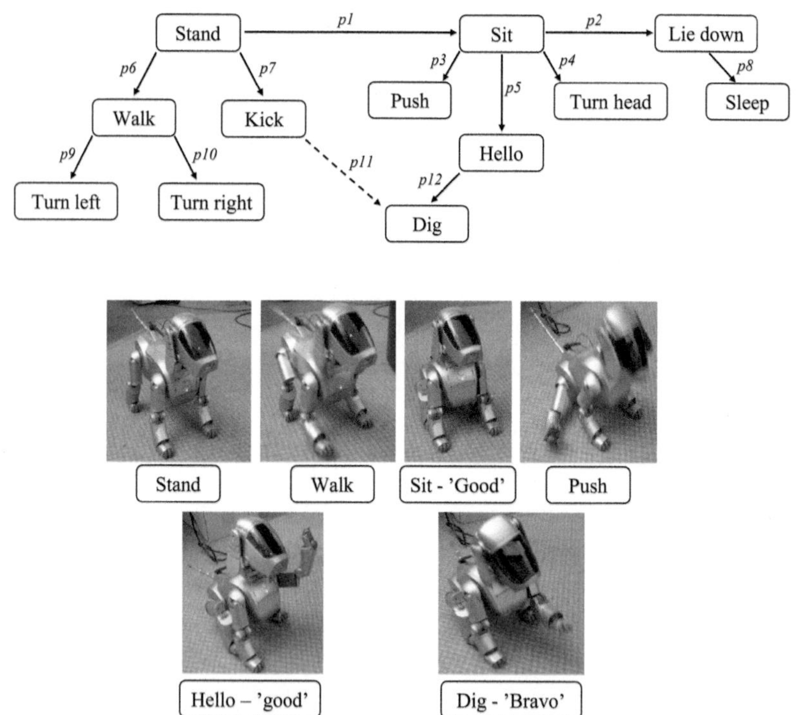

Figure 2.30 AIBO robot learns how to display an action on human command (see text for explanation) (from Frédéric Kaplan).

2.6.6 Conclusions, prospects, and questions

This section provided a concise overview of four crucial social skills: attachment, communication, social learning, and cooperation. While there could be additional significant abilities influencing the performance of an ethorobot, substantial success is likely achievable when these foundational skills are well-developed. The practical utility of each skill has been demonstrated in controlled laboratory environments; however, their effectiveness in real-world contexts remains relatively unexplored. Presently, there exists no robot possessing a complete version of these four essential skills.

Attachment is essential for strengthening and maintaining social connections among ethorobots and human partners, and the system can be accommodated to specific situations by adjusting its behavioural styles. Engaging in prosocial behaviours might trigger attachment responses in partners, a dynamic that can prove valuable during the initial encounter between the ethorobot and a human. The endowment of ethorobots with attachment skills seems to be inevitable in the future (see Box 3.24; Kovács et al., 2011; Niitsuma et al., 2014).

Despite extensive research in the field, the non-verbal communication of social robots remains relatively basic. This research has, however, underscored the significance of these skills, particularly in relation to attachment and cooperation. It is important for ethorobots to develop a distinctive yet compatible communicative system akin to that of humans. This could potentially require a reciprocal learning of communicative behaviours, but this should be regarded as serving as an investment in supporting the relationship. At present, the embodiment of robots provides little flexibility for non-verbal communication. Adding 'head' or 'arm' or other types of appendages with added degrees of freedom could increase the versatility of communicative interactions.

Social learning constitutes a natural way of acquiring knowledge from more experienced peers within a group. It involves a relatively seamless process that also serves to safeguard the ethorobot from potentially detrimental scenarios. Possessing such skills can significantly enhance learning efficacy, particularly when observers gather information pertaining to objects, the surrounding environment, and potential courses of action. While at the beginning, the emphasis was on imitative skills in robots, there are many different ways to gain knowledge by observing others. This could be especially useful if the embodiment of the robot differs from that of a human.

Collaboration stands as the cornerstone of HRI. However, at present the achievement of robust cooperation between ethorobots and humans is hindered by technological constraints. The challenge in crafting collaborative interactions with humans lies in ensuring that, ultimately, working alongside robots genuinely ease humans' workload. Currently, in the field, such robots tend to present more challenges than contribute to solutions. Therefore, a precise task analysis is necessary to identify specific segments of collaborative interactions where the robots' contribution can be effective. For instance, there is a widespread vision that social robots could play significant roles in elderly care. Although the nature of such work is highly intricate, social robots realistically cannot fully take on the role of 'care' in its entirety.

Nonetheless, this does not rule out the possibility that they can successfully perform very specific and well-defined tasks in collaboration with human staff.

The vision is for ethorobots to embody a robust iteration of these capabilities; however, it is worth noting that even with today's technology, simple solutions can yield positive outcomes. One might posit that even fundamental versions of these skills could contribute synergistically to the achievements of ethorobots.

2.6.6.1 A summary: how to design a cooperative ethorobot?

1 Conduct a thorough behavioural examination of the collaborative task and the robot's functionality;
2 Analyse the interaction sequence and establish action coupling between partners;
3 Determine the role of the ethorobot, including its task representation (partial or full);
4 Check all aspects of the environment in which the robot and the humans interact and ensure that the robot has maximum mobility and autonomy;
5 Select an appropriate embodiment aligned with expected physical requirements for task execution and human interaction;
6 Develop a comprehensive task analysis plan that encompasses the majority of potential interactions during collaboration;
7 Define communicative actions to enhance interaction by leveraging the capabilities offered by the chosen embodiment.

This list serves only as a guide, and even if adhered to, iterations among the steps, particularly in selecting the most suitable embodiment, will inevitably be necessary. Another overlooked aspect is the testing of the robot under realistic circumstances and leveraging the obtained results for continuous improvement. This involves making adjustments to the robot's embodiment or re-evaluating the tasks it can perform. It is crucial to recognise and always be aware that the apparent simplicity of a task for a human and the unpredictability of the environment pose significant challenges for these robots.

2.7 Ethical considerations: ethorobots are machines

Moral principles guide human behaviour and shaping how people should interact with others and the environment. These principles emerge through negotiations among adult members of society, occurring at various societal tiers and unfolding as a complex, iterative process. Participation in these negotiations requires individuals with comparable and aligned cognitive abilities and societal status. These principles typically mirror the outcomes of these negotiations and are frequently codified in laws and regulations. They are profoundly influenced by historical legacies and cultural dynamics and may change over time. While some facets of moral behaviour are broadly accepted across the whole human society, others are rooted in local customs and traditions (e.g., Russell, 1954; Valentini, 2023).

In the context of the present concise discussion, it is important to acknowledge that moral principles may seem arbitrary when applied to entities not engaged in the negotiation process. Examples of this include children or animals. In these instances, adult humans superimpose their own third-person perspectives when formulating moral principles of conduct for them including their moral rights as members of the society.

As moral principles take shape and are employed among group members with shared status, the demarcation between those who belong to the group and those who do not becomes significant. The ethorobotic approach offers a clear delineation, as all such robots belong to distinct species and are thus categorised as 'others'. This is important due to humans' predisposition to be sensitive in identifying group membership based on appearance and behaviour (Everett et al., 2015). Thus, ethorobots pose no complications when it comes to make such decisions. Ethorobots can participate in human group activities but despite being a member of the groups they do not (and should not) have the same moral status as humans. However, this can pose substantial challenges in the case of robots that closely resemble and behave like humans (e.g., humanoids, androids).

The insights gained by developing moral principles applicable to animals can serve as a valuable starting point. However, an essential distinction must be acknowledged. Ethorobots do not conform to principles governing living beings; consequently, from a societal and moral standpoint, they are machines and should be treated as such. Consequently, the moral principles imposed by society for social robots should be perceived as operational guidelines to be implemented by manufacturers. This perspective entails that manufacturers bear the responsibility if these robots cause harm (see also Guerra et al., 2022).

It is worth noting that human society has encountered analogous situations before, as seen in the case of novel drugs or transportation innovations such as cars and airplanes. Consequently, manufacturers should develop a comprehensive process to ensure these robots adhere to all stipulated regulations. A parallel can be drawn with guide dogs that aid their visually impaired owners in navigation. These dogs undergo a selection process to assess their behavioural aptitude. Subsequently, they participate in an extensive and purposefully tailored training. Finally, they have to pass a complex exam testing their blind leading skills under natural conditions. This comprehensive process guarantees that certified dogs adeptly guide their owners, resulting in relatively few significant mishaps. A comparable systematic approach can be envisioned for social robots as well. Social robots ought to be conceived with an array of safety features and subjected to rigorous testing before their introduction to the market.

A further complicating problem about social robots is that they might replace humans and that people could develop strong emotional bonds with them, possibly affecting their relationships with other humans (Lutz et al., 2019). The issue of displacement has been a recurring topic in relation to machines. However, historical evidence suggests that the automation of industries over time does not necessarily result in long-term job loss for people. The roles that social robots could assume generally involve tasks that alleviate individuals from demanding, sometimes very boring, labour, which is frequently characterised by repetitive routines. This means that with appropriate planning humans should have more time for each other if robots take over part of their work. In addition, new type of jobs will also emerge in connection with developing, constructing, maintaining, etc. these robots.

The concept of replacement can prove problematic if it leads to a significant reduction in human interactions with fellow humans (Sharkey & Sharkey, 2012). However, when it comes to ethorobots, it is rather straightforward to argue that substituting humans with these robots does not equate to a replacement for genuine human-to-human connections; instead, it addresses specific issues with specific solutions. This distinction becomes less clear in the context of robots designed to mimic humans.

Engaging with ethorobots can readily align with moral principles to uphold healthy social dynamics among humans. For instance, ethorobots could serve as valuable tools for educating children living with autism. Nonetheless, interventions of this kind should be constrained by time limits, much akin to the concept of 'dosage' applied in the case of drugs. In general, the introduction of any social robot in large numbers should be planned in advance with considering various possible scenarios.

Numerous ethical dilemmas warrant discussion, yet the field of 'roboethics' is poised to evolve in tandem with the rise of social robots (Veruggio & Operto, 2006). Additionally, a spectrum of interconnected modern technologies, such as internet and mobile phone usage, presents other pertinent concerns (e.g., privacy), which are not explored within this context. Given the current capabilities of contemporary robots, certain ethical queries may remain pending definitive resolution. However, those intrigued by these matters might find valuable insights within references like Salem et al. (2015a), Fosch-Villaronga et al. (2020), Lutz et al. (2019), and Sharkey (2017).

References

Abdai, J., Baño Terencio, C., & Miklósi, Á. (2017). Novel approach to study the perception of animacy in dogs. *PLoS One*, *12*(5), e0177010. https://doi.org/10.1371/journal.pone.0177010

Abdai, J., Baño Terencio, C., Pérez Fraga, P., & Miklósi, Á. (2018). Investigating jealous behaviour in dogs. *Scientific Reports*, *8*, 8911. https://doi.org/10.1038/s41598-018-27251-1

Abdai, J., Bartus, D., Kraus, S., Gedai, Z., Laczi, B., & Miklósi, Á. (2022). Individual recognition and long-term memory of inanimate interactive agents and humans in dogs. *Animal Cognition*, *25*, 1427–1442. https://doi.org/10.1007/s10071-022-01624-6

Abdai, J., Gergely, A., Petró, E., Topál, J., & Miklósi, Á. (2015). An investigation on social representations: Inanimate agent can mislead dogs (*Canis familiaris*) in a food choice task. *PLoS One*, *10*(8), e0134575. https://doi.org/10.1371/journal.pone.0134575

Abdai, J., & Miklósi, Á. (2023). After 150 years of watching: Is there a need for synthetic ethology? *Animal Cognition*, *26*(1), 261–274. https://doi.org/10.1007/s10071-022-01719-0

Abdai, J., Uccheddu, S., Gácsi, M., & Miklósi, Á. (2022). Exploring the advantages of using artificial agents to investigate animacy perception in cats and dogs. *Bioinspiration and Biomimetics*, *17*(6), 065009. https://doi.org/10.1088/1748-3190/ac93d9

Abdi, J., Al-Hindawi, A., Ng, T., & Vizcaychipi, M. P. (2018). Scoping review on the use of socially assistive robot technology in elderly care. *BMJ Open*, *8*(2). https://doi.org/10.1136/bmjopen-2017-018815

Ainsworth, M. D. S., Blehar, M. C., Waters, E., & Wall, S. N. (1978). *Patterns of attachment: A psychological study of the strange situation*. Erlbaum.

Åkesson, S., Boström, J., Liedvogel, M., & Muheim, R. (2014). Animal navigation. In L.-A. Hansson & S. Åkesson (Eds.), *Animal movement across scales* (pp. 151–178). Oxford University Press. https://doi.org/10.1093/acprof:oso/9780199677184.003.0009

Ali, H., Bhansali, S., Köksal, I., Möller, M., Pekarek-Rosin, T., Sharma, S., Thebille, A.-K., Tobergte, J., Hübner, S., Logacjov, A., Özdemir, O., Parra, J. R., Sanchez, M., Surendrakumar, N. S., Alpay, T., Griffiths, S., Heinrich, S., Strahl, E., Weber, C., & Wermter, S. (2019). Virtual or physical? Social robots teaching a fictional language through a role-playing game inspired by Game of Thrones. In M. A. Salichs, S. S. Ge, E. Ivanova Barakova, J.-J. Cabibihan, A. R. Wagner, Á. Castro-González, & H. He (Eds.), *Social Robotics. ICSR 2019. Lecture Notes in Computer Science* (pp. 358–367). Springer. https://doi.org/10.1007/978-3-030-35888-4_33

Al-Shawaf, L., Conroy-Beam, D., Asao, K., & Buss, D. M. (2016). Human emotions: An evolutionary psychological perspective. *Emotion Review*, *8*(2), 173–186. https://doi.org/10.1177/1754073914565518

Alt, M., Meyers, C., & Ancharski, A. (2012). Using principles of learning to inform language therapy design for children with specific language impairment. *International Journal of Language and Communication Disorders*, *47*(5), 487–498. https://doi.org/10.1111/j.1460-6984.2012.00169.x

Amirova, A., Rakhymbayeva, N., Yadollahi, E., Sandygulova, A., & Johal, W. (2021). 10 years of human-NAO interaction research: A scoping review. *Frontiers in Robotics and AI*, *8*, 744526. https://doi.org/10.3389/frobt.2021.744526

Anderson, C. J., & Prentice, N. M. (1994). Encounter with reality: Children's reactions on discovering the Santa Claus myth. *Child Psychiatry and Human Development*, *25*(2), 67–84. https://doi.org/10.1007/BF02253287

Anderson, D. J., & Adolphs, R. (2014). A framework for studying emotions across species. *Cell*, *157*(1), 187–200. https://doi.org/10.1016/j.cell.2014.03.003

Andics, A., Gábor, A., Gácsi, M., Faragó, T., Szabó, D., & Miklósi, Á. (2016). Neural mechanisms for lexical processing in dogs. *Science*, *353*(6303), 1030–1032. https://doi.org/10.1126/science.aaf3777

Andriella, A., Siqueira, H., Fu, D., Magg, S., Barros, P., Wermter, S., Torras, C., & Alenyà, G. (2021). Do I have a personality? Endowing care robots with context-dependent personality traits. *International Journal of Social Robotics*, *13*(8), 2081–2102. https://doi.org/10.1007/s12369-020-00690-5

Annett, J., & Stanton, N. A. (2000). *Task analysis*. Taylor & Francis.

Antle, M. C., & Silver, R. (2016). Circadian insights into motivated behavior. In E. H. Simpson & P. D. Balsam (Eds.), *Behavioral neuroscience of motivation. Current topics in behavioral neurosciences: Vol. 27* (pp. 137–169). Springer. https://doi.org/10.1007/7854_2015_384

Arkin, R. C. (1998). *Behavior-based robotics*. MIT Press.

Aronson, E., Willerman, B., & Floyd. Joanne. (1966). The effect of a pratfall on increasing interpersonal attractiveness. *Psychonomic Science*, *4*(6), 227–228. https://doi.org/10.3758/BF03342263

Asgharian, P., Panchea, A. M., & Ferland, F. (2022). A review on the use of mobile service robots in elderly care. *Robotics*, *11*(6), 127. https://doi.org/10.3390/robotics11060127

Atkinson, J. W. (1964). *An introduction to motivation*. Van Nostrad.

Aureli, M., Fiorilli, F., & Porfiri, M. (2012). Portraits of self-organization in fish schools interacting with robots. *Physica D*, *241*(9), 908–920. https://doi.org/10.1016/j.physd.2012.02.005

Barrett, L. (2012). Why behaviorism isn't satanism. In J. Vonk & T. K. Shackelford (Eds.), *The Oxford handbook of comparative evolutionary psychology* (pp. 17–38). Oxford University Press. https://doi.org/10.1093/oxfordhb/9780199738182.013.0002

Bartneck, C., Kanda, T., Ishiguro, H., & Hagita, N. (2009). My robotic doppelgänger - A critical look at the uncanny valley theory. *Proceedings of the 18th IEEE International Symposium on Robot and Human Interactive Communication, RO-MAN2009*, 269–276. https://doi.org/10.1109/ROMAN.2009.5326351

Bartneck, C., Kulić, D., & Croft, E. (2008). Measurement instruments for the anthropomorphism, animacy, likeability, perceived intelligence, and perceived safety of robots. In *Proceedings of the Metrics for Human-Robot Interaction Workshop in affiliation with the 3rd ACM/IEEE International Conference on Human-Robot Interaction (HRI 2008)* (Vol. 471). https://doi.org/10.1007/s12369-008-0001-3

Bartneck, C., van der Hoek, M., Mubin, O., & Al Mahmud, A. (2007). "Daisy, Daisy, give me your answer do!" Switching off a robot. *HRI '07: Proceedings of the ACM/ IEEE International Conference on Human-Robot Interaction*, 217–222. https://doi.org/10.1145/1228716.1228746

Basalla, G. (1988). *The evolution of technology*. Cambridge University Press.

Bassili, J. N. (1976). Temporal and spatial contingencies in the perception of social events. *Journal of Personality and Social Psychology*, *33*(6), 680–685. https://doi.org/10.1037/0022-3514.33.6.680

Bekey, G., Ambrose, R., Kumar, V., Lavery, D., Arthur Sanderson, A., Wilcox, B., Yuh, J., & Zheng, Y. (2008). *Robotics: State of the art and future challenges*. Imperial College Press.

Bell, A. M. (2007). Future directions in behavioural syndromes research. *Proceedings of the Royal Society B: Biological Sciences*, *274*(1611), 755–761. https://doi.org/10.1098/rspb.2006.0199

Belpaeme, T., Kennedy, J., Ramachandran, A., Scassellati, B., & Tanaka, F. (2018). Social robots for education: A review. *Science Robotics*, *3*, eeat5954. https://doi.org/10.1126/scirobotics.aat59

Berridge, K. C. (2004). Motivation concepts in behavioral neuroscience. *Physiology and Behavior*, *81*(2), 179–209. https://doi.org/10.1016/j.physbeh.2004.02.004

Berridge, K. C. (2018). Evolving concepts of emotion and motivation. *Frontiers in Psychology*, *9*, 1647. https://doi.org/10.3389/fpsyg.2018.01647

Berridge, K. C., & Kringelbach, M. L. (2015). Pleasure systems in the brain. *Neuron*, *86*(3), 646–664. https://doi.org/10.1016/j.neuron.2015.02.018

Binetti, N., Harrison, C., Coutrot, A., Johnston, A., & Mareschal, I. (2016). Pupil dilation as an index of preferred mutual gaze duration. *Royal Society Open Science*, *3*(7), 160086. https://doi.org/10.1098/rsos.160086

Biswas, S., Ali, I., Chakrabortty, R. K., Turan, H. H., Elsawah, S., & Ryan, M. J. (2022). Dynamic modeling for product family evolution combined with artificial neural network based forecasting model: A study of iPhone evolution. *Technological Forecasting and Social Change*, *178*, 121549. https://doi.org/10.1016/j.techfore.2022.121549

Blumberg, B. M. (2002). D-Learning: What learning in dogs tells us about building characters that learn what they ought to learn. In G. Lakemeyer & B. Nebel (Eds.), *Exploring artificial intelligence in the new millenium* (pp. 37–67). Morgan Kaufman Publishers.

Boden, M. A. (1996). *The philosophy of artificial life*. Oxford University Press.

Bowlby, J. (1969). *Attachment and loss: Vol. 1: Attachment*. Basic Books.

Breazeal, C. (2002). Regulation and entrainment in human-robot interaction. *The International Journal of Robotics Research*, *21*(10–11), 883–902. https://doi.org/10.1177/0278364902021010096

Breazeal, C. (2003a). Emotion and sociable humanoid robots. *International Journal of Human Computer Studies*, *59*(1–2), 119–155. https://doi.org/10.1016/S1071-5819(03)00018-1

Breazeal, C. (2003b). Toward sociable robots. *Robotics and Autonomous Systems*, *42*(3–4), 167–175. https://doi.org/10.1016/S0921-8890(02)00373-1

Breazeal, C., & Brooks, R. (2005). Robot emotion: A functional perspective. In J.-M. Fellous & M. A. Arbib (Eds.), *Who needs emotions? The brain meets the robot* (pp. 271–310). Oxford University Press. https://doi.org/10.1093/acprof:oso/9780195166194.003.0010

Breazeal, C., Kidd, C. D., Lockerd Thomaz, A., Hoffman, G., & Berlin, M. (2005). Effects of nonverbal communication on efficiency and robustness in human-robot teamwork. *2005 IEEE/RSJ International Conference on Intelligent Robots and Systems*, 708–713. https://doi.org/10.1109/IROS.2005.1545011

Breazeal, C., & Scassellati, B. (2002). Robots that imitate humans. *Trends in Cognitive Sciences*, *6*(11), 481–487. https://doi.org/10.1016/S1364-6613(02)02016-8
Broekens, J., Heerink, M., & Rosendal, H. (2009). Assistive social robots in elderly care: A review. *Gerontechnology*, *8*(2), 94–103. https://doi.org/10.4017/gt.2009.08.02.002.00
Brooks, R. A. (1986). A robust layered control system for a mobile robot. *IEEE Journal of Robotics and Automation*, *2*, 14–23.
Bryson, J. J. (2018). Patiency is not a virtue: The design of intelligent systems and systems of ethics. *Ethics and Information Technology*, *20*(1), 15–26. https://doi.org/10.1007/s10676-018-9448-6
Burgard, W., Cremers, A. B., Fox, D., Haehnel, D., Lakemeyer, G., Schulz, D., Steiner, W., & Thrun, S. (1998). Interactive museum tour-guide robot. *Proceedings of the National Conference on Artificial Intelligence*, 11–18.
Burghart, C. R., Mikut, R., & Holzapfel, H. (2007). Cognition-oriented building blocks of future benchmark scenarios for humanoid home robots. In *Proceedings of AAAI Workshop Evaluating Architectures for Intelligence*.
Burgoon, J. K. (1992). Applying a comparative approach to nonverbal expectancy violations theory. In J. Blumler, K. E. Rosengren, & J. M. McLeod (Eds.), *Comparatively speaking: Communication and culture Across space and time* (pp. 53–69). Sage.
Burgoon, J. K. (2016). Expectancy violations theory. In C. R. Berger, M. E. Roloff, S. R. Wilson, J. P. Dillard, J. Caughlin, & D. Solomon (Eds.), *The international encyclopedia of interpersonal communication* (pp. 1–9). John Wiley and Sons. https://doi.org/10.1002/9781118540190.wbeic0102
Burns, R., Jeon, M., & Park, C. H. (2018). Robotic motion learning framework to promote social engagement. *Applied Sciences*, *8*, 241. https://doi.org/10.3390/app8020241
Cabanac, M. (1971). Physiological role of pleasure. *Science*, *173*(4002), 1103–1107. https://doi.org/10.1126/science.173.4002.1103
Cabanac, M. (1992). Pleasure: The common currency. *Journal of Theoretical Biology*, *155*, 173–200. https://doi.org/10.1016/S0022-5193(05)80594-6
Cahya, D. E., Ramakrishnan, R., & Giuliani, M. (2019). Static and temporal differences in social signals between error-free and erroneous situations in human-robot collaboration. In M. A. Salichs, S. S. Ge, E. I. Barakova, J.-J. Cabibihan, A. R. Wagner, Á. Castro-González, & H. He (Eds.), *11th International Conference, ICSR 2019* (Vol. 11876, pp. 189–199). Springer International Publishing. https://doi.org/10.1007/978-3-030-35888-4_18
Cameron, D., Fernando, S., Collins, E. C., Millings, A., Szollosy, M., Moore, R., Sharkey, A., & Prescott, T. (2017). You made him be alive: Children's perceptions of animacy in a humanoid robot. In M. Mangan, M. Cutkosky, A. Mura, P. F. M. J. Verschure, T. Prescott, & N. Lepora (Eds.), *Biomimetic and Biohybrid Systems. Living Machines 2017. Lecture Notes in Computer Science* (pp. 73–85). Springer International Publishing AG. https://doi.org/10.1007/978-3-319-63537-8_7
Campos, A., Port, J. D., & Acosta, A. (2022). Integrative hedonic and homeostatic food intake regulation by the central nervous system: Insights from neuroimaging. *Brain Sciences*, *12*(4), 431. https://doi.org/10.3390/brainsci12040431
Cañamero, D. (1997). Modeling motivations and emotions as a basis for intelligent behavior. *AGENTS '97: Proceedings of the First International Conference on Autonomous Agents*, 148–155. https://doi.org/10.1145/267658.267688
Cangelosi, A., & Schlesinger, M. (2015). *Developmental robotics: From babies to robots*. The MIT Press. https://doi.org/10.7551/mitpress/9320.001.0001
Capitain, S., Miklósi, Á., & Abdai, J. (2023). Influence of reward and location on dogs' behaviour toward an interactive artificial agent. *Scientific Reports*, *13*(1), 1093. https://doi.org/10.1038/s41598-023-27930-8
Carp, S. B., Santistevan, A. C., Machado, C. J., Whitaker, A. M., Aguilar, B. L., & Bliss-Moreau, E. (2022). Monkey visual attention does not fall into the uncanny valley. *Scientific Reports*, *12*(1), 11760. https://doi.org/10.1038/s41598-022-14615-x

Cavallo, F., Semeraro, F., Fiorini, L., Magyar, G., Sinčák, P., & Dario, P. (2018). Emotion modelling for social robotics applications: A review. *Journal of Bionic Engineering*, *15*(2), 185–203. https://doi.org/10.1007/s42235-018-0015-y

Celiktutan, O., & Gunes, H. (2015). Computational analysis of human-robot interactions through first-person vision: Personality and interaction experience. *Proceedings of the 24th IEEE International Symposium on Robot and Human Interactive Communication*, 815–820. https://doi.org/10.0/Linux-x86_64

Chemero, A. (2023). LLMs differ from human cognition because they are not embodied. *Nature Human Behaviour*, *7*(11), 1828–1829. https://doi.org/10.1038/s41562-023-01723-5

Chen, H., Park, H. W., & Breazeal, C. (2020). Teaching and learning with children: Impact of reciprocal peer learning with a social robot on children's learning and emotive engagement. *Computers and Education*, *150*. https://doi.org/10.1016/j.compedu.2020.103836

Chen, H., Russell, R., & Nakayama, K. (2010). Crossing the 'uncanny valley': Adaptation to cartoon faces can influence perception of human faces. *Perception*, *39*(3), 378–386. https://doi.org/10.1068/p6492

Chen, Y., Wu, F., Shuai, W., Wang, N., Chen, R., & Chen, X. (2015). KeJia robot – An attractive shopping mall guider. In A. Tapus, E. André, J. C. Martin, F. Ferland, & M. Ammi (Eds.), *Social Robotics. ICSR 2015. Lecture Notes in Computer Science* (Vol. 9388). Springer.

Chipman, A. D., & Edgecombe, G. D. (2019). Developing an integrated understanding of the evolution of arthropod segmentation using fossils and evo-devo. *Proceedings of the Royal Society B: Biological Sciences*, *286*, 20191881. https://doi.org/10.1098/rspb.2019.1881

Choi, Y., & Luo, Y. (2023). Understanding preferences in infancy. *Wiley Interdisciplinary Reviews: Cognitive Science*, *14*(4), e1643. https://doi.org/10.1002/wcs.1643

Christaller, T., Decker, M., Gilsbach, J.-M., Hirzinger, G., Lauterbach, K., Schweighofer, E., Schweitzer, G., & Sturma, D. (2001). *Robotik: Perspektiven für menschliches Handeln in der zukünftigen Gesellschaft*. Springer-Verlag Berlin Heidelberg.

Chuah, S. H. W., & Yu, J. (2021). The future of service: The power of emotion in human-robot interaction. *Journal of Retailing and Consumer Services*, *61*, 102551. https://doi.org/10.1016/j.jretconser.2021.102551

Cochrane, T. (2009). Eight dimensions for the emotions. *Social Science Information*, *48*(3), 379–420. https://doi.org/10.1177/0539018409106198

Collins, E. C., Millings, A., & Prescott, T. J. (2013). Attachment to assistive technology: A new conceptualisation. In *Assistive technology research series. Volume 33: Assistive technology: From research to practice* (pp. 823–828). IOS Press. https://doi.org/10.3233/978-1-61499-304-9-823

Cruz-Sandoval, D., Morales-Tellez, A., Sandoval, E. B., & Favela, J. (2020). A social robot as therapy facilitator in interventions to deal with dementia-related behavioral symptoms. *ACM/IEEE International Conference on Human-Robot Interaction*, 161–169. https://doi.org/10.1145/3319502.3374840

Danaher, J. (2017). Should we be thinking about sex robots? In J. Danaher & N. McArthur (Eds.), *Robot sex: Social and ethical implications* (pp. 3–14). MIT Press.

Danaher, J. (2020). Welcoming robots into the moral circle: A defence of ethical behaviourism. *Science and Engineering Ethics*, *26*(4), 2023–2049. https://doi.org/10.1007/s11948-019-00119-x

Dario, P. (2005). Biorobotics. *Journal of the Robotics Society of Japan*, *23*(5), 552–554. https://doi.org/10.7210/jrsj.23.552

Das, D., Rashed, M. G., Kobayashi, Y., & Kuno, Y. (2015). Supporting human-robot interaction based on the level of visual focus of attention. *IEEE Transactions on Human-Machine Systems*, *45*(6), 664–675. https://doi.org/10.1109/THMS.2015.2445856

Dautenhahn, K. (2007). Socially intelligent robots: Dimensions of human-robot interaction. *Philosophical Transactions of the Royal Society B: Biological Sciences*, *362*(1480), 679–704. https://doi.org/10.1098/rstb.2006.2004

Dautenhahn, K., Ogden, B., & Quick, T. (2002). From embodied to socially embedded agents – Implications for interaction-aware robots. *Cognitive Systems Research*, *3*, 397–428. https://doi.org/10.1016/S1389-0417(02)00050-5

Dawkins, M. S. (1988). Behavioural deprivation: A central problem in animal welfare. *Applied Animal Behaviour Science*, *20*, 209–225. https://doi.org/10.1016/0168-1591(88)90047-0

Dawkins, R. (1982). *The extended phenotype*. Oxford University Press.

de Gelder, B. (2009). Why bodies? Twelve reasons for including bodily expressions in affective neuroscience. *Philosophical Transactions of the Royal Society B: Biological Sciences*, *364*(1535), 3475–3484. https://doi.org/10.1098/rstb.2009.0190

De Jaegher, H., Di Paolo, E., & Gallagher, S. (2010). Can social interaction constitute social cognition? *Trends in Cognitive Sciences*, *14*(10), 441–447. https://doi.org/10.1016/j.tics.2010.06.009

de Vignemont, F. (2020). Bodily awareness. In E. N. Zalta (Ed.), *The Stanford Encyclopedia of Philosophy* (Fall 2020). Metaphysics Research Lab, Stanford University. https://plato.stanford.edu/archives/fall2020/entries/bodily-awareness/

de Waal, F. B. M. (2011). What is an animal emotion? *Annals of the New York Academy of Sciences*, *1224*(1), 191–206. https://doi.org/10.1111/j.1749-6632.2010.05912.x

Delmerico, J., Mintchev, S., Giusti, A., Gromov, B., Melo, K., Horvat, T., Cadena, C., Hutter, M., Ijspeert, A., Floreano, D., Gambardella, L. M., Siegwart, R., & Scaramuzza, D. (2019). The current state and future outlook of rescue robotics. *Journal of Field Robotics*, *36*(7), 1171–1191. https://doi.org/10.1002/rob.21887

Deng, E., Mutlu, B., & Matarić, M. J. (2019). Embodiment in socially interactive robots. *Foundations and Trends® in Robotics*, *7*(4), 251–356. https://doi.org/10.1561/2300000056

Di Giacomo, D., Ranieri, J., D'Amico, M., Guerra, F., & Passafiume, D. (2019). Psychological barriers to digital living in older adults: Computer anxiety as predictive mechanism for technophobia. *Behavioral Sciences*, *9*, 96. https://doi.org/10.3390/bs9090096

Di Giorgio, E., Lunghi, M., Vallortigara, G., & Simion, F. (2021). Newborns' sensitivity to speed changes as a building block for animacy perception. *Scientific Reports*, *11*, 542. https://doi.org/10.1038/s41598-020-79451-3

Diel, A., Weigelt, S., & MacDorman, K. F. (2022). A meta-analysis of the uncanny valley's independent and dependent variables. *ACM Transactions on Human-Robot Interaction*, *11*(1), 1. https://doi.org/10.1145/3470742

Dingemanse, N. J., & Wolf, M. (2010). Recent models for adaptive personality differences: A review. *Philosophical Transactions of the Royal Society B: Biological Sciences*, *365*(1560), 3947–3958. https://doi.org/10.1098/rstb.2010.0221

Dubal, S., Foucher, A., Jouvent, R., & Nadel, J. (2011). Human brain spots emotion in non-humanoid robots. *Social Cognitive and Affective Neuroscience*, *6*(1), 90–97. https://doi.org/10.1093/scan/nsq019

Duffy, B. R. (2006). Fundamental issues in social robotics. *The International Review of Information Ethics*, *6*, 31–36. https://doi.org/10.29173/irie137

Dyke, C. (1988). *The evolutionary dynamics of complex systems*. Oxford University Press.

Eble, G. J. (2005). Morphological modularity and macroevolution: Conceptual and empirical aspects. In W. Callebaut & D. Rasskin-Gutman (Eds.), *Modularity: Understanding the development of evolution of complex natural systems* (pp. 221–238). MIT Press. https://doi.org/10.13140/2.1.1116.5120

Eibl-Eibesfeldt, I. (1989). *Human ethology*. Aldine de Gruyter.

Ekman, P., & Davidson, R. J. (1994). *The nature of emotion: Fundamental questions*. Oxford University Press.

Endsley, M. R. (1995). Toward a theory of situation awareness in dynamic systems. *Human Factors*, *37*(1), 32–64. https://doi.org/10.1518/001872095779049543

Endsley, M. R. (2015). Situation awareness misconceptions and misunderstandings. *Journal of Cognitive Engineering and Decision Making*, *9*(1), 4–32. https://doi.org/10.1177/1555343415572631

Engwall, O., & Lopes, J. (2022). Interaction and collaboration in robot-assisted language learning for adults. *Computer Assisted Language Learning*, *35*(5–6), 1273–1309. https://doi.org/10.1080/09588221.2020.1799821

Everett, J. A. C., Faber, N. S., & Crockett, M. (2015). Preferences and beliefs in ingroup favoritism. *Frontiers in Behavioral Neuroscience*, *9*, 15. https://doi.org/10.3389/fnbeh.2015.00015

Falck, A., Labouret, G., Izard, V., Wertz, A. E., Keil, F. C., & Strickland, B. (2020). Core cognition in adult vision: A surprising discrepancy between the principles of object continuity and solidity. *Journal of Experimental Psychology: General*, *149*, 2250–2263. https://doi.org/10.1037/xge0000785

Fang, B., Jia, S., Guo, D., Xu, M., Wen, S., & Sun, F. (2019). Survey of imitation learning for robotic manipulation. *International Journal of Intelligent Robotics and Applications*, *3*(4), 362–369. https://doi.org/10.1007/s41315-019-00103-5

Faragó, T., & Miklósi, Á. (2012). Cellphone evolution - applying evolution theory to an info-communication system. *2012 IEEE 3rd International Conference on Cognitive Infocommunications (CogInfoCom)*, 763–768. https://doi.org/10.1109/CogInfoCom.2012.6421953

Faria, J. J., Dyer, J. R. G., Clément, R. O., Couzin, I. D., Holt, N., Ward, A. J. W., Waters, D., & Krause, J. (2010). A novel method for investigating the collective behaviour of fish: Introducing 'Robofish'. *Behavioral Ecology and Sociobiology*, *64*(8), 1211–1218. https://doi.org/10.1007/s00265-010-0988-y

Feil-Seifer, D., Skinner, K., & Matarić, M. J. (2007). Benchmarks for evaluating socially assistive robotics. *Interaction Studies*, *8*(3), 423–439. https://doi.org/10.1075/is.8.3.07fei

Fellous, J.-M., & Arbib, M. A. (2005). *Who needs emotions? The brain meets the robot*. Oxford University Press.

Ferrey, A. E., Burleigh, T. J., & Fenske, M. J. (2015). Stimulus-category competition, inhibition, and affective devaluation: A novel account of the uncanny valley. *Frontiers in Psychology*, *6*, 249. https://doi.org/10.3389/fpsyg.2015.00249

Fink, J., Bauwens, V., Kaplan, F., & Dillenbourg, P. (2013). Living with a vacuum cleaning robot: A 6-month ethnographic study. *International Journal of Social Robotics*, *5*, 389–408. https://doi.org/10.1007/s12369-013-0190-2

Fitter, N. T., Funke, R., Pulido, J. C., Eisenman, L. E., Deng, W., Rosales, M. R., Bradley, N. S., Sargent, B., Smith, B. A., & Mataric, M. J. (2019). Socially assistive infant-robot interaction: Using robots to encourage infant leg-motion training. *IEEE Robotics and Automation Magazine*, *26*(2), 12–23. https://doi.org/10.1109/MRA.2019.2905644

Flaherty, C. F. (1982). Incentive contrast: A review of behavioral changes following shifts in reward. *Animal Learning & Behavior*, *10*(4), 409–440. https://doi.org/10.3758/BF03212282

Flanagan, J. R., & Beltzner, M. A. (2000). Independence of perceptual and sensorimotor predictions in the size-weight illusion. *Nature Neuroscience*, *3*(7), 737–741. https://doi.org/10.1038/76701

Floreano, D., Husbands, P., & Nolfi, S. (2008). Evolutionary robotics. In B. Siciliano & O. Khatib (Eds.), *Springer handbook of robotics* (pp. 1423–1451). Springer. https://doi.org/10.1007/978-3-540-30301-5_62

Fong, T., Nourbakhsh, I., & Dautenhahn, K. (2003). A survey of socially interactive robots. *Robotics and Autonomous Systems*, *42*(3–4), 143–166. https://doi.org/10.1016/S0921-8890(02)00372-X

Fosch-Villaronga, E., Lutz, C., & Tamò-Larrieux, A. (2020). Gathering expert opinions for social robots' ethical, legal, and societal concerns: Findings from four international workshops. *International Journal of Social Robotics*, *12*(2), 441–458. https://doi.org/10.1007/s12369-019-00605-z

Frith, C. (2009). Role of facial expressions in social interactions. *Philosophical Transactions of the Royal Society B: Biological Sciences*, *364*(1535), 3453–3458. https://doi.org/10.1098/rstb.2009.0142

Fugazza, C., & Miklósi, Á. (2014). Should old dog trainers learn new tricks? The efficiency of the Do as I do method and shaping/clicker training method to train dogs. *Applied Animal Behaviour Science*, *153*, 53–61. https://doi.org/10.1016/j.applanim.2014.01.009

Fukuhara, A., Gunji, M., & Masuda, Y. (2022). Comparative anatomy of quadruped robots and animals: A review. *Advanced Robotics*, *36*(13), 612–630. https://doi.org/10.1080/01691864.2022.2086018

Gácsi, M., Szakadát, S., & Miklósi, Á. (2013). Assistance dogs provide a useful behavioral model to enrich communicative skills of assistance robots. *Frontiers in Psychology*, *4*, 971. https://doi.org/10.3389/fpsyg.2013.00971

Gallagher, S. (2001). The practice of mind: Theory, simulation or primary interaction? *Journal of Consciousness Studies*, *8*(5–7), 83–108. https://www.researchgate.net/publication/233674767

Gallese, V., & Goldman, A. (1998). Mirror neurons and the simulation theory of mind-reading. *Trends in Cognitive Sciences*, *2*(12), 493–501. https://doi.org/10.1016/S1364-6613(98)01262-5

Gallistel, C. R. (1990). *The organization of learning*. MIT Press.

Gambino, A., Fox, J., & Ratan, R. A. (2020). Building a stronger CASA: Extending the computers are social actors paradigm. *Human-Machine Communication*, *1*(1), 71–85. https://doi.org/10.30658/hmc.1.5

Gao, Y., & Huang, C. M. (2022). Evaluation of socially-aware robot navigation. *Frontiers in Robotics and AI*, *8*, 721317. https://doi.org/10.3389/frobt.2021.721317

Garcia-Haro, J. M., Oña, E. D., Hernandez-Vicen, J., Martinez, S., & Balaguer, C. (2021). Service robots in catering applications : A review and future challenges. *Electronics*, *10*, 47. https://doi.org/10.3390/electronics10010047

Gavoille, N., & Hazans, M. (2022). Personality traits, remote work and productivity. *IZA Discussion Paper*, 15486. https://doi.org/10.2139./ssrn.4188297

Gergely, A., Abdai, J., Petró, E., Kosztolányi, A., Topál, J., & Miklósi, Á. (2015). Dogs rapidly develop socially competent behaviour while interacting with a contingently responding self-propelled object. *Animal Behaviour*, *108*, 137–144. https://doi.org/10.1016/j.anbehav.2015.07.024

Gergely, A., Faragó, T., Galambos, Á., & Topál, J. (2017). Differential effects of speech situations on mothers' and fathers' infant-directed and dog-directed speech: An acoustic analysis. *Scientific Reports*, *7*(1), 13739. https://doi.org/10.1038/s41598-017-13883-2

Gergely, G., & Watson, J. S. (1999). Early social-emotional development: Contingency perception and the social biofeedback model. In P. Rochat (Ed.), *Early social cognition: Understanding others in the first months of life* (pp. 101–136). Lawrence Erlbaum Associates.

Gibson, E. J. (1988). Exploratory behavior in the development of perceiving, acting, and the acquiring of knowledge. *Annual Review of Psychology*, *39*, 1–41. https://doi.org/10.1146/annurev.ps.39.020188.000245

Gildert, N., Millard, A. G., Pomfret, A., & Timmis, J. (2018). The need for combining implicit and explicit communication in cooperative robotic systems. *Frontiers Robotics AI*, *5*, 65. https://doi.org/10.3389/frobt.2018.00065

Goetz, J., Kiesler, S., & Powers, A. (2003). Matching robot appearance and behavior to tasks to improve human-robot cooperation. *The 12th IEEE International Workshop on Robot and Human Interactive Communication, 2003. Proceedings. ROMAN 2003*, 55–60. https://doi.org/10.1109/ROMAN.2003.1251796

Goldman, A. I. (1989). Interpretation psychologized. *Mind & Language*, *4*(3), 161–185. https://doi.org/10.1111/j.1468-0017.1989.tb00249.x

Goldman, A. I. (2012). Theory of mind. In E. Margolis, R. Samuels, & S. P. Stich (Eds.), *The Oxford handbook of philosophy of cognitive science* (pp. 402–424). Oxford University Press. https://doi.org/10.1093/oxfordhb/9780195309799.013.0017

Goodrich, M. A., & Schultz, A. C. (2007). Human-robot interaction: A survey. *Foundations and Trends in Human-Computer Interaction*, *1*(3), 203–275. https://doi.org/10.1561/1100000005

Gosling, S. D., & John, O. P. (1999). Personality dimensions in nonhuman animals: A cross-species review. *Current Directions in Psychological Science*, *8*(3), 69–75. https://doi.org/10.1111/1467-8721.00017

Grammer, K., Schiefenhövel, W., Schleidt, M., Lorenz, B., & Eibl-Eibesfeldt, I. (1988). Patterns on the face: The eyebrow flash in crosscultural comparison. *Ethology*, *77*, 279–299. https://doi.org/10.1111/j.1439-0310.1988.tb00211.x

Guerra, A., Parisi, F., & Pi, D. (2022). Liability for robots I: Legal challenges. *Journal of Institutional Economics*, *18*, 331–343. https://doi.org/10.1017/S1744137421000825

Hall, E. T. (1966). *The hidden dimensions*. Doubleday.

Halloy, J., Sempo, G., Caprari, G., Rivault, C., Asadpour, M., Tâche, F., Saïd, I., Durier, V., Canonge, S., Amé, J. M., Detrain, C., Correll, N., Martinoli, A., Mondada, F., Siegwart, R., & Deneubourg, J. L. (2007). Social integration of robots into groups of cockroaches to control self-organized choices. *Science, 318*(5853), 1155–1158. https://doi.org/10.1126/science.1144259

Hanson, D. (2005). Expanding the aesthetic possibilities for humanoid robots. *Proceedings of the 5th IEEE-RAS International Conference on Humanoid Robots*, 24–31.

Haring, K. S., Watanabe, K., Velonaki, M., Tossell, C. C., & Finomore, V. (2018). FFAB – The form function attribution bias in human robot interaction. *IEEE Transactions on Cognitive and Developmental Systems*, *10*(4), 843–851. https://doi.org/10.1109/TCDS.2018.2851569

Harnad, S. (1990). The symbol grounding problem. *Physica D: Nonlinear Phenomena*, *42*(1–3), 335–346. https://doi.org/10.1016/0167-2789(90)90087-6

Haslam, N. (1994). Mental representation of social relationships: Dimensions, laws, or categories? *Journal of Personality and Social Psychology*, *67*(4), 575–584. https://doi.org/10.1037/0022-3514.67.4.575

Hassenstein, B. (1965). *Biologische kybernetik: eine elementare Einführung*. Heidelberg Quelle & Meyer.

Hazan, C., & Shaver, P. (1987). Romantic love conceptualized as an attachment process. *Journal of Personality and Social Psychology*, *52*(3), 511–524. https://doi.org/10.1037/0022-3514.52.3.511

Hecht, J., & Horowitz, A. (2015). Seeing dogs: Human preferences for dog physical attributes. *Anthrozoös*, *28*(1), 153–163. https://doi.org/10.2752/089279315X14129350722217

Hecht, J., Miklósi, Á., & Gácsi, M. (2012). Behavioral assessment and owner perceptions of behaviors associated with guilt in dogs. *Applied Animal Behaviour Science*, *139*(1–2), 134–142. https://doi.org/10.1016/j.applanim.2012.02.015

Hecker, J. P., & Moses, M. E. (2013). An evolutionary approach for robust adaptation of robot behavior to sensor error. *GECCO '13 Companion: Proceedings of the 15th Annual Conference Companion on Genetic and Evolutionary Computation*, 1437–1444. https://doi.org/10.1145/2464576.2482724

Heider, F., & Simmel, M. (1944). An experimental study of apparent behavior. *The American Journal of Psychology*, *57*(2), 243–259. https://doi.org/10.2307/1416950

Henrich, J., Heine, S. J., & Norenzayan, A. (2010). The weirdest people in the world? *Behavioral and Brain Sciences*, *33*(2–3), 61–83. https://doi.org/10.1017/S0140525X0999152X

Heyes, C. (2000). Evolutionary psychology in the round. In C. Heyes & L. Huber (Eds.), *The evolution of cognition* (pp. 3–22). MIT Press.

Hieida, C., & Nagai, T. (2022). Survey and perspective on social emotions in robotics. *Advanced Robotics*, *36*(1–2), 17–32. https://doi.org/10.1080/01691864.2021.2012512

Hiolle, A., Cañamero, L., Davila-Ross, M., & Bard, K. A. (2012). Eliciting caregiving behavior in dyadic human-robot attachment-like interactions. *ACM Transactions on Interactive Intelligent Systems*, *2*(1), 1–24. https://doi.org/10.1145/2133366.2133369

Hiolle, A., Lewis, M., & Cañamero, L. (2014). Arousal regulation and affective adaptation to human responsiveness by a robot that explores and learns a novel environment. *Frontiers in Neurorobotics*, *8*, 17. https://doi.org/10.3389/fnbot.2014.00017

Hoffmann, L., Derksen, M., & Kopp, S. (2020). What a pity, Pepper! How warmth in robots' language impacts reactions to errors during a collaborative task. *ACM/IEEE International Conference on Human-Robot Interaction*, 245–247. https://doi.org/10.1145/3371382.3378242

Hoffmann, M. (2021). Body models in humans, animals, and robots: Mechanisms and plasticity. In Y. Ataria, S. Tanaka, & S. Gallagher (Eds.), *Body schema and body image: New directions* (pp. 152–180). Oxford University Press. https://doi.org/10.1093/oso/9780198851721.003.0010

Holland, J. H. (1992). Complex adaptive systems. *Daedalus. A New Era in Computation*, *121*(1), 17–30.

Holland, O. (2003). The first biologically inspired robots. *Robotica*, *21*(4), 351–363. https://doi.org/10.1017/S0263574703004971

Holz, D., Iocchi, L., & van der Zant, T. (2013). Benchmarking intelligent service robots through scientific competitions: The RoboCup@Home approach. In *AAAI Spring Symposium Series*.

Hommel, B. (2022). GOALIATH: A theory of goal-directed behavior. *Psychological Research*, *86*(4), 1054–1077. https://doi.org/10.1007/s00426-021-01563-w

Honig, S., & Oron-Gilad, T. (2018). Understanding and resolving failures in human-robot interaction: Literature review and model development. *Frontiers in Psychology*, *9*, 861. https://doi.org/10.3389/fpsyg.2018.00861

Hortensius, R., Hekele, F., & Cross, E. S. (2018). The perception of emotion in artificial agents. *IEEE Transactions on Cognitive and Developmental Systems*, *10*(4), 852–864. https://doi.org/10.1109/TCDS.2018.2826921

Hsiao, H. S., Chang, C. S., Lin, C. Y., & Hsu, H. L. (2015). "iRobiQ": The influence of bidirectional interaction on kindergarteners' reading motivation, literacy, and behavior. *Interactive Learning Environments*, *23*(3), 269–292. https://doi.org/10.1080/10494820.2012.745435

Hudlicka, E. (2004). Beyond cognition: Modeling emotion in cognitive architectures. *Proceedings of the International Conference on Cognitive Modelling*, 118–123. https://www.researchgate.net/publication/221548280

International Federation of Robotics - IFR. (2023). *Industrial Robots*. https://ifr.org/industrial-robots

Jackson, R. B., & Williams, T. (2021). A theory of social agency for human-robot interaction. *Frontiers in Robotics and AI*, *8*, 687726. https://doi.org/10.3389/frobt.2021.687726

Jentsch, E. (1997). On the psychology of the uncanny (1906). *Angelaki - Journal of the Theoretical Humanities*, *2*(1), 7–16. https://doi.org/10.1080/09697259708571910

Jochum, E., Vlachos, E., Christoffersen, A., Nielsen, S. G., Hameed, I. A., & Tan, Z. H. (2016). Using theatre to study interaction with care robots. *International Journal of Social Robotics*, *8*(4), 457–470. https://doi.org/10.1007/s12369-016-0370-y

Johnson, C. M. (2001). Distributed primate cognition: A review. *Animal Cognition*, *3*(4), 167–183. https://doi.org/10.1007/s100710100077

Jonas, E., & Kording, K. P. (2017). Could a neuroscientist understand a microprocessor? *PLoS Computational Biology*, *13*(1), e1005268. https://doi.org/10.1371/journal.pcbi.1005268

Kahn, P. H., Friedman, B., Pérez-Granados, D. R., & Freier, N. G. (2006). Robotic pets in the lives of preschool children. *Interaction Studies*, *7*(3), 405–436. https://doi.org/10.1075/is.7.3.13kah

Kahn, P. H., Kanda, T., Ishiguro, H., Freier, N. G., Severson, R. L., Gill, B. T., Ruckert, J. H., & Shen, S. (2012). 'Robovie, you'll have to go into the closet now': Children's social and moral relationships with a humanoid robot. *Developmental Psychology*, *48*(2), 303–314. https://doi.org/10.1037/a0027033

Kaisler, S. H., & Madey, G. (2009). *Complex adaptive systems: Emergence and self-organization*. Lecture. Hawaii International Conference on Systems Sciences 42 (HICSS 42). https://www3.nd.edu/~gmadey/Activities/CAS-Briefing.pdf

Kanda, T., Shiomi, M., Miyashita, Z., Ishiguro, H., & Hagita, N. (2009). An affective guide robot in a shopping mall. *HRI '09: Proceedings of the 4th ACM/IEEE International Conference on Human Robot Interaction*, 173–180. https://doi.org/10.1145/1514095.1514127

Kaplan, F., Oudeyer, P.-Y., Kubinyi, E., & Miklósi, Á. (2002). Robotic clicker training. *Robotics and Autonomous Systems*, *38*, 197–206.

Kasuga, H., & Ikeda, Y. (2020). Gap between owner's perceptions and dog's behaviors toward the same physical agents: Using a dog-like speaker and a humanoid robot. *HAI 2020-Proceedings of the 8th International Conference on Human-Agent Interaction*, 96–104. https://doi.org/10.1145/3406499.3415068

Khan, A. S., Anwar, A. A., & Azeem, K. (2021). Bio-inspired motivational architecture for autonomous robots using glucose and insulin theories. *Evolutionary Intelligence*, *14*(4), 2039–2050. https://doi.org/10.1007/s12065-020-00489-3

Kim, B. H., Jo, S., & Choi, S. (2020). A-Situ: A computational framework for affective labeling from psychological behaviors in real-life situations. *Scientific Reports*, *10*, 15916. https://doi.org/10.1038/s41598-020-72829-3

Kim, H., Kwak, S. S., & Kim, M. (2008). Personality design of sociable robots by control of gesture design factors. *Proceedings of the 17th IEEE International Symposium on Robot and Human Interactive Communication*, 494–499.

Kim, J. T., & Shim, H. S. (2007). *The characteristics of the Myers-Briggs Type Indicator*. Assessta Publishers.

Kim, K., Boelling, L., Haesler, S., Bailenson, J., Bruder, G., & Welch, G. F. (2018). Does a digital assistant need a body? The influence of visual embodiment and social behavior on the perception of intelligent virtual agents in AR. *Proceedings of the 2018 IEEE International Symposium on Mixed and Augmented Reality, ISMAR 2018*, 105–114. https://doi.org/10.1109/ISMAR.2018.00039

Kishi, T., Kojima, T., Endo, N., Destephe, M., Otani, T., Jamone, L., Kryczka, P., Trovato, G., Hashimoto, K., Cosentino, S., & Takanishi, A. (2013). Impression survey of the emotion expression humanoid robot with mental model based dynamic emotions. *2013 IEEE International Conference on Robotics and Automation (ICRA)*, 1663–1668. https://doi.org/10.1109/ICRA.2013.6630793

Klingenberg, C. P. (2014). Studying morphological integration and modularity at multiple levels: Concepts and analysis. *Philosophical Transactions of the Royal Society B: Biological Sciences*, *369*(1649), 20130249. https://doi.org/10.1098/rstb.2013.0249

Kompatsiari, K., Bossi, F., & Wykowska, A. (2021). Eye contact during joint attention with a humanoid robot modulates oscillatory brain activity. *Social Cognitive and Affective Neuroscience*, *16*(4), 383–392. https://doi.org/10.1093/scan/nsab001

Kompatsiari, K., Ciardo, F., De Tommaso, D., & Wykowska, A. (2019). Measuring engagement elicited by eye contact in human-robot interaction. *2019 IEEE/RSJ International Conference on Intelligent Robots and Systems (IROS)*, 6979–6985. https://doi.org/10.1109/IROS40897.2019.8967747

Konidaris, G., & Barto, A. (2006). An adaptive robot motivational system. *Lecture Notes in Computer Science (Including Subseries Lecture Notes in Artificial Intelligence and Lecture Notes in Bioinformatics)*, *4095 LNAI*, 346–356. https://doi.org/10.1007/11840541_29

Konok, V., Gigler, D., Bereczky, B. M., & Miklósi, Á. (2016). Humans' attachment to their mobile phones and its relationship with interpersonal attachment style. *Computers in Human Behavior*, *61*, 537–547. https://doi.org/10.1016/j.chb.2016.03.062

Konok, V., Kosztolányi, A., Rainer, W., Mutschler, B., Halsband, U., & Miklósi, Á. (2015). Influence of owners' attachment style and personality on their dogs' (*Canis familiaris*) separation-related disorder. *PLoS One*, *10*(2), e0118375. https://doi.org/10.1371/journal.pone.0118375

Konok, V., Pogány, Á., & Miklósi, Á. (2017). Mobile attachment: Separation from the mobile phone induces physiological and behavioural stress and attentional bias to separation-related stimuli. *Computers in Human Behavior*, *71*, 228–239. https://doi.org/10.1016/j.chb.2017.02.002

Korcsok, B., Faragó, T., Ferdinandy, B., Miklósi, Á., Korondi, P., & Gácsi, M. (2020). Artificial sounds following biological rules: A novel approach for non-verbal communication in HRI. *Scientific Reports*, *10*, 7080. https://doi.org/10.1038/s41598-020-63504-8

Korcsok, B., Konok, V., Persa, G., Faragó, T., Niitsuma, M., Miklósi, Á., Korondi, P., Baranyi, P., & Gácsi, M. (2018). Biologically inspired emotional expressions for artificial agents. *Frontiers in Psychology*, *9*, 1191. https://doi.org/10.3389/fpsyg.2018.01191

Kory-Westlund, J. M., & Breazeal, C. (2019). A long-term study of young children's rapport, social emulation, and language learning with a peer-like robot playmate in preschool. *Frontiers in Robotics and AI*, *6*, 81. https://doi.org/10.3389/frobt.2019.00081

Koski, S. E. (2014). Broader horizons for animal personality research. *Frontiers in Ecology and Evolution*, *2*, 70. https://doi.org/10.3389/fevo.2014.00070

Kotseruba, I., & Tsotsos, J. K. (2020). 40 years of cognitive architectures: Core cognitive abilities and practical applications. *Artificial Intelligence Review*, *53*(1), 17–94. https://doi.org/10.1007/s10462-018-9646-y

Kouroupa, A., Laws, K. R., Irvine, K., Mengoni, S. E., Baird, A., & Sharma, S. (2022). The use of social robots with children and young people on the autism spectrum: A systematic review and meta-analysis. *PLoS One*, *17*(6), e0269800. https://doi.org/10.1371/journal.pone.0269800

Kovács, S., Gácsi, M., Vincze, D., Korondi, P., & Miklósi, Á. (2011). A novel, ethologically inspired HRI model implementation: Simulating dog-human attachment. *2nd International Conference on Cognitive Infocommunications.*

Krämer, N. C., Eimler, S., von der Pütten, A., & Payr, S. (2011). Theory of companions: What can theoretical models contribute to applications and understanding of human-robot interaction? *Applied Artificial Intelligence*, *25*(6), 474–502. https://doi.org/10.1080/08839514.2011.587153

Krause, J., Winfield, A. F. T., & Deneubourg, J.-L. (2011). Interactive robots in experimental biology. *Trends in Ecology & Evolution*, *26*(7), 369–375. https://doi.org/10.1016/j.tree.2011.03.015

Krupenye, C., & Call, J. (2019). Theory of mind in animals: Current and future directions. *Wiley Interdisciplinary Reviews: Cognitive Science*, *10*(6), e1503. https://doi.org/10.1002/wcs.1503

Kubinyi, E., Miklósi, Á., Kaplan, F., Gácsi, M., Topál, J., & Csányi, V. (2004). Social behaviour of dogs encountering AIBO, an animal-like robot in a neutral and in a feeding situation. *Behavioural Processes*, *65*(3), 231–239. https://doi.org/10.1016/j.beproc.2003.10.003

Kubinyi, E., Pongrácz, P., & Miklósi, Á. (2010). Can you kill a robot nanny?: Ethological approach to the effect of robot caregivers on child development and human evolution. *Interaction Studies*, *11*(2), 214–219. https://doi.org/10.1075/is.11.2.06kub

Kunold, L., & Onnasch, L. (2022). A framework to study and design communication with social robots. *Robotics*, *11*(6), 129. https://doi.org/10.3390/robotics11060129

Lakatos, G., Gácsi, M., Konok, V., Brúder, I., Bereczky, B., Korondi, P., & Miklósi, Á. (2014). Emotion attribution to a non-humanoid robot in different social situations. *PLoS One*, *9*(12), e114207. https://doi.org/10.1371/journal.pone.0114207

Lakatos, G., Janiak, M., Malek, L., Muszynski, R., Konok, V., Tchon, K., & Miklósi, Á. (2014). Sensing sociality in dogs: What may make an interactive robot social? *Animal Cognition*, *17*, 387–397. https://doi.org/10.1007/s10071-013-0670-7

Lakin, J. L., Jefferis, V. E., Cheng, C. M., & Chartrand, T. L. (2003). The chameleon effect as social glue: Evidence for the evolutionary significance of nonconscious mimicry. *Journal of Nonverbal Behavior*, *27*(3), 145–162. https://doi.org/10.1023/A:1025389814290

Lazzeri, N., Mazzei, D., Zaraki, A., & De Rossi, D. (2013). Towards a believable social robot. In N. F. Lepora, A. Mura, H. G. Krapp, P. F. M. J. Verschure, & T. J. Prescott (Eds.),

Biomimetic and Biohybrid Systems. Living Machines 2013. Lecture Notes in Computer Science: Vol. 8064. Springer.

LeDoux, J. E. (2012). Evolution of human emotion: A view through fear. In M. A. Hofman & D. Falk (Eds.), *Progress in brain research: Vol. 195* (pp. 431–442). https://doi.org/10.1016/B978-0-444-53860-4.00021-0

Lee, I. (2021). Service robots: A systematic literature review. *Electronics*, *10*(21), 2658. https://doi.org/10.3390/electronics10212658

Lee, M. K., Kiesler, S., Forlizzi, J., Srinivasa, S., & Rybski, P. (2010). Gracefully mitigating breakdowns in robotic services. *2010 5th ACM/IEEE International Conference on Human-Robot Interaction (HRI)*. https://doi.org/10.1109/HRI.2010.5453195

Leslie, A. M., Friedman, O., & German, T. P. (2004). Core mechanisms in 'theory of mind'. *Trends in Cognitive Sciences*, *8*(12), 528–533. https://doi.org/10.1016/j.tics.2004.10.001

Letunic, I., & Bork, P. (2021). Interactive tree of life (iTOL) v5: An online tool for phylogenetic tree display and annotation. *Nucleic Acids Research*, *49*(W1), W293–W296. https://doi.org/10.1093/nar/gkab301

Lewkowicz, D. J., & Ghazanfar, A. A. (2012). The development of the uncanny valley in infants. *Developmental Psychobiology*, *54*(2), 124–132. https://doi.org/10.1002/dev.20583

Lin, V., Yeh, H. C., & Chen, N. S. (2022). A systematic review on oral interactions in robot-assisted language learning. *Electronics*, 11(2), 290. https://doi.org/10.3390/electronics11020290

Little, A. C., & Roberts, S. C. (2012). Evolutionary psychology evolution, appearance, and occupational success. *Evolutionary Psychology*, *10*(5), 782–801. https://doi.org/10.1177/147470491201000050

Liu, B., Tetteroo, D., & Markopoulos, P. (2022). A systematic review of experimental work on persuasive social robots. *International Journal of Social Robotics*, *14*(6), 1339–1378. https://doi.org/10.1007/s12369-022-00870-5

Looser, C. E., & Wheatley, T. (2010). The tipping point of animacy: How, when, and where we perceive life in a face. *Psychological Science*, *21*(12), 1854–1862. https://doi.org/10.1177/0956797610388044

Lorenzini, M., Lagomarsino, M., Fortini, L., Gholami, S., & Ajoudani, A. (2023). Ergonomic human-robot collaboration in industry: A review. *Frontiers in Robotics and AI*, *9*, 813907. https://doi.org/10.3389/frobt.2022.813907

Lutz, C., Schöttler, M., & Hoffmann, C. P. (2019). The privacy implications of social robots: Scoping review and expert interviews. *Mobile Media and Communication*, *7*(3), 412–434. https://doi.org/10.1177/2050157919843961

MacDorman, K. F., & Ishiguro, H. (2006). The uncanny advantage of using androids in cognitive and social science research. *Interaction Studies*, *7*(3), 297–337. https://doi.org/10.1075/is.7.3.10mac

Mangin, O., Roncone, A., & Scassellati, B. (2022). How to be helpful? Supportive behaviors and personalization for human-robot collaboration. *Frontiers in Robotics and AI*, *8*, 725780. https://doi.org/10.3389/frobt.2021.725780

Maniscalco, U., Storniolo, P., & Messina, A. (2022). Bidirectional multi-modal signs of checking human-robot engagement and interaction. *International Journal of Social Robotics*, *14*(5), 1295–1309. https://doi.org/10.1007/s12369-021-00855-w

Maris, M., & te Boekhorst, R. (1996). Exploiting physical constraints: Heap formation through behavioral error in a group of robots. *Proceedings of IEEE/RSJ International Conference on Intelligent Robots and Systems. IROS '96*, 1655–1660. https://doi.org/10.1109/IROS.1996.569034

Marshall-Pescini, S., Ceretta, M., & Prato-Previde, E. (2014). Do domestic dogs understand human actions as goal-directed? *PLoS One*, *9*(9), e106530. https://doi.org/10.1371/journal.pone.0106530

Matheson, E., Minto, R., Zampieri, E. G. G., Faccio, M., & Rosati, G. (2019). Human-robot collaboration in manufacturing applications: A review. *Robotics*, *8*(4), 100. https://doi.org/10.3390/robotics8040100

Mathur, M. B., & Reichling, D. B. (2016). Navigating a social world with robot partners: A quantitative cartography of the Uncanny Valley. *Cognition*, *146*, 22–32. https://doi.org/10.1016/j.cognition.2015.09.008

Maturana, H. R., & Varela, F. J. (1987). *The tree of knowledge: The biological roots of human understanding*. New Science Library/Shambhala Publications.

Mauss, I. B., McCarter, L., Levenson, R. W., Wilhelm, F. H., & Gross, J. J. (2005). The tie that binds? Coherence among emotion experience, behavior, and physiology. *Emotion*, *5*(2), 175–190. https://doi.org/10.1037/1528-3542.5.2.175

McCrae, R. R., & John, O. P. (1992). An introduction to the five-factor model and its applications. *Journal of Personality*, *60*(2), 175–215. https://doi.org/10.1111/j.1467-6494.1992.tb00970.x

McFarland, D. (1989). Goals, no-goals and own goals. In A. Montefiore & D. Noble (Eds.), *Goals, no-goals and own goals: A debate on goal-directed and intentional behaviour*. Routledge.

Melson, G. F., Kahn, P. H., Beck, A. M., & Friedman, B. (2009). Robotic pets in human lives: Implications for the human - Animal bond and for human relationships with personified technologies. *Journal of Social Issues*, *65*(3), 545–567. https://doi.org/10.1111/j.1540-4560.2009.01613.x

Mendl, M., Neville, V., & Paul, E. S. (2022). Bridging the gap: Human emotions and animal emotions. *Affective Science*, *3*(4), 703–712. https://doi.org/10.1007/s42761-022-00125-6

Mennie, N. R., Hayhoe, M. M., Stupak, N., & Sullivan, B. T. (2005). Sources of information for catching balls. *Journal of Vision*, *5*(8), 383–383. https://doi.org/10.1167/5.8.383

Merritt, M. (2021). *Minding dogs: Humans, canine companions, and a new philosophy of cognitive science*. The University of Georgia.

Miklósi, Á. (2007). *Dog behaviour, evolution, and cognition* (1st ed.). Oxford University Press.

Miklósi, Á. (2015). *Dog behaviour, evolution, and cognition* (2nd ed.). Oxford University Press.

Miklósi, Á., Korondi, P., Matellán, V., & Gácsi, M. (2017). Ethorobotics: A new approach to human-robot relationship. *Frontiers in Psychology*, *8*, 958. https://doi.org/10.3389/fpsyg.2017.00958

Miłkowski, M. (2018). From computer metaphor to computational modeling: The evolution of computationalism. *Minds and Machines*, *28*(3), 515–541. https://doi.org/10.1007/s11023-018-9468-3

Mirnig, N., Stollnberger, G., Miksch, M., Stadler, S., Giuliani, M., & Tscheligi, M. (2017). To err is robot: How humans assess and act toward an erroneous social robot. *Frontiers in Robotics and AI*, *4*, 21. https://doi.org/10.3389/frobt.2017.00021

Mitra, S., & Crawley, E. (2014). Effectiveness of self-organised learning by children: Gateshead experiments. *Journal of Education and Human Development*, *3*(3), 79–88. https://doi.org/10.15640/JEHD.V3N3A6

Mohapatra, R. K., Panda, S., & Acharya, U. R. (2014). Study on activity pattern and incidence of stereotypic behavior in captive tigers. *Journal of Veterinary Behavior: Clinical Applications and Research*, *9*(4), 172–176. https://doi.org/10.1016/j.jveb.2014.04.003

Moll, H., Pueschel, E., Ni, Q., & Little, A. (2021). Sharing experiences in infancy: From primary intersubjectivity to shared intentionality. *Frontiers in Psychology*, *12*, 667679. https://doi.org/10.3389/fpsyg.2021.667679

Moore, A., Wozniak, M., Yousef, A., Barnes, C. C., Cha, D., Courchesne, E., & Pierce, K. (2018). The geometric preference subtype in ASD: Identifying a consistent, early-emerging phenomenon through eye tracking. *Molecular Autism*, *9*(1), 19. https://doi.org/10.1186/s13229-018-0202-z

Moreno, A., Etxeberria, A., & Umerez, J. (2008). The autonomy of biological individuals and artificial models. *Biosystems*, *91*(2), 309–319. https://doi.org/10.1016/j.biosystems.2007.05.009

Mori, M. (1970). The uncanny valley. *Energy*, *7*, 33–35.

Morovitz, M., Mueller, M., & Scheutz, M. (2017). Animal-robot interaction: The role of human likeness on the success of dog-robot interactions. *1st International Workshop on Vocal Interactivity In-and-between Humans, Animals and Robots*.

Morton, E. S. (1977). On the occurrence and significance of motivation-structural rules in some bird and mammal sounds. *The American Naturalist*, *111*(981), 855–869. https://doi.org/10.1086/283219

Mou, Y., Shi, C., Shen, T., & Xu, K. (2020). A systematic review of the personality of robot: Mapping its conceptualization, operationalization, contextualization and effects. *International Journal of Human-Computer Interaction*, *36*(6), 591–605. https://doi.org/10.1080/10447318.2019.1663008

Muro, C., Escobedo, R., Spector, L., & Coppinger, R. P. (2011). Wolf-pack (*Canis lupus*) hunting strategies emerge from simple rules in computational simulations. *Behavioural Processes*, *88*(3), 192–197. https://doi.org/10.1016/j.beproc.2011.09.006

Mutlu, B., Duff, M., & Turkstra, L. (2019). Social-cue perception and mentalizing ability following traumatic brain injury: A human-robot interaction study. *Brain Injury*, *33*(1), 23–31. https://doi.org/10.1080/02699052.2018.1531305

Naderi, S., Miklósi, Á., Dóka, A., & Csányi, V. (2001). Co-operative interactions between blind persons and their dogs. *Applied Animal Behaviour Science*, *74*(1), 59–80. https://doi.org/10.1016/S0168-1591(01)00152-6

Nass, C., & Moon, Y. (2000). Machines and mindlessness: Social responses to computers. *Journal of Social Issues*, *56*(1), 81–103. https://doi.org/10.1111/0022-4537.00153

Nass, C., Steuer, J., & Tauber, E. R. (1994). Computers are social actors. *Proceedings of the SIGCHI Conference on Human Factors in Computing Systems*, 72–78. https://doi.org/10.1145/191666.191703

Nazir, T. A., Lebrun, B., & Li, B. (2023). Improving the acceptability of social robots: Make them look different from humans. *PLoS One*, *18*(11), e0287507. https://doi.org/10.1371/journal.pone.0287507

Niitsuma, M., Takahashi, Y., & Takahashi, S. (2014). Improvement in mutual interaction between robot and person for attachment behavior of robot. *IECON 2014–40th Annual Conference of the IEEE Industrial Electronics Society*, 4022–4027. https://doi.org/10.1109/IECON.2014.7049104

Nocentini, O., Fiorini, L., Acerbi, G., Sorrentino, A., Mancioppi, G., & Cavallo, F. (2019). A survey of behavioral models for social robots. *Robotics*, *8*(3), 54. https://doi.org/10.3390/robotics8030054

Okita, S. Y., & Schwartz, D. L. (2006). Young children's understanding of animacy and entertainment robots. *International Journal of Humanoid Robotics*, *3*(3), 393–412. https://doi.org/10.1142/S0219843606000795

Oudeyer, P.-Y., Kaplan, F., Hafner, V. V, & Whyte, A. (2005). The playground experiment: Task-independent development of a curious robot. *Proceedings of the AAAI Spring Symposium on Developmental Robotics*. http://playground.csl.sony.fr/.

Panksepp, J. (2005). Affective consciousness: Core emotional feelings in animals and humans. *Consciousness and Cognition*, *14*(1), 30–80. https://doi.org/10.1016/j.concog.2004.10.004

Panksepp, J. (2010). Affective consciousness in animals: Perspectives on dimensional and primary process emotion approaches. *Proceedings of the Royal Society B: Biological Sciences*, *277*(1696), 2905–2907. https://doi.org/10.1098/rspb.2010.1017

Panksepp, J. (2011). The basic emotional circuits of mammalian brains: Do animals have affective lives? *Neuroscience and Biobehavioral Reviews*, *35*(9), 1791–1804. https://doi.org/10.1016/j.neubiorev.2011.08.003

Paul, E. S., & Mendl, M. T. (2018). Animal emotion: Descriptive and prescriptive definitions and their implications for a comparative perspective. *Applied Animal Behaviour Science*, *205*, 202–209. https://doi.org/10.1016/j.applanim.2018.01.008

Peer, M., Brunec, I. K., Newcombe, N. S., & Epstein, R. A. (2021). Structuring knowledge with cognitive maps and cognitive graphs. *Trends in Cognitive Sciences*, *25*(1), 37–54. https://doi.org/10.1016/j.tics.2020.10.004

Penney, L. M., David, E., & Witt, L. A. (2011). A review of personality and performance: Identifying boundaries, contingencies, and future research directions. *Human Resource Management Review*, *21*(4), 297–310. https://doi.org/10.1016/j.hrmr.2010.10.005

Pereira, G. A. S., Pimentel, B. S., Chaimowicz, L., & Campos, M. F. M. (2002). Coordination of multiple mobile robots in an object carrying task using implicit communication. *Proceedings of the 2002 IEEE International Conference on Robotics & Automation*, 281–286. https://doi.org/10.1109/ROBOT.2002.1013374

Pezzulo, G., Baldassarre, G., Butz, M., Castelfranchi, C., Butz, M. V, & Hoffmann, J. (2006). An analysis of the ideomotor principle and TOTE. In M. V Butz, O. Sigaud, G. Pezzulo, & G. Baldassarre (Eds.), *Proceedings of the Third Workshop on Anticipatory Behavior in Adaptive Learning Systems (ABiALS 2006)*. https://www.researchgate.net/publication/249781497

Phillips, E., Zhao, X., Ullman, D., & Malle, B. F. (2018). What is human-like?: Decomposing robots' human-like appearance using the Anthropomorphic roBOT (ABOT) database. *HRI '18: Proceedings of the 2018 ACM/IEEE International Conference on Human-Robot Interaction*, 105–113. https://doi.org/10.1145/3171221

Pierce, C. S. (1998). *The Essential Peirce: Selected Philosophical Writings (1893–1913): Vol. 2*. Indiana University Press.

Pitron, V., & de Vignemont, F. (2017). Beyond differences between the body schema and the body image: Insights from body hallucinations. *Consciousness and Cognition*, *53*, 115–121. https://doi.org/10.1016/j.concog.2017.06.006

Plass, J. L., & Kalyuga, S. (2019). Four ways of considering emotion in cognitive load theory. *Educational Psychology Review*, *31*(2), 339–359. https://doi.org/10.1007/s10648-019-09473-5

Plutchik, R. (1980). A general psychoevolutionary theory of emotion. In *Theories of emotion* (pp. 3–33). Academic Press. https://doi.org/10.1016/b978-0-12-558701-3.50007-7

Plutchik, R. (2001). The nature of emotions: Human emotions have deep evolutionary roots, a fact that may explain their complexity and provide tools for clinical practice. *American Scientist*, *89*(4), 344–350. https://www.jstor.org/stable/27857503

Pollini, A. (2009). A theoretical perspective on social agency. *AI and Society*, *24*(2), 165–171. https://doi.org/10.1007/s00146-009-0189-2

Polverino, G., Soman, V. R., Karakaya, M., Gasparini, C., Evans, J. P., & Porfiri, M. (2022). Ecology of fear in highly invasive fish revealed by robots. *IScience*, *25*(1), 103529. https://doi.org/10.1016/j.isci.2021.103529

Pool, E., Sennwald, V., Delplanque, S., Brosch, T., & Sander, D. (2016). Measuring wanting and liking from animals to humans: A systematic review. *Neuroscience and Biobehavioral Reviews*, *63*, 124–142. https://doi.org/10.1016/j.neubiorev.2016.01.006

Prescott, T. J., & Robillard, J. M. (2021). Are friends electric? The benefits and risks of human-robot relationships. *IScience*, *24*(1), 101993. https://doi.org/10.1016/j.isci.2020.101993

Primack, B. A., Shensa, A., Sidani, J. E., Whaite, E. O., Lin, L. yi, Rosen, D., Colditz, J. B., Radovic, A., & Miller, E. (2017). Social media use and perceived social isolation among young adults in the U.S. *American Journal of Preventive Medicine*, *53*(1), 1–8. https://doi.org/10.1016/j.amepre.2017.01.010

Quinn, L. K., Schuster, L. P., Aguilar-Rivera, M., Arnold, J., Ball, D., Gygi, E., Holt, J., Lee, D. J., Taufatofua, J., Wiles, J., & Chiba, A. A. (2018). When rats rescue robots. *Animal Behavior and Cognition*, *5*(4), 368–379. https://doi.org/10.26451/abc.05.04.04.2018

Rabb, N., Law, T., Chita-Tegmark, M., & Scheutz, M. (2022). An attachment framework for human-robot interaction. *International Journal of Social Robotics*, *14*(2), 539–559. https://doi.org/10.1007/s12369-021-00802-9

Ragni, M., Rudenko, A., Kuhnert, B., & Arras, K. O. (2016). Errare humanum est: Erroneous robots in human-robot interaction. *2016 25th IEEE International Symposium on Robot and Human Interactive Communication (RO-MAN)*, 501–506. https://doi.org/10.1109/ROMAN.2016.7745164

Rahwan, I., Cebrian, M., Obradovich, N., Bongard, J., Bonnefon, J. F., Breazeal, C., Crandall, J. W., Christakis, N. A., Couzin, I. D., Jackson, M. O., Jennings, N. R., Kamar, E., Kloumann, I. M., Larochelle, H., Lazer, D., McElreath, R., Mislove, A., Parkes, D. C., Pentland, A. 'Sandy', Roberts, M. E., Shariff, A., Tenenbaum, J. B., & Wellman, M. (2019). Machine behaviour. *Nature*, *568*(7753), 477–486. https://doi.org/10.1038/s41586-019-1138-y

Raptopoulou, A., Komnidis, A., Bamidis, P. D., & Astaras, A. (2021). Human–robot interaction for social skill development in children with ASD: A literature review. *Healthcare Technology Letters*, *8*(4), 90–96. https://doi.org/10.1049/htl2.12013

Reis, H. T., Collins, W. A., & Berscheid, E. (2000). The relationship context of human behavior and development. *Psychological Bulletin*, *126*(6), 844–872. https://doi.org/10.1037/0033-2909.126.6.844

Riek, L. (2012). Wizard of Oz studies in HRI: A systematic review and new reporting guidelines. *Journal of Human-Robot Interaction*, *1*(1), 119–136. https://doi.org/10.5898/jhri.1.1.riek

Riener, R., Rabezzana, L., & Zimmermann, Y. (2023). Do robots outperform humans in human-centered domains? *Frontiers in Robotics and AI*, *10*, 1223946. https://doi.org/10.3389/frobt.2023.1223946

Ritzmann, R. E., Quinn, R. D., & Fischer, M. S. (2004). Convergent evolution and locomotion through complex terrain by insects, vertebrates and robots. *Arthropod Structure and Development*, *33*(3), 361–379. https://doi.org/10.1016/j.asd.2004.05.001

Robert Jr., L. P., Alahmad, R., Esterwood, C., Kim, S., You, S., & Zhang, Q. (2020). A review of personality in human-robot interactions. *Foundations and Trends® in Information Systems*, *4*(2), 107–212. https://doi.org/10.1561/2900000018

Rodseth, O. J., & Vagia, M. (2020). A taxonomy for autonomy in industrial autonomous mobile robots including autonomous merchant ships. *IOP Conference Series: Materials Science and Engineering*, *929*(1). https://doi.org/10.1088/1757-899X/929/1/012003

Rogers, L. J., & Andrew, R. J. (2002). *Comparative vertebrate lateralization*. Cambridge University Press.

Rogers, Y. (2005). New theoretical approaches for human-computer interaction. *ARIST: Annual Review of Information Science and Technology*, *38*(1), 87–143. https://doi.org/10.1002/aris.1440380103

Rohlfing, K. J., Altvater-Mackensen, N., Caruana, N., van den Berghe, R., Bruno, B., Tolksdorf, N. F., & Hanulíková, A. (2022). Social/dialogical roles of social robots in supporting children's learning of language and literacy — A review and analysis of innovative roles. *Frontiers in Robotics and AI*, *9*, 971749. https://doi.org/10.3389/frobt.2022.971749

Rolls, E. T. (1999). *The brain and emotions*. Oxford University Press.

Romano, D., Donati, E., Benelli, G., & Stefanini, C. (2019). A review on animal–robot interaction: From bio-hybrid organisms to mixed societies. *Biological Cybernetics*, *113*(3), 201–225. https://doi.org/10.1007/s00422-018-0787-5

Roncone, A., Mangin, O., & Scassellati, B. (2017). Transparent role assignment and task allocation in human robot collaboration. *2017 IEEE International Conference on Robotics and Automation (ICRA)*, 1014–1021. https://doi.org/10.1109/ICRA.2017.7989122

Russell, B. (1954). *Human society in ethics and politics*. Routledge.

Russell, J. A. (1980). A circumplex model of affect. *Journal of Personality and Social Psychology*, *39*(6), 1161–1178. https://doi.org/10.1037/h0077714

Russell, J. A. (2003). Core affect and the psychological construction of emotion. *Psychological Review*, *110*(1), 145–172. https://doi.org/10.1037/0033-295X.110.1.145

Sage, K. D., & Baldwin, D. (2010). Social gating and pedagogy: Mechanisms for learning and implications for robotics. *Neural Networks*, *23*(8–9), 1091–1098. https://doi.org/10.1016/j.neunet.2010.09.004

Salem, M., Lakatos, G., Amirabdollahian, F., & Dautenhahn, K. (2015a). Towards safe and trustworthy social robots: Ethical challenges and practical issues. In A. Tapus, E. André, J.-C. Martin, F. Ferland, & M. Ammi (Eds.), *Social Robotics. ICSR 2015. Lecture Notes in Computer Science* (Vol. 9388). Springer. https://doi.org/10.1007/978-3-319-25554-5

Salem, M., Lakatos, G., Amirabdollahian, F., & Dautenhahn, K. (2015b). Would you trust a (faulty) robot?: Effects of error, task type and personality on human-robot cooperation and trust. *ACM/IEEE International Conference on Human-Robot Interaction, 2015-March*, 141–148. https://doi.org/10.1145/2696454.2696497
Samsonovich, A. V. (2010). Toward a unified catalog of implemented cognitive architectures. In A. V Samsonovich, K. R. Jóhannsdóttir, A. Chella, & B. Goertzel (Eds.), *Biologically inspired cognitive architectures 2010: Proceedings of the first annual meeting of the BICA Society* (pp. 195–244). IOS Press. https://doi.org/10.3233/978-1-60750-660-7-195
Sayigh, L. S., El Haddad, N., Tyack, P. L., Janik, V. M., Wells, R. S., & Jensen, F. H. (2023). Bottlenose dolphin mothers modify signature whistles in the presence of their own calves. *Proceedings of the National Academy of Sciences of the United States of America*, *120*(27), e2300262120. https://doi.org/10.1073/pnas.2300262120
Scassellati, B., Boccanfuso, L., Huang, C. M., Mademtzi, M., Qin, M., Salomons, N., Ventola, P., & Shic, F. (2018). Improving social skills in children with ASD using a long-term, in-home social robot. *Science Robotics*, *3*(21), eaat7544. https://doi.org/10.1126/SCIROBOTICS.AAT7544
Schaal, S. (1999). Is imitation learning the route to humanoid robots. *Trends in Cognitive Sciences*, *3*(6), 233–242. https://doi.org/10.1016/S1364-6613(99)01327-3
Scherer, K. R., Dan, E. S., & Flykt, A. (2006). What determines a feeling's position in affective space? A case for appraisal. *Cognition and Emotion*, *20*(1), 92–113. https://doi.org/10.1080/02699930500305016
Scholl, B. J., & Tremoulet, P. D. (2000). Perceptual causality and animacy. *Trends in Cognitive Sciences*, *4*(8), 299–309. https://doi.org/10.1016/S1364-6613(00)01506-0
Scholtz, J. (2003). Theory and evaluation of human robot interactions. *Proceedings of the 36th Annual Hawaii International Conference on System Sciences, HICSS 2003*. https://doi.org/10.1109/HICSS.2003.1174284
Schweitzer, G. (2003). Robotics – Chances and challenges of a key science. *17th International Congress of Mechanical Engineering (COBEM 2003)*.
Searcy, W. A., & Nowicki, S. (2010). *The evolution of animal communication: Reliability and deception in signaling systems*. Princeton University Press. https://doi.org/10.1515/9781400835720
Seyama, J., & Nagayama, R. S. (2007). The uncanny valley: Effect of realism on the impression of artificial human faces. *Presence: Teleoperators and Virtual Environments*, *16*(4), 337–351. https://doi.org/10.1162/pres.16.4.337
Sfetsos, P., Stamelos, I., Angelis, L., & Deligiannis, I. (2006). Investigating the impact of personality types on communication and collaboration-viability in pair programming - An empirical study. In P. Abrahamsson, M. Marchesi, & G. Succi (Eds.), *Extreme Programming and Agile Processes in Software Engineering. XP 2006. Lecture Notes in Computer Science* (Vol. 4044, pp. 43–52). Springer. https://doi.org/10.1007/11774129_5
Shagrir, O. (2010). Brains as analog-model computers. *Studies in History and Philosophy of Science*, *41*(3), 271–279. https://doi.org/10.1016/j.shpsa.2010.07.007
Shagrir, O. (2018). The brain as an input–output model of the world. *Minds and Machines*, *28*(1), 53–75. https://doi.org/10.1007/s11023-017-9443-4
Sharkey, A. (2017). Can robots be responsible moral agents? And why should we care? *Connection Science*, *29*(3), 210–216. https://doi.org/10.1080/09540091.2017.1313815
Sharkey, A., & Sharkey, N. (2012). Granny and the robots: Ethical issues in robot care for the elderly. *Ethics and Information Technology*, *14*(1), 27–40. https://doi.org/10.1007/s10676-010-9234-6
Sharkey, A., & Sharkey, N. (2021). We need to talk about deception in social robotics! *Ethics and Information Technology*, *23*(3), 309–316. https://doi.org/10.1007/s10676-020-09573-9
Sharkey, N., & Sharkey, A. (2010). The crying shame of robot nannies: An ethical appraisal. *Interaction Studies*, *11*(2), 161–190.
Shettleworth, S. J. (2010). *Cognition, evolution, and behavior* (2nd ed.). Oxford University Press.

Shiomi, M., Nakagawa, K., Shinozawa, K., Matsumura, R., Ishiguro, H., & Hagita, N. (2017). Does a robot's touch encourage human effort? *International Journal of Social Robotics*, *9*(1), 5–15. https://doi.org/10.1007/s12369-016-0339-x

Silverstein, P., Westermann, G., Parise, E., & Twomey, K. (2019). New evidence for learning-based accounts of gaze following: Testing a robotic prediction. *2019 Joint IEEE 9th International Conference on Development and Learning and Epigenetic Robotics (ICDL-EpiRob)*, 302–306. https://doi.org/10.1109/DEVLRN.2019.8850716

Slaughter, V. (2015). Theory of mind in infants and young children: A review. *Australian Psychologist*, *50*(3), 169–172. https://doi.org/10.1111/ap.12080

Smith, D. J. (2005). *Reliability, maintainability and risk: Practical methods for engineers* (7th ed.). Butterworth-Heinemann.

Spatola, N. (2019). Switch off a robot, switch off a mind? How dualist and computational philosophy of mind predict attitudes and behaviours towards robots in human-robot interaction (HRI). *HAI 2019- Proceedings of the 7th International Conference on Human-Agent Interaction*, 194–199. https://doi.org/10.1145/3349537.3351897

Stanton, N. A. (2006). Hierarchical task analysis: Developments, applications, and extensions. *Applied Ergonomics*, *37*, 55–79. https://doi.org/10.1016/j.apergo.2005.06.003

Steckenfinger, S. A., & Ghazanfar, A. A. (2009). Monkey visual behavior falls into the uncanny valley. *Proceedings of the National Academy of Sciences of the United States of America*, *106*(43), 18362–18366. https://doi.org/10.1073/pnas.0910063106

Steel, R. P., Bishop, N. C., & Taylor, I. M. (2021). The relationship between multidimensional motivation and endocrine-related responses: A systematic review. *Perspectives on Psychological Science*, *16*(3), 614–638. https://doi.org/10.1177/1745691620958008

Steinfeld, A., Fong, T., Kaber, D., Lewis, M., Scholtz, J., Schultz, A., & Goodrich, M. (2006). Common metrics for human-robot interaction. *HRI 2006: Proceedings of the 2006 ACM Conference on Human-Robot Interaction*, 33–40. https://doi.org/10.1145/1121241.1121249

Stoytchev, A., & Arkin, R. C. (2004). Incorporating motivation in a hybrid robot architecture. *Journal of Advanced Computational Intelligence and Intelligent Informatics*, *8*(3), 100–105.

Strohmann, T., Siemon, D., Khosrawi-Rad, B., & Robra-Bissantz, S. (2023). Toward a design theory for virtual companionship. *Human-Computer Interaction*, *38*(3–4), 194–234. https://doi.org/10.1080/07370024.2022.2084620

Sweeney, P. (2023). Trusting social robots. *AI and Ethics*, *3*(2), 419–426. https://doi.org/10.1007/s43681-022-00165-5

Tanaka, F., Cicourel, A., & Movellan, J. R. (2007). Socialization between toddlers and robots at an early childhood education center. *Proceedings of the National Academy of Sciences of the United States of America*, *104*(46), 17954–17958. https://doi.org/10.1073/pnas.0707769104

Tanaka, F., & Matsuzoe, S. (2012). Children teach a care-receiving robot to promote their learning: Field experiments in a classroom for vocabulary learning. *Journal of Human-Robot Interaction*, *1*(1), 78–95. https://doi.org/10.5898/jhri.1.1.tanaka

Taylor, H. O., Cudjoe, T. K. M., Bu, F., & Lim, M. H. (2023). The state of loneliness and social isolation research: Current knowledge and future directions. *BMC Public Health*, *23*(1), 1049. https://doi.org/10.1186/s12889-023-15967-3

Téglás, E., Gergely, A., Kupán, K., Miklósi, Á., & Topál, J. (2012). Dogs' gaze following is tuned to human communicative signals. *Current Biology*, *22*(3), 209–212. https://doi.org/10.1016/j.cub.2011.12.018

ten Tusscher, K. H. W. J. (2013). Mechanisms and constraints shaping the evolution of body plan segmentation. *European Physical Journal E*, *36*(5), 54. https://doi.org/10.1140/epje/i2013-13054-7

TenHouten, W. D. (2017). From primary emotions to the spectrum of affect: An evolutionary neurosociology of the emotions. In A. Ibáñez, L. Sedeño, & A. M. García (Eds.), *Neuroscience and social science: The missing link* (pp. 141–168). Springer International Publishing. https://doi.org/10.1007/978-3-319-68421-5

Terzioğlu, Y., Mutlu, B., & Şahin, E. (2020). Designing social cues for collaborative robots: The role of gaze and breathing in human-robot collaboration. *ACM/IEEE International Conference on Human-Robot Interaction*, 343–357. https://doi.org/10.1145/3319502.3374829

Thomaz, A. L., & Cakmak, M. (2009). Social learning mechanisms for robots. *International Symposium on Robotics Research (ISRR)*.

Tinbergen, N. (1952). 'Derived' activities; their causation, biological significance, origin, and emancipation during evolution. *The Quarterly Review of Biology*, *27*(1), 1–32. https://doi.org/10.1086/398642

Tomasello, M. (2008). *Origins of human communication*. MIT Press.

Topál, J., Miklósi, Á., & Csányi, V. (1998). Attachment behaviour in the dogs (*Canis familiaris*): A new application of the Ainsworth's (1969) Strange Situation Test. *Journal of Comparative Psychology*, *112*(3), 219–229. https://doi.org/10.1037/0735-7036.112.3.219

Trappes, R., Nematipour, B., & Krohs, U. (2022). Introduction to niches and mechanisms in ecology and evolution. *Biology and Philosophy*, *37*(6), 55. https://doi.org/10.1007/s10539-022-09890-x

Trevarthen, C. (2011). What is it like to be a person who knows nothing? Defining the active intersubjective mind of a newborn human being. *Infant and Child Development*, *20*(1), 119–135. https://doi.org/10.1002/icd.689

Trevarthen, C., & Aitken, K. J. (2001). Infant intersubjectivity: Research, theory, and clinical applications. *Journal of Child Psychology and Psychiatry and Allied Disciplines*, *42*(1), 3–48. https://doi.org/10.1017/S0021963001006552

Turkle, S. (2011). *Alone together: Why we expect more from technology and less from each other*. Basic Books.

Vaesen, K. (2012). The cognitive bases of human tool use. *Behavioral and Brain Sciences*, *35*(4), 203–218. https://doi.org/10.1017/S0140525X11001452

Valentini, L. (2023). *Morality and socially constructed norms*. Oxford University Press.

Vernon, D. (2019). The architect's dilemmas. In M. I. Aldinhas Ferreira, J. S. Sequeira, & R. Ventura (Eds.), *Cognitive architectures* (pp. 59–70). Springer Nature. https://doi.org/10.1016/j.neuron.2015.09.046

Vernon, D., Metta, G., & Sandini, G. (2007). A survey of artificial cognitive systems: Implications for the autonomous development of mental capabilities in computational agents. *IEEE Transactions on Evolutionary Computation*, *11*(2), 151–180. https://doi.org/10.1109/TEVC.2006.890274

Veruggio, G., & Operto, F. (2006). Roboethics: A bottom-up interdisciplinary discourse in the field of applied ethics in robotics. *International Review of Information Ethics*, *6*, 3–8. www.i-r-i-e.net

Vijay, R., Kapuria, A., Datta, C., Dubey, G., Sharma, C., & Taank, G. (2012). Ecosystem of a shopping mall robot - Neel. *9th International Conference on Ubiquitous Robots and Ambient Intelligence (URAI)*, 120–125. https://doi.org/10.1109/URAI.2012.6462949

von Uexküll, J. (1928). *Theoretische biologie*. Springer Verlag.

von Uexküll, J. (1982). The theory of meaning. *Semiotica*, *42*(I), 25–82. https://doi.org/10.1515/semi.1982.42.1.25

Vulpe, A., Crăciunescu, R., Drăgulinescu, A. M., Kyriazakos, S., Paikan, A., & Ziafati, P. (2021). Enabling security services in socially assistive robot scenarios for healthcare applications. *Sensors*, *21*, 6912. https://doi.org/10.3390/s21206912

Walia, M., Lukmanji, T., Farrell, R., & Green, J. R. (2010). Towards development of a robotic guide dog. *33rd Conference of the Canadian Medical and Biological Engineering Society*. https://doi.org/10.13140/2.1.1952.0965

Walters, M. L., Syrdal, D. S., Dautenhahn, K., te Boekhorst, R., & Koay, K. L. (2008). Avoiding the uncanny valley: Robot appearance, personality and consistency of behavior in an attention-seeking home scenario for a robot companion. *Autonomous Robots*, *24*(2), 159–178. https://doi.org/10.1007/s10514-007-9058-3

Wang, S., Lilienfeld, S. O., & Rochat, P. (2015). The uncanny valley: Existence and explanations. *Review of General Psychology*, *19*(4), 393–407. https://doi.org/10.1037/gpr0000056

Wang, S., & Rochat, P. (2017). Human perception of animacy in light of the uncanny valley phenomenon. *Perception, 46*(12), 1386–1411. https://doi.org/10.1177/0301006617722742

Wang, Z., Mülling, K., Deisenroth, M. P., Ben Amor, H., Vogt, D., Schölkopf, B., & Peters, J. (2013). Probabilistic movement modeling for intention inference in human-robot interaction. *International Journal of Robotics Research, 32*(7), 841–858. https://doi.org/10.1177/0278364913478447

Warren, W. H. (2019). Non-Euclidean navigation. *Journal of Experimental Biology, 222*, jeb187971. https://doi.org/10.1242/jeb.187971

Watzlawick, P., Beavin, J., & Jackson, D. (1967). *Pragmatics of human communication*. W.W. Norton.

Webb, B. (2000). What does robotics offer animal behaviour? *Animal Behaviour, 60*(5), 545–558. https://doi.org/10.1006/anbe.2000.1514

Webb, B. (2006). Transformation, encoding and representation. *Current Biology, 16*(6), 184–185. https://doi.org/10.1016/j.cub.2006.02.034

Weiser, M. (1991). The computer for the 21st century. *Scientific American, 265*(3), 94–105. https://doi.org/10.1145/329124.329126

Wellman, H. M. (2018). Theory of mind: The state of the art. *European Journal of Developmental Psychology, 15*(6), 728–755. https://doi.org/10.1080/17405629.2018.1435413

Wiener, J., Shettleworth, S., Bingman, V. P., Cheng, K., Healy, S., Jacobs, L. F., Jeffery, K. J., Mallot, H. A., Menzel, R., & Newcombe, N. S. (2011). Animal navigation: A synthesis. In R. Menzel & J. Fischer (Eds.), *Animal thinking: Contemporary issues in comparative cognition* (pp. 51–76). MIT Press.

Williams, B. (2016). An introduction to robotics. In *ME 4290/5290 Mechanics and Control of Robotic Manipulators. Mechanical Engineering, Ohio University*. https://sopantalekar.files.wordpress.com/2019/06/1introrob.pdf

Wilson, D. S., Clark, A. B., Coleman, K., & Dearstyne, T. (1994). Shyness and boldness in humans and other animals. *Trends in Ecology and Evolution, 9*(11), 442–446. https://doi.org/10.1016/0169-5347(94)90134-1

Windhager, S., Bookstein, F. L., Grammer, K., Oberzaucher, E., Said, H., Slice, D. E., Thorstensen, T., & Schaefer, K. (2012). 'Cars have their own faces': Cross-cultural ratings of car shapes in biological (stereotypical) terms. *Evolution and Human Behavior, 33*(2), 109–120. https://doi.org/10.1016/j.evolhumbehav.2011.06.003

Windhager, S., Slice, D. E., Schaefer, K., Oberzaucher, E., Thorstensen, T., & Grammer, K. (2008). Face to face: The perception of automotive designs. *Human Nature, 19*(4), 331–346. https://doi.org/10.1007/s12110-008-9047-z

Winkle, K., Senft, E., & Lemaignan, S. (2021). LEADOR: A method for end-to-end participatory design of autonomous social robots. *Frontiers in Robotics and AI, 8*, 704119. https://doi.org/10.3389/frobt.2021.704119

Wisspeintner, T., van der Zant, T., Iocchi, L., & Schiffer, S. (2009). RoboCup@Home: Scientific competition and benchmarking for domestic service robots. *Interaction Studies, 10*(3), 392–426. https://doi.org/10.1075/is.10.3.06wis

Wong, A. L., Goldsmith, J., Forrence, A. D., Haith, A. M., & Krakauer, J. W. (2017). Reaction times can reflect habits rather than computations. *ELife, 6*, e28075. https://doi.org/10.7554/eLife.28075

Wood, L. J., Zaraki, A., Robins, B., & Dautenhahn, K. (2021). Developing Kaspar: A humanoid robot for children with autism. *International Journal of Social Robotics, 13*(3), 491–508. https://doi.org/10.1007/s12369-019-00563-6

Xiao, J., Catrambone, R., & Stasko, J. (2003). Be quiet? Evaluating proactive and reactive user interface assistants. *IFIP TC13 International Conference on Human-Computer Interaction*. https://www.researchgate.net/publication/27521266

Xu, J., Broekens, J., Hindriks, K., & Neerincx, M. A. (2015). Mood contagion of robot body language in human robot interaction. *Autonomous Agents and Multi-Agent Systems, 29*(6), 1216–1248. https://doi.org/10.1007/s10458-015-9307-3

Yamazaki, R., Nishio, S., Nagata, Y., Satake, Y., Suzuki, M., Kanemoto, H., Yamakawa, M., Figueroa, D., Ishiguro, H., & Ikeda, M. (2023). Long-term effect of the absence of a companion robot on older adults: A preliminary pilot study. *Frontiers in Computer Science*, *5*, 1129506. https://doi.org/10.3389/fcomp.2023.1129506

Zhang, J., Li, S., Zhang, J. Y., Du, F., Qi, Y., & Liu, X. (2020). A literature review of the research on the Uncanny Valley. *Lecture Notes in Computer Science (Including Subseries Lecture Notes in Artificial Intelligence and Lecture Notes in Bioinformatics)*, *12192 LNCS*, 255–268. https://doi.org/10.1007/978-3-030-49788-0_19

Ziemke, T. (2003). What's that thing called embodiment? *Proceedings of the Annual Meeting of the Cognitive Science Society*, 1305–1310.

Ziemke, T. (2008). On the role of emotion in biological and robotic autonomy. *BioSystems*, *91*(2), 401–408. https://doi.org/10.1016/j.biosystems.2007.05.015

Ziemke, T., & Sharkey, N. E. (2001). A stroll through the worlds of robots and animals: Applying Jakob von Uexküll's theory of meaning to adaptive robots and artificial life. *Semiotica*, *134*(1/4), 701–746. https://doi.org/10.1515/semi.2001.050

Zimmermann, P., Maier, M. A., Winter, M., & Grossmann, K. E. (2001). Attachment and adolescents' emotion regulation during a joint problem-solving task with a friend. *International Journal of Behavioral Development*, *25*(4), 331–343. https://doi.org/10.1080/01650250143000157

Złotowski, J. A., Sumioka, H., Nishio, S., Glas, D. F., Bartneck, C., & Ishiguro, H. (2015). Persistence of the uncanny valley: The influence of repeated interactions and a robot's attitude on its perception. *Frontiers in Psychology*, *6*, 883. https://doi.org/10.3389/fpsyg.2015.00883

3

AN INTRODUCTION TO ROBOT CONSTRUCTION

Bence Ferdinandy, Péter Telkes, Judit Abdai, and Ádám Miklósi

3.1 Introduction

This chapter offers an overview of robot construction to both biologists and engineers from a broad engineering and ethological perspective. For biologists, we aim to provide a concise summary of the available technical solutions for robot construction, as explored in works such as Matarić (2007), Siciliano & Khatib (2008), and Correll et al. (2021). For engineers, we look into the integration of biological and ethological knowledge into modern robotics, a concept known as ethorobotics (Chapter 2), in order to generate a new generation of robots, called ethorobots. In-depth technical details are covered in dedicated books and papers on robotics (e.g., Arkin, 1998; Arkin et al., 2003; Correll et al., 2021; Matarić, 2007), as keeping pace with the ever-evolving developments in hardware and software would be a formidable challenge.

In contrast, this chapter directs its attention to the possibilities presented by contemporary, state-of-the-art technological advancements and the boundaries that accompany them. We aim to draw a parallel between the facets of robot construction and the biological phenomena explored in preceding chapters, shedding light on the advantages and constraints of specific robotic engineering solutions. Comparing biological and artificial agents is a challenging task, given that evolution had billions of years to generate potential solutions, whereas the field of robotics is still in its infancy (Figure 3.1). Despite the wide array of technical developments, including, for example, soft robotics, we anticipate that, in the foreseeable future, robots will primarily be fashioned from rigid materials like metal and plastic, powered by electric motors as actuators, and dependent on electricity as their primary energy source. While there is undoubtedly room for technological improvement, the choice of hardware also imposes significant limitations (Correll et al., 2021; Craig, 2018; McKinnon, 2016). The trend to mimic biological agents is regarded as a highly valuable asset in robotics, offering important insights for both biology and technology. This mimicking not

DOI: 10.4324/9781003182931-4

Figure 3.1 Taking specific inspiration from nature is an important asset of robotics. Bioinspired robotics aims to develop analogue solutions by focusing on a specific skill of an animal species. Baines et al. (2022) suggest that it is possible to design a robot that can accommodate itself to various environments. To this end, they took inspiration from three similar reptile species living in water and/or on land and designed a robot that is able to change its morphology as required. The Amphibious Robotic Turtle has a unified structural and actuation systems which allows him to move efficiently in different terrestrial and aquatic environments. (a) Galapagos giant tortoise (*Chelonoidis niger*) and the green sea turtle (*Chelonia mydas*) served as animal models; (b) Side view (scale bar, 30 mm): The limb of the robot can accommodate to different environments by changing its shape and stiffness; (c) Isometric and top perspectives of the robot are depicted, with the joint arrangement illustrated in the inset. The joints, usually covered with rubber bellows (omitted for clarity), facilitate movements along the θ, ϕ, and α axes, representing forward and backward, up and down, and angle-of-attack motions, respectively (figure from Baines et al., 2022).

only enhances our comprehension of the biological processes, but the technological realisation may lead to an improvement in performance if some biological constraints can be by-passed or eliminated. There is no single ‘best solution’, and artificial agents typically take the form of hybrids, incorporating components inspired by biomimicry alongside other elements rooted in purely artificial mechanisms. On the contrary, autonomy, encompassing the capability for unrestricted movement in wide range of environments and the freedom from external support or supervision, should be regarded as a critical constraint for the development of ethorobots that interact with humans. All ethorobots are expected to be autonomous and mobile as much as possible, and this chapter takes on this aim (or constraint) when presenting advises for construction.

Contemporary biological and robotic agents exhibit distinct disparities in terms of energy optimisation requirements. Biological agents invest significant time and effort in acquiring the necessary energy to sustain optimal functioning, whereas this is not a concern for robots, as they have a readily available source of electricity 'for free'. While this advantage eases the construction of robots, it also renders certain capabilities unrealistic for practical, widespread use.

Robots encounter a challenge akin to that faced by living agents: how to optimise the utilisation of the technological resources available onboard. There exists a trade-off between solving a problem successfully by relying on the minimal necessary technology and achieving speed, reliability, and precision (Correll et al., 2021). Real-world problems in robotics differ conceptually from typical issues confronted by artificial intelligence (AI), as they involve inherent unreliability in both sensing and execution. This situation can be ameliorated by methods such as the utilisation of multiple input sources (e.g., sensor fusion, see below) or acquiring multiple feedback loops regarding the executed actions. Incorporating uncertainty into the construction and execution of behaviour can also be a beneficial approach.

Living organisms have developed robust and flexible mechanisms, and yet these attributes have not received adequate consideration in the realm of autonomous robotics. Industrial robots are expected to perform their tasks without errors, repeatedly, for millions of cycles. While such reliability may not be demanded of ethorobots, which have the potential for alternative mechanisms to compensate for errors, a certain level of performance must still be attained. The challenge of benchmarking remains unresolved, despite several attempts (see Section 2.2.4). Without rigorous testing in real-world environments and under natural conditions, ethorobots may fall short of their potential to become useful tools for humans.

The primary objective of this chapter is to stimulate the creativity of ethorobot developers by highlighting the wide-ranging functional solutions found in living beings and drawing parallels with contemporary trends in robotics. When confronted with specific challenges, roboticists frequently seek inspiration from animals and tend to emulate the functionalities of living organisms. We aspire to broaden this approach into area of social robotics, fostering a novel perspective on these agents as exemplified by ethorobotics.

This chapter adheres to the triadic division of the mental machinery of problem solving, involving sensing/perceiving, action, and mental architecture/central mechanisms (Section 1.4). Despite the organisational framework observed in robots, these subsystems are not as distinctly separated in living beings as their box-like depictions in figures might imply. For example, Ward and Stapleton (2012) also contend that actions in living beings cannot be strictly categorised as solely cognitive or affective; instead, they represent a blended form encompassing elements of both states.

3.2 An overview of sensing skills

For survival, embodied agents must possess the capacity to accommodate to changes in their environment. In both artificial and organic agents, the input for this process is provided by sensors or sensory organs, respectively. Sensors are typically defined as physical devices that measure various physical or chemical quantities, while the primary role of sensory organs is to convert diverse environmental stimuli (such as light waves, molecules, etc.) into electrical signals transmitted to the central nervous system (Matarić, 2007).

Sensory ecology investigates how different animal species process information collected by their sensory receptors, built on the evolutionary understanding that species are finely tuned to their ecological niches, and sensory receptors play a pivotal role in this process (Stevens, 2013). Animals possess a variety of sensory receptors that have evolved in accordance with the physical properties of information most crucial for their survival. Robots require similar devices to navigate within their niches, resulting in numerous similarities as well as distinctions between robotic sensors and the sensory organs found in animals.

It is equally important to distinguish between sensation and perception. In animals, sensation occurs within sensory organs, involving the activation of sensory machinery by environmental stimuli (e.g., light, sound, or pressure). This is followed by specific primary processing and the transmission of the acquired input to the brain. Perception, on the other hand, encompasses the brain's mechanisms that use mental representations to choose, organise, and process sensory inputs. Unlike sensation, perception is viewed as an active process. The same sensations received by the same sensory organs can be perceived differently based on the perceptual processes, which also encompass learning and experience (Box 3.1).

Box 3.1 Visual illusions

Illusions reveal the importance of perception in visual sensing. Different types of illusions may rely on different aspects of visual perception. For example, some illusions capitalise the rules of perception on representing depth (Figure 3.2a). Upon seeing the two converging lines, the mind automatically simulates a 3rd dimension (depth) to the image, due to which the vertical stripe on the top is judged to be longer than the one on the bottom. In typical situations, this judgement is important for the mind to know that objects further away are actually not smaller than those staying closer. Other types of illusions rest on the perceptual skill to look for specific patterns in visual scenes. Depending on what the viewer 'wants' to see, either a young or an old woman shows up in the picture (Figure 3.2b). Falling for illusions depends on the perceptual apparatus and how the mind evaluates visual scenarios. This also depends on evolutionary and ecological factors, and thus we share some of our illusions with animals, while differences also exist.

Figure 3.2 Two classic visual illusions: (a) capitalising on depth perception (photo by Alex Heuvink, Unsplash); and (b) ambiguous image of a young or an old woman.

In animals, sensory organs and central mechanisms provide a clear separation between sensation and perception; however, this demarcation is less distinct in robotics. As a result, in robotics, the term 'perception' is often employed to describe the entire process. While some aspects of detecting environmental physical changes may differ in robots compared to animals, there are numerous shared challenges that all sensors or sensory organs must address.

1 Essentially, any physical change can theoretically be detected through some form of mechanism. Nevertheless, the true challenge lies in ascertaining whether the gathered information provides any advantages to the agent and whether a sensor can be designed both effectively and with adequate sensitivity to make precise measurements or detections as required. For example, typical earthworms have very basic visual sensors, being able to discriminate light and dark but the complexity of their visual sensory organs increases in species with higher mobility and more complex predatory behaviour (Harley & Asplen, 2018; see also Table 3.1). For reasons of efficiency, robots should be equipped with sensors that provide the necessary capabilities but not more.
2 In the animal kingdom, five so-called fundamental senses are widely recognised: vision, hearing, olfaction, taste, and touch. While there are variations in the sensory range within which these organs operate, in the case of most complex animal species, all these sensor types are employed in some capacity. This implies that, irrespective of their specific environment, social robots could also gain advantages from possessing optimal combination of the five sensory capabilities.
3 In animals, sensory receptors, or more precisely, receptor cells, serve as the fundamental units for sensation. For instance, in the context of vision, a receptor cell, known as a photoreceptor, contains a specific light-sensitive protein (e.g., rhodopsin), which becomes activated when exposed to light. Sensory organs responsible for vision, like the eye, are typically composed of varying numbers of these photoreceptors, ranging from just a few to potentially millions. This

structural similarity extends to artificial sensors as well. In cameras, the photosensitive surface comprises specific light-sensitive elements situated at precise locations, responding to incident photons. The quantity of these units, whether in an animal or a camera, directly influences their sensitivity to visual input.

4 Sensor resolution, or measurement resolution, refers to the smallest detectable change in the input of the measured quantity. In the case of a sensor with digital output, resolution typically corresponds to the numerical precision of the digital output. The resolution of a sensory organ depends on various specific factors related to the sensor and typically requires experimental determination. For the basic senses, the resolution may also adhere to the Weber–Fechner law (Box 1.3), which posits that the perceived difference in a stimulus is proportional to its initial magnitude. This law, however, usually does not apply to artificial sensors, as they tend to have a uniform resolution along their entire measurement range. The resolution of a sensor is therefore usually only a function of the measurement range itself (having a larger range generally results in having a lower resolution), given a certain sensor technology.

The precision of a sensor is defined by the consistency and repeatability of its measurements when the measurable feature remains constant. It assesses the sensor's ability to provide consistent results under the same conditions. Accuracy quantifies the disparity between the measurement provided by a sensor and the true or actual value of the feature being measured. It gauges how closely the sensor's measurements align with the real values, which may include systematic errors or biases.

5 The majority of animal sensory organs are passive, meaning they are activated by specific environmental input. In contrast, certain animals possess active sensory systems. This means that they generate a physical change in the environment, and the subsequent consequences of this change are detected afterwards. These active sensor systems typically comprise an emitter and a detector working in close synchrony. For example, bats emit ultrasonic sound waves and then sense the echoes of these waves bouncing off objects in their environment for navigation, a process known as echolocation. Similar analogue devices inspired by nature are also used in robotics (see below).

6 Sensors can be deployed proactively to search for specific patterns or inputs in the environment. In this case, mental representations play a pivotal role in determining what the sensor should be seeking (see 'search image' in Section 1.11.6) This search process is initiated by the agent and contributes to effective sensing, enabling the agent to actively engage with its surroundings and gather the information it requires.

7 The primary function of a sensor is to selectively respond to those aspects of physical changes that carry significance for the agent. This process is often referred to as the signal-noise problem, where the goal is to distinguish the meaningful input (signal) from the irrelevant or unwanted input (noise). The signal-to-noise ratio is a measure that defines the ratio of the power of the signal to the power of the background noise, a ratio higher than 1:1 indicates more signal than noise. While noise is commonly associated with acoustic backgrounds, similar detection

challenges arise in all other sensory modalities. For example, in vision, it might involve spotting a small bird on a tree amidst a background of leaves, or in olfaction, it could be the act of detecting a specific odour amid various other scents. Improving the signal-to-noise ratio can be achieved through processes like perception that involve changes in attention (Section 1.11.6) or by activating specific mental representations, enhancing the agent's ability to discern and focus on the meaningful signals amidst the noise.

8 Sensors are typically highly selective in their ability to detect specific types of physical inputs. This selectivity makes them efficient at capturing particular information but may also lead to the agent ignoring other forms of input. To address these limitations, one type of sensor can be complemented by others that provide different and complementary measurements. This is why both animals and robots employ various types of sensors to gather a wide range of information from their environments (see Table 3.1).

9 In contrast to noise, which represents the presence of unwanted or irrelevant input, errors in sensor measurements stem from specific limitations of the sensor. These limitations can arise from changes in the sensor's physical properties and unexpected environmental effects. In the context of animals, ageing can compromise the functionality of sensory organs. For instance, as animals age, their vision, hearing, or other sensory functions may deteriorate, leading to a decline in sensory acuity. Additionally, external factors like fog on the road can interfere with the performance of visual systems, causing errors in the sensation of the environment. Analogue processes may occur in artificial sensors also, as the materials wear off.

10 Sensor fusion occurs when the mind or the artificial architecture integrates input from various sensors to construct a complex representation of the environment, facilitating the decision-making process. Sensor fusion is highly beneficial because sensors often offer complementary, and at times overlapping, environmental information. For example, combining vision and hearing can lead to improved localisation of prey compared to relying on just one type of sensor.

11 Sensors are inundated with substantial volumes of environmental stimulation, and the processing and storage of this information can impose a significant burden on perceptual and central mechanisms. To address this challenge, a common approach is to implement specific filtering at the sensor level, which involves the selective extraction of pertinent information while discarding less relevant data. For instance, in the case of colour cameras used for vision, not all the acquired input needs to be processed and stored Eliminating certain elements, like redundant colour information, can reduce the data size and expedite the processing time, making the system more efficient and manageable.

12 Calibration is indeed an active process aimed at enhancing the accuracy and precision of a sensory system. In animals, this calibration process can take on various forms, often involving both developmental and learning mechanisms. These processes contribute to fine-tuning the sensor's functionality and the executive processes. For example, by repeated actions to peck at small objects, newly hatched chicks increase rapidly their hitting accuracy during the first few days of their life.

Table 3.1 Summary table on sensor types including the advantages and disadvantages for robotic applications. Accelerometers, gyroscopes, and magnetometers are often combined into a single sensor, called IMU (Inertial Measurement Unit). This provides the robot with information about its orientation in the reference frame of the world relative to gravity vector, and relative change in speed (based on Shangari et al., 2017). Importantly, engineers should conduct a very thorough analysis when designing the 'sensor-cocktail' for a robot with assigned to a specific task. In addition to considering the performance parameters, the advantages and disadvantages of each specific sensor should be taken into account, guided by a detailed task analysis (Section 2.4)

	Physical property	*Sensing technology*	*Advantage*	*Disadvantage*
Sensing the surrounding	Light level	Visible spectrum camera	- Rich in contents - Easy to interpret - Low price - No interference problems with the environment	- Limited performance in darkness
		Infrared camera	- Ability to measure temperature - Independent of light source	- Sensitive to weather conditions - Sensitive to environmental conditions - Limited range
	Smell	Chemical sensors	- Detection of chemicals - Low concentration	- Limited to specific chemical compounds
	Sound	Microphones	- Low cost	- Limited usability in noisy environment
	Temperature	Thermal camera	- Wide field of view - Suitable for multiple materials - Insensitive to light conditions	- Limited accuracy - High cost - Sensitive to environmental conditions
	Pressure	Pressure gauges	- Low cost	- Limited accuracy - Wear-and-tear - Sensitive to temperature and vibration

(*Continued*)

Table 3.1 (Continued)

	Physical property	*Sensing technology*	*Advantage*	*Disadvantage*
Navigation in space	Distance	Ultrasound	- Wide detection angle - Low cost	- Accuracy decreases at longer distances - Sensitive to environmental conditions
		Radar	- Images can be acquired day or night - No strong limitations in different environmental conditions	- High cost
		Infrared	- Fast response time - Low power consumption	- Accuracy decreases at longer distances - Sensitive to temperature
		Sonar	- Measures distance and speed of target object	- Difficult to interpret output signal - High price - Sensitive to weather conditions
		Laser	- High accuracy in lateral and longitudinal directions	- High cost
	Magnetism	Compasses	- Lightweight - Low power consumption - Low cost - Global reference frame	- Sensitive to magnetic interference - Limited precision
	Altitude	Altimeters	- Accurate and reliable measurement	- Sensitive to weather conditions

(*Continued*)

Table 3.1 (Continued)

	Physical property	*Sensing technology*	*Advantage*	*Disadvantage*
Sensing its own state/motion in space	Acceleration	Accelerometers	- Low cost - Low power consumption - Accurate and stable information over time	- Limited to measure linear acceleration - Sensitive to the changes in the orientation of the sensor
	Rotational speed	Gyroscopes	- Ability to measure rotational motion - Influenced less by changes in orientation	- Inaccurate in long-term measurements - High consumption power
	Rotation angle	Encoders	- High accuracy - Ability to measure multiple rotations - Long-term reliability due to lack of wear-and-tear - Information about the position and direction of the change	- High cost - Sensitive to environmental conditions
		Potentiometers	- Low cost - Low power consumption	- Limited lifespan - Decreased accuracy over time - Limited rotation angles - Limited accuracy
	Inclination	Inclinometers	- Compact and lightweight - Low power consumption - Operates properly in different environmental conditions	- Limited accuracy - Limitations in measuring rapid changes - Limitations in measuring in complete 3D

(*Continued*)

Table 3.1 (Continued)

Physical property	Sensing technology	Advantage	Disadvantage
	Gyroscopes	-	-
Strain	Strain gauges	- Highly sensitive - Accurate and reliable measurement	- Sensitive to temperature - Sensitive to environmental conditions
Contact	Bump	- Low cost - Durable - Low power consumption	- Limited sensitivity - Susceptible to false triggers
	Switch	- Reliable	- Limited sensitivity - Wear-and-tear; limited long-term reliability

3.3 Internal sensors

3.3.1 Measuring kinematic parameters: proprioceptors

Specific internal receptors (proprioceptors) are responsible for detecting changes within the body and/or between parts of the body by being sensitive to various inputs such as pressure and mechanical effects. The main function of proprioceptors is to support movement by providing regular measurements that reflect the changes in body position and relationship between the body and the environment. Typical mechanical proprioceptors are located in the muscles, joints, and tendons and measure the changes in the length of the muscles and the rate of change (e.g., Box 3.2). Integrated sensations of proprioceptor inputs of the whole body are processed to stabilise position and control movement. More complex bodies have to rely on more sophisticated systems of proprioceptors (Proske & Gandevia, 2012).

Robots can also obtain such inputs about their actual body conformation. The analogue function of an organic proprioceptors is fulfilled by so-called encoders (Correll et al., 2021), but the mechanisms applied are very different. Encoders can be divided into incremental (relative, used primarily in mobile robotics) and absolute encoders (mainly used in robot manipulators). Encoders employ either a magnetic or optical beacon that rotates alongside the motor and is detected by a suitable sensor counting each pass-through. The quadrature encoder, a prevalent type in robotics, is specifically an optical encoder. It utilises a pattern rotating with the motor and an optical sensor capable of recognising black/white transitions. Knowing the count equivalent to a full turn allows the sensor data to be leveraged for calculating the distance covered by the wheel (Box 3.2).

Extending the same principle, encoders can measure rotation angles at joints or any rotational axis. Joint position and speed measurements contribute to determining the position, orientation, and velocity of a robotic arm, leg, or head through kinematic equations derived from the mechanical structure. When a torsion spring is affixed to the axis of an encoder, the spring torque can be computed from the rotational angle using the spring's characteristics, providing a relationship to the force acting on the manipulator.

Box 3.2 Analogue mechanisms for proprioception in animals and robots

Both animals and robots have to know the relative position of their body parts to organise ambulation or movement in any medium. This function is achieved by proprioceptors in animals, and so-called encoders in robots (Figure 3.3).

a Proprioceptors typically measure the stretching of a muscle. Afferent axons weave around a section of small bunch of specific (intrafusal) muscle fibres forming a spiral. The intrafusal fibres change their length in parallel to the stretching or contraction of the muscles. The cell membrane of the afferent axons (Ia afferents) contains stretch-sensitive ion channels activation of which triggers an electric signal. The signal about the changes in muscle length reaches the brain along two pathways starting with the afferent neurons. The cerebellum receives input via the spinocerebellar tracts for continuous updates on the body's actual conformation. A different neural (dorsal column-medial lemniscus) pathway provides the input via the thalamus to the cerebral cortex for the coordination of voluntary action planning.

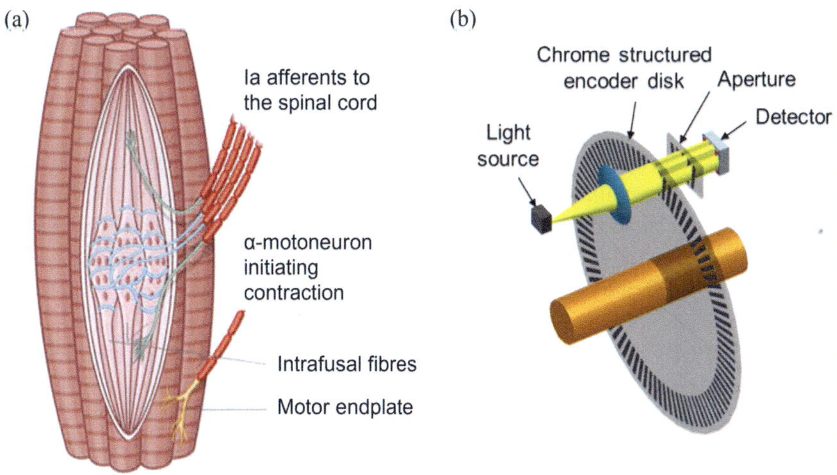

Figure 3.3 (a) Schematic representation of a muscle spindle (figure from Vega & Cobo, 2021), and (b) an optical encoder (figure from Seybold et al., 2019).

b Interestingly, the same function is achieved in robots by very different means, most often using a form of active sensing. In general, an encoder converts the angular or linear position of a shaft or object into an electrical signal that can be read and processed by a control system. The physical basis for an encoder can be very different. The most popular optical encoders use a light-emitting device and a photo sensor. A disc (code wheel) with small holes is fixed to the moving shaft. As the wheel turns, the light gets through the holes of the wheel (black stripes along the circumference of the wheel) and activates the photosensor (detector). This way the rotation of the wheel is converted to electrical pulses. Knowing the number of holes allows to determine a full turn, and thus the distance moved by the wheel.

3.3.2 Measuring dynamic parameters

In vertebrates, together with the network of proprioceptors, the vestibular system is responsible for keeping the balance and detecting speed and acceleration of the body. It consists of two vesicles and three semicircular canals oriented at the pitch, roll, and yaw axes and cavities close to their base filled with a specific fluid (endolymph) (Figure 3.4a).

The two vesicles, the saccule and utricle each have a macula of hair cells in vertical and horizontal planes, respectively. A gelatinous membrane with embedded otoliths rests on these cells. Linear acceleration changes pressure on the otoliths, displacing the processes on hair cells and triggering membrane depolarization. This arrangement makes the receptors a highly sensitive detection system for both vertical and horizontal linear acceleration. In parallel, the bending of the hairs of the sensor cells located in the semicircular canals as response to the fluid displacement maps the movement of the head in space. The generated electric signal is transmitted to the brain, representing the rotational movements (angular velocity) in all three dimensions. One vestibular system is functioning on each side of the head. These have to work in concert, and thus turning the head in one direction activates the hairy cell receptors at one side and inhibits the corresponding ones on the other side (Raphan & Cohen, 2002).

In robotics, accelerometers rely on different mechanisms to measure acceleration. A cost-effective and practical approach involves utilising the MEMS (Micro-Electro-Mechanical Systems) capacitance-based acceleration measurement method. This method operates on a foundational principle: when a specific mass and set of springs are in place, the displacement of the mass becomes proportional to the accelerations experienced by the system.

Within MEMS, one end of the springs is fixed to the comb capacitor plate, while the other end is connected to the seismic mass. When an external force acts on the sensor, it induces motion in the seismic mass, altering the separation distance between the plate (fixed electrode) and the mass (moving electrode). Consequently, this change in distance results in a modification of the capacitance of the system (Figure 3.4).

Assuming knowledge of the spring constant, mass of the seismic element, and the initial equilibrium separation distance between the arms of the capacitor, the

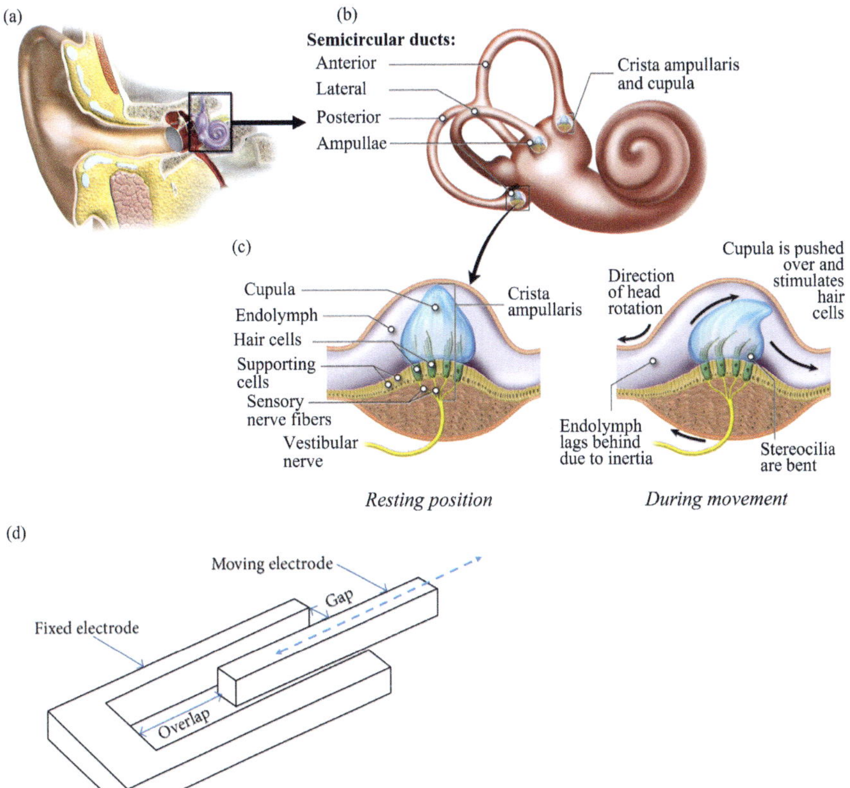

Figure 3.4 Ways of measuring the acceleration of the body in vertebrates and robots. In vertebrates the vestibular apparatus is responsible for supplying information on the body and head movements (a) the location of the vestibular apparatus in the human ear (figure by Blausen.com staff, 2014) (b) The simplified structure of the vestibular apparatus; the three semicircular ducts play the key role in sensing movement in the 3D space. The snail-house like structure on the left is part of the hearing system, see also Figure 3.8 (figure by Cenveo). (c) The mechanic decoding of the movement (black arrows) is initiated by the movement of the copula in the endolymph. This is sensed by the stereocilia of the hair cells and an electric signal is generated. The vestibular nerve (yellow) transmits this signal to the brain stem (figure by Cenveo). (d) The same function is achieved by Micro-Electro-Mechanical Systems (MEMS) which can measure capacity of the system affected by the distance between the moving and fixed electrode. The forces generated during the movement (dashed arrow) affect the position of the electrode (see text for details) (figure from Megat Hasnan et al., 2014).

acceleration of the system can be determined by measuring the applied voltage according to predetermined values. To measure acceleration in any direction, a single accelerometer typically incorporates at least one capacitive sensor along each of the three axes. As MEMS sensors are constructed from silicon using the same principal technology as integrated circuits, these sensors are commonly integrated into a single microchip with signal conditioning and processing circuitry that transforms the measured small voltages and electric charges into digital values which can be sent to microcontrollers via common communication protocols. This integration facilitates their straightforward use in various technical applications.

Gyroscopes are used to measure angular velocities. In robotics, the most commonly used type is the vibrating structure gyroscope, which has a mass that is forced to vibrate along an axis. If the structure is being rotated, the Coriolis force will act on the mass, and by measuring this force, the angular velocity can be calculated with a preliminary knowledge of elements of the system. As one such structure can only measure angular velocity around one axis, in most sensors three vibrating masses are used for the three axes. In practice, these devices are manufactured in small sizes as MEMS structures and are usually integrated with accelerometers in the same chips.

These inner receptor systems also play an important role in maintaining body orientation and stability during movements. For example, the vestibulo-ocular reflex ensures that the animal can keep its focus of vision on a specific location during movements (Box 3.3). The reflex stabilises the eyes' line of vision even when the head is making small movements to either sides or up and down. This control is very important to obtain a stable image projected onto the retina. Similar control systems are also needed for autonomous moving robots, especially when they are moving on legs on uneven surfaces.

3.4 External sensors

External sensors provide the agent with diverse information gleaned from the environment. While the challenge for internal sensors is relatively universal across various organisms, the array of external sensors exhibits substantial variation contingent upon the species' specific ecological niche. External sensors not only differ in their capacity to sense various changes in environmental states (e.g., waves, chemical or mechanical alterations) but also in the range of sensing specific types of input.

The evolution of sensory capacity presents numerous instances of homology, where the same structure undergoes specific modifications as species diverge

Box 3.3 The function of the vestibulo-ocular reflex

Stabilising vision is important for being able to collect smooth visual images for mental processing. Even in a stationary position, the head is making short compensatory movements to balance, and these movements may become even larger when the whole body is moving. The vestibulo-ocular reflex ensures that the animal can hold focus on a visual target despite continuous head movements (Figure 3.5). Even small head turns are noticed by the receptor cells in the semicircular canal (ductus semicircularis) because the fluid inside starts to move into the opposite direction (see also Figure 3.4a). This triggers excitatory signals (red) to the muscles on one side of the eyes and inhibitory signals (blue) to the muscles on the other side. The contraction of the relevant muscles makes the eyes moving in the opposite direction (compensatory eye movements; white arrows at the top).

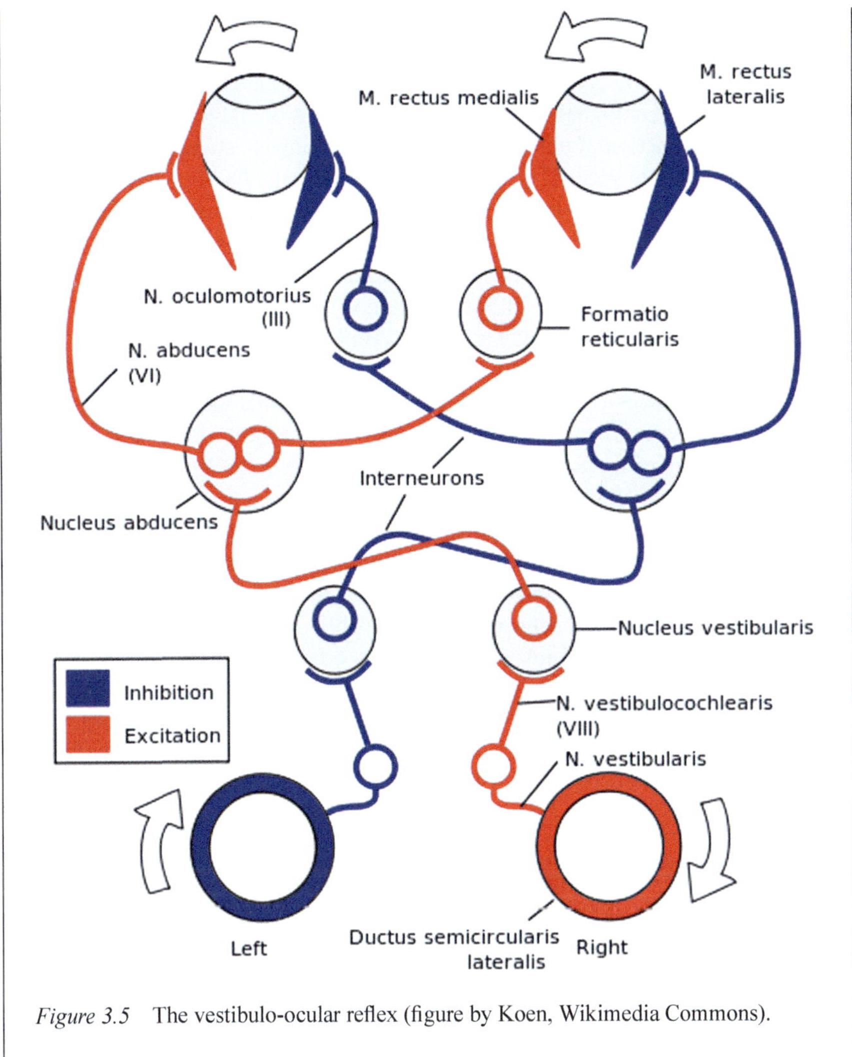

Figure 3.5 The vestibulo-ocular reflex (figure by Koen, Wikimedia Commons).

(e.g., the vertebrate eye). Additionally, there are instances of convergent processes, where similar sensory inputs are achieved by sensors with different structures, such as the ears of grasshoppers and mice. Even in such cases, the underlying physical mechanisms tend to be remarkably similar.

3.4.1 Visual sensors – eyes

The phenomenon known as light is a result of electromagnetic radiation at various wavelengths. Although there is a continuous spectrum of these radiations, they are generally categorised into several types based on wavelength, progressing from low to high frequencies: radio waves, microwaves, infrared (IR) light (heat), visible light, ultraviolet, X-rays, and gamma rays. While human technology enables the detection of all types of electromagnetic waves, living organisms are typically limited to sensing only a small fraction of this spectrum. This limitation is determined by the ability of photons to activate specific molecules in receptor cells. Photons with insufficient energy, such as radio waves, have no effect, while those with excessively high energy, like X-rays, can potentially damage biological structures. As a result of the presence of specific receptor cells, typical eyes are generally sensitive to electromagnetic waves falling within the range of 350–720 nm, with some variations influenced by other factors. This particular range is termed visible light.

Simple eyes consist of a lens positioned near the surface of a cavity, directing electromagnetic waves ('light') to a layer of receptor cells after undergoing some filtration. Some types of eyes feature a pupil (a hole) in front of the lens, regulating the overall amount of light entering the eye (Box 3.4).

The sensation experienced by the organism is determined by the specific tuning of the receptor cells. In the most basic scenario, these eyes only sense the quantity of light reaching the receptors, allowing for a rudimentary distinction between varying levels of illumination. Depending on the lighting conditions and the presence of nearby objects, these visual sensors may be capable of detecting blurred images of objects and discerning their movement (Land & Fernald, 1992).

Box 3.4 The evolution and the structure of a complex eye detecting light waves

a Animal eyes have evolved over tens of millions of years, exhibiting a myriad of variations in specific structures at various levels of complexity (Gehring, 2014; Figure 3.6a). Typically, six levels of the construction are discriminated: (1) Eye-spot: presence of photosensitive receptors; (2) Eye pit: photosensitive receptors are located in a depression; (3) Pinhole eye: the hole has only a small opening; (4) The enclosed chamber of the eye is filled with a transparent liquid like specific substance; (5) Lens evolve for focusing the light beam; and (6) Iris evolves for regulating the amount of the incoming light.

b There is an interesting similarity between the basic structure of the most complex eyes and mechanical cameras (Figure 3.6b). Both incorporate a lens that projects the image onto the retina or film, respectively. The diaphragm in the camera and the iris in the eye regulate the amount of light reaching the film/retina by adjusting the aperture/pupil size (as indicated by the arrow). However, the iris is positioned in front of the lens, whereas the diaphragm is situated behind the lens. The lens is also responsible for determining the sharpness of the image. In the eye, this is accomplished by altering the thickness of the lens (see arrow), while in the camera this is

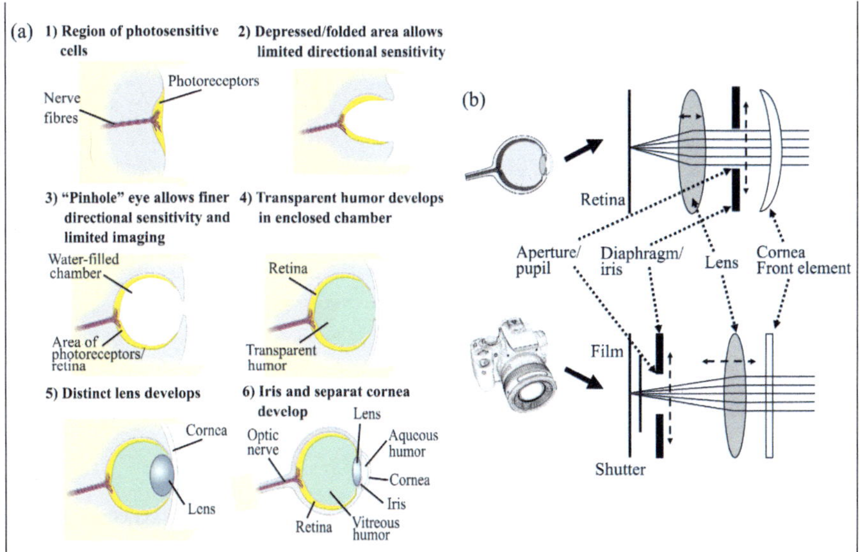

Figure 3.6 (a) The evolution of the animal eye (figure by Matticus78, Wikimedia Commons). (b) Comparison of detecting light waves by a complex eye and a camera (camera picture from PickPik.com).

achieved by moving the lens forward and backward (see arrow) to obtain a sharp image on the film. Additionally, cameras feature a shutter that governs the duration of exposure to incoming light, a component notably absent in the eye.

Three crucial features contribute to significantly enhancing information processing in the eyes. First, the shape of the lens can be altered (accommodation) through a set of muscles, ensuring that incoming light from objects at various distances is focused precisely on the layer of receptor cells. The more receptor cells involved in the sensing process, the more detailed a picture of the environment can be generated. Additionally, specialisation may arise among receptor cells. One type (rods) is exceptionally sensitive and can be activated by a single photon, enabling vision in low-light conditions. Other types of receptor cells (cones) are less sensitive but are specialised for a narrow range of electromagnetic frequencies, facilitating colour vision. Humans possess one type of rods and three types of cones specialised for blue, green, and red light, respectively. In contrast, birds and fish typically have four types of cones (see Land & Fernald, 1992).

A sophisticated hierarchical system, consisting of a specific network of neurons (the retina), manages the inputs from these four types of photoreceptors (Figure 3.7). Parallel processing of the incoming visual input is handled by bipolar cells and ganglion cells, with horizontal cells and amacrine cells contributing lateral connections

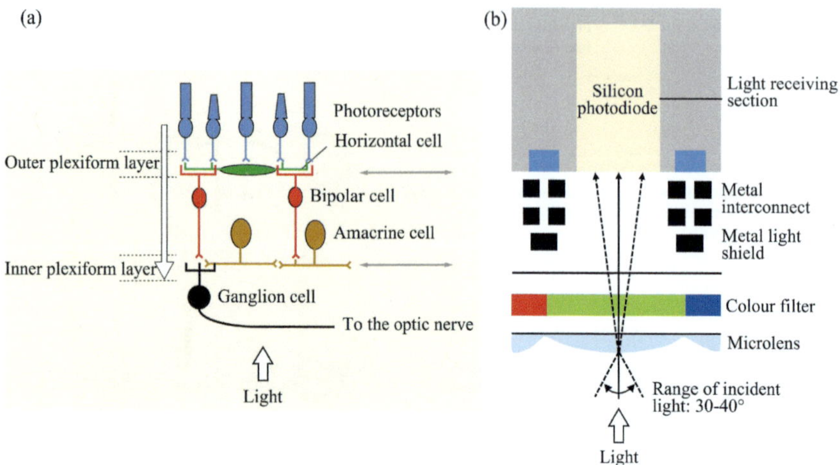

Figure 3.7 Light-sensitive structures; retina versus photosensors. (a) The retina consists of ten specific layers that first convert light into electrical signals and then encode motion changes, shapes, and colour of the visual information that is transmitted to the brain via the ganglion cells (figure from Lagnado, 1998). (b) Artificial light sensors are composed of an array of very small sensor elements (CMOS). At each location there is a microlens that focuses light on the sensor, after which a colour filter is placed, transmitting only one of the red, green or blue light waves of the spectrum. At the bottom of the sensor element is a photodiode which converts incoming photons into electrons. Between the colour filter and the photodiode there is a layer of metal interconnects, which contain necessary electronics for the operation of the sensor.

within this network. The retina serves as the initial stage for information processing (early vision), followed by more advanced computation at higher levels in the brain.

The receptive field, considered the fundamental unit of vision, comprises nearby photoreceptors connected to the same ganglion cell by bipolar cells situated between the receptor cells and the ganglion cells. These receptive fields exhibit a crucial characteristic known as the centre-surround feature. In half of the receptive fields, circularly positioned photoreceptors are organised with an ON-centre and an OFF-surround, while in other receptive fields, the arrangement is reversed. ON-centres and ON-surrounds are excitatory, contrasting with the inhibitory OFF-centres and OFF-surrounds. Horizontal cells maintain this relationship by inhibiting the surrounding photoreceptor cells when the centre photoreceptor cell is activated. This phenomenon enhances visual acuity by increasing contrast and sharpness, aiding in the identification of object outlines and other essential visual elements (see also Figure 3.8).

Furthermore, one type of ganglion cells exhibits low spatial resolution, high sensitivity, and high temporal resolution, providing no colour information. In contrast, other types display the opposite characteristics in all these features. Information from the latter type of ganglion cells is primarily utilised for recognising scenes and objects, while the former contributes to the analysis of physical aspects such as

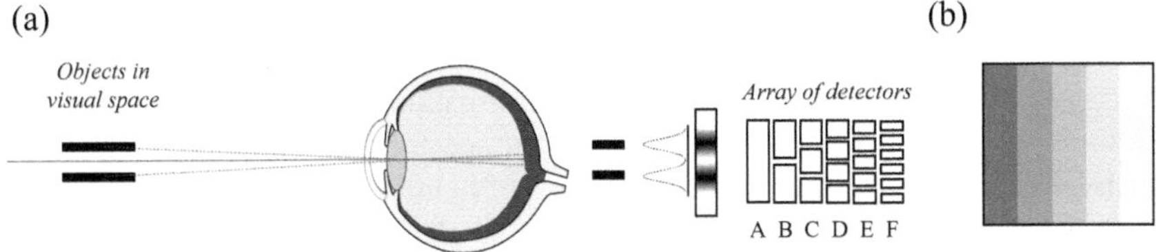

Figure 3.8 (a) Visual acuity refers to the spatial resolving capacity of the visual system, reflecting the eye's ability to discern fine details. Visual acuity depends on factors such as diffraction, and photoreceptor density in the eye. In addition, visual acuity changes with refractive error, illumination, contrast, and the specific location on the retina being stimulated. The accompanying figure illustrates two black lines and their point spread function at the back of the eye. The resolution of these lines relies on the distribution of retinal cones across the retina. The hypothetical arrays from A to F represent detectors with different number of sensors from 1 to 6. It has been suggested that to distinguish between the two lines, at least one inactive sensor must be positioned between two stimulated ones. Consequently, the detectors A and B do not resolve the two lines, in contrast to the detectors C to F, where the two dots would be distinguishable (drawing based on Kolb et al., 1995). (b) The experience of the Match band illusion is based on the amplification of the contrast along the edges where slightly different shades of grey meet. Lateral inhibition and centre-surround receptive fields are also contributing to this effect.

distance and movement. Following spatial summation, the output of the ganglion cells is transmitted by the optic nerve to the brain, arriving at different neural areas that facilitate parallel processing of the actual information (Figure 3.7).

The human retina contains approximately 1 million ganglion cells, gathering information from about 100 million photoreceptor cells. However, the connection ratio is not uniform. In the centrally positioned fovea, one photoreceptor cell is connected to one ganglion cell. As one moves towards the periphery, the ratio increases, with 100–1000 rods possibly providing input for a single ganglion cell. Consequently, the fovea is the region associated with the best vision, offering high resolution (and the greatest acuity), while the outer parts of the eye contribute to an expanded field of view with lower levels of acuity. Thus, there is a trade-off in the eye between the resolution of a scene and the extent of the scene that can be viewed at any given time. This arrangement also dictates that the eye should be positioned in a way that the crucial aspects of the scene fall onto the fovea.

3.4.1.1 Specific visual features detected at the sensory level

The specific organisation of the retina serves for early information processing, detecting simple visual phenomena, and providing the brain with selectively reduced or structured input. Rough calculations suggest that the retina may transmit approximately 10% of the information it gathers from the environment to the brain. However, the brain is not merely a passive receiver of input from the retina; it can also proactively influence low-level visual processing. Here we refer only to a few such mechanisms; for more details, see Daw (2012), Joselevitch (2008), and Kolb et al. (1995).

Edge Detection: ganglion cells transmit information about relative differences to the brain, making the recognition of edges in the scene easier. Ganglion cells indicate relatively higher differences at the boundaries of luminance differences, exemplified by the Mach Bands illusion. While the brain also plays a role in the perceptual process, this function is rooted in retinal processing.

Movement Detection: various mechanisms have been proposed for the retina's ability to detect moving objects. A central assumption is that the retina identifies a moving object when a horizontally positioned, direction-sensitive amacrine cell (see Figure 3.7) receives simultaneous stimulation from two distant bipolar cells. This co-occurrence of stimulation is possible only if the input from the initially stimulated bipolar cell is transmitted after a delay, a process supported both anatomically and physiologically. The activated amacrine cell then transmits a signal to the corresponding ganglion cell. In summary, ganglion cells positioned in the direction of movement are sequentially activated, and the integration of retinal coding occurs at a higher neural level in the brain.

Flickering Detection: this phenomenon is based on the time required for a receptor cell to react again to a light stimulus, known as recovery time. Recovery time differs for rods and cones and is influenced by external conditions. The shortest recovery time, termed critical fusion frequency, can be determined experimentally. When a light source flashes more frequently than the critical fusion frequency, the viewer perceives it as continuous.

Box 3.5 Perception of depths

Stereopsis, the process enabling the perception of the surrounding environment in three dimensions, involves comparing the same scene from distinct lines of sight. Stereopsis may be present in animals having two or more eyes, but binocular eyes do not automatically come with the ability of depth perception. In scenarios (a) and (b) (Figure 3.9), the apple is focalised on the fovea, while the orange, positioned to the left of the fovea in both eyes, creates an angle α in the left eye and β in the right eye. Consequently, the retinal disparity remains consistent: the absolute disparity of the apple is 0, and that of the orange is $\alpha-\beta$, representing the relative disparity between the two objects. However, due to the varying eye positions (less convergence in A, more pronounced convergence in B), the spatial locations of the objects differ significantly in the two cases. In both instances, the inference of the orange being closer can be drawn from the relative disparity, yet determining the absolute distance to either object necessitates knowledge of the vergence angle (V1, V2). This information can potentially be derived from signals from the eye muscles, or from the geometric projection of objects in the visual scene onto the two eyes (see Nityananda & Read, 2017).

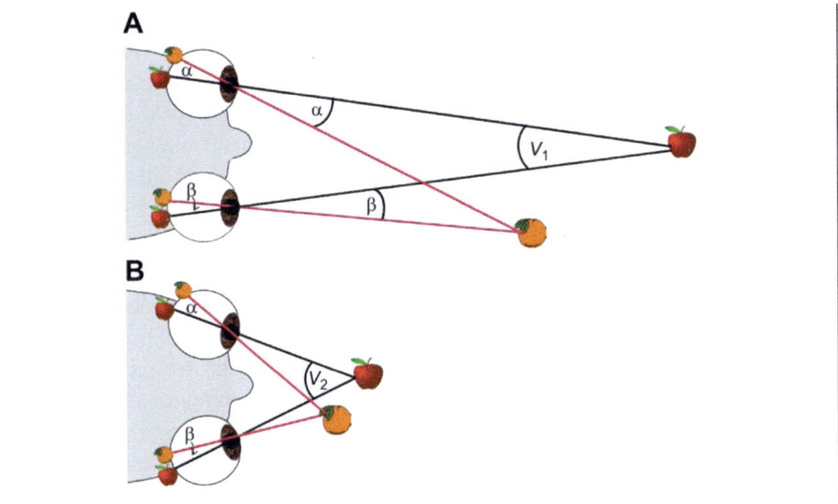

Figure 3.9 Stereopsis in mobile eyes (figure from Nityananda & Read, 2017).

3.4.1.2 Depth vision for estimating distance

The eye relies on various cues to estimate the relative distances between different parts of a scene. The contribution of diverse processes is advantageous because decisions often need to be made rapidly (e.g., during a threat or escape), and there can be variations in the available information. Higher brain centres also play a crucial role by providing memory-based information for scene analysis.

When viewing the same scene from different angles with both eyes, the brain can use the disparity in the images (stereopsis) to calculate the distance of an object based on triangulation. As two objects fall on different locations of the retina, the disparity, or the calculation of the angles (Box 3.5), allows for determining which object is closer.

Relative depth can also be assessed with just one eye if the brain has prior experience with typical scenes. For instance, if two identical objects appear different in size, the smaller one is perceived as further away (Box 3.1). This mechanism operates effectively when the viewer has prior knowledge of the object sizes. Additionally, depth perception can be estimated by moving the eye across the scene, causing the images of closer objects to move more rapidly across the retina.

3.4.2 Cameras as robotic eyes

Robots require sophisticated and resilient visual capabilities to operate effectively in the human-made environment. Cameras in robots serve the same function as eyes, sharing many features but also presenting significant differences (Box 3.4) (Szeliski, 2022).

Most cameras feature an adjustable aperture as a physical device, akin to pupils, regulating the amount of light entering the camera through its variable diameter. The light then traverses a lens system, typically comprising various lenses, forming a complex optical system. This complexity aims to minimise optical errors, such as spherical and chromatic aberrations, ensuring a higher-quality image. Certain elements of the lens system can be adjusted along the optical axis, either manually or with an electric motor, to focus incoming light rays on the digital sensor. This functionality mirrors the accommodation of the lens in a biological eye, contributing to maintaining a sharp image (Box 3.4)

Modern cameras are equipped with digital sensors, which can be of the CCD (charge-coupled device) or CMOS (complementary metal-oxide semiconductor) type (Durini, 2014). Both sensor types comprise a rectangular matrix of small photodiodes constituting the pixels of the sensor. These photodiodes convert incoming photons into electric charge, subsequently converted into a digital value (i.e., a number) representing the light intensity at the specific pixel. The primary distinction between the two sensor types lies in the process of handling the accumulated electric charge. In a CCD sensor, the charge must be transferred to the sensor's edge for conversion into digital data, whereas a CMOS sensor has a dedicated converter for each pixel. Similar to the adaptation of the retina in a biological eye, the sensitivity of a digital sensor can be adjusted by altering the amplification of the accumulated electric charge (Figure 3.7).

In both types of sensors, each pixel is of the same type, indicating that there is no variation in their functionalities as seen with the rods and cones in a biological eye. For colour vision, most cameras employ a colour filter array. Each pixel is equipped with its colour filter, typically in red, green, or blue, causing the pixel to detect only the wavelengths of light that can pass through its filter. In practical terms, colour filters are often organised in a 'Bayer arrangement', featuring twice as many green filters as red or blue to replicate the characteristics of the human eye (Kolb et al., 1995).

The output of the image sensor can be saved on a hard drive as a raw image file, which directly contains data as obtained from the sensor itself. Although this format has lossless data from the moment of the image capture, it has a notably high file size that requires large amount of both storage and bandwidth. Most cameras are able to output raw files, but an additional disadvantage of this format is that it contains direct sensor data, and it is usually more difficult to use as direct input for image processing algorithms. As most engineering use cases require the actual image itself, the raw data is usually converted into either an image or a video stream by the camera, so that it can be viewed and processed on a computer more easily.

The camera transmits the data from each pixel on the sensor to a computer chip for processing into an image. While the human eye boasts approximately 130 million neurons in the retina (as mentioned earlier), the optic nerve responsible for carrying these sensor signals to the brain has only 1.2 million fibres. This implies that less than 10% of the retina's data is conveyed to the brain at any given moment. In comparison, higher-end photography cameras often feature sensors with a few tens of millions of pixels. However, when capturing video (as opposed to still images), the practical resolution of cameras is typically lower, usually below 10 million pixels. Common resolutions include Full HD (1,920 × 1,080 pixels) and 4K (3,840 × 2,160 pixels). In cases where a complete image-type vision is not needed, individual photodiodes can also be used to measure the presence or intensity of light.

3.4.2.1 Detecting simple visual features

As mentioned before, modern digital cameras gather information as a 2D matrix of pixels, where usually each pixel has three intensity values, one for each colour channel (red, green, blue). As many cases in robotics are concerned with higher-level uses of images (e.g., face detection or object detection), computer algorithms are deployed for extracting such features. Some of the simpler tasks performed are edge detection and filtering of the noise in the image.

Convolution is the mathematical operation that is the backbone of many signal-processing techniques, including computer vision. It takes two mathematical functions as its input and multiplies their respective values while 'sliding' one function along the other. In image processing, one of the functions is the 2D image itself, while the other one is often referred to as a 'convolution window', or window function. To calculate the result of convolution at a given pixel, the window function is placed at the given location, and the pixel values of the image are multiplied by the values of the convolution window, then the products are added (Khan et al., 2022).

A special case must be defined for pixels on the edges of the image, as the convolution window will hang off from the image area. A common solution for this problem is 'zero-padding', which simply assumes that all pixels outside the image have zero values for all colour channels.

3.4.2.2 Gaussian smoothing and edge detection with convolution

Both edge detection and Gaussian smoothing (a type of noise filtering) rely on the convolutional operation, but they use different convolution windows. The Gaussian smoothing window is the following:

$$1/10 * \begin{pmatrix} 1 & 1 & 1 \\ 1 & 2 & 1 \\ 1 & 1 & 1 \end{pmatrix}$$

During convolution, the above matrix is placed such that the pixel for which the new value is being calculated is at the centre of the matrix. The other elements of the matrix represent the weights with which the surrounding pixels are counted. As it can be seen, this matrix simply averages the surrounding pixels, while weighing the current one twice as much as the others. The result of the averaging is multiplied by 1/10 to normalise the overall sum of all weights in the convolution window to 1, so that the convolution does not scale the pixel values.

For edge detection, the so-called 'Sobel-kernel' is used. There are two kernels, or convolution windows for this operation, as one is used for detecting vertical edges, and the other for horizontal ones.

$$s_x(x,y) = \begin{pmatrix} -1 & 0 & 1 \\ -2 & 0 & 2 \\ -1 & 0 & 1 \end{pmatrix}$$

$$s_y(x,y) = \begin{pmatrix} 1 & 2 & 1 \\ 0 & 0 & 0 \\ -1 & -2 & -1 \end{pmatrix}$$

3.4.2.3 Motion vision

As digital cameras have no inherent capability for detecting motion (in contrast to eyes), it is detected by computer vision techniques using differences between subsequent images. A method for this is calculating the so-called 'optical flow', which is the distribution of the perceived motion of the different brightness areas in the image. The result of the calculation is a vector field representing the estimated velocities of the image points, which can be related to the actual motion of the projected points on the sensor's plane.

As this method requires additional algorithms, it can be computationally expensive. There is ongoing research about devices using retina-inspired hardware with specifically designed artificial neural networks for more efficient motion detection, but currently these devices are not yet available for general robotics uses (Zhang et al., 2022).

3.4.2.4 Depth vision

Similarly, to biological vision, a single image from a single point of view does not provide enough information to determine the distance of objects from the camera, but there are several ways to overcome this in robotics (Lakshmanan et al., 2021).

Stereo vision is when two cameras record the scene at the same time, while motion parallax refers to using the movement of one camera relative to the scene in order to record the scene from slightly different angles. Both of these mechanisms are present and used in biological vision as well. In computer vision and robotics, RGBD cameras are also used, which usually have a time-of-flight (TOF)-based measurement system and are able to gather depth information directly. Optical flow can also be used for depth estimation, building upon the fact that objects closer to the camera appear to move faster than those further away.

A problematic point in stereo vision is merging the information from the two viewpoints into a common knowledge base. For any calculations or depth estimations, points between the two images must be matched first, which is not a straightforward task, and it comes with certain challenges. A common algorithm for solving this problem is template matching, in which the algorithm selects a feature (that is, a set of pixels) in one of the images and scans the other image for the area that matches best. It is assumed that both the optical properties and the relative positions of the cameras are known, and that the cameras lie in the same horizontal plane. As template matching has its own difficulties (such as choosing a suitable feature size and defining a good metric for similarity between features), algorithms for stereo vision are a topic of ongoing research (Kok & Rajendran, 2019).

3.4.3 Active visual sensors

Laser, abbreviation for Light Amplification by Stimulated Emission of Radiation, is an artificial source of light that has been invented in 1960 and is widely used in many engineering applications due to its unique properties (Albustanji et al., 2023; Heckman, 2022).

A laser creates light by first exciting the atoms of a so-called active medium by the means of a pumping mechanism, and then triggering a stimulated emission which releases a photon of a given frequency and phase. This process happens inside the laser cavity, which is bounded by a mirror on both sides. One of the mirrors is fully reflective, while the other one has a small transmission value, which lets part of the photons exit the cavity, and therefore emit laser light. Compared to other light sources, laser has a very narrow spectrum, and from the perspective of practical usage, it can usually be considered monochromatic. Due to its operation principle, laser is spatially coherent, and therefore it can be focused onto a small surface area, possessing a high energy density. There are no natural sources that produce laser; thus, this type of light is less sensitive to interference from environmental sources.

Lidar, an acronym for light detection and ranging, is an optical measurement technology and device that uses lasers. Lidar works in a TOF principle by emitting a laser beam in a given direction and measuring the time until the reflected beam returns to the device (see also sonars below). The distance to the surface can be calculated using the speed of light in air. One laser beam is only capable of measuring the distance in one direction, therefore lidar sensors usually scan multiple directions either in a 2D or a 3D space.

Reflective optosensors are another type of active light-based sensors, which consist of an emitter and a receiver part, but they can be built into the same housing. Their main operating principle is that the emitter emits light with known characteristics (usually in the IR wavelength), and the receiver measures some property of it, such as its intensity or the round-trip time of the beam. In the simplest of the cases, such a sensor can be used to detect the presence of an object by detecting reflected light. A similar concept is used when detecting markings, such as lines on the floor or a wall: the lines have a different reflectivity at the wavelength of the emitted light, and therefore produce a sudden change in measured light intensity which can be used for controlling the motion of the robot. The same idea is used for rotary encoders, which are devices that are used to measure the angular position of a rotating shaft. These sensors have a disk fixed on the shaft, which has markings that alternate between having light-reflecting and absorbing properties. An optical sensor detects the different parts of the disk, and therefore measures the change in the angular position of the shaft. By using more than one set of markings, the direction of rotation or the absolute angular position can also be determined (see also above encoders).

A more advanced type of sensor built on the same principle is the so-called TOF sensor. These sensors use laser beams as the emitted light and measure their round-trip time, that is, the time it takes for light to travel to the detected object and arrive at the receiver after a reflection. Given the speed of light, the measurement can be used to calculate the distance between the sensor and the detected object. The TOF principle is used extensively by lidar sensors.

3.4.4 Bioinspired sensors for vision

3.4.4.1 Distributed vision systems

An alternative strategy for enhancing the effectiveness of the visual system involves increasing the number of visually sensitive receptors. Various clades of animals have evolved distributed vision systems as a solution. For example, certain species of marine molluscs, like chitons, possess numerous eye-like sensors distributed across various parts of their bodies (Chappell & Speiser, 2023). These animals seem capable of integrating information from these sensors, utilising them to evade threats from predators. This capacity relies on their ability to improve visual resolution through redundant sampling of visual input from the distributed eyes, characterised by overlapping visual fields. While the deployment of multiple cameras is gaining popularity for navigating large objects like cars (Häne et al., 2017), the concept of utilising many simpler visual sensors distributed on the body has not yet been introduced into mobile robotics.

3.4.4.2 Infrared sensation

Heated objects emit IR light that is invisible to humans and most animal species. However, a few groups of animals, such as snakes and vampire bats, have evolved sensors capable of detecting warm objects at a distance based on IR radiation. This ability proves particularly useful in the dark when conventional vision is ineffective. Snakes, for instance, possess a specific membrane that absorbs IR radiation, leading to an increase in temperature. Thermosensory cells detect this change and transmit the information to the central nervous system (Bleckmann et al., 2004). Another type of IR receptor, more specific to the spectrum of radiation, was discovered by Schmitz and Bleckmann (1998). In the case of fire beetle (*Melanophila*) species, they have a photomechano-receptor where specific protein molecules begin to vibrate when struck by IR radiation of a particular spectrum. The movement of these molecules is then detected by mechanoreceptor cells and further processed by the brain.

Artificial sensors employ different mechanisms to measure IR radiation (Adarsh et al., 2016). In engineering, non-imaging IR sensors can be categorised as active or passive. Active sensors consist of an emitter and a receiver, measuring the intensity of the reflection of their own emitted IR light. These sensors are commonly used as distance sensors. On the contrary, passive IR (PIR) sensors do not emit light; instead, they rely on the emitted IR radiation from objects, animals, or humans in their field of view. Detecting changes in light intensity, PIR sensors are often employed as motion detectors, providing binary output. They find applications in security alarms or light switches due to their non-image-forming nature.

Imaging IR sensors essentially function as digital cameras tuned for IR detection in the electromagnetic spectrum. Regular cameras, sensitive to near-IR light, typically include IR filters to block these wavelengths and produce natural-looking images. However, in specific cases, IR filters may be intentionally omitted, allowing the camera to capture images in the IR range. This technique is commonly used in night vision cameras, where a dedicated IR light source illuminates the scene, enabling the camera to capture reflections from objects even in the absence of visible light.

IR technology is employed in thermal cameras, which detect IR radiation emitted from objects in the environment and calculate their temperature based on the amount of thermal energy. Due to the longer wavelength of IR light compared to visible light, IR sensors utilise larger sensor elements, resulting in lower resolution than regular cameras, typically in the range of tens of thousands of pixels.

3.4.4.3 Eyes on tentacles

Many snail species exhibit a unique sensory adaptation featuring light-sensitive receptors at the tips of bilateral tentacles. Unlike conventional eyes fixed to the body, these specialised structures enable snails to move their eyes independently without requiring body movement (Eakin & Brandenburger, 1975). However, these eyes have limited visual capabilities, primarily used for detecting the intensity and direction of light, with some evidence suggesting a potential for object detection to some extent.

In principle, eyes or other sensors could be situated at the end of a tendon-like extension, facilitating independent movement of the body. In an engineered solution, the use of fibre optic cables eliminates the necessity of placing sensors directly at the tip. These cables are designed to ensure total internal reflection, allowing light to travel through without dissipation, making them particularly valuable in medical applications. In medical practice, for instance, only the thin cable needs to be inserted into the body, while the components related to image detection can remain outside (Carmo & Ribeiro, 2013).

3.4.5 The advantages and disadvantages of using visual sensors

Vision is one of the main sensory abilities in the animal kingdom. During evolution, animals have featured sensors that also reflect the ecological challenges of the particular species because there is trade-off between the amount of information provided by the sensor and the cost of maintaining the sensor and the mental structures for evaluating the input. Processing of visual information may require a lot of mental effort. Thus, the complexity of visual sensors has been adjusted to the species needs by selective forces. The most sophisticated cameras may provide immense amount of data as input making the data processing too slow and too costly. One possible solution is that only a fraction of the incoming data is passed on the evaluating system, or some kind of input is not considered at all, e.g., using b/w camera instead of a colour camera.

There are also issues with analysing visual data because present architectures are far from the analytic complexity of the animal brain. Thus, even multiple camera systems controlled by sophisticated software tools do not reach the robustness of vision present in insects or fishes. Interference from changes in light intensity, shadow, or various background effects may make vision relatively unreliable in mobile robots.

Typically, visual sensors, eyes, are located on a mobile head which also means that they cannot be moved independently. In robotics, there is the possibility to obtain visual input from various parts of the body by using optic cables (see above). The fixed positions and the visibility of the eyes on the head evolved other mechanisms, as recognising the eyes (and the face) as belonging to another animate being. Heads and eyes oriented at social partner for a short time may indicate attention, interest but extended watching may be a signal for threat (Section 1.11.6).

Interestingly, many social robots have fake eyes, that is, the cameras are located at different places. This unnatural placement could become a problem because the human partner likely orients at the non-functional eyes during interaction and thus the real camera may have difficulties to pick up facial or bodily actions of the other.

3.4.6 Hearing sensors – ears

Sound is a wave that induces vibrations in mediums such as gas, liquid, or solid. When air vibrates, it leads to two outcomes. First, air particles move in a specific direction, and second, this movement results in localised, fluctuating changes in air pressure. Some of these sound waves may arise from the movement of living beings, suggesting that animals could have gained an advantage by evolving mechanoreceptors sensitive to such inputs. The crucial components of these mechanoreceptors, known as hair cells, include fine hairs located at the top of the cell. These receptor cells exhibit high sensitivity to hair deflection in space, translating these subtle movements into analogue electrical signals (Pickles, 2008).

In proximity to the sound source, particles in motion can directly displace these extremely light hairs. This phenomenon occurs in insects with cerci located at the rear-most segments, allowing them to detect rapid air vibrations typically caused by potential threats, such as an approaching rat. This sensory input aids the animal in initiating an escape response (e.g., Libersat & Camhi, 1988).

Other mechanosensory organs respond to the pressure component of sound waves, utilising a specific membrane that moves when there are changes in air pressure on either side (Figure 3.10). Interestingly, the concept of using a membrane has independently evolved several times in different animals, and such membranes are also significant components of artificial sensors like microphones (see below).

Unlike vision, the sense of hearing is not widely distributed among animals. Vertebrates and insects have the most well-developed hearing senses. Species can be characterised by their hearing range, determined by various morphological factors (e.g., the shape and size of the outer, middle, and inner ear) and physiological factors (e.g., sensitivity of the hair cells), often in association with environmental challenges. The sound spectrum is somewhat arbitrary divided into three sections based on the human hearing range. Infrasound ranges from 0 to 20 Hz, acoustic (human audible) sounds fall between 20 and 20,000 Hz, and sound frequencies higher are described as ultrasound. There is a variation in hearing range of animals, for example, whales hear

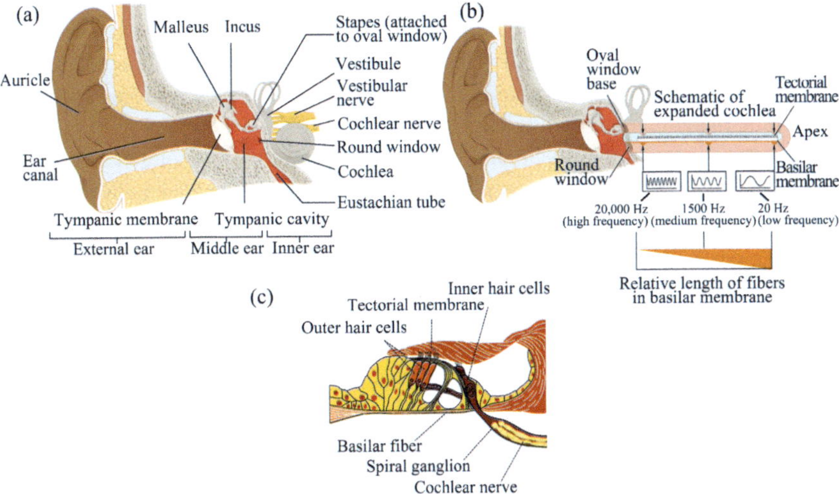

Figure 3.10 Overview of the hearing apparatus in humans. (a) Cross-section showing the position of the middle and inner ear, and (b) mechanical depiction of the cochlea indicating the effect of the sound waves relied by the auditory ossicles (malleus, incus, stapes) on the movement of the basal membrane. The main function of these small bones is to overcome the large difference in acoustic impedance between air and fluid (in the cochlea). High-pitched sounds (approx. 20 kHz) are detected at the lower part of the basal membrane, closer to the oval window (figures from Betts et al., 2013; Access for free at https://openstax.org/books/anatomy-and-physiology/pages/1-introduction). (d) The Corti organ (figure by Madhero88, Wikimedia Commons): The movement of the hair of the receptor cells (yellow arrow) provides the mechanical input for decoding the sound frequency. The non-linearity of the hearing systems has several sources (Eguíluz et al., 2000): (1) Frequency perception is not linearly related to the physical frequency of sound waves. Instead, it follows a logarithmic scale. Doubling of the frequency does not result in a perceived doubling of pitch. For example, an increase from 100 Hz to 200 Hz is perceived as a more significant change in pitch than an increase from 1,000 Hz to 2,000 Hz. (2) Perception of intensity has a logarithmic relationship with the physical intensity of a sound wave. As intensity increases, the perceived loudness does not increase at the same rate (see also the Weber–Fechner law, Box 1.3). (3) Nonlinearities emerge also in the cochlea. In the tonotopically organised basilar membrane different regions respond more strongly to specific frequencies. (4) Auditory nerve fibres exhibit a compression effect as the firing rate of auditory nerve fibres does not increase linearly with the sound intensity. Non-linear interactions occur in the masking of one sound by another. A loud sound may mask a quieter sound at a nearby frequency, making it less perceptible.

sounds between 20 and 100,000 Hz, and dogs and mice perceive sounds between 40 and 40,000 Hz and 1,000 and 91,000 Hz, respectively (Pickles, 2008).

In the vertebrate ear (Figure 3.10), the membrane (ear drum) is connected to a fluid-filled tube by three rigid bones (malleus, incus, and stapes). This way the movement of the membrane is transmitted to the fluid in which the hairs of the receptor cells sense any movement. This tube (cochlea) has a shape similar to a snail's house with a stretched membrane in the middle. The vibration of the fluid makes the (basilar) membrane swing. This movement bends the hairs of receptor cells, sitting on the membrane, at different degrees, providing the mechanical input for auditory sensation. Due to differences in the thickness and width of the membrane, the peak amplitude of sound wave frequencies shows a specific spatial distribution. High-frequency sounds resonate at the base of the membrane, while low-frequency sounds are received at the tip of the cochlea (tonotopic organisation). Naturally, complex sound waves consisting of a mixture of different frequencies produce a complex movement across the whole length of the tube.

The main role of the inner hair cells is to transmit signals to the brain, while the outer hair cells perform a vital function in modulating the sound waves that reach the innermost part of the inner ear. The outer hair cells function as envelope detectors, allowing them to precisely detect changes in sound intensity and adjust these fluctuations before relaying them to the inner ear. This mechanism acts as a protective measure for the inner hair cells, preventing them from being overwhelmed by extremely loud sounds. This process closely resembles automatic gain control, a technique used in various devices designed for receiving and transmitting sound waves to achieve dynamic range compression (DRC). In essence, this process dampens loud sounds while amplifying quiet ones, reducing the disparity between the peak volume of the loudest sound and the lowest volume of the quietest sound. Damage to the outer hair cells can result in a notable decrease in the amplitude of vibrations transmitted to the inner hair cells (Pickles, 2008).

3.4.6.1 Hearing acuity

There are several measures of hearing acuity. One aspect relates to the ability to discriminate frequencies, and humans show the best performance for such pitch discrimination in the 1,000 to 4,000 Hz range which is also important for speech perception. Loudness perception is measured by modulating the amplitude of the sound at specific frequencies. The threshold of hearing, or the softest sound that can be detected by the average human ear, is typically around 0 decibels (dB) at 1,000 Hz. The detection of a gap between two sounds is also a form of hearing acuity. Gap detection thresholds are about 1 ms for white noise, for pure tones the gap threshold increases with decreasing frequency from 2.3 ms at 8,000 Hz to 22.5 ms at 200 Hz (Shailer & Moore, 1983).

3.4.6.2 Selective hearing

Hearing ability faces challenges when detecting signals in a noisy environment, especially when the noise is similar to the signal, akin to trying to listen to a specific conversation in the midst of many discussions at a party –a phenomenon known as the cocktail party effect (Bee & Micheyl, 2008). Animals also encounter similar challenges; for instance, female toads must locate conspecific males amidst the calls of various species (Burmeister, 2017).

The initial task for the auditory system involves segregating sounds from a continuous acoustic stream, as visualised in a sonogram (Box 3.6). It is crucial to note that since all acoustic signals reaching the ear combine into an auditory scene, any sound heard at a specific amplitude or frequency could originate from multiple independent sources. Hence, the brain requires specific algorithms to separate vocal outputs from different callers.

Box 3.6 Solving the cocktail-party problem by natural and artificial minds

Listening selectively to a specific sound source is a very intriguing ability of animals. Although most experience has been gathered about human hearing, the skill also exists in non-humans. Selective hearing nick-named as the cocktail-party effect is involved when the listener aims to focus (listen to the speech) of one of many speakers who provide an interfering background noise. Selective hearing is also at work when humans listen to music and can follow the sounds emitted by specific instruments. McDermott (2009) described the challenges of such situations both from the perspective of a human and an artificial hearing system, noting the significant advantage of the former.

The spectrogram depicted in the figure shows a specific utterance (upper row) and the same utterance mixed with one additional speech signal from a different speaker (lower row). The mixture simulates the auditory experience as if the additional speaker was speaking at the same volume as the target speaker but was positioned twice as far from the listener, mimicking a scenario resembling a cocktail party. Spectrograms were generated using a filter bank with bandwidths and frequency spacing akin to those found in the human ear. Each pixel in the spectrogram represents the root mean square amplitude of the signal within a specific frequency band and time window. Notably, the spectrogram excludes local phase information, as listeners are generally insensitive to it.

The grayscale indicates attenuation (in dB) relative to the maximum amplitude of all pixels in all spectrograms, allowing for the comparison of grey levels across different spectrograms. The acoustic cues thought to contribute to sound segregation are highlighted in the spectrogram of the target speech (top row). On the right, the spectrogram is identical to the one on the left, except for the addition of overlaid colour masks. Pixels labelled green indicate areas where the original target speech signal is more than −50 dB, but the mixture level is at least 5 dB higher. Pixels labelled red denote areas where the target was quieter, and the mixture had an amplitude of more than −50 dB. The listener may rely on the statistical regularities in the sound stream, for example, sounds consisting of specific frequencies start and end at the same point in time, and even if these critical events sometimes overlap with other sounds, the majority of these repetitions (based also on former experience) can be heard and predicted.

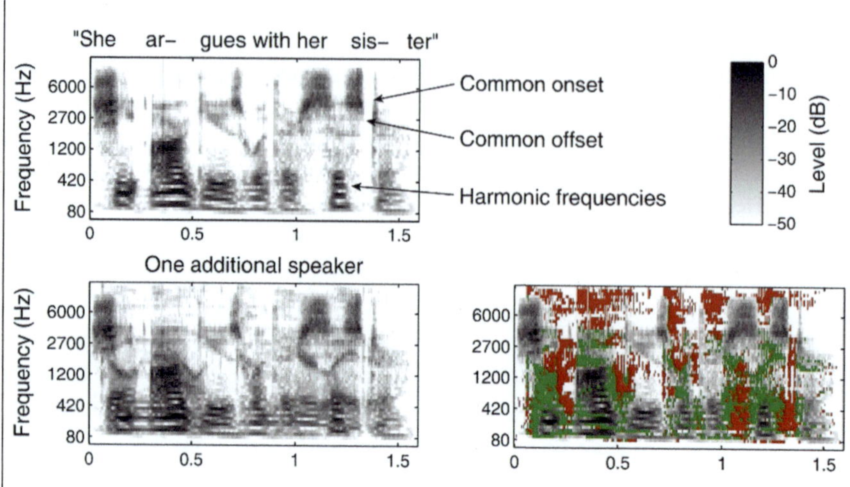

Figure 3.11 The spectrograms recorded from the focal speaker (a), a second partner (b) and a combined spectrogram of the two speakers (modified from McDermott, 2009).

Dealing with the cocktail party problem is a huge challenge for artificial systems. Deployment of multiple microphones located at different distances provides usable input for architectures based on independent component analysis. The varying distances between the speakers and the microphones result in different weighted combination of the sources. If the number of microphones and the speakers is the same rudimentary assumptions about the statistical properties of speakers met the requirements to derive the source signals from the mixtures.

While successful, this method is different from the mechanism in living agents because the latter has only two ears for the same job, and typically one ear is enough for selective hearing. While natural systems may rely on statistical methods, comparable to independent component analysis, they must also capitalise on other solutions (Figure 3.11).

McDermott (2009) highlights two distinct mental mechanisms. Bottom-up processes focus on detecting statistical regularities in the sound stream. First, specific frequency ranges preferred by an individual likely starting and ending at the same time. Second, individual voices also exhibit unique frequency patterns, including amplitude changes at specific points in time. Third, harmonics (integer multiples of a fundamental frequency) present in many vocalisations offer valuable input for statistical sound analyses. These mechanistic processes are complemented by top-down cognitive inputs based on experience. In humans, familiarity with the voice of the conversing partner or knowledge of the conversation topic aids in eavesdropping on specific communicative interactions.

3.4.6.3 Directional hearing

Directional hearing is determined by the bilateral positions of the ears on the body, typically located on the head, at varying distances from the sound source. Consequently, the sound arrives faster in one ear than the other, allowing for the measurement of interaural time differences and intensity variations. Physical properties suggest that low-frequency sounds are best detected by measuring the time difference between the waves reaching the ears, while for higher frequencies, intensity differences offer a more accurate estimation (Pickles, 2008). The brain processes these differences, and the subject learns the corresponding direction of the sound source. Notably, this method may not be effective when the source is equidistant from both ears, but such situations are rare, and minor head movements to either side alter the distance of the ears from the source (Van Wanrooij & Van Opstal, 2004).

3.4.7 Microphones as robotic ears

Microphones share the same basic physical principle with biological ears for sensing sound: they have a membrane (often referred to as a diaphragm) that vibrates as a consequence of changes in air pressure. The different types of microphones differ in how they convert this vibration into measurable electric signals, with the two most common types being the condenser microphone and the dynamic microphone (Eargle, 2005).

Condenser microphones have a backplate placed close to the diaphragm, which together form a capacitor (Figure 3.12). As the diaphragm moves due to changes in air pressure, the capacitance of the capacitor changes, and this change is converted into an electric signal. Since the mass of the diaphragm is usually low, it can easily be moved by air, and therefore this type of microphone can be made very sensitive. In order to measure capacitance, a voltage is required across the two plates forming the capacitor, meaning that this type of microphone requires an active power source (Eargle, 2005).

Dynamic microphones use electromagnetic induction as their main working principle. They feature a coil attached to the diaphragm, and a permanent magnet that creates a magnetic field. When the diaphragm vibrates, the coil starts moving within the magnetic field, and (as per Faraday's law) this induces an electromotive force and thus a current in the coil, which is then converted into an electric signal. As a coil is attached to the diaphragm, dynamic microphones are heavier and less sensitive compared to condenser microphones, especially at higher frequencies. However, they are also more durable, and able to withstand high sound pressure, and they also do not require an external power supply.

3.4.7.1 Hearing acuity

In computer-based audio processing, the ability to discriminate frequencies is built upon a mathematical integral transform, called the Fourier transform. According to the mathematical properties of this operation, the frequency resolution is limited by

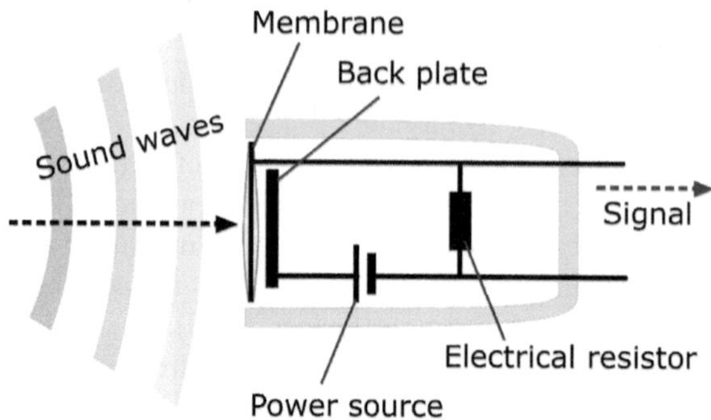

Figure 3.12 The microphone is an analogue to the hearing organ. Similar to biological ears, microphones also feature a membrane (often called the front plate or the diaphragm) for sensing pressure variations of the medium. The front and the back plates form a capacitor, the capacitance of which varies with the changing distance between the plates. When connected to a voltage source, such as a battery, this results in a change in accumulated electric charge, which then can be measured by electric circuitry. Opposed to hearing organs, microphones do not feature spatial separation of frequencies (no tonotopic lay out), as the diaphragm moves according to a superposition of the vibrations of all frequencies. This means that measuring separate frequencies requires software algorithms, namely, an implementation of Fourier transform, but this also allows for other types of digital signal processing to take place. Microphones can be designed more specifically to the desired use-case and frequency response compared to biological ears, but generally they offer a significantly smaller dynamic range. (Figure by Kevin, Wikimedia Commons)

the length of the processed input audio sequence and is given as $\Delta f = 1/T$, where T denotes the length of the audio that is being processed at once.

The lowest loudness that can be recorded with a microphone is generally limited by the microphone's self-noise that is generated either mechanically or in the electrical circuitry. The corresponding sound pressure level varies between microphones, and it is a design parameter when choosing a microphone for a given application (Eargle, 2005).

3.4.7.2 Selective hearing

Selecting relevant information from an incoming audio stream in artificial hearing faces similar challenges to those of the biological one. When recognising certain sounds, such as human speech, artificial systems usually rely on the signal being more intense than the noises in the environment. Noise filtering algorithms are used widely, and they are often implemented as frequency filters that amplify or attenuate sounds of selected frequencies. This approach does not require too much computational power, but it can only be used if there is a clear separation in frequency between the noise and the useful part of the sound signal.

An additional challenge in artificial hearing is that extracting knowledge about the topics of conversation is a computationally intensive process that relies on speech recognition, and it is usually not fast enough to be used for contextual aid in hearing. The performance of speech recognition deteriorates quickly with an increased noise level, and it is nearly impossible to separate different speakers in a noisy environment (Han et al., 2019).

3.4.7.3 Directional hearing

Directional hearing in robotics is handled in a similar fashion to its biological equivalent. A single microphone is only able to produce a single audio signal that contains no spatial information, which implies that for locating a sound source, a configuration of at least two microphones is needed. For detecting the direction, the used techniques are similar to the ones used in biological ears, that is, measuring the differences in time and intensity between the two microphones (Hwang et al., 2011). As with biological hearing, some processing of sound signals is needed to obtain spatial information, which requires computational power and advanced algorithms. In real-world scenarios, calculating a precise direction can often be problematic due to the noisiness of the environment and measurement imperfections, and finding suitable algorithms is a topic of active research.

Like biological ears, microphones may also have different sensitivities in different directions, which is often referred to as directionality or polar pattern. Omnidirectional microphones have a uniform sensitivity in all directions, while unidirectional microphones, for example, are tuned to be more sensitive to sounds coming from only one direction (usually the front) and suppress sounds from the sides (Eargle, 2005).

3.4.8 Active acoustic sensors

A few clades of animal species have independently evolved an active acoustic sensor system that can be used in similar functions to vision: detecting objects in the environment both for navigation and for interaction with them (Au, 1997). These alternative sensors are especially useful under conditions of low visibility, such as at night or underwater. For example, both bats (at night) and dolphins (underwater) can detect distance to objects, their size, texture, and velocity all through characteristics of their reflected echoes (e.g., Harley & DeLong, 2008). These active sensors consist of an emitter that sends out sound waves at regular intervals, and a sensor that is sensitive to the sound waves reflected from the objects. These rapid clicks are generated by the nasal sacs located in the dolphins' forehead. The emission of clicks has its own dynamics with short silence periods between two sounds and is also influenced by the emitter's interest in exploring the surrounding field or objects in it (e.g., prey). Both bats and dolphins rely on ultrasounds that usually fall between 20 and 200 kHz. The use of this frequency range can be advantageous because most animals do not hear in that range and there are no natural sources of ultrasound (Box 3.7).

The minimum audible angle for dolphins is comparable to terrestrial species (between 0.7 and 4° depending on the sound frequency and testing method) (Renaud & Popper, 1975). Thus, they can use the sonar relatively efficiently for underwater orientation and for localisation of underwater objects.

The minimum inter-sound interval for being able to hear two subsequent vocalisations as distinct is relatively long for dolphins (264 ms compared to 1–5 ms in humans). Dolphins are not able to separate pulses sent with shorter intervals. Since dolphins need to analyse the echo of the pulses, it is important that sent and repulsed sound waves do not interfere. They seem to manipulate the inter-sound interval during search and often the echoes were received before the next pulse was emitted. Dolphins probably integrate information obtained across click trains (Altes et al., 2003).

Bats can detect distance to objects, their size, texture, and velocity by the means of echolocation, and studies have revealed similar ability of dolphins. They can discriminate less than one mm difference in wall thickness of otherwise identical aluminium cylinders. It is claimed that within the range of 1–100 m, echolocating dolphins are superior to any human-made sonar for object detection and discrimination (Au & Simmons, 2007). Interestingly, they may also listen to clicks emitted by other dolphins close by and recognise the object based on this kind of information (c.f., eavesdropping, Xitco & Roitblat, 1996).

Box 3.7 Sonar use in dolphins and the problem of reflections

The echolocation involves emitting sound waves (blue) and sensing the echoes that bounce back (dark grey), providing information about the surrounding objects (Figure 3.13). There are however different types of reflections that make the analysis of the incoming signal difficult. Some part of the reflections may travel towards the dolphin, while others get dispersed in the environment. These reflections, however, may also reach other non-specific targets, and these 'secondary' reflections (black) may also interfere with the primary reflections from the target object. The specular reflection provides a specific challenge when sound waves reflect off smooth and hard surfaces, such as the surface of the water, producing strong, coherent echoes. These echoes can be challenging to interpret, as they may not accurately represent the location or nature of an object. Dolphins employ several strategies to deal with the specular reflection problem:

1. Dolphins adjust the characteristics of their echolocation clicks based on the environment. They may modify the frequency, amplitude, or duration of the clicks to minimise the impact of specular reflections.
2. They emit echolocation clicks at different angles, not just directly ahead. By doing so, they can receive echoes from multiple directions, allowing them to gather more information about the surroundings and differentiate between echoes caused by specular reflections and those from actual objects.
3. By moving their heads, they can alter the angle of their echolocation beams, reducing the impact of echoes from specular reflections and enhance their ability to perceive objects in the environment.

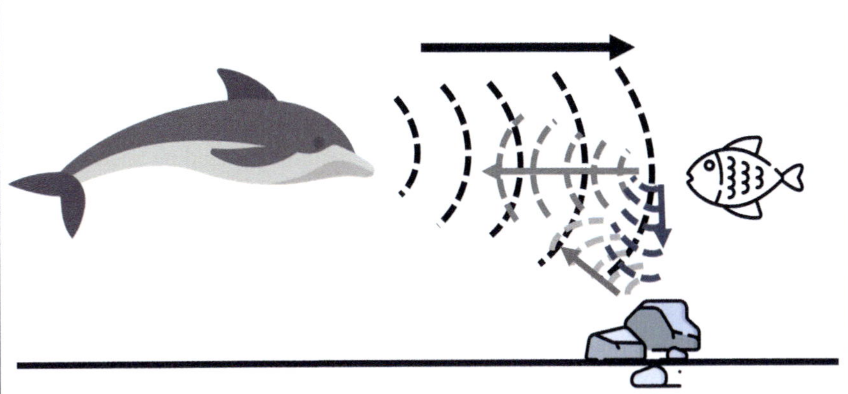

Figure 3.13 Sonar use in dolphins (icons from Freepik, Flaticon).

4 Emitting rapid sequences or click trains, they gather more information over time, allowing them to distinguish between echoes from specular reflections and those from objects with complex shapes.

Robots face similar problems when they use sonars to navigate in anthropogenic environments. While dolphins have only a single sonar, robots could be equipped with an array of sonars and also multiple sensors.

The sonar technology was independently invented by engineers at the beginning of the last century. In contrast, animal sonars were only discovered 30 years later. The term 'sonar' can refer to both active and passive devices. Passive ones just detect noises from other sources. This limits the usability, however. Thus, active ones are more widespread in robotics (e.g., Carelli & Freire, 2003), although biological sensors are vastly more sophisticated than current artificial ones.

Just like in the case of dolphins or bats, active sonars work by emitting acoustic pluses and then listening to their echoes. By measuring the time between emission and reception, the distance to the object can be calculated using the speed of sound in the medium. By using multiple sound beams and additional measurements (such as beam energy), a more detailed picture can be created of the surroundings.

Sonars can use a wide range of frequencies, from a few 1,000 Hz to 1,000 kHz frequencies, but most devices use sounds in the ultrasound range, with pulse lengths of a few 10 ms. There is a general trade-off between range and resolution, as lower frequencies offer higher maximum range, while higher frequencies can provide more details with a higher resolution. Low-resolution fish-finders have maximum ranges of up to a 1,000 m, while high-resolution imaging sonars are usually limited to 10–50 m. Simpler sonars use a constant frequency signal, while more advanced devices deploy a so-called chirp signal which has a varying frequency, allowing for more detailed signal processing and Doppler-effect compensation (Christ & Wernli, 2014; Lurton, 2002).

3.4.9 Bioinspired sensors for hearing

While sound waves are typically sensed by specific membranes, such as the eardrum in mammals, research has also revealed that the body or parts of the body, like the skull, may also resonate with acoustic waves (Stenfelt & Goode, 2005). Five distinct processes of bone-conducted sounds have been identified, including sound entering the external ear canal, inertia of the middle ear ossicle, cochlear fluid inertia, cochlear wall compression, and pressure transmission originating from the cerebrospinal fluid. Bone-conducted sounds interact with each other through vectorial integration, and a complex pattern of stimuli reaches the cochlea (Freeman et al., 2000). Sensing bone-based sounds may have a significant perceptual effect in addition to airborne sounds. Depending on the size of the resonating structure, these sounds may have a larger amplitude (appear to be louder) or differ in tonality, varying according to the individual characteristics of the perceiver.

Vibro-tactile devices have been developed for humans to aid hearing as a complementing solution to more well-known cochlear implants. Fletcher et al. (2023) reported that haptic actuators can transmit speech through vibration at multiple frequencies to the human wrist, and this technology has the potential to improve human hearing at a relatively low cost.

3.4.10 The advantages and disadvantages of using acoustic sensors

Despite the production of noise by larger-bodied creatures during their activities, the prevalence of sensory apparatus to detect such stimuli is significantly less widespread in the animal kingdom compared to vision. The evolution of the vertebrate ear underwent several improvements after the emergence of fishes (Manley, 2017), in parallel less significant changes can be observed in the evolution of visual sensory organs. Interestingly, the evolution of hearing seems to be limited to arthropods among invertebrates and evolved independently multiple times, resulting in varying levels of functional complexity (Warren & Nowotny, 2021).

Sound waves can convey additional information about the environment, which becomes crucial in situations where poor light conditions or physical obstacles render vision ineffective. In specific scenarios, hearing may require less computing power than vision, particularly in tasks like individual recognition. The utilisation of acoustic sensors could offer a cost-effective means of enhancing an agent's sensing performance. However, it is essential to acknowledge that hearing performance diminishes with increasing background noise, presenting challenges for acoustic sensors in anthropogenic environments.

At a certain level of complexity, hearing capabilities can be comparable to visual abilities, and the integration of both types of information (sensor fusion) may result in a robust sensory capacity in social robots. Active hearing (sonars) could be often an alternative to vision, however complex environments may provide a challenge, how to deal with reflections that make the analysis of the incoming signal difficult.

3.4.11 Smelling and tasting sensors – noses and tongues

In biological organisms, chemosensation provides a means to sense potentially attractive or harmful objects or living beings, such as food, potential mates, or predators, based on their chemical composition and sensing the molecules emitted by them. Chemical senses may also be utilised to detect environmental events, such as smoke, or assist in navigation. The involvement of chemical senses in sensing the environment depends on the ecology of the species, but some mechanisms to detect molecules are shared by many species in the animal kingdom. In vertebrates, olfaction and gustation typically occur mainly in well-defined anatomical structures called the 'nose' and the 'tongue', respectively. In invertebrates, however, chemosensation can occur at different places on the body. For example, bees and ants have such receptors on their antennae, snails' and slugs' olfactory organs are on their tentacles, and worms and many aquatic invertebrates possess chemoreceptors distributed across their body surface or in specific regions (Hansson & Stensmyr, 2011).

Both the sense of smell and taste allow organisms to react to volatile molecules (odorants) in the environment, but gustation can also reveal chemical aspects of non-volatile, liquid compounds. The overall design and the chemical and neural mechanisms of olfaction and tasting are very similar across animals, but there are also some differences. Additionally, while olfaction can be regarded as a truly independent sensory system, the perception of taste (at least in investigated mammals) also depends on olfactory input. It is estimated that the sense of smell is responsible for about 80% of what is tasted (Spence, 2015).

3.4.12 Olfaction

Olfaction involves the detection of the chemical composition of volatile substances, such as gases, through direct contact between volatile molecules and sensor molecules (proteins). In terrestrial animals, olfaction is closely associated with breathing, while in fish, olfaction detects chemicals dissolved in water. Typically, volatile chemical compounds reach olfactory sensors through inhalation, and they are partially removed during exhalation. This passive process is often complemented by active inhalation (sniffing), aiming to increase the amount and concentration of chemical compounds reaching the sensors for better recognition performance. Sniffing involves increasing the frequency or depth of inhalation, or both (Mainland & Sobel, 2006).

Olfaction relies on specific chemoreceptor cells located in the olfactory epithelium (OE). These cells have several small filaments (micorvilli) that extend into the mucous fluid covering the epithelium. Therefore, molecules must first enter the mucous material on the epithelium before coming into contact with the protein receptors located in the micorvilli's membrane (Figure 3.14). This mucous fluid provides the appropriate molecular and ionic environment for the interaction between the odorant and the protein receptors. In humans, 7–8 molecules must interact with these protein receptors for the receptor cell to generate an electric signal, whereas, for example, in female

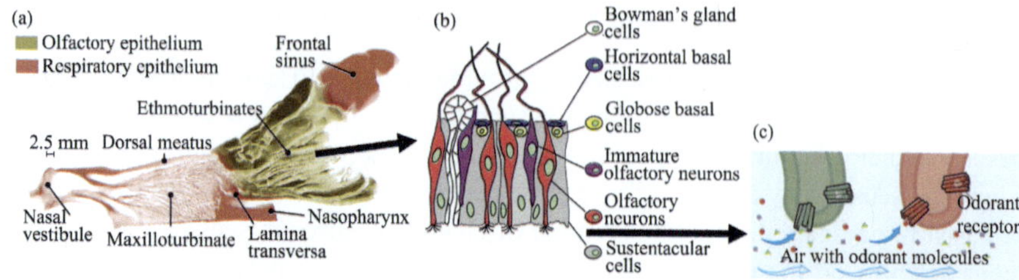

Figure 3.14 Anatomy of the olfactory system: (a) Illustrated in a sagittal section of the dog's skull are two distinct areas: the respiratory region and the olfactory region. The olfactory region comprises scroll-like ethmoturbinates adorned with the olfactory epithelium. These intricately folded structures not only provide an expansive surface for the epithelium but also impede substantial amounts of inhaled air (figure from Lawson et al., 2012) (b) The molecules from the inhaled air become caught in the mucus, where the cilia of the olfactory cells are immersed (figure from Lavoie et al., 2017) (c) The initiation of the olfactory signal occurs as these molecules encounter a suitable protein receptor in the cell membranes of the cilia (figure from Nobel Prize Organization (2004). © The Nobel Assembly at Karolinska Institutet, https://www.nobelprize.org/).

silk moth (*Bombyx mori*), a single molecule sensed can activate the receptor cell, and approach behaviour is initiated when about 1% of all odour receptors are aroused (Kaissling & Priesner, 1970).

In vertebrates, electric impulses are relayed to the olfactory nerve, which establishes synapses with other cells in the glomeruli located in the olfactory bulb (OB; Laska, 2017; Box 3.7). The epithelium of humans contains approximately 20 million olfactory receptor cells, while this number is between 125 and 300 million for dogs. The area of the epithelium correlates with the overall size of the nose, but inside, it also covers a complex folded laminar structure, resulting in a much-extended surface ready to interact with the odorants entering the nose. The overall size of the epithelium is 3–5 m^2 in humans, 6–10 cm^2 in rats, and 60–170 cm^2 in dogs (depending on the breed). Interestingly, olfactory cells are renewed every one to two months, in contrast to other receptor cells for vision or hearing.

Each olfactory receptor cell expresses only one type of receptor protein, sensitive to only a few odorants. This sensitivity is determined by the structure of the protein molecule coded by a specific gene. Recent studies indicate that humans possess 400–450 functional genes, rats 1,200–1,500, and dogs 800–1,000, resulting in around 400 different types of olfactory receptor cells in humans (e.g., Ache & Young, 2005). Each type of protein receptor responds to only a small range of chemicals, causing a pure compound entering the nose to elicit a varied and distributed response by olfactory receptor cells. Different types of protein receptors may be activated, and the strength of this activation can vary. The number of reacting receptors is greater at higher concentrations of the odorant. Thus, an odour elicits a patterned response from a huge number of olfactory receptor cells, with particular odorants represented by different and unique combinations of activated receptors.

This mechanism is supported by the specific spatial arrangement of receptors (Box 3.8). Each odorant receptor type is localised in one specific zone in the OE bilaterally, situated among receptors showing different sensitivities. The output of similar receptors converges on only a few glomeruli at fixed locations. There is a map-like organisation in the OB that facilitates the recognition of chemicals of specific natures. Ultimately, based on the spatial and temporal activity of a unique set of glomeruli in the OB, a specific neural code is generated for that particular odorant. This complex network of connections explains how a relatively small number of specific receptors can be used to distinguish and recognise thousands of different compounds and mixtures, leading to a relatively large individual variability in the sensation of the same chemical stimuli (Imai et al., 2010).

Box 3.8 The anatomical arrangement of the glomerular map in the olfactory bulb in mice (based on Manzini & Korsching, 2011)

The cross-section of the mice skull shows the location of the OE at the back of the nasal cavity (Figure 3.15a). This structure is connected to the OB by a bunch of axons crossing a porous bone structure. Axons from the olfactory receptor neurons (ORNs) (sitting in the OE), which express the same type of olfactory (protein) receptors (ORs), are connected to the same glomeruli in the OB. Input from many thousands of such ORNs reach the same glomerulus (Figure 3.15b). In mammals, the number of glomeruli falls between 1,100 and 2,400 depending on the species, humans have between 1,100 and 1,200. Interestingly, rodents have two glomeruli assigned for each olfactory receptor type on each side of the OB. Already at the level of OE specific OSNs are present in one of two zones (dorsomedial (D) and ventrolateral (V) zones). ORs can be categorised phylogenetically into classes I and II, on the basis of homology of deduced amino acid sequences. Most ORs belonging to class I are dominantly expressed in zone D, while ORs of class II are present in both zones. Olfactory input is segregated in the OB as a chemotopic map. Thus, different classes of chemical compounds (see list) are activating topographically different anatomical structures (Figure 3.15c).

Despite the specificity of the ORs, various chemical compounds may activate non-selective ORs to a lesser extent, contingent upon their actual concentration. Assuming that these effects also exhibit temporal variations (as the molecules are travelling along the OE), the olfactory input manifests as a dynamically changing map-like spatial activity pattern in the OB.

Crucially, studies involving mice and fruit flies reveal close similarities in the functional organisation of the olfactory system, including the map-like representation of chemical compounds. While the actual molecular mechanisms underpinning this structure appear to differ, it suggests a case for evolutionary convergence. This suggests that mimicking the overall structure and mechanisms of animal olfactory sensors could enhance the performance of electronic noses (Hurot et al., 2020).

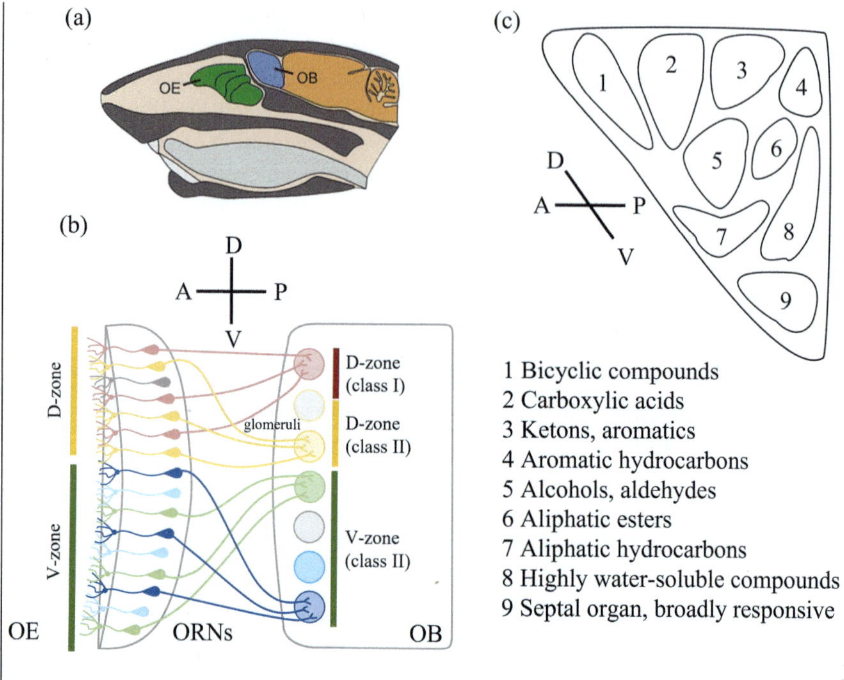

Figure 3.15 (a) The cross-section of the skull of a mouse (figure from Lohr et al., 2014). (b) The axons of receptor neurons expressing the same receptor converge onto common glomeruli, and (c) the different classes of chemical compounds are activating topographically different anatomical structures; A, anterior; V, ventral; P, posterior; D, dorsal (figures drawn based on Manzini & Korsching, 2011).

3.4.12.1 Olfactory fatigue

Continuous exposure to an odorant or a mix of odours decreases sensitivity to the same smell for a short duration. This process ensures that olfaction is ready to detect new odours in the environment despite constant exposure to background odours (Dalton, 2000). The magnitude of the effect depends on various factors, including the nature and dose of the odorant, the duration of exposure, but generally, this kind of fatigue lasts longer than in other sensory systems. It should also be noted that, in contrast to vision and hearing, the offset of the stimulus from the receptor takes much longer because the molecules of the odorant dilute at a much slower pace in the nasal cavity.

The decrease in sensitivity is regulated both at the receptor level and in central areas of the brain. At the receptor level, two rapid and one slower molecular mechanism is responsible for the decreased sensitivity, which is based on changes in the properties of the protein receptors (Zufall & Leinders-Zufall, 2000).

3.4.12.2 Detection threshold

In principle, the detection threshold can be a single molecule, but this is achieved only under specific conditions (see above). Nevertheless, a crucial function of the olfactory system is to detect the presence of a compound at the lowest possible concentration. The performance depends on several factors, including the number and type of sensitive receptors, the chemical structure of the compound (e.g., molecules with longer carbon chains are easier to detect than those with shorter chains), and other ambient and subjective variables (Laska, 2017).

For example, in the case of amyl acetate, humans (detection at 0.00011 ppm) appear to be more sensitive than pigtail macaques (*Macaca nemestrina*) (0.14 ppm), but dogs (0.0000114 ppm) can detect even lower concentrations. In the case of proline, spider monkeys (*Ateles*) (0.002 ppm) outperform dogs (0.023 ppm) (Wackermannová et al., 2016). The literature often categorises mammalian species as being 'macrosmatic' or 'microsmatic'; however, there is little evidence that such simple dichotomies make ecological sense.

3.4.12.3 Olfaction-based navigation

The spatial distribution of volatile or solvable chemicals in nature provides animals with cues to navigate in their environment and find specific targets, such as prey or location. Assuming that the odour concentration is highest at its source, following a trace is possible if the chemoreceptors can detect relatively small changes in odour gradients. Evidence for olfaction-based navigation comes, for example, from Atlantic salmons (*Salmo salar*), which find their spawning locations based on odour cues (Bett & Hinch, 2016). Predators also utilise their prey's odour trail to track them. Experimental evidence in dogs revealed that they could only follow a path if the odour trail was arranged along a specific concentration difference, increasing from the source to the target (Hepper & Wells, 2005).

3.4.13 Gustation

Gustation provides an additional chemical sense, often allowing more direct contact with a specific target. In vertebrates, taste primarily involves judging the safety of food and discerning edible from toxic substances (Snijders et al., 2021). Gustatory receptors for this function are found in the mouth, such as on the tongue or mouth palate. However, in insects, similar receptors have also been detected on the antennae, legs, and wings, serving to facilitate the removal of potentially dangerous chemical compounds (King & Gunathunga, 2023).

In vertebrates, taste receptor cells are specialised epithelial cells located in the so-called taste buds on the tongue. These cells have microvilli at their apex that come into contact with chemical compounds (tastants), forming the basis for taste sensation. Generally, taste receptor cells have a lifespan of only around ten days and are continuously replaced by new ones.

Research on taste sensing has agreed on the existence of five types of different receptors. Originally, specific receptors for sweet, bitter, salty, and sour have been revealed, but there appears to be a fifth taste receptor for umami (Chandrashekar et al., 2006). Taste detection relies on two different mechanisms. One type is similar to how olfactory receptors work, where the tastant activates the protein receptor molecule (localised in the cell membrane), leading to an electric signal produced by the sensory cell. This mechanism applies to substances with a sweet, bitter, or umami taste. The alternative mechanism comes into play when salty or sour compounds reach the surface of the receptor cell. In this case, the cellular electric signal is generated after small, charged molecules enter the cell through specific protein channels located in the cell membrane.

The molecular receptors for these five tastes are relatively specific, allowing for the separate sensing and recognition. Thus, gustatory sensing follows the labelled line principle, assuming that one receptor reacts to a very narrow range of stimuli, which is then sent by the nerves for central processing. Tasting and hearing seem to share such a mechanism.

3.4.14 Artificial chemosensation: e-noses and e-tongues

Equipment for measuring the presence of specific molecules was developed relatively late compared to visual or acoustic sensors, emerging in the 1950s. The primary challenge was constructing machinery capable of transforming physical and/or chemical processes into measurable electrical signals. Most electronic noses, or e-noses, use chemical sensor arrays that react to volatile compounds upon contact: the adsorption of volatile compounds on the sensor surface induces specific physical changes. The electronic interface records this response and transforms the signal into a digital value (for a review, see Cheng et al., 2021).

The sensor part of the e-nose comprises a sampling unit and a detection unit. The sampling unit ensures that the chemical or medium (typically air) containing the chemical reaches the detection unit, where the interaction between the sensor and the chemical compound(s) takes place. Over time, various sensor types have been developed, differing in the physical process capitalised for the measurement (Dymerski, 2017; Figure 3.16).

The utilization of a specific physico-chemical mechanism depends on many factors, including both performance spectrum and cost (Dymerski, 2017). For example, the oxygen species on the surface of metal-oxide-semiconductor sensors interact with the chemical compounds, resulting in a change in conductance. Conducting polymer sensors consist of a substrate stretching between two gold-plated electrodes. The substrate is coated with a specific conducting polymer that changes its conductance during interaction. In piezoelectric sensors, the mass of the piezoelectric sensor coating changes due to the absorption of a chemical compound, leading to corresponding changes in the resonant frequency of the sensor. The variations in sensor types make each kind of e-nose very specific because the output measure depends on the interaction between the specific sensor and the chemical compound. The sensors present various advantages and disadvantages, such as sensitivity to temperature, humidity,

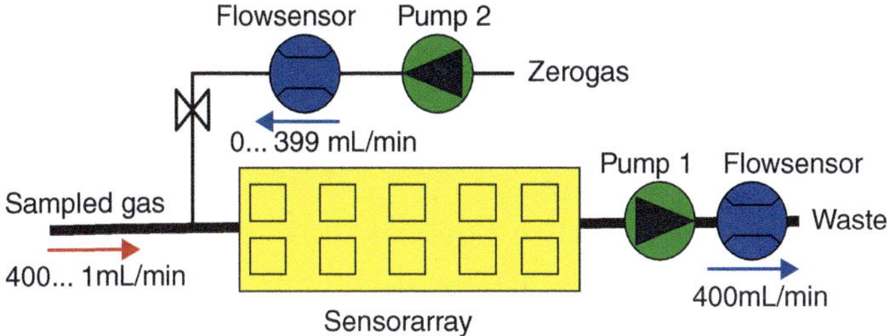

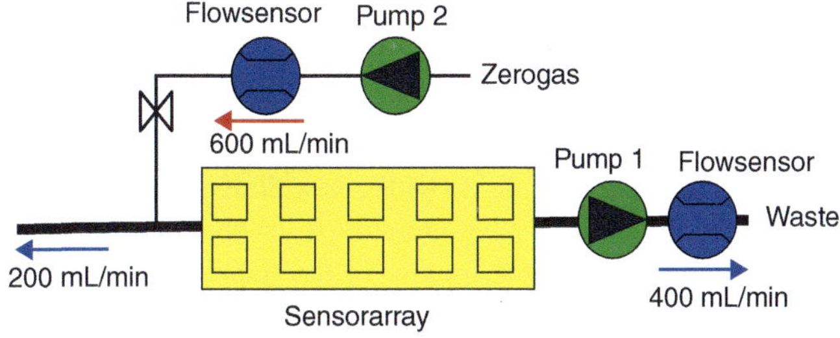

Figure 3.16 Schematic diagrams of the gas flow during the electronic nose measurements (from on Macías et al., 2013). The measurement phase of the electronic nose consists of two distinct stages. (a) During the injection phase, the volatiles from the sample are conveyed to the sensor chamber for analysis by the sensor array. Pump 1 sucks the sample gas compounds through the sensor array, while pump 2 introduces filtered reference air into the sensor array. This setup facilitates dilution, preventing sensor saturation caused by, for instance, elevated concentrations of ethanol. (b) In the cleaning phase, all remnants of volatiles must be purged from the electronic nose to prevent interference with the upcoming measurement. The cleaning air flow from pump 2 is utilised to flush the system. The higher flow rate of pump 2 results in the inversion of the original gas flow direction at the inlet. Hence, akin to animals, the act of smelling is an active process that entails transporting airborne compounds to the sensor and subsequently removing them as thoroughly as possible.

selectivity, and response time. Some sensors are also more sensitive to sensor poisoning, where the interaction with the chemical compound damages the sensor. For example, conducting polymer sensors can operate at ambient temperatures and are more resistant to sensor poisoning compared to metal-oxide-semiconductor sensors (Kalita et al., 2015; Karakaya et al., 2020).

Specific sensors may detect a chemical compound; thus, an array of sensors is applied to reveal the presence of mixed odours. The construction of an e-nose begins with deciding the character and number of chemical compounds to be detected. Next, the type and number of olfactory sensors must be determined based on their specificity, sensitivity, etc. The larger the number of different odour sources that need discrimination, the more sensors are needed. Most e-noses include 5–32 sensors, limiting the range of detection capabilities. In practice, e-noses successfully detect only specific groups of chemical compounds and their mixtures (Karakaya et al., 2020).

In many cases, these devices are very sensitive and can detect one molecule among millions (ppm) or billions (ppb), but it must be ensured that these molecules reach the sensor. Thus, many e-noses rely on active sensing, similar to animals – they actively inhale the volatile substance ('sniffing').

While e-noses are used to analyse the chemical composition of volatile fractions, e-tongues are deployed for the chemical analysis of solutions. This also means that e-noses have a more general capacity because all kinds of objects may release odours as volatile molecules (albeit in small amounts), while e-tongues are limited to detecting odours in liquids. The general physico-chemical mechanisms used for sensing are very similar for e-noses and e-tongues; thus, no further details are provided here (see Dymerski, 2017).

3.4.15 The advantages and disadvantages of using chemosensors

In many respects, chemosensors can provide a new dimension for collecting environmental information. Locations, living individuals, or objects detected by vision and audition also possess a chemical character. So far, social robots have not frequently been equipped with chemosensors, particularly because other sensing capabilities are easier and cheaper to implement. Most chemosensors also occupy more space and need to be close to the source of the odour, although this could be addressed by active inhalation ('sniffing') through a moveable long tube. Despite this, there are several types of e-noses designed to detect the presence of humans in hazardous conditions (e.g., after earthquakes), potentially taking over the role of rescue dogs at some future point in time (Brown et al., 2023).

Current e-noses excel at detecting a well-specified but narrow range of chemical compounds and their mixes. Since most equipment has only 32 sensors compared to the more than 20 million sensors in the human nose, they have an obvious upper limit for recognising many odours at present. Sensor poisoning is also a typical problem that affects the utility of these sensors more frequently compared to microphones and cameras. Consequently, just like in living beings, olfactory sensors in e-noses need to be replaced more frequently.

E-noses have the advantage that the process of perception, i.e., the recognition of substances, relies on simpler computing procedures (e.g., neural networks) compared to analysing vision or audition. Training e-noses may also require smaller amounts of data compared to vision or audition, but this aligns with their limitations regarding recognition and also that training is very sensor specific.

In general, gustation can be seen as a complementary and somewhat more specific process to olfaction. At present, the role of taste sensing is less clear in social robots because its main function is closely connected to feeding in many animal species. In the long term, it may turn out that some gustatory abilities could be useful for ensuring the safety of the robot itself, such as analysing water quality in aquatic robots.

3.4.16 Magnetosensors

The Earth's magnetic field has consistently provided directional information since the evolution of animals, despite its gradual and periodic changes in orientation. However, animal species have not fully capitalised on this possibility, as the search for magnetosensing abilities has yet to lead to a breakthrough. Although the theory of animals possessing a magnetic sense was proposed over 120 years ago, researchers have recently reached a consensus on this phenomenon, though the specific organ for this ability remains elusive.

From a purely physical perspective, it is evident why magnetic cues may offer advantages. They remain (relatively) constant over time, allowing animals to accommodate changes occurring over longer time scales, can be sensed everywhere and always, and have a directional component (Granger et al., 2020). Thus, the Earth's magnetic field provides potential ways for orientation and navigation (Lohmann et al., 2007). The functioning of a magnetic compass has been demonstrated in many bird species, such as pigeons, which rely on directional orientation to return home from distant locations (Wiltschko & Wiltschko, 2015, 2005). Others use the map-like features of the magnetic field to calculate their actual position and change direction when necessary, during long-distance migration (Lohmann et al., 2022). Through various experimental scenarios, Lohmann and colleagues have discovered that loggerhead turtles (*Caretta caretta*) can use positional magnetic information to navigate in the Atlantic Ocean and, for example, locate warm currents.

Despite these findings, confined so far to a few species, the understanding of the mechanism of magnetic sensing remains elusive. Three main hypotheses have been put forward: electromagnetic induction, the use of magnetite, or chemical magnetoreception. However, all of these hypotheses are subject to criticism (see Box 3.9).

Box 3.9 Different hypotheses for magnetosensing in animals

Several authors have described three primary mechanisms proposed to elucidate animals' sensitivity to the magnetic field. The following brief summary is founded on their comprehensive explanations (Lohmann, 2010).

Magnetite

Magnetotactic bacteria possess the remarkable ability to align and navigate along the Earth's magnetic field lines, enabling them to reach preferred environmental conditions. This capability is facilitated by magnetosomes located within these cells, which are comprised of multiple single-domain magnetite (Fe_3O_4) crystals. Magnetite has been found in some magnetically sensitive animals, e.g., salmon and rainbow trout in a region of the nose near a nerve that responds to magnetic stimuli. These crystals collectively possess an ample magnetic moment and can modify their shape upon alignment with the geomagnetic field. If connected to other cytoskeletal filaments, this alteration in shape may impact the cell membrane. In various animal species, analogous magnetites could play a role in employing an intracellular magnetite-based compass, associated with mechanically sensitive cation channels. The activation of these channels enables ions to flow across the cell membrane, generating electrical signals (Figure 3.17a).

Electromagnetic induction

In the presence of a magnetic field, electrically conductive objects, such as the bodies of animals, experience the migration of negatively and positively charged particles to

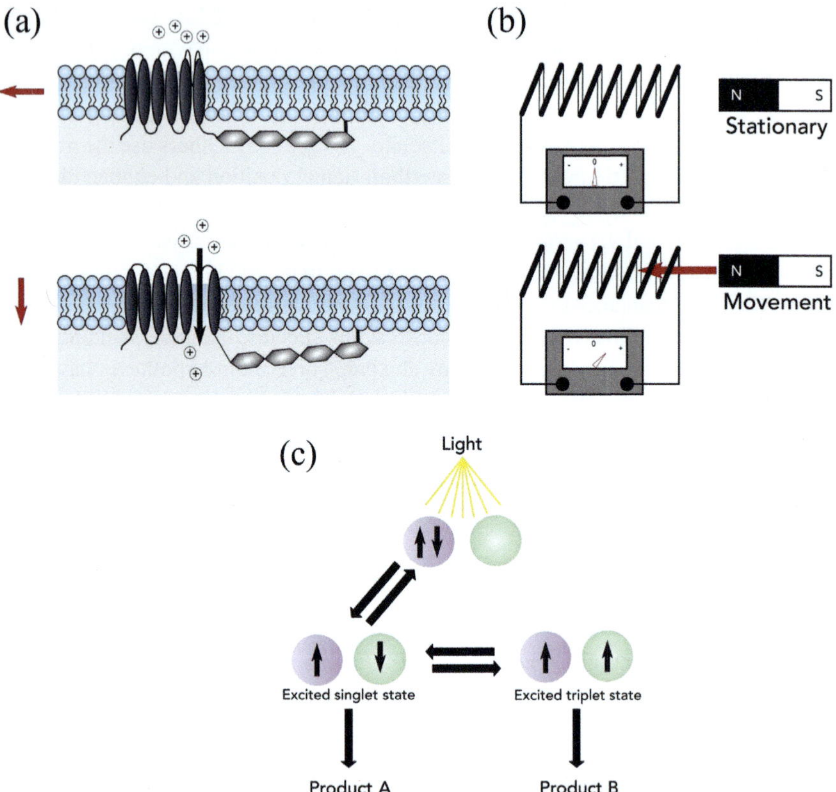

Figure 3.17 (a) Magnetite crystals are linked to a mechanosensitive cation channel. The orientation of the Earth's magnetic vector (red arrows) applies torque to the channel, resulting in either its closure or opening. (b) A moving magnetic field can generate a voltage in a conductive material, which can be quantified using a galvanometer. (c) Light initiates an electron transfer between two molecules, leading to a singlet or triplet state. Magnetic fields can influence the interconversion of these states, potentially modifying the ratio of the produced products A and B. (Figure from Nimpf & Keays, 2022).

opposite sides. This results in a constant voltage, determined by the object's speed and direction relative to the magnetic field. Some animals, such as sharks and rays, may rely on this mechanism if they can measure electric current. They possess very sensitive electroreceptors which may be able to detect the voltage drop of the induced current that arises when they swim through Earth's magnetic field. However, doubts exist regarding whether small electric currents of this nature can be sensed by these animals (Figure 3.17b).

Photochemical magnetoreception

Photosensitive molecules, activated by light, generate radical pairs. If the magnetic field alters the ratio of these states, it may initiate a cascade of downstream events. Light-sensitive proteins, such as cryptochromes, found in plants, insects, birds, and mammals, could function as receptors in this process. Despite substantial evidence supporting this mechanism, there are also controversial issues. For instance, it remains unclear when the signals are strong enough to be detected in a noisy background (Figure 3.17c).

Reviewing the present stand of magnetosensing in animals, Nimpf and Keays (2022) take on a cautious approach indicating that more controlled experiments and more critical interpretations are needed. Researchers need an open mind to reveal the receptors for magnetic sense in animals.

It also appears that animals possessing magnetic sensing abilities still rely on other senses for navigation, utilising multiple compasses for their travel. For example, the bogong moth (*Agrotis infusa*) seems to regularly sample its magnetic compasses to ensure it is moving in the right general direction but relies predominantly on visual cues for ongoing travels (Dreyer et al., 2018). Similar mechanism is also present in pigeons (Wiltschko & Wiltschko, 2005).

The seemingly rare involvement of the magnetic sense in animal navigation has led scientists to seek explanations. One serious limitation could be the weakness of the magnetic cue at the surface of the Earth, and such a weak signal has an even lower chance of stimulating physical or chemical changes in biological tissue or organic molecules (Johnsen et al., 2020). This situation may also explain why many animals show a relatively long latency (compared to other senses), ranging from seconds to minutes or even hours, to respond to changes in magnetic stimulation because they need to integrate inputs from various inner sensors. Behavioural experiments also reveal that magnetic orientation manifests more at the population level; that is, the reactions of individuals show larger variability in comparison to other senses (Johnsen et al., 2020).

3.4.17 Artificial magnetosensors

Artificial magnetosensors serve a similar purpose in robotics as in the animal world, providing the orientation of the robot relative to the Earth's magnetic field. They are also employed to measure the orientation of electric motors and axes of actuators (Lenz & Edelstein, 2006).

The primary working principle of most artificial magnetosensors is the Hall effect. When a conductor is placed perpendicular to an external magnetic field, the moving charge carriers experience a force that is perpendicular to both their velocity and the magnetic field. This results in a measurable voltage on the conductor, proportional to both the current in the conductor and the external field. Therefore, the strength of the magnetic field in that direction can be determined by measuring the voltage. Since this technique can only sense the magnetic field along one axis, three sensor pieces are required for a complete measurement of the field. Hall-effect sensors can be easily integrated into a single chip, making them commonly incorporated into 9-axis integrated measurement units (IMUs) along with an accelerometer and a gyroscope (Khan et al., 2021).

The main reason for using magnetosensors is that they can provide absolute orientation information using the Earth's magnetic field as reference, similar to a compass. This is in contrast to accelerometers and gyroscopes, as they are unable to provide information about orientation around the yaw axis. In measuring orientation, magnetosensors are able to reach an accuracy of a few degrees on their own, or few tenths of degrees when using sensor fusion algorithms with data from gyroscopes (Tang et al., 2015).

3.4.18 Conclusions on sensing for ethorobotics

The input from sensors is a critical factor in determining the operational range of any ethorobot. It is not only the type of sensor that matters but also how the data is processed to extract meaningful information about both the robot's internal and external states, along with the changes since the last measurement. This process needs to be rapid and efficient, optimising the robot's resources.

All sensors come with practical advantages and constraints, as well as technical limitations based on the current state of technology. While technological limitations may decrease over time, certain physical constraints, like those related to light and sound, remain. Living beings have addressed this issue by evolving different types of sensors, employing sensor fusion, and relying on the cumulative inputs from all sensors (see Box 3.10).

Box 3.10 Combining information from different sensors improves the robots' chances to navigate successfully in the environment

Sensor fusion refers to techniques using information from multiple sensor inputs to create a better perception of the environment. The main reason for using sensor fusion is that each individual sensor has its own uncertainty, and their errors can propagate to a level that is not suitable for the given use-case. When combining data from multiple sensors (which can be either of the same or different types), their combined uncertainty can be reduced, reaching a higher accuracy. Sensor fusion also allows for using several sensors of the same type, introducing redundancy to the system, making it more resistant against sensor failure.

For example, localisation in robotics is a very typical situation in which sensor fusion is used. A mobile robot could have encoders for its wheels to sense their rotation, and it could also be equipped with an IMU that senses acceleration and rotational velocity (Figure 3.18). In this case, the encoders are considered accurate for slow motions but would lose accuracy if the robot moves quickly or the wheels start sliding. In contrast, the IMU usually has a static error that accumulates over time but is able to capture quick changes in motion. With using sensor fusion algorithms, the advantages of these sensor types can be combined while minimising the overall measurement errors.

Sensor fusion has its own difficulties. These include choosing a suitable algorithm and estimating the accuracy of the sensors, which plays a key role in determining the degree to which each measurement will be trusted. As a particular implementation may include different types of sensors, this introduces different sensor characteristics and time delays, which also need to be taken into consideration.

In practical applications, the Kalman filter is one of the most widely used sensor fusion techniques (Huggi et al., 2021).

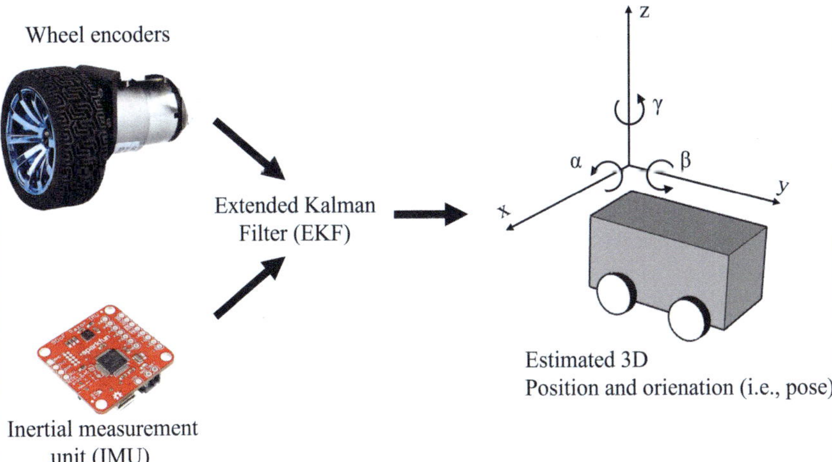

Figure 3.18 Wheel encoders measure the rotation of the wheels, while the IMU measures acceleration and rotational velocity of the robot body. Using sensor fusion algorithms such as an Extended Kalman Filter, these measurements can be combined to provide an estimate of the robot position and orientation in the 3D space.

While one might compare the efficiency of vision and audition based on light waves and sound waves, respectively, it is conceivable that even at a lower level of complexity, deploying different types of sensors could be more advantageous than relying on one type with high levels of ability. Research in analysing the trade-offs in such situations is lacking, and there are few architectures capitalising on the synergic effects of combining inputs from various sensors for decision making. For instance, incorporating relatively simple vision, audition, and olfaction could significantly enhance a robot's action range both spatially and temporally.

The choice of sensor also dictates the need for necessary computation, which can be a limiting factor for autonomous and interactive robots expected to perform with everything on board. This necessitates careful planning of data collection, data selection and reduction, and processing, including data storage for future use. Ethical considerations related to storing and handling sensitive data should also be taken into account, ensuring data safety and preventing data leakage to unauthorised third parties.

Perceptual analysis of sensor inputs often directly influences inner states due to their significance. In animals, emotional states fulfil this role and provide support for organising and maintaining appropriate behaviour. This link is often missing from robotic architectures, making them fundamentally different from living counterparts. Some of these connections are predetermined biologically, such as a fearful reaction to a growling-type sound. These prewired connections may also act as a kind of crystallisation point for the organisation of further experiences collected via sensory interaction with the environment.

3.5 Mechanisms for movement and actions

Evolution has produced a wide range of means to travel on the ground, in water, or in air. In spite of this, technology has mostly used different kinds of solutions including wheels or thrust engines, probably because the natural mechanisms were too sophisticated to be mimicked by early developers. This scenario is slowly changing, for example, by the construction of legged robots.

This does not mean that copying of biological mechanisms may be the optimal way to go in social robotics. Studying the behaviour of animals may provide important insight of how to design embodied robots but there could be many situations when the technical solution should be favoured or only some aspects of the biological mechanism are utilised in the artificial agents.

All devices on a robot that have an effect on the surroundings are named collectively as effectors. Legs, hands, or wheels are typical examples of effectors. There is always a specific mechanism that makes effectors work represented by actuators. In animals, actuators are typically muscles and tendons, while in machines electric motors, hydraulic or pneumatic cylinders play a similar role (Box 3.11).

Degrees of freedom (DoFs) describe how parts of an effector can move independently from each other. In the real world, physical objects may have six DoFs in relation to their three axes (forward–backward, to either side, and up–down) and the rotation along these axes. Rotations around different axes of the body or effector are

Box 3.11 Characteristics of electric motors used as actuators

Although using electric motors is widespread in robotics, there are some areas where biological actuators of animals perform better. Biological structures are more flexible in general, and it allows them to absorb external forces during movement easily. This is difficult to achieve with electric motors, as their mechanical structure is inherently rigid (Figure 3.19). When designing and building arms (or legs), engineers have to decide along many trade-offs (e.g., Kavalieros et al., 2022).

1. Metal is much heavier than biological tissue, this dictates the size of the electromotors which are relatively heavy themselves, to be able to move the arm, the gripper, and the grasped object.
2. Electric motors themselves are usually also relatively bulky, meaning that they require considerable space at the joints. Their weight is also concentrated in the same area where the actual motion takes place, resulting in relatively high acceleration forces and a need for a suitable mechanical support structure.
3. They have to exert the necessary force to move the object around an axis (joints) that is referred to as torque. The required torque increases with the mass to be moved, but also with the length of the arm.
4. Electromotors also have to deal with inertia, that is, the tendency of objects not to change their state in the absence of any forces. In addition, all movements should be executed smoothly, without jerks, trembling, and with high precision. Stronger electromotors would be more advantageous, but they are also heavier, more expensive and make the robot appear and move clumsier.
5. The characteristics of most electric motors are such that they perform best at relatively high rotational speeds and low torque loads, which is usually the opposite of what is needed in robotic applications. Most applications therefore feature a reduction drive which reduces the rotational speed and increases the output torque, but it also adds to the complexity of the actuation system and increases the weight and volume of the actuator.

Figure 3.19 A typical servomotor used for controlling movements in robots (photo by Kitmondo PPM, Flickr).

called roll (around front-to-back axis), yaw (around vertical axis), and pitch (around side-to-side axis). If the human wrist is fixed, then it has two DoFs (pitch and yaw).

The comparison of biological and today's artificial agents shows that there are differences with regard to the relationship between body structure and function (Fukuhara et al., 2022). Typically, animal body parts or extremities are typically multifunctional and have more DoFs as the corresponding artificial actuators in robots. Thus, animals can exhibit more complex modes of action also because there is a complex interaction between different body parts and different functions of the body are achieved by changing its conformation but not the basic structure.

Physics characterises movement in space in three parallel ways. Kinematics describes the movement displayed by an object in a specific environment with reference to time, distance, and speed without considering causal factors of the movement. Dynamics is concerned with modelling the forces that make an object to move, to stop, or get into contact with others, including acceleration. Finally, there could be a need to take into consideration the rapid changes in acceleration (jerking) which make the movement 'shaky' and less smooth.

3.5.1 What do we expect from robot behaviour?

No matter how fancy a social robot looks or what kind of sophisticated machinery it is driven by, at the end its utility depends on the performance achieved by interacting with humans. People may show some tolerance but in real life (outside the scientists' laboratory), any (social) robot must perform well meeting high expectations. From the first introduction of the Rumba vacuum cleaner in 2002, it took more than 10–15 years before these robots gained wider acceptance and found their way into human homes. Despite being specifically designed for the job, it is quite clear that their performance differs still on many magnitudes from that of a cleaning person.

The media and marketing efforts typically exaggerate the potential effectiveness of social robots. Thus, one should not wonder that people may form even greater expectancies towards these robots, because it is assumed that they may share some aspects of our social lives. Given the limited technical ability of contemporary social robots, any skill they possess should be as reliable as possible. Humans may tolerate some error rates (Section 2.2.4), but roboticists should not exploit this situation.

Thus, when making a choice for the technical solution (hardware), the expected functionality of the robot should be studied in detail. Ethorobots should be physically stable (stability) and robust in their performance (being able to correct errors). Behaviourally, ethorobots should show animate movement as much as possible, with smooth transitions from one action to another, and also with reaction times that fit to human interaction.

3.5.1.1 Bodies to be moved

The nature of the embodiment determines how the agent is able to move and also what kind of problems it can solve while navigating in a complex environment.

Autonomy: in this context, autonomy can be defined as the robot's attribute to perform in the natural anthropogenic environment by relying solely on features offered by its own embodiment. Autonomous robots may also need to be familiarised with novel locations and have to learn about the specific features of their new environment, but their capacity does not depend on specific input from the surroundings. An alternative situation is offered by intelligent (smart) spaces (Hashimoto, 2003), which are endowed with embedded systems that take over wholly or partially the function of sensing and computation from the functioning robots. The advantage of such smart spaces is that the robot development is less constrained by the hardware needed on board, but such robots are bound to stay in that specific environment and their functioning heavily relies on the stability of wireless communication between the agent and the outer computation technology. Using smart spaces can be attractive for specific solutions, but reliance only on the robot's own skills may provide more stable functioning.

Rigidity: most of today's social robots have a one-piece body covered by some rigid material, like plastic or metal. The central part of this body (trunk) is sometimes connected to extremities. Assuming a cylinder-shaped body, these robots can move only in spaces that are about 10%–20% wider than their diameter allowing for safe passing through. Animals need much smaller spaces, some can cross through slots that are narrower than their width because in addition to more precise navigation skills, their bodies can change shape and tolerate some compression also. This feature is mimicked partly by soft robots where units of the robot are connected by flexible joints, and the trunk may also consist of such connected units. Soft robots are made of more elastic materials which make the interaction with humans safer and more pleasant (Yasa et al., 2023). Such soft robots have large potential in future use, but they also need specific actuation systems for generating forces (e.g., pneumatic artificial muscles, Section 3.4.2), and complex architecture has to be in place to control their movement and action.

Stability: agents need to maintain their body orientation during movement despite potential environmental perturbations. With some exceptions, in biological agents, the body orientation is determined by the head in relation to the body both in static and dynamic situations.

Stability of the robot is ensured if its centre of gravity (CoG) falls within the area (polygon of support) that is defined by its ground contact points. If for any reason the CoG gets outside this area, then the robot starts to fall over. Stability can be maintained over longer time scale by one of the two possibilities. One can construct a body that is resistant to most environmental effects, and it is ensured that the CoG stays within the polygon of support. This can be achieved by embodiments with weights placed close to the ground (e.g., the batteries of the robot), or if they have a relatively broad basis supported by more legs or wheels (Figure 3.20). This represents a state of stable balance, but in extreme situations, the reinstatement of this stability is possible only by external assistance. This design is used in many wheeled robots.

In other robots, the support polygon is relatively small, and thus the CoG moves frequently outside the preferred area during actions. These robots are continuously in an unstable balance state (especially when they move) and need a dynamic action system that is able to maintain the robot's balance either by moving the support polygon

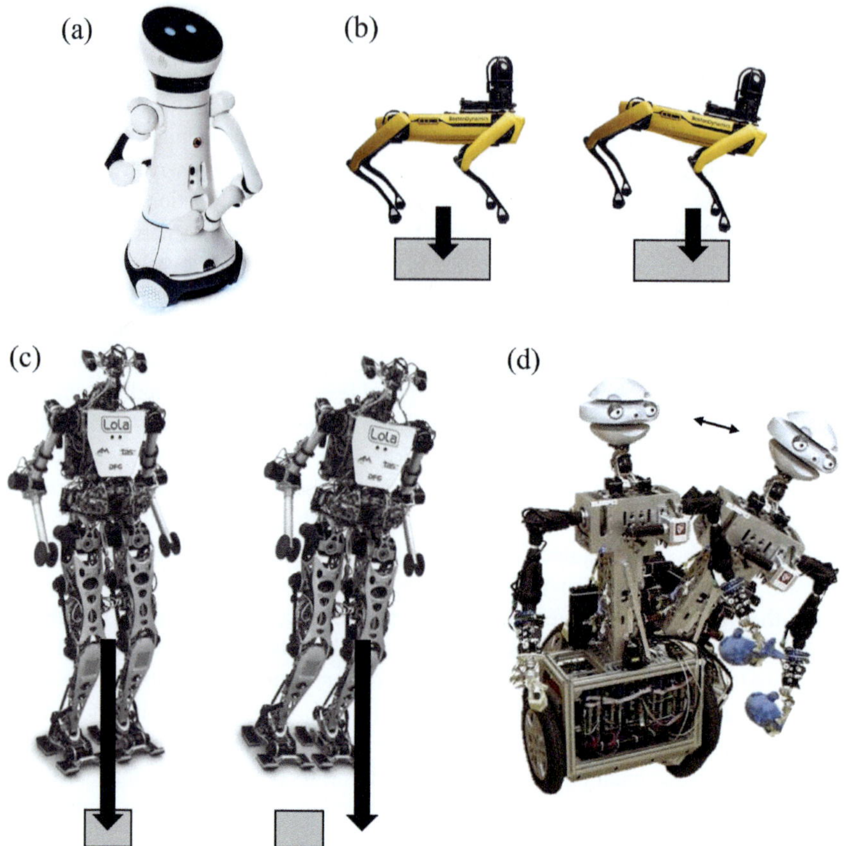

Figure 3.20 Stable and unstable balance in robots. Some robots have a large polygon of support that ensures their stability in the case of wide-ranging environmental effects, and typically they also lack mechanisms to reinstate their stability: (a) Care-O-bot and similar devices achieve stability by placing the heavy batteries close to the bottom (photo from Bodenhagen et al., 2019); (b) Four or more legs also provide stability by covering a large polygon of support (photo from Brosque & Fischer, 2022). (c) Two-legged (photo from Buschmann et al., 2013) or (d) two-wheeled (photo from Kędzierski et al., 2013) robots have a small polygon of support therefore they need a mechanism for regaining stability after losing balance or fall.

under CoG or moving the CoG over the polygon (Figure 3.20). Relying on a complex system of muscles and tendons biological agents are well-prepared for this task but the constant balancing of artificial agents could be a challenge (Mikolajczyk et al., 2022). Importantly, such unstable balancing robots consume more energy compared to those with a stable balance. As a byproduct, the small compensating movements produced during the balancing make such robots appear more animate (Section 2.4.8) that could be useful in some interactive contexts. Two-legged or two-wheeled robots are the typical examples for this category (Klemm et al., 2019; Box 3.12).

Box 3.12 A wheeled-legged jumping robot: Ascento

Klemm et al. (2019) developed a robot that aims to combine the advantages of motion on wheels and legs. Ascento (Figure 3.21a) consists of a body, housing all electronics of the robot, and two individually actuated legs, that are moved by hip motors in the body. The legs have been designed using topology optimisation for weight reduction, and they end in wheels actuated by hub motors. Structurally, each leg is made of a three-bar linkage that allows the wheels to move on a nearly vertical trajectory, while the inner joint of the leg contains a torsion spring that reduces the control efforts of the hip motor. Although the legs are actuated, they do not perform a human-like stepping motion but are used only for jumping and fall recovery, meaning that the controllers of the robot are also designed for performing these latter motions.

High-level motion goals are sent to the robot by a user via a remote controller, allowing for easy control. The actual motion control of the robot uses model-based control, for which a mechanical model of the robot was created as a two-wheeled inverted

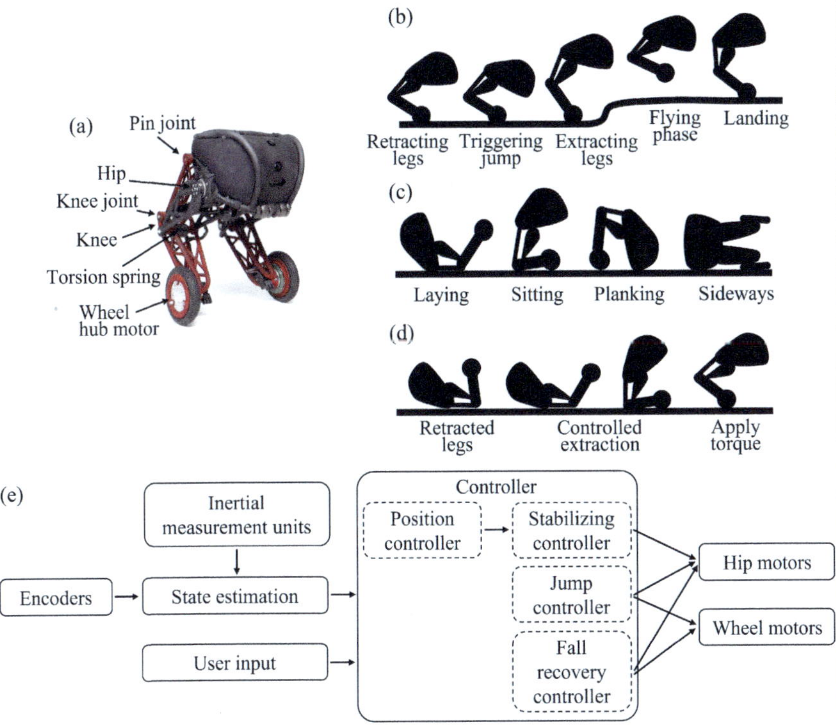

Figure 3.21 (a) The prototype of Ascento (figure from Romanova-Bolshakova & Poleschikov, 2023). (b) Jumping sequence of Ascento. (c) Ascento can fall over to four stable positions. (d) Ascento is able to stand up from laying in four steps. (e) Overview of the controller architecture (based on Klemm et al., 2019).

pendulum. To compensate for the movement of the hips, and thus the varying height of the structure, ten different models had been defined for different leg positions, and the controller uses linear interpolation between them.

The entire controller (Figure 3.21e) consists of three sub-controllers, each for a different operation state of the robot. As a two-wheeled robot is inherently unstable, a stabilising controller is used for maintaining upright position when driving or standing still. There is a dedicated controller for executing a predefined jump sequence, allowing the robot to handle obstacles. A graphical user interface allows for changing the height and velocity of the jump (Figure 3.21b), making it possible to perform different jumps under different circumstances. Ascento can stand up and get into either of four stable resting positions (Figure 3.21c), for which a fall recovery controller has been implemented (Figure 3.21d). As jumping and recovering require special motion sequences, these two controllers overwrite drive controllers for the duration of the manoeuvre.

Testing of Ascento has revealed that it can perform both jumping and fall recovery, and it can also endure both momentarily and long-lasting disturbances. Ascento allows for further research in motion control, such as tilting the robot body inwards for high-speed turns or balancing the robot on one leg.

Manoeuvrability: biological agents were selected for the most optimal behaviour both in using the available space and manipulating objects. Such skills can be denoted as manoeuvrability, that is, when the agent (or a particular actuator) can solve a physical problem in the most efficient way. This could be conceptualised by investigating what is the smallest change that leads to the expected end state of the machine. Technically, manoeuvrability has to do with the availability of DoFs and their control with regard to speed. For wheeled robots, the degree of manoeuvrability refers to the overall DoF that a robot manipulates by changing wheels' speed (direct mobility) and wheels' orientation (indirect mobility) (Correll et al., 2021). In principle, the manoeuvrability of a system reaches its maximum value if all DoFs can be controlled independently from each other (controllable DoF; cDoF). Uncontrolled DoFs decrease the manoeuvrability of the system. There are three possibilities (Correll et al., 2021; Matarić, 2007):

1 *Holonomic agents*: the total number of DoFs equals the total number of cDoFs. Robots having three wheels that can change moving direction and speed independently are typical examples. They can turn around without displacement in the horizontal plane that is, the centre of their rotation falls at the centre of their horizontal body plan. Many legged animals (and robots) can also move this way. Being holonomic means also more effective use of small spaces that can be especially advantageous in case of danger. In principle (functionally), many legged animals can behave as holonomic agents, but they may not have the fine control to show such behaviour. Interestingly, even animals with holonomic behaviour seem

to have a preference for non-holonomic movements in general, probably because they prefer that the movement takes place in the direction of their head/face.

2 *Nonholonomic agents*: if the total number of DoF is larger than the cDoF, then the agent's manoeuvrability is constrained. The centre of rotation of such agents falls outside their body plan when turning around on the ground. Thus, they need more space for turning back and/or have to make more complicated moves to achieve the same goal. Cars are typical examples of nonholonomic agents since they are not able to move sideways.

3 *Redundant agents*: in some cases, the cDoF is larger than the total number of DoFs. The human arm (shoulder, elbow, wrist) is a good example of a redundant system. Having seven DoF with independent control, not only can it manipulate an object in a very efficient way, but it can have many different solutions for the same kind of manipulation. This redundancy from an engineering perspective increases its manoeuvrability and makes new solutions to problems possible.

3.5.2 Moving on the ground: trunk, neck, and tail

The prevalent methods for traversing solid surfaces involve either wheels or legs, although alternative approaches exist, exemplified by the slithering movement of snakes.

From a phylogenetic perspective, the conventional legged body structure has independently evolved at least twice, occurring in both vertebrates and invertebrates (see also Section 2.4.3). Both systems have developed various features conducive to walking behaviour, providing valuable insights for robotics in constructing agents endowed with analogous abilities (Fukuhara et al., 2022).

3.5.2.1 The trunk

This constitutes the core region of the body, essential for ensuring the overall stability of the animal or agent. It houses vital organs or hardware and serves as the central linkage connecting all extremities. In the animal kingdom, trunks are commonly oriented horizontally, offering increased stability at the expense of manoeuvrability. In social robotics, vertical trunks are more commonly deployed. This design choice ensures that the robot can engage in face-to-face interactions with human partners and presents a less cumbersome appearance.

The trunk also plays a crucial role in determining the overall rigidity or flexibility of the body. Insect thorax ('trunks') is composed of tightly connected segments, forming a relatively rigid structure that restricts lateral flexibility during movement. Conversely, vertebrate trunks exhibit a notable degree of flexibility. This is achieved by a segmented structure, the spine (backbone) which, however, is located inside the trunk. The vertebrae (making up the backbone) are connected to each other by joints, which are supported by longitudinal filaments running in parallel with the spine and shorter ones connecting adjacent vertebrae. Importantly, this arrangement allows for bending both along the horizontal and vertical plane but at the same time it also limits the extent of the bending movements. Gross changes are achieved by making

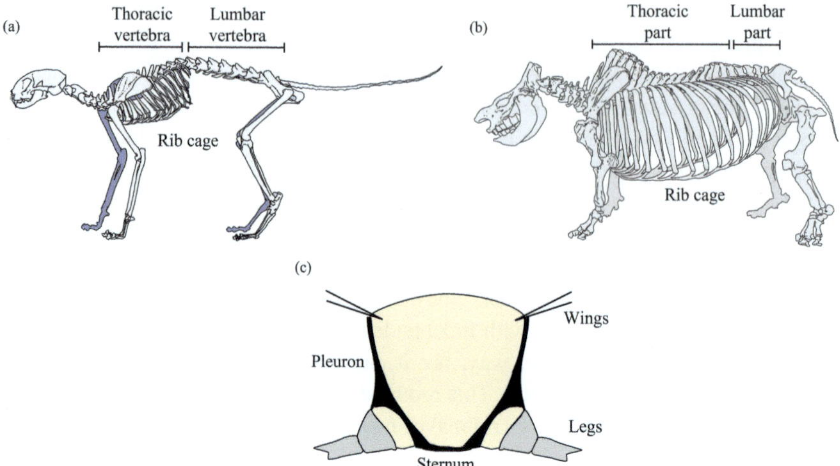

Figure 3.22 Lateral view of flexible (a, cat) and inflexible (b, rhino) trunks (from Fukuhara et al., 2022). In cats, the vertebrae are linked by specialised intervertebral discs that function as cushions between the bones of the spine, providing support for shock absorption. Strong muscles involved in moving the spine and the elasticity associated with these discs enable larger and faster movements in all directions, facilitating a swift righting reflex in the event of a fall. Functionally, most contemporary social robots have an insect-like trunk which does not have any structure like a backbone. The thorax of insects (cross-section, c), which is functionally similar to the trunk, consists of rigid outside walls (exoskeleton) to which the legs and wings are attached.

relatively small movements at any individual vertebrae. Thanks to this feature some types of trunks can add a lot flexibility to the whole body (Figure 3.22). In the case of vertically oriented trunks similar movements support bending to the side and bending forward. In vertebrates, the backbone also connects the head to the trunk by the neck, provides a sticking point for attaching further extremities (e.g., legs), and also forms the tail (see below).

While the presence of the spine determines the length (or height) of the trunk in vertebrates, there is the possibility of manipulating the length of the trunk by specific mechanisms in social robots. For example, hydraulic mechanisms can change the height of the robot to adjust it to the height of the actual human partner. However, such manipulation could be limited because the trunk houses often important hardware for the robot.

3.5.2.2 Neck

The neck is a specific section of the spine, providing a flexible connection between the trunk and the head. Together with the head, the neck may play a role also in balancing and the action radius of the neck determines the actual position of the head in space. The flexibility of the neck is also determined by the number of the connecting vertebrae in animals which allows for more varied movement then achieved by most social robots. Having a flexible neck allows the agent to change the orientation of the head without moving the body, and the neck has also to perform coordinated actions when the head is used to grasp or manipulate objects (e.g., for eating or transporting).

The movement of the neck and the head may affect the position of CoG, and thus compensatory actions could be necessary to keep the CoG over the support polygon. There is also some evidence that the nodding behaviour may store elastic energy reducing the energetic cost of walking if the head and neck are moving up and down with suitable timing (Zhang et al., 2016).

3.5.2.3 Tail

Descriptively tails are defined as post-anal appendages or extremities (Schwaner et al., 2021). Tails have evolved several times independently for various functions, often having multiple roles in animal behaviour. The convergent nature of their history is also underlined by diverse anatomy. For example, in insects, tails evolved as an extension of their abdomen, while in vertebrates, the tail is the prolonged part of the backbone. The tail of terrestrial animals has three non-exclusive main functions. It can be used as a counterweight for balancing during movements, tails may be deployed as graspers or tails can also serve communicative functions (see below).

The role of tail in maintaining stability has been investigated in several species. Shield et al. (2021) observed how the tail moves when cheetahs perform a rapid deceleration. The detailed behavioural analysis and modelling revealed that the tail movements help the animal to perform a safe braking by stabilising the body. Similar functions have been associated with the movements of the forelimb or arm. Fukushima et al. (2021) studied the rightening behaviour of squirrels during unexpected falls. First, the squirrels typically aimed to stabilise their body orientation for landing. Next, they continuously spun the tail to slow down the body rotation before the start of the landing. The researchers built a bioinspired mechanical model of squirrels and showed that the tail has indeed a stabilising role during falls.

3.5.3 Moving on the ground: legs

Legs are defined as specific extremities for moving the body above the ground. This function of legs is achieved by several structures depending on the evolutionary history of the species. Some extremities with leg function have been also modified during animal evolution. For example, the wings of the birds have evolved from the forelegs of ancient reptiles. Looking at the animal kingdom, there seem to be two widespread basic designs. The quadruped mechanism evolved (approximately 350–400 million years ago) in tetrapods (Capdevila & Belmonte, 2000) allowing for movement on the hard terrestrial surface, and parallel in time or from somewhat earlier, there is also evidence for insects moving by six legs (hexapod), which has become a very successful paradigm for walking (Garwood & Edgecombe, 2011). Both mechanisms received considerable interest from both biologists and roboticists because they differ in various ways not just in the number of legs. Not surprisingly, specific attention has been devoted to biped solutions of ambulation, featured dominantly in birds and humans (Fukuhara et al., 2022).

3.5.3.1 Optimal number of legs

There is no general rule for the optimal number of legs. There is a trade-off between stability, manoeuvrability, and cost of maintenance. The increased number of legs enhances stability, while having too many legs may infer with flexibility. In the evolution of tetrapods, the number of legs remained constant meaning that four legs can provide a robust solution. In other animal clades where the number of (potential) legs increased over 10, there has been an evolutionary trend to modify these extremities for other functions, suggesting that the problem of locomotion can be safely solved by fewer legs. Thus, four to six legs provide the required stability on the ground for any agent while in case of bipedal support, there is a need for specific control systems to maintain the balance (Figure 3.23).

3.5.3.2 Multi-functionality of legs

It should be also noted that many legs support other functions then locomotion in animals. As the evolution of new extremities is constrained by phylogenetic, developmental, and genetic reasons, the modification of existing body parts provides an easier alternative. Legs also have new functions, for example, legs can be used for grabbing, pushing, or holding objects, manipulating objects which is also reflected in their form and represents a compromise between the original function and the new one.

3.5.3.3 Structures of legs

Fukuhara et al. (2022) distinguish three functional sections of the tetrapod leg. The proximal part refers to the joints (shoulder and hip) connecting the upper part of the leg to the trunk. The intermediate part (elbows, wrists, knees, and ankles) provides a flexible connection between the rigid elements (e.g., femur, humerus), finally, the distal part is the interface between the leg and the ground (e.g., foot). Functionally, the same division may be also applied to the arthropod leg while the actual mechanics is partially different.

Proximal leg parts: in tetrapods, there seems to be a major difference in the way how the shoulder and hip connect the legs to the trunk. While the hindleg is firmly connected to a ball-and-socket joint (pelvis) that supports the direct translation of the propulsive forces to make the body move, shoulders typically provide a more flexible connection. The upper end of the humerus is embedded in a cavity made up from several bones and is fixed to the trunk by a complex system of muscles and tendons. This loose coupling increases the potential action range of the foreleg which can also be used in functions other than locomotion (Figure 3.24).

In most robots (and in arthropods), there is no major structural difference in the connecting joints of forelegs and hindlegs; moreover, the orientation of the femur

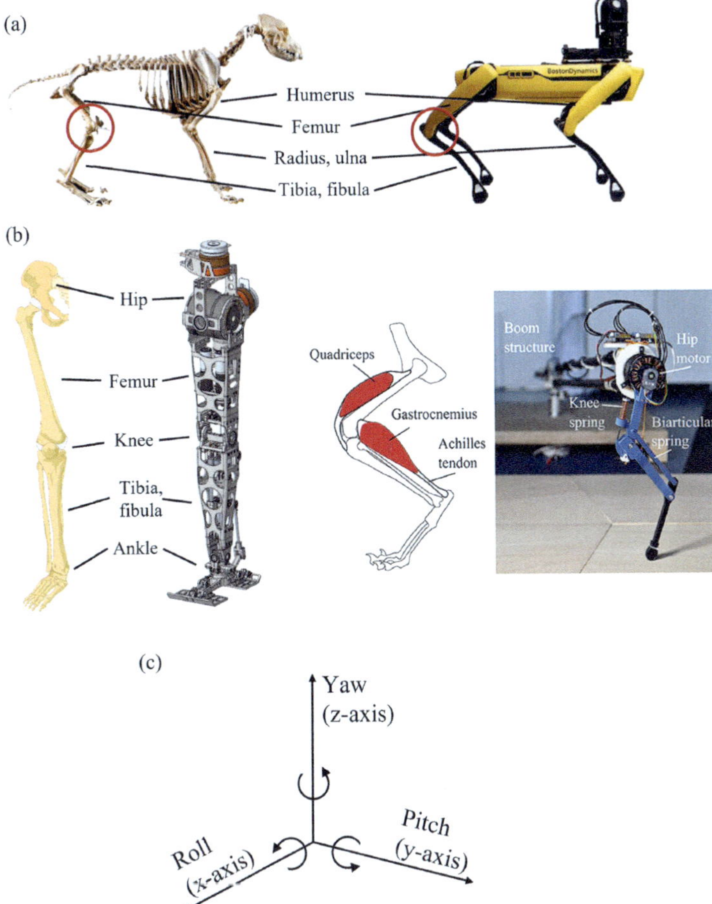

Figure 3.23 (a) The comparative morphology of a dog and a robot dog (Spot by Boston Dynamics) shows both similarities and differences (photos by Museum of Veterinary Anatomy FMVZ USP, Wikimedia Commons; Brosque & Fischer, 2022). Note that the forelegs and hindlegs of the robot dogs have different structure and orientation and conform to the foreleg of the dog. In the dog, the upper part (femur) of the hindleg and the knee is oriented forward. There are also four-legged robots that mimic the orientation of dog legs. The dog's shank consists of two longer bones (tibia and fibula) while robots have only a single unit connecting the knee and the ankle. Robot dogs often have to functional 'ankle' in contrast to dogs or other mammals. (b) The solution for moving the leg by actuators is very different for animals and robots. The muscles connecting distal points of the bony structure and using elastic force by contracting the muscle fibres. The contracting and expanding muscles can generate large amount of force in contrast to servomotors which have a limited torque by design. A simplified illustration of a bird's leg is presented alongside a comparable representation of a robot leg for comparative analysis. (Photos from Ruppert & Badri-Spröwitz, 2019; Zhu & Thomas, 2023). (c) Pitch, yaw, and roll are the three dimensions of movement when an object moves through a medium.

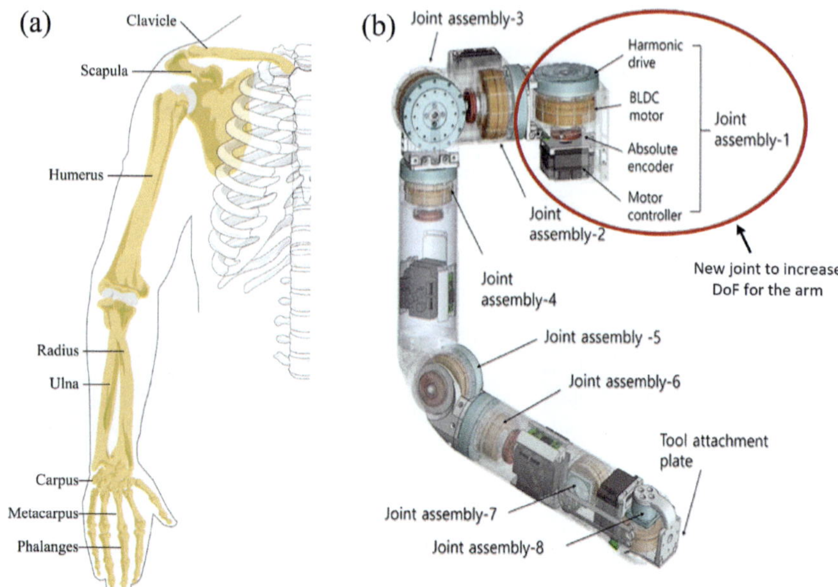

Figure 3.24 A comparison between a human and robot arm (proximate leg). There has been an evolutionary trend for the foreleg to become more flexible and part of this change was supported by alterations in the shoulder. This increased flexibility is achieved (among other changes) by the head of the humerus forming only a loose connection with the relative swallow glenoid cavity on the scapula. The connection is maintained by a specific set of muscles (not shown) allowing for much larger action range of the humerus (a). This difference allowed that over the evolution the forearm could be deployed for other functions but walking. For example, such arms are better in manipulating objects, self-grooming, but also assist in mitigating the consequences of falling. Flexible shoulders allow for further functional gains in biped animals. (b) Robotic shoulders are approaching the complexity of human shoulders with servomotors acting as actuators instead of muscles. Lee et al. (2017) constructed a bioinspired robotic shoulder that mimics the behaviour much closer (figure from Lee et al., 2017). To make the artificial shoulder more flexible they added a further joint close to the trunk increasing the three degrees of freedom (DoF) of natural shoulders to four DoFs. This is a case for analogy when the same function is achieved by (partly) different structures.

and humerus is also parallel in many popular quadruped robots (e.g., dog-like Spot robot), in contrast to their biological counterparts (e.g., cats and dogs). It should be mentioned that the natural orientation of the legs and their joints seem to result in a more stable body structure.

Intermediate leg parts: in tetrapods, this section plays a key role in initiating and controlling the movement of the leg. The muscles cross over two joints, where they generate and restrict their motion. The gastrocnemius muscle and the Achilles tendon bends the knee joint and extends the ankle joint, while the semitendinosus muscle stretches the hip joint and bends the knee joint. The resultant motion depends on the combination of activated muscles. This arrangement has been mimicked in many robotic designs by replacing muscles with springs and other elastic materials. This coupling brings about several benefits for the animal or the robot. First, the elastic elements in the limb can store and release energy. Second, this flexible arrangement contributes to resisting and recovering from disturbances. Third, the direct mechanical connection between the two joints reduces the number of actuators needed for effective movement (Fukuhara et al., 2022). This structure seems to be specific to vertebrates.

Distal leg parts: this part in tetrapods (and anthropods) has a variable structure depending on many ecological factors and whether the leg has also other functions. The utility of the foot is often enhanced by the inclusion of further elements connected to each other by joints. These can serve as stabilising contact with the ground and allow for smoother adjustment to unexpected changes on the surface. Increased flexibility of the foot can be deployed in tasks requiring manipulations. Today's robots utilise distal leg parts mostly for movement, and only rarely do they have multiple functions.

3.5.3.4 Bipedal humans and robots

One of the greatest challenges in robotics is to mimic the way how humans move. Walking on two legs represents a complicated hardware problem, as well as a need for sophisticated software control (Mikolajczyk et al., 2022). Despite this, once constructed, bipedal robots may provide advantages because potentially they have high levels of manoeuvrability, require small space for their actions and are also fit for the existing human environment.

There are two different approaches for making bipedal robots. Some engineers aim to build a robot that practically copies all aspects of the human body including the hardware and motion control. Others take a more practical view and look for the minimum capabilities the robot must have to perform human-like bipedal movements, while still being able to carry out tasks a human would do.

The stiff-standing human body is typically in a very sensitive instable balance state because the CoG can easily move out of the horizontal area of the support polygon. The movement of the upper body or manipulating objects by the hand has always an effect on the location of the CoG.

In humans, this instable balance is maintained by continuous and synchronised skeletomuscular adjustment of posture (Skoyles, 2006). Instability occurs within 100 ms when the CoG moves outside the support polygon. Since the typical reaction time is longer, some preparatory mechanisms should be in place which can initiate rapidly adequate compensatory actions. These actions depend on both the ongoing events and on the actual body position. Such mechanisms presume the existence of a detailed body schema (Section 2.5.6). Motor reflexes play also a role which rely on the spring-like property of the neuromuscular system. Finally, cognitive processes may also elicit balancing response based on earlier experience.

Bipedal robots also utilise several mechanisms to maintain control of the upright position (Mikolajczyk et al., 2022; Box 3.13). Researchers implemented software that is prepared for imbalance occurring. Accelerometers and gyroscopes estimate the change of the robot's position and estimate the movement speed and direction of CoG that is followed by correcting and compensating body movements to move the CoG back over the support polygon. Walking or running demands additional calculations and control algorithms, as the robot has to keep its balance dynamically while only one or even none of its legs touches the ground.

Movement planning itself is usually done in two stages: first, a walking pattern generation, which calculates the footstep positions, and then a balance controller which stabilises the robot along the trajectory. Most kinematic models for bipedal robots are based on the inverted pendulum, which is a statically unstable structure and therefore it can be used for developing control algorithms and structures for bipedal robots. There exist multiple kinematic models, for example, a model used for planning a running motion uses a so-called spring-loaded inverted pendulum, which approximates the dynamic bending of the knee of humans with a virtual spring that absorbs and releases kinetic energy when the foot touches the ground (Zhu & Thomas, 2023). As control algorithms usually use sensor inputs from multiple sensors, sensor fusion techniques such as Kalman filters play an integral role in bipedal robots.

Box 3.13 Multi-sensor feedback controller

The block diagram of a multi-sensor feedback controller for stable bipedal walking (Figure 3.25; for details, see Chen et al., 2015) indicates the body state feedback control algorithm and foot posture corrector.

The coloured lines show sensor data coming from the three types of sensors (six-axis inertial measurement units (IMUs), one two-axis inclinometer, encoders) on the robot. The task of the body state estimator is to combine ('fuse') these data and calculate the best estimation of the position and orientation of the robot body. Importantly, this estimation is used as a reference for the feedback of the controller, and multiple data sources improve the accuracy of the estimation. There are several types of algorithms for state estimation, with different versions of the Kalman filter being the most popular.

The Zero-Moment Point (ZMP) is involved in the walking pattern generation, and its reference value comes from the stable walking pattern created in the first step of the control algorithm. The ZMP is used to calculate a reference for the CoG, and the body state controller calculates the required changes of the robot body to keep the CoG (and thus the ZMP) on the desired trajectory.

Inverse kinematics is the problem of calculating joint angles from a given state of the end-effector on a robot. For bipedal robots, it calculates leg joint values from the body state, and its output is an array of rotational angle reference values for each joint in both legs that should be used for the desired movements of the robot. In addition to this

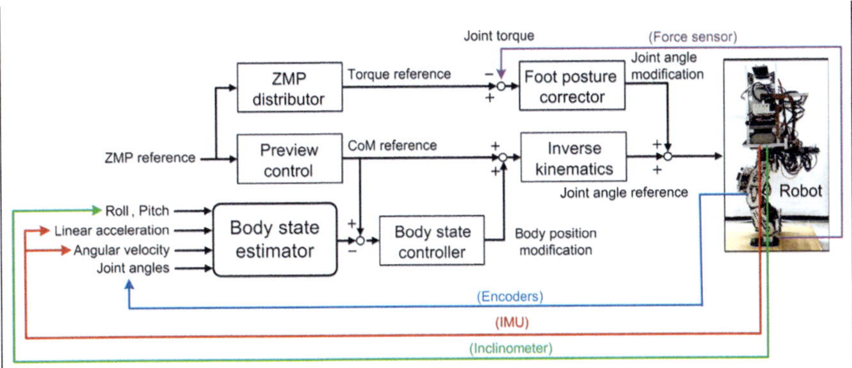

Figure 3.25 Block diagram of the multi-sensor feedback controller (figure from Chen et al., 2015).

fundamental control algorithm, a torque control strategy is also applied, as unevenness of the ground causes a bumpy contact with the feet, and this needs to be compensated. The ZMP distributor calculates a reference force profile, to which actual forces from the joints are compared to. In case there is a difference, the foot posture corrector module changes the joint angles to compensate for the differences.

3.5.4 Moving on the ground: wheels

While wheeled vehicles are commonplace for everyone, they represent a huge difference to the way how biological agents move. Wheels rolling on axes was one of the important innovations of humans taking place possible around 3,500 BCE (before the contemporary era). The first wheels were probably utilised in the potteries in ancient Mesopotamia for shaping plates, jugs, or pots. The application of wheels on chariots emerged when horses or donkeys were trained for the job of pulling. The structure of wheels has not changed for long, the next significant development took place more than thousand years later with the invention of the spokes.

Today's technology is using one of the four types of wheels, depending on the expected functionality (Correll et al., 2021; see below):

1 Standard wheel rotates on an axel bearing has one DoF (excluding the rotation around its contact point with the ground);
2 Castor wheel rotates around an offset steering joint with two DoFs;
3 Swedish (mecanum) wheel has a series of obliquely attached rollers to the whole circumference of its rim. Thus, it has three DoFs and makes movement in all directions possible (omnidirectional);
4 Ball or spherical wheels are placed in a specific a socket, which allows movement in many directions with three DoFs. This solution is only rarely used because the drive solution is complicated.

How to choose the best-wheeled structure?

Wheeled robots are very attractive because they are easy to build and repair and cost effective and the hardware is typically easy to control. They perform also well in typical anthropogenic environments in factory buildings, offices, or many other places where the ground surface is even.

The choice of wheeled drive depends on the many factors (e.g., Campion & Chung, 2008). For example, the manoeuvrability of the wheeled drive used in cars is very poor, and thus this solution does not work for robots. Steerability represents the number of steering wheels that can be oriented independently for steering the robot. The total number of DoFs that can be achieved by a robot depends on the possibility to change the speed of the wheels (direct mobility) and their orientation (indirect mobility). Finally, the stability refers to the robot's ability to maintain its orientation by placing its CoG above the support polygon (Section 3.5.1.1).

With regard to mobility and steerability engineers distinguish the following types of wheeled vehicles (based on Campion & Chung, 2008; Figure 3.26), but note that

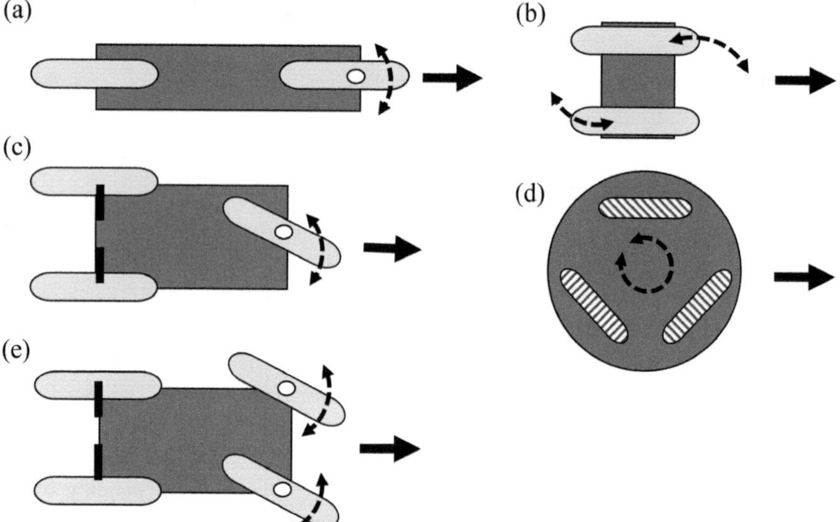

Figure 3.26 Wheels can be organised in diverse configurations, each presenting its own set of advantages and disadvantages. Prior to devising the driving mechanism, it is crucial to take into account both the specific role of the robot and the constraints imposed by the environment. (a) Two wheels each after another are applied rarely because they represent a very instable 'bicycle-like' robot. (b) Two wheels with a common axis and differential drive (wheels can be rotated independently). While this construction is also instable, there are such balancing robots that have a rapid compensatory mechanism for shifts in centre of mass (see also Figure 3.20d). (c) A differential drive system with two wheels and a castor wheel. (d) Three motorised Swedish wheels. (e) Two steered wheels at the front and two fixed wheels on a shared axis at the rear. This construction closely resembles that found in conventional cars (see also Campion & Chung, 2008).

other categorisations are also possible based on the kinematics and dynamics properties of the respective mathematical models:

Two-wheeled structures

There exist two versions of two-wheeled robots: the wheels can be positioned either one after another in a bicycle-like configuration, or they can be placed on a common axis.

1 Two wheels each after another: often the rear wheel is driven, and the front wheel is steered. This configuration is rarely used in practice, as it is not able to maintain stability in a standing position.
2 Two wheels with a common axis with differential drive: when two wheels are aligned on a shared axis, a differential drive is employed for robot propulsion. To ensure stability, the robot's CoG must be positioned over the common axis of the wheels. This setup forms an inverted pendulum structure, demanding continuous dynamic stabilisation to maintain an upright position – similar to the stabilisation required for bipedal robots or a standing human. The necessity for ongoing stabilisation can be mitigated by adding one or two passive caster or spherical wheels to the structure (see below).

Three-wheeled structures

Utilising three wheels imparts static stability to the robot, given that the three contact points establish a plane allowing the robot to remain statically upright without requiring active balancing. The sole condition is that the robot's CoG must be positioned over the triangle defined by the contact points of the wheels. While enhancing stability is achievable by incorporating additional wheels, doing so necessitates the implementation of a suspension system for each wheel. This precaution is taken to prevent the creation of a statically overdetermined system when the robot operates on a surface that is not perfectly flat.

1 Two wheels with a common axis with differential drive, and a third point of contact: Two independently driven wheels (a differential drive system) are supplemented by a (usually unpowered) caster wheel not on the same axis. Such robot can move forwards or backwards on a trajectory of an arbitrary radius and rotate around the centre point of the axis of the differentially driven wheels.
2 Two connected traction wheels on a common axis (with a differential gear in between) and one steered free wheel or two free wheels on a common axis and one steered traction wheel: these arrangements mimic the setup found in rear-wheel drive and front-wheel drive cars, each featuring only one steered wheel. Consequently, a robot configured in this manner can move in a forward or backward direction along a trajectory of any radius but lacks the capability to execute a rotation in place.
3 Three motorised Swedish (mecanum) or spherical wheels are all powered and typically positioned in the shape of an equilateral triangle. A robot designed with this arrangement is omnimobile, allowing movement in any direction and orientation. An advantage of this setup is its relatively straightforward mechanical

design, given the absence of steered wheels. However, this simplicity comes at the expense of having more intricate individual wheels.

Four-wheeled structures

The most common design among four-wheeled robots resembles that of car-like drive in robots. This layout typically incorporates two steered wheels at the front and two fixed wheels on a shared axis at the rear. The powered wheels can be located either in the front, resembling front-wheel drive cars, or at the rear, resembling rear-wheel drive cars.

In both scenarios, synchronous rotation of the steered wheels is crucial to maintaining a constant centre of rotation, thereby preventing skidding. Additionally, to facilitate varied rotational speeds, the non-steered wheels on the common axis should be equipped with a differential gear for the same reason.

3.5.5 Legs or wheels?

Looking at the contemporary scenario of social robots, wheeled solutions are more popular. This may not represent the preferences of the engineers rather the lower cost and easier implementation of wheeled technology. In addition, one should keep an open mind about possible solutions by legs, wheels, or tracked locomotion, including hybrid structures (e.g., Rubio et al., 2019; Box 3.14).

Box 3.14 Quadrupedal robot ANYmal by Bjelonic et al. (2021)

Bjelonic and colleagues (2021) developed a wheeled-legged robot capable of traversing various terrains (Figure 3.27a). On smooth surfaces, the robot achieves high speeds by rolling on its wheels.

There are different designs that aim to combine the advantages of wheeled and legged robots with a different number and arrangement of wheels, which also influences the choice of and requirements from the motion controller. In the case of a four-wheeled robot, as its mechanical structure is statically stable, the most important aspect is the generation of the gait sequence, i.e., calculating the timings of the contact and lift-off times for each of the robot's legs in order to increase the speed and decrease the overall energy consumption.

The motion controller for this wheeled-legged robot is a whole-body controller, which makes it possible to find complex motions for the actuators, and it consists of three controllers in a cascaded structure (Figure 3.27b). The whole motion controller receives as input a reference velocity or trajectory that is global for the robot and describes the desired high-level motion plan that the robot body should follow. Only after that can the motion planning of the individual legs take place.

First, the gait sequence generator calculates the gait timings and sequences over a predefined time horizon. If this module is able to create aperiodic timings, the robot' movements can be more tailored to its environment, allowing for a lower energy use for moving.

Next, the model-predictive control (MPC) calculates the velocities and angles of the joints in the robot's legs, and thus creates a motion plan with a trajectory from the gait sequence. An MPC controller uses a dynamic model of the controlled system (hence its name), and some type of cost function to optimise; in the case of a legged robot the

(a)

(b)

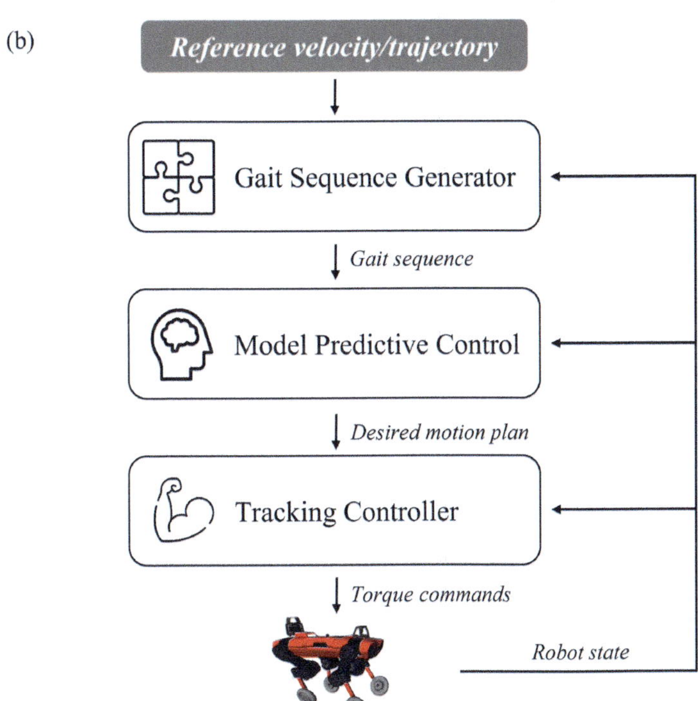

Figure 3.27 (a) Quadrupedal robot ANYmal moves on different surfaces and stepping down (figure from Bjelonic et al., 2022). (b) Overview of the motion control of the ANYmal robot (icons from Freepik, Flaticon; based on Bjelonic et al., 2021).

controller must have a model of the robot's (and its legs') kinematics and aim to minimise the errors to the pre-calculated trajectory of the robot body.

The final step is to calculate the torques and actuator efforts that should be used to achieve the demanded trajectories; this is what the tracking controller does. The whole controller makes the calculations over some defined time horizon and therefore the commanded motions optimise the robot's behaviour not just at an instant but over a short duration of time.

Stability: to be fair, one should always compare robots having the same number of effectors (legs or wheels). While two-legged or two-wheeled robots are similarly instable, the maintenance of a stable orientation can be achieved easier in the latter ones. In contrast, while both four-legged and four-wheeled robots are statically stable to the same degree, legged robots have an advantage over wheeled ones because they can manipulate the position of the CoG more flexibly by changing the length of their legs. This can be helpful when the robot has to overcome steep slopes.

Mobility: wheeled robots can be much faster than legged ones on smooth surfaces, but this may not be the case on rough terrain. Even small heaps (e.g., thick carpets or doorsteps) can be a challenge for many types of wheeled robots having typical size of wheels. Wheels with larger diameter could make robots to overcome such obstacles, but this solution may have other disadvantages. Legged robots can step over hindrances, which are impossible for a wheeled agent. Legged robots may also have a greater chance of crossing gaps by either stepping over or jumping. Although there are some clever solutions that make wheeled robots jump (see Klemm et al., 2019), a crossing by deploying legs is more controllable and can be designed with higher precision.

Animacy: the moving pattern of legged and wheeled robots is clearly different. While both types can mimic some aspects of animated behaviour, for example, showing self-propelledness (see Section 2.4.8), legged robots are more similar to biological agents in the way how they balance during movement. It is possibly easier to anthropomorphise legged robots, but this can make them also eerier. Eeriness can also emerge if the robot legs/arms are differently designed from that of the living counterpart. The upper part of the hindlegs in many four-legged robots orients backwards (rather than forward). This may add to their strange look because their knee bends in the opposite direction than what one observes in dogs or cats (Figure 3.22a).

The continuous small balancing movements of the two-legged or wheeled robots can also contribute to them appearing livelier. While such movements are not necessary in robots having at least three wheels, technical solutions for legged agents with four or more effectors cannot avoid investing some energy to maintain their balance, especially if they aim to avoid resisting sudden environmental perturbations.

Flexibility: most of the legs and all wheels are very specialised effectors. Biological agents have also evolved legs with additional functionalities (see also Fukuhara et al., 2022). The forelegs of monkeys are not only used for mobility, but they are

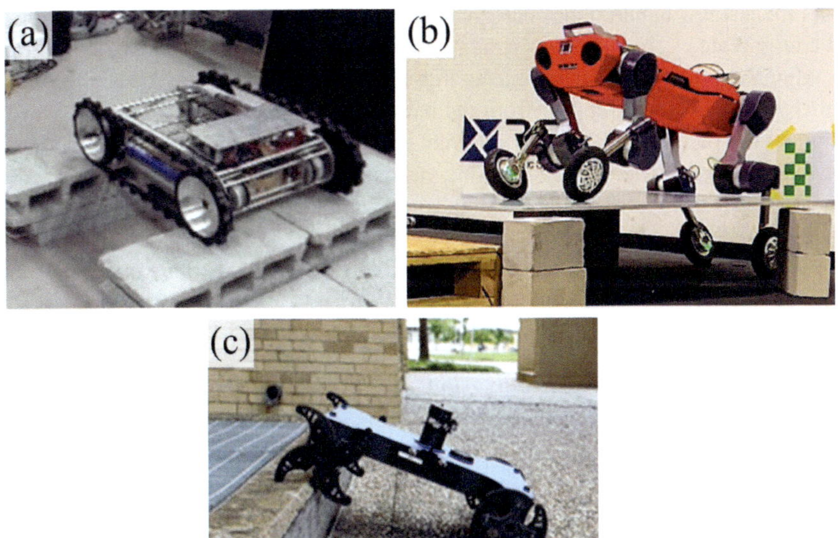

Figure 3.28 Robots with different artificial (non-natural) locomotion systems. (a) This tracked robot has a moveable centre of gravity which can be moved longitudinally to maintain stability (figure from Bruzzone et al., 2022); (b) a hybrid wheeled-legged robot aims to combine the advantages of the two locomotion systems (figure from Bjelonic et al., 2022); (c) this wheeled-legged robot can automatically switch legs to wheels or vice versa as the terrain changes (figure from Zheng et al., 2023).

also effective grabbers (hands) for manipulating objects. Thus, in the longer run, multifunctional legs could bear with some advantage.

Modern engineering may also allow merging different solutions. In this regard, technology has some advantages over evolution because new solutions can be invented and tested rapidly. Thus, despite the challenging structure, researchers put forward solutions to wheeled-legged robots having increased mobility (Bjelonic et al., 2021). This robot has wheels instead of a foot attached to the legs. Thus, it can step over larger obstacles if needed while moving fast on even terrain (Box 3.14). Alternative, partially convergent solutions are offered by tracked robot systems (Figure 3.28; see, e.g., Bruzzone et al., 2022).

3.5.6 Manipulating objects

The anthropogenic environment is full of various types of objects designed for being manipulated by humans. The design of these objects is aligned to human manipulation skills including some historical and cultural preferences. The human forearm with hands and fingers, as 'graspers' is a unique evolutionary construct that differs also significantly from that of other apes. The typical use of hands has also changed over the last few hundred thousand years, for example, movements can be controlled at a finer scale.

Humans often invent new objects that may rely on a new way of acting, while other activities may disappear from the human motor repertoire. The typical way of flipping the pages of a book is a different movement from scrolling pages on a cell phone (Marzke & Marzke, 2000), This means that most objects are not designed to be optimal for being manipulated by others than humans, so, despite its significance handling of objects by robots presents a huge challenge. This means that robots with clever manipulation skills have a long way to go, nevertheless there could be some simpler applications that prove to be useful.

3.5.6.1 Basic algorithm of object manipulation

There are few standard steps that should be followed by any agent that has the goal to interact with an object to achieve a specific goal. These steps are not necessarily executed one after the other rather they should overlap extensively to make the action time effective (Gong et al., 2023). The specific example for a grasping action is as follows:

1 Localisation of the object in space and recognising its physical properties. This happens mainly through vision, much rarer by active hearing (echolocation: Section 3.4.8) or touch.
2 Bringing the gripper into contact with the object and making sure that the grasping action is executed in the expected manner. This may also involve a specific interface (e.g., arm) that connects the gripper to the body and brings it into the right position. During these actions the CoG may also change, and thus the robot needs to execute stabilising motions. This could be more important for robots with two-legged or wheeled dynamic balancing state.
3 Based on the input from the object and the execution of the action the robot should ensure that the grasp is firm and safe for the entire handling procedure. The grasp should not damage the object but also should exert the necessary torque for prevention of dropping.
4 Next, the robot should have a plan for the displacement of the object, including information about the new location, and have to execute a safe release at the goal. Changes in CoG may need to be compensated also.

All four steps should be accompanied by several feedback mechanisms that ensure that the actions had the desired effects, and in case of errors corrective interventions are executed.

3.5.6.2 Types of object manipulation

From a behavioural point of view, there are many ways to manipulate an object but for obvious reason robotics is most concerned with grasping. Hitting, kicking, pulling, and tearing are important ways of handling objects. With grasping representing one of the most complex type of actions, it may turn out that in some scenarios simpler actions also suffice. Thus, manipulation consists of any action(s) that aim to change the location or orientation or other character (e.g., shape) of an object by physical contact. Below, there is an incomplete list of the most important types of manipulation (see also Gong et al., 2023).

Pushing objects: pushing can be defined as an interaction during which the robot exerts horizontal force on an object that moves along on the ground. To maintain the object on a predetermined track, often some kind of pliers are utilised. In principle, objects can be pushed by any part of the body, which can be brought into stable physical contact with the object, such as the body itself, legs, or even heads or trunks. Pushing could be useful to move light and/or wheeled objects, although the control and the precision of the action can be problematic. Rigo et al. (2023) introduced a four-legged robot that is able to execute planar pushing of large and heavy objects to specific locations by relying on a novel hierarchical model of predictive control. Pushing could be applied also to gain more space or can be part of agonistic interaction, like in many animal species.

Kicking objects: kicking can be considered being different from pushing because the short duration of contact between the robot and the object. The robot transfers its momentum during a brief contact that determines both the acceleration and the direction of the object's movement. Like in the case with pushing, several body parts can be used for kicking but applying the appropriate direction of force with regard to the goal is the most challenging aspect. Kicking may be useful when there is a need to rapidly remove a dangerous object from vicinity (including hurting the other in agonistic interactions) or in specific sport interactions, like soccer.

Grasping objects: grasping denotes the process of making a firm physical contact with an object for an extended duration typically to execute further manipulations. In addition, grasping can be also used to stabilise the agent's balance. Animals have evolved many different types of gripper devices. Apart from the primate hand, mouth can be also regarded as a potential gripper. Birds use their beaks to manipulate objects in several ways, while other species use their legs (octopuses) or trunk (elephants) or other altered appendages (e.g., tail) for handling objects. In animals, similar grippers function by the active movement of both 'fingers' or lower and upper parts of the beak in birds. This solution makes the gripping action faster. Grippers with two fingers/mandibles/pincers are the most common apart from the hand (paw) with five fingers which is present in several mammals (e.g., rodents, primates). While grasping seems to be a relatively simple action, it relies on challenging control mechanisms (Correll et al., 2021) that may also explain why their respective grasping behaviour is genetically supported in most species and then also practiced during development.

A grasping (more specifically grasping and holding) mechanism has to estimate gravity and apply opposite forces by also relying on friction using appropriate torque on the object. The interaction of these factors is represented by complex functions, where inputs from the objects can make significant differences. For example, the grasping of an empty or full transparent plastic bottle should be executed differently

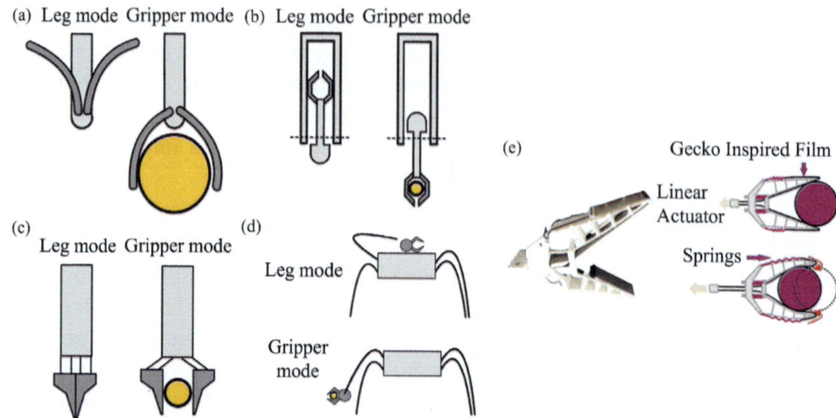

Figure 3.29 Schematic drawings (a to d) of various structures of robot manipulators with leg and gripper functions (from Gong et al., 2023). (e) This gripper utilises adhesive films to grasp and securely hold objects of diverse shapes and textures. Employing these adhesive films results in a 2.6x increase in pull-out force on rough surfaces compared to using soft rubber (Hashizume et al., 2019) (figure from Hernandez et al., 2023).

from grasping an empty or full glass bottle of the same size because differences in rigidity, weight, and friction.

Typical 'hand-like' grippers consist of at least two opposable fingers where at least one of them can be moved actively. Relying on the emerging friction between the fingers and the object is utilised that the closing of the fingers eventually holds the object firmly. There is also a trade-off between the number of fingers and successful task execution. More fingers increase the contact surface, and the necessary torque can be distributed better making the grasping action more reliable but too many fingers can make the computing more complicated. It is actually quite challenging to develop a strong grip that does not harm the object. Using elastic surfaces, like rubber, could be useful to increase friction but may at the same time decrease stability of the object held.

Gong et al. (2023) review several examples of grasping with legs. In this case, legs have a double function, as they are involved both in walking and manipulation of objects (Figure 3.29). Grasping can be executed by one leg, more legs, or with all legs involved. Such ideas also present novel challenges for coordinating the movement of the legs, the action space needed to carry out the manipulation, and controlling for the balance of the robot. Although, in principle, technology can offer various combinations of such solutions, their effectiveness at present is questionable.

3.5.6.3 Interfaces between bodies and grippers

If grippers are attached to the body directly then they cannot be moved independently, that is, body movements are needed to put the gripper in the actual position for the manipulation. Animals evolved specific interfaces that allow for independent movement of body and gripper such as arms, necks, or trunks. These interfaces typically consist also of several independent joints and thus increase the DoFs of the manipulating action.

Human-like robotic effectors having three joints which represent the shoulder, elbow, and wrist may achieve comparable functionality as the human arm. Although, despite their sophistication, present-day robotic arms are still constrained in their action space and are unable to copy all possible movements of a human arm (Gulletta et al., 2020) (see also above).

Although the focus of action is on the manipulation by the gripper, the proper control of the arms and feedback of their movement is essential for successful task completion. Arms have their own action space and specifically artificial arms may need larger obstacle-free area for executing their movements than biological ones. To ensure good performance, specific architectures are needed that can deal with forward and inverse kinematics, providing mathematic calculations for the appropriate movements in space (Box 3.15).

Box 3.15 Forward and backward kinematics for action planning

Kinematics is fundamental in robotics as it provides the mathematical tools and methodologies needed to understand, plan, and control the motion of robotic systems. Forward kinematics involves calculating the position and orientation of the end-effector (tool centre point, TCP) based on the current values of the joint axes. It is used to determine the pose of the robot's end-effector in relation to the starting point of the kinematic chain. Inverse kinematics is a method used in robotics to determine the joint configurations that enable a robotic arm or mechanism to reach a specific position and orientation. It involves finding the joint angles or parameters that position the end-effector (e.g., the robot's hand) at a desired location.

The difference between forward kinematics and inverse kinematics lies in the calculations they involve and the purpose they serve in robotics (see also Correll et al., 2021).

Forward kinematics

1. Calculates the position and orientation of an end-effector using the variables of the joints and linkages;
2. Determines the end-effector's position and orientation based on the given positions, angles, and orientation of the joints and linkages;
3. Useful for understanding the geometry and motion of a robot's mechanism, but it requires a comprehensive understanding of the variables of the joints and linkages connected to the end-effector.

Inverse kinematics

1. Calculates the variables of the set of joints and linkages connected to the end-effector, given the position and orientation of the end-effector;
2. Reverses the process of forward kinematics, allowing the robot to move its joints to achieve the desired position and orientation of the end-effector;
3. Used to reveal how to position the actuators to achieve the desired end-effector position and orientation;
4. Can be more complex than forward kinematics, as it involves a set of circular functions feeding back into each other.

Consider a humanoid robot arm with six DoFs. In forward kinematics, one would calculate the position and orientation of the end-effector using the joint angles and linkages. In inverse kinematics, one would calculate the joint angles required to achieve a specific position and orientation of the end-effector. Let's consider a simple example of a robotic arm with two rotational joints (two-link planar robot, see Figure 3.30). Each joint can be rotated, and the end-effector is located at the tip of the second link.

Parameters

- Length of the first link: L1;
- Length of the second link: L2;
- Joint angles: θ_1 (angle of the first joint), θ_2 (angle of the second joint).

Forward Kinematics: The forward kinematics problem involves determining the position of the end-effector given the joint angles. It can be expressed as follows:

$$x = \mathrm{L1}\cos(\theta_1) + \mathrm{L2}\cos(\theta_1 + \theta_2),$$
$$y = \mathrm{L1}\sin(\theta_1) + \mathrm{L2}\sin(\theta_1 + \theta_2)$$

In this example, if one knows the joint angles (θ_1 and θ_2), then one can calculate the (x, y) coordinates of the end-effector.

Inverse Kinematics: Solving an inverse kinematics problem means finding the joint angles given the desired (x, y) coordinates of the end-effector. This problem requires determining the values of θ_1 and θ_2 that satisfy the equations. This process can be more complex than forward kinematics, especially for robots with more DoFs.

In this case, the equations for θ_1 and θ_2 are derived from the inverse trigonometric functions. The solution involves considering the geometry of the robot and applying trigonometric relationships.

$$\theta_2 = \mathrm{atan2}\ (y - \mathrm{L1}\sin(\theta_1)/\mathrm{L12},\ x - \mathrm{L1}\cos(\theta_1)\ /\mathrm{L2}),$$
$$\theta_1 = \mathrm{atan2}\ (y, x) - \mathrm{atan2}\ (\mathrm{L2}\sin(\theta_2)/\mathrm{L1} + \mathrm{L2}\cos(\theta_2),\ \mathrm{L2}\cos(\theta_2)/\mathrm{L1} + \mathrm{L2}\cos(\theta_2))$$

These equations provide the joint angles needed to position the end-effector at the desired (x, y) coordinates.

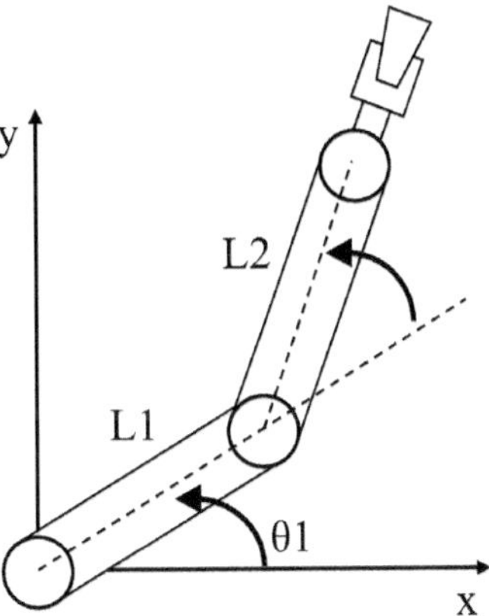

Figure 3.30 A simple arm model for calculating forward and backward kinematics. Large open circles represent joints, L1 and L2 are the length of arm sections, θ_1 and θ_2 are angles referring to the arm movements.

Inverse kinematics plays an important role in robotics for path planning, motion control, and programming robot manipulators to perform specific tasks by specifying the desired end-effector positions and orientations, that is, it deals with the problem of figuring out how a robot's joints should move to achieve a specific position and orientation of its end-effector (e.g., the robot's hand or tool).

3.5.6.4 Grasping without fingers: adhesive solutions

Many species of animals are able to walk or run on vertical surfaces or fix their body to objects to avoid being driven away by water current or wind blow. In order to solve such problems animals evolved several different adhesive solutions that rely either or both on physical adhesion (e.g., vacuum) or chemical adhesion (e.g., organic glue compounds). For example, clingfishes (Gobiesocidae) use a specific suction disc to adhere onto different kinds of objects or surfaces (Green & Barber, 1988). The clingfishes apply many mechanisms to achieve this effect, one of which is to suck out water from below the disc after contacting the surface to create a zone with lower pressure. Sandoval et al. (2019) copied this mechanism to build a biomimetic sucking disc for holding different types of objects both in air and underwater (Figure 3.31).

Similar solution could be highly effective to replace or complement grasping with fingers. The pressure formed between the disc and the surface can be adjusted to allow for grabbing and release and the contact between the gripper and the object during this kind of grip can be firmer than that between the fingers. The effect can be enhanced by using several suction discs on a gripper. The disc margins (rim) sealing off the central low-pressure zone from the outside can be made of materials that ensure a proper adherence independent from the nature of the surface. There is less

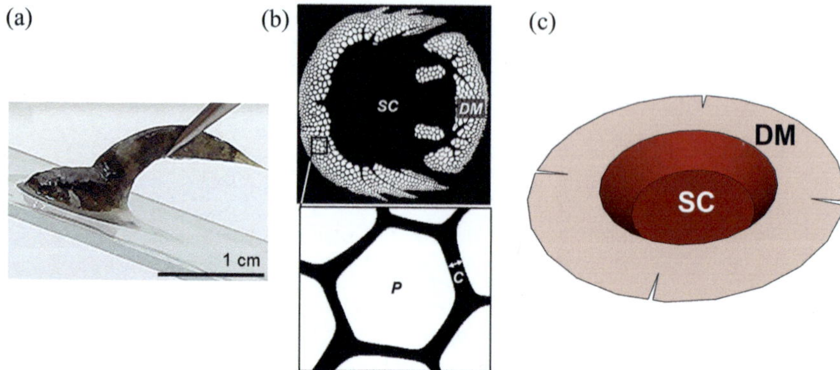

Figure 3.31 The suction disc of the clingfish inspires technology of adhesion in robots. (a) Clingfish (*Gobiesox maeandricus*) living at the coast of Southern California. Scalebar, 1 cm. (b) Binary image of the suction disc, showing the papillae; DM indicates the disc margin, SC the suction chamber, P the papillae, and C the channels formed between the papillae (panels (a) and (b) are reprinted with permission from *Sandoval et al.* (2020) *Toward Bioinspired Wet Adhesives: Lessons from Assessing Surface Structures of the Suction Disc of Intertidal Clingfish*. Copyright 2020 American Chemical Society). (c) Schematic of one artificial suction disk prototype: radially symmetric suction disc with four slits in the disc margin. The soft layer of the disc margin (DM) is composed of EcoFlex 00–30. The suction chamber (SC) is composed of Dragon Skin 20 (not to scale) (drawn based on Sandoval et al., 2019).

danger that the target object gets harmed by such grips, and the suction method is especially advantageous when the friction is low.

3.5.7 Alternative actuators for controlling movements

Robotic arms appear often more robust in size and construction compared to an animal and human arm with equivalent performance. One main reason for this is the difference in the actuators. In typical robots, the arms are moved by electromotors (see Box 3.11), although over the years many other types of actuators have been developed which have the potential to mimic the function of muscles.

Muscles and tendons are elastic structures that play a key role in moving body parts along the joints exerting the needed torque. The active movement is created by the muscles by changing their length that is transmitted by the tendons which are adhered to the bones. The contraction of the muscles takes place upon an electric signal received from the nervous system. This elicits a series of cellular and biochemical changes. According to the core process, molecular changes release energy that is used to alter the conformation of a specific protein molecule (myosin) making it shorter. The micromovements of these tiny molecules add together and make the muscle contract at the macroscale (Frontera & Ochala, 2015).

One alternative solution has been to use pneumatic (based on air) or hydraulic (based on fluids) machinery in which changes in pressure provide the required contraction and relaxation for making the joint move (e.g., Tavakoli et al., 2008). However, such structures also come with many specific constraints and have never been adopted widely in social robotics.

There are also emerging biomimetic solutions that aim to copy the biological mechanisms by using elastic materials, which can be shortened or relaxed by specific triggers (Mirvakili & Hunter, 2018). These artificial muscles have undergone significant improvements over the last ten years and may become realistic contenders for replacing electromotors (Box 3.16). Higueras-Ruiz et al. (2022) provide a very detailed review comparing the material and functional features of biological and artificial muscles. Measuring specific performance metrics (e.g., power per mass, response time, and efficiency), the values obtained for artificial muscles fall within the range of that shown by their biological counterpart. The development of artificial muscles also faces some challenges. For example, there is a need for improving the control of the muscles to provide a controlled and coordinated movement, especially when responding to unexpected ambient stimuli.

Box 3.16 Biological muscles instead of servomotors?

There is a trend to replace servomotors as actuators in robotics, by introducing the artificial counterpart of biological muscles (Figure 3.32). In a recent comparative work by Liang et al. (2020), it has been revealed that specific artificial actuators show often better performances than biological muscle from the aspect of single actuation parameters but overall, they do not reach the performance of biological muscles (Table 3.2).

Actuators compared:

Pneumatic artificial muscles mimic the function of biological muscles using compressed air or other gases. Pneumatic artificial muscles operate by introducing pressurised air or gas into their chambers. The increased pressure causes the muscle to contract, and when the pressure is released, the muscle returns to its original state through passive elongation or extension.

Dielectric elastomer actuators operate based on the electromechanical properties of dielectric elastomer materials. These materials can deform when subjected to an electric field. When a voltage is applied, the elastomer experiences electrostatic pressure, causing it to deform and change shape.

Conducting polymers (also known as intrinsically conducting polymers or synthetic metals) are a class of organic polymers that exhibit electrical conductivity. When an electrical potential is applied, ions (either cations or anions) from the surrounding electrolyte migrate into or out of the polymer matrix. The migration of ions causes the polymer to either swell or contract, leading to reversible change in volume. The actuation of conductive polymer-based artificial muscles can be precisely controlled by adjusting the electrical potential applied to the material.

Twisted fibre actuators are composed of polymer fibres that exhibit a helical structure. They undergo reversible and controllable expansion or contraction resulting in twisting motions (coiling or uncoiling) in response to external stimuli, such as changes in temperature or humidity or in the presence of certain chemicals.

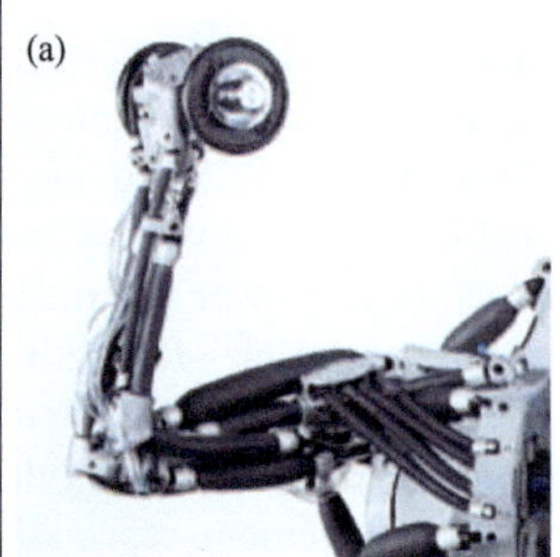

Figure 3.32 (a) Festo's robot arm is designed with a skeletal structure resembling that of a human arm, complete with bones and muscles. Thirty pneumatic muscles control the movement of the artificial bone structure. Similar to the human anatomy, this structure includes ulna and radius bones, hand and finger bones, a shoulder ball joint, and a shoulder blad (figure from Petre, 2012). (b) The KUKA robot depicted is effortlessly writing text on paper. However, such robots are also commonly utilised in the automotive and electronics industries, typically operating in physical separation from human workers (Photo by Marc Wathieu, Flickr).

Table 3.2 Actuation performances of biological and artificial muscles. There are many ways to compare actuators, the present evaluation is based on selected data taken from Liang et al. (2020) and see also for specific references therein

Actuators	*Strain (%)*	*Stress (mPa)*	*Strain rate (%/s)*	*Work density (kJ/m3)*	*Specific power kW/kg*	*Efficiency (%)*
Biological muscle	40	0.35	50	40	0.28	40
Pneumatic artificial muscles	25	1.16	800	200	10	49
Dielectric elastomer actuators	380	7.7	450	3,400	3.6	80
Conducting polymers	10	34	12	100	0.15	1
Twisted fibre actuators	49	22	50	5.3	5.3	2

Definitions of the parameters used:

Maximum actuation strain: The maximum deformation ratio (expressed as %) that an actuator is capable of performing divided by the initial length of the actuator (measured under non-loaded conditions).

Maximum actuation stress: The maximum stress ('effective work output') that an actuator can perform is divided by the cross-section area of the actuator (measured under blocked force conditions, in megapascal, mPa).

Strain rate: The change in strain per unit time (s) that indicates how fast the muscle can expand or shrink in response to the motion commands delivered from the central controller.

Work density: The work (kilo Joule, kJ) generated in the actuator per cycle and normalised by the volume, indicating the amount of work that can be exerted by the actuator in a limited space (m3).

Specific power: It is the ratio of the power (kilo Watt, kW) that is generated by the actuator to its mass.

Efficiency: It corresponds to the ability of actuators to convert the input energy to output mechanical energy (expressed in %).

3.5.8 Signalling actions

The communication function of many social robots is enhanced by displaying various actions that may involve the whole body or specific body parts, such as the head or arm (Section 1.8). Often these body parts and their actions are specifically designed for communication only in the absence of any other utility. For example, the NAO robot can perform various gestures with its arm but is unable to use the arm for any other purpose (Section 4.1). This situation presents a contrast to biological agents in which communicative signals typically evolved secondary to some primary (practical) function. Many extremities, such as legs or tails, have multiple functions. They are used in various ways for interacting with the environment but in specific situations, they can also act as signals. Very general rules involved in signalling, which increase the signal-to-noise ratio, have specific consequences on the evolution of a signalling action. At the same time signals have to have the potential to represent the actual inner state of signaller that also constrains their form. The signalling ability of social robots has been given much attention but much of the insight gained by ethologist has been neglected, partly because the focus has been to somehow replicate the human behaviour (Kunold & Onnasch, 2022).

From a technical perspective, visual signals emerge as the result of some bodily activity involving muscle contraction and relaxation, but signalling activity should not interfere with the ongoing activity of the body. These challenges can be solved by making typical actions to a signal by small modifications without impairing the basic functionality or by confining signals to extremities that are not in use at the time of the signalling event. For example, dogs use a play bow to invite group mates to play. The signal involves lowering the front half of its body while keeping its hindquarters and rear end elevated. This relatively small change in body conformation does not jeopardise the stability of the dog's body posture. However, dogs also wag their tail very actively that they can do because the tail has no functional role in that situation (Byosiere et al., 2016).

In general, the complexity of the signals can be increased by evolving multi-functioning body parts, involving actually non-functional body parts, both of which increase the flexibility of the body and increase the DoFs of the whole system. Many ethologists share the notion that species living in more complex social groups have evolved more complex signalling systems (Peckre et al., 2019) that allow the sender to express finer differences of its inner state and the receiver has also the opportunity to attend these important minute details.

3.5.8.1 Gestures – actions of the body and body parts

In line with the above, the body of a frog-like creature has a lower signalling capacity than a body of a mice because the higher behavioural flexibility of the latter. It follows that, given the phylogenetic constraints, species could differ in their potential to display communicative signals, and under some scenarios, there could be a pressure to evolve and/or modify body parts that serve communication specifically. This may explain the huge divergence in extremities of animal species and their role in communication, such as modified legs, tails, ears, and antennae.

It should be added that despite large variability of visual signalling capacity in animals, most species display relatively low number of signals (typically less than 60–80) (Bradbury, 2011). Comparisons are different to make in the absence of detailed data but, for example, chimpanzees and humans, who share the basic morphological features of arm, hand, and fingers, seem to show a marked difference in the number of visual gestures displayed. Many morphologically simple gestures such as pointing or waving are missing from the chimpanzees' signalling repertoire, while both species stretch out their arm for request and use their hands for communicative touches (Patricelli & Hebets, 2016).

Both humanoids and androids were designed to use their body for signalling inner states. These skills depend on the flexibility of the embodiment, robots with moveable arms or legs have more opportunities to communicate with gestures, often referred to erroneously as 'body language'. Three types of gestures can be discriminated:

1 *Whole bodily gestures indicating inner states:* inner emotional states are often reflected by the conformation of the body, and these physical changes have gained a communicative function during evolution (c.f., ritualisation see Section 1.8.2). This is true for human and non-human animals as well, and there are also some general rules describing the changes in physical appearance. For example, animals displaying threats make their body silhouette appear as big as possible in the opponents' visual field, and the opposite changes happen in case of fear. A four-legged robot may appear as being aggressive with stretched legs and a stiff body while it displays fear in a crouched position on the ground with its legs tucked in. Thus, both for zoomorph, humanoid, and android robots there is the possibility to convey their inner states by appropriate changes in body conformation. Importantly,

ethologist have developed a catalogue (ethogram) for a number of relatively stereotyped behaviours or behavioural signals related to inner states of specific animal species (e.g., Stanton et al., 2015). In contrast, the situation is more complicated in the case of humans, because there is no general behaviourally validated catalogue for human body gestures. Engineers planning the gestural behaviour of their robots are left to rely on their own, sometimes subjective insights (Zabala et al., 2021), or may get input from actors (Kishi et al., 2013).

Instead of discreate pre-programmed actions, some researchers design an affective architecture for controlling the emotional displays using valence, arousal (intensity), and stance as independent scales (Section 2.5.5, Box 2.15). These models have the advantage that they support also the display of blended or mixed emotional signals (Beck et al., 2010), and also the outputs can drive various kinds of embodiments as also shown by Bretan et al. (2015) (see also Box 3.17).

Box 3.17 A generalised model of displaying emotions on embodied or virtual agents

A Darwinian approach, focused on a pertinent and specific set of features for expressing emotions, has revealed that human subjects are capable of discerning the inner states of abstract, faceless, robot-like entities (Bretan et al., 2015). The researchers suggest a set of relevant body features for differentiating the behaviours that have the potential to display emotional states on agents with various embodiments.

(1) *Posture Height*: Showing the relative distance between the height of a person's chest and height of the waist, that is, how erect or crouched a person's torso is; (2) *Shoulder Height*: Indicating the relative distance between the waist and the shoulders; (3) *Arm Position*: The location of the arms in relation to the rest of the body (e.g., 'at the sides' vs 'in front of the face'); (4) *Gaze (Head Position)*: Referring to the attention of the person by the position of the head and the direction of looking; (5) *Body Activation*: Motion type displayed by the body and arms exhibit (slow vs fast) including (a) up and down activation (b) left and right activation (c) rotational activation (twisting of the body and arms); (6) *Head Activation*: Motion type displayed by the head; (a) positive head activation (nodding up and down) (b) negative head activation (shaking left and right); (7) *Volatility or Periodicity*: Variation in movement oscillation in terms of amplitude, rate, and regularity (highly periodic movement indicates a motion with a low volatility); (8) *Exaggeration*: The range of motions exhibited by each DoF (smaller vibrations vs massive convulsions).

Participants demonstrated the ability to recognise six basic emotional displays (see Figure 3.33a) better than chance. The encoding of affective behaviour was enhanced by animated presentations as opposed to static displays and by utilising an embodied device rather than a video demonstration.

This research lends additional support to the ethorobotic approach, asserting that humans can attribute emotional states to a physically constrained robot even in the absence of a face, provided that the structure and dynamics of affective displays adhere to the general rules of communication (see Section 1.8).

Similar outcomes were achieved by displaying emotional states on an arbitrary virtual object (Korcsok et al., 2018; see Figure 3.33b). Hungarian and Japanese participants were exposed to a short movie featuring the six emotional states presented by an arbitrary virtual agent. The emotional signals were constructed based on ethological findings, manipulating the movement, rotation, size, and colour of the agent (see

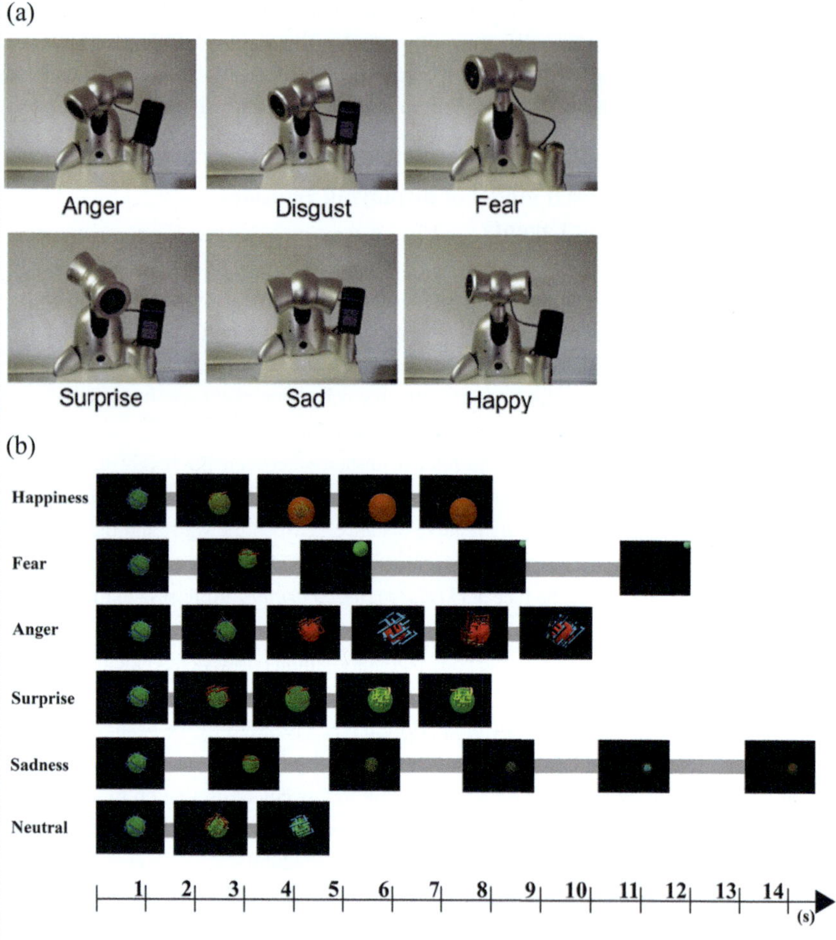

Figure 3.33 Behavioural attributes based on ethological insights in relation to emotional displays including ritualised behaviours, and some cultural influences (see also Section 1.8; and for details Bretan et al., 2015). (a) Static poses of Shimi robot representing six basic emotions (figure from Bretan et al., 2015), and (b) the dynamic changes of the artificial agent (following the same starting state) representing six emotions (figure from Korcsok et al., 2018).

references in Korcsok et al., 2018). Although the recognition ratios of Hungarian and Japanese participants were comparable, cultural differences also surfaced (Table 2.11).

In summary, this line of research suggests that robots lacking many human-like or zoomorph features can acquire the capacity to meaningfully convey their emotional states. Through specific planning, robots with simpler designs or arbitrary morphologies due to functional constraints can be imbued with affective behaviour.

2 *Gestures referring to external events:* head turning, gaze alternation, and pointing (with arm and/or finger) may display the sender's tendency to draw the receiver's attention to some specific aspect of the environment. Pointing is a social gesture, as it is typically not displayed when alone. Head turning may be used for collecting information for oneself as well as a communicative signal. While head movements are shared by many animal species which have a relatively flexible neck, so far pointing with arms has only been described in humans (although animals socialised with humans display the ability to react to the pointing gesture) (Cooperrider & Mesh, 2022). These behaviours can be significant to achieve joint attention between the interactants which facilitates coordinated actions and cooperation.

   In comparison to the multifunctional head-turning behaviour, pointing is a very specific ritualised referential signal that can also be transferred to other extremities such as legs or trunks (Kita, 2003). To carry out these actions, the robot must know the location of the object or event in space and execute the behaviour in a way that the direction of the gesture points at the right location. Industrial applications can provide a good practice for such problems.

3 *Contact gestures*: the physical contact (touch) adds a further dimension to the effectiveness of gestures. Examples include gentle or more forceful touches of the other's body by the whole body or various body parts, such as hands or legs. Apart from the physical contact, these gestures have other important characteristics that should be taken into account. This includes the pressure put on the body of the other, the dynamics of the action, the location of contact, and temperature of the contacting surface.

The endowment of social robots with the ability to shake hands provides a good example for the challenges (Prasad et al., 2022). Despite its simplicity to be carried out by humans, the three main behavioural components of handshake (reaching, grasping, and shaking) rely on a complex hardware and architecture. According to Prasad et al. (2022), to achieve a robust handshaking skill in robots, there is a need to increase context awareness (e.g., gaining more information about the partner and its inner state), the robot should have a better coupling between the force exerted by the human and the force expressed by it, making the grasping and shaking more natural. However, from an ethorobotic perspective handshaking is dispensable action for social robots because it does not exist in many human cultures, and there are other ways to express similar behavioural tendencies or attitudes, such as bowing. Unfortunately, there are also no field studies, and no data to show whether alternative behaviours are or are not as effective as handshaking. In addition, the absence of physical contact between partners could also be advantageous because it can decrease the chance of disease transmission (Gravina et al., 2020).

While it may turn out that handshaking is an unnecessary skill for social robots, touching could have important functionality in educational or medical situations (Eckstein et al., 2020; Hoffmann & Krämer, 2021) in parallel to the effect of therapy animals. There have been numerous attempts to observe the outcome of robotic touch on humans, investigating the influence of the type of action (e.g., patting and stroking (Zheng et al., 2020), speed of movement (Zamani et al., 2020), and temperature (Block & Kuchenbecker, 2019)). In short, patting, lower speed and warmer temperatures were preferred by the human subjects. Thus, social robots may provide emotional support through touch, but there is very little verified evidence for positive effect of touch by therapy animals.

Expandable features of the body could be also used to enhance human-robot interaction by making the robot more acceptable (Hedayati et al., 2022). These appendages can have different designs and may function as attention-getting devices, complementing communication and allowing for playful interaction.

3.5.8.2 Gestures – actions of the head and face

Gesturing by facial muscles is typical only for a relatively small number of species and evolved mostly among vertebrates. Facial muscles evolved primarily for moving the jaws and controlling eye movements. Their role of taking part in communicative facial gestures emerged much later in mammals (see Diogo & Santana, 2017). While the platypus (*Ornithorhynchus anatinus*) has only ten facial muscles, detailed anatomical descriptions have revealed that rodents and all primates, including humans, share about 20–24 facial muscles. Despite the similar number, there are specific variations in mammalian species, some muscles were lost (e.g., muscles moving the ear), and some other were gained (e.g., muscles moving the mouth and the eye region) during the evolution of specific clades that could be also explained by their particular feeding ecology or even social habits. It has been also hypothesised (Diogo & Santana, 2017), especially for primates, that facial colours (and probably also longer fur) may blur the expression facial gestures, and thus the effectiveness of facial gesturing, as a signal, is greater in species with pale facial skin. Having no colouration and no facial fur, humans represent a special case for increasing complexity of facial gestures but interestingly this is achieved by comparable number of muscles, probably by changes in the neuromotor control.

Parallel research suggests that the signal components of the body and face follow similar rules (Oosterhof & Todorov, 2008; Tzschaschel et al., 2022). The inner state of the sender is mirrored on the body, and the primate facial gestures could be interpreted as redundant to the evolutionary more ancient signals shaped by the body as a whole.

Artificial applications of human facial expressions have a broad scale stretching from simplified graphic emojis to complex hardware in robots' head designed to mimic human emotions on the face. Both humanoids and androids have been equipped with such capabilities.

For example, Kobian-R is a humanoid robot that has a specifically designed head for displaying human-like emotions (Kishi et al., 2013). Controlled by a three-dimensional emotional architecture, Kobian-R is constructed to show seven Ekmanian emotions, such as happiness, fear, sadness, etc. The robot's face is composed of a fixed middle face area with nose and independently moveable mouth, forehead, eyes with eyelids and eye browns. These metal (and rubber) units are controlled by several electromotors which all together allow for 21 DoFs. Experimental subjects were able to recognise some of the emotional states displayed by Kobian (Zecca et al., 2009).

Androids were also endowed with the capacity to display emotions on their face (for overviews, see Faraj et al., 2021; Li et al., 2023; Rawal & Stock-Homburg, 2022). The hardware of these android differs mainly in the number of actuators installed in the role of the facial muscles, the type of actuators (electromotors, pneumatic artificial muscles), and the material used for the skin (silicone, urethan resin). For example, the robot named Eva has 12 servomotors to manipulate a silicone facial silicon skin and can achieve 15 mm displacement (Faraj et al., 2021). Using the Ekmanian emotional model, the appropriate control of specific servomotors elicits all six basic emotional displays (Box 3.18).

Box 3.18 Constructing a humanoid face having affective displays

There have been many projects aiming to construct a human-like face for robots that is able to express various emotions. One relatively recent attempt was reported by Faraj and colleagues (2021) who constructed an android robot ('Eva') that emulates human facial expressions, head movements, and speech through the use of 25 muscles (substituted by servomotors), including 12 facial muscles that can produce a maximum skin displacement of 15 mm (Figure 3.34 and Table 3.3).

Table 3.3 The numbers represent the action units ('muscles') involved in the expression of the basic emotions.

Emotion	*Action units*
Happy	(6+12)
Sad	(1+4+11+15)
Surprised	(1+2+5+26)
Afraid	(1+2+4+5+20+26)
Disgusted	(9+15+16+26)
Angry	(4+5+7+10+26)

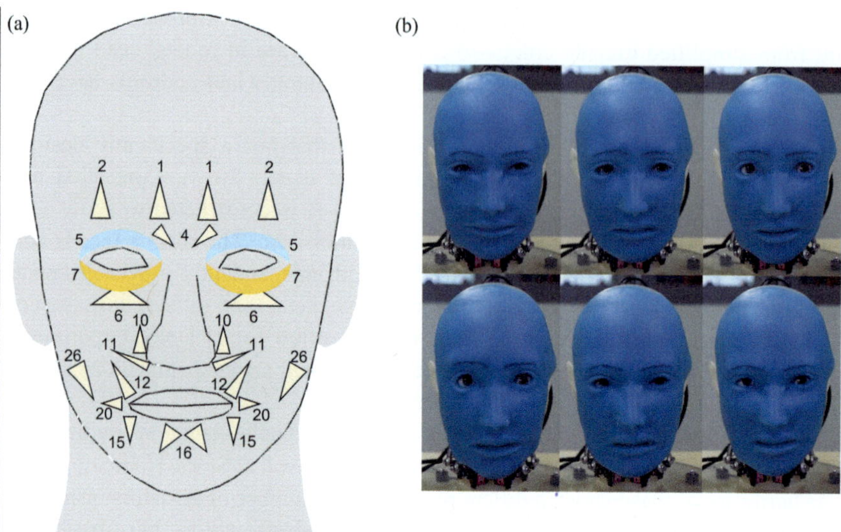

Figure 3.34 (a) The muscles and their movements have been defined based on human facial coding system (Freitas-Magalhães, 2012) that also determines which muscles contribute to the display of a specific emotional state (see Table 3.3). 1- Inner brow raiser; 2- Outer brow raiser; 4- Brow lowerer; 5- Upper lid raiser; 6- Cheek raiser; 7- Lid tightener; 10- Upper lip raiser; 11- Nasolabial deepener; 12- Lip corner puller; 15- Lip corner depressor; 16- Lower lip depressor; 20- Lip stretcher; 26- Jaw drop (based on Faraj et al., 2021). (b) The mask is made from blue-coloured silicone (Smooth-On EcoFlex 00–30) that has similar material properties to the human skin, and can be programmed to show a wide range of facial movements, including basic emotional displays, such as happy, sad, surprised, afraid, disgusted and angry (from left to right) (figure from Faraj et al., 2021). Triangular symbols in light colours indicate the location of servomotors within the head, with the apex of the triangle pointing in the direction of movement.

Building of such structures is an interesting engineering challenge, and such humanoid heads will improve even more in time. It is however an open question whether these machines will be utilised in social robotics. Note that this head is only superficially similar to a human head because apart from showing similar facial movements it lacks all fundamental functions like eating, seeing, hearing, smelling, and being sensitive to touch. Twenty-five servomotors are needed for manipulating the surface which probably make some noise and can get overheated (there is a cooling fan also in the head). It is likely that ‘Eva’ may elicit some ‘uncanny’ feelings (see Section 2.3.3) in the human partners. Thus, it could be more advantageous (and perhaps also cheaper) to utilise ‘artificial’ designs (see Box 3.13) for displaying emotional states.

3.5.8.3 Vocalisation

In animals, acoustic signals are typically the result of some muscle activity. For example, crickets produce their chirping sound by rubbing the bottom of one of their wings against the upper part of the other. The emerging friction causes vibration that generates audible sound waves in the air.

Broadly speaking, the head and the face are also involved in the vocalisations in many other animals, and thus sound production is also accompanied by changes in the body conformation. These observable actions may become a visual component of the dominantly acoustic communication.

Many terrestrial vertebrates utilise their respiratory system for producing sounds (Box 3.19). The muscles in the lung push air into the larynx that causes vibration in the vocal folds. The tension and the thickness of the vocal folds regulated by muscles determine the acoustic quality of the sound. The vibrations are also modified in the mouth when they pass through mobile obstacles like the tongue and teeth (Ladich & Winkler, 2017). Thus, an important aspect of such vocalising systems is the involvement of several body parts controlled by specific neural input. This process ensures that the sound produced is the function of the individual's actual inner state. This tight relationship between the actual physical body shape and state and the output is specific feature of acoustic communication in animals. In sharp contrast, today's robots use either pre-recorded sounds or generate sounds electrically that do not have a direct, inherent connection to the state of the agent. Although the effects can be mimicked by appropriate programming, this provides only a superficial similarity.

There has been some interest in building vocal tracks that correspond to that of humans. For example, Nhu and Sawada (2017) constructed a 'talking' system in which electromotors were used to modify the shape of the vocal folds and the vocal cavity. They were able to produce various vowels across a wide range of frequencies by adjusting the vocal system using electromotors. Although, impressive, such solutions are still very different from that functioning in biological organisms, like mammals, because there is no direct connection between the actual physical state of the body and the sound emitted, but at least such applications make the sound production more realistic. While human-like talking could still provide a challenge for such systems, similar concepts could be utilised to emit simpler vocalisations present in many other species or implemented in ethorobots.

Box 3.19 The vocal apparatus

A simplified and generalised model of the source-filter theory (Fant, 1960) divides the vocal apparatus into two functional units:

1 *The Source*: The energy required for sound production is generated by the chest's breathing muscles, pushing air from the lungs into the bronchial tubes. Sound production occurs in the larynx, composed of cartilaginous elements. In a resting state, the vocal cords permit the free flow of air. During sound production, specific muscles

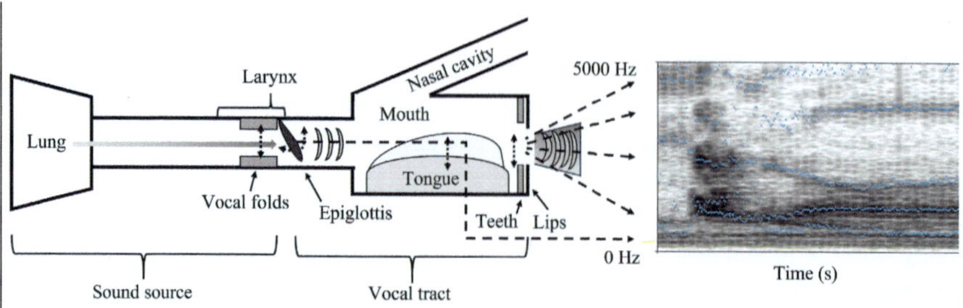

Figure 3.35 The main anatomical structures involved in sound production. Double-headed arrows indicate possible movements (opening and closing) at the vocal cords, epiglottis, and teeth. The complex movements of the tongue also influence the airway's shape. Dashed lines to the spectrogram illustrate the contributions of the source and the filter to the acoustic structure of the sound.

pull the vocal cords into a closed position, blocking the airflow. The increased pressure eventually allows the air to break through the stretched vocal cords, after which the cords are relaxed again. Subsequent muscle contractions initiate a new cycle of sound waves. This activity gives rise to the fundamental frequency (F0) of the produced sound.

2 *Vocal Track (Filter)*: Leaving the larynx, sound waves traverse the upper part of the larynx, throat, and oral and nasal cavities before reaching the environment. The filter essentially acts as an elastic-walled tube filled with air, possessing resonance frequencies that can easily move particles in the column. This creates a band filter effect, enhancing certain frequency ranges (formants) and muting others in the sound wave spectrum produced in the larynx (Figure 3.35).

The fundamental frequency (F0) is defined by the opening and closing frequency of the vocal cords and is also influenced by the vocal cord tension, and heavier and thicker structures typically produce lower sounds. Formant frequencies are influenced by vocal tract length, with longer tracts emitting lower-frequency formants.

The source-filter theory provides a clear background for the fundamental difference between animals and robots in vocalisation. In animals, there exists a strong physiological connection between the body's state, controlled by the nervous system, and the mental state influencing nuances in sound. In contrast, robots produce vocalisations independently of embodiment, utilising specific electrical systems.

3.5.8.4 An ethorobotic approach to gesturing behaviour

Gesturing behaviour underwent significant development in social robotics during the last 20 years. Improvements in hardware and novel approaches to the architecture enhanced the way how robots can display expressive behaviours. There are two different tendencies that are difficult to reconcile when there is an aim for building human-like robots.

For obvious reasons, one very confident strategy is to build robots that are able to communicate like humans. While the strive for doing so in understandable as human complexity provides a challenge of roboticists, efforts to make robots act like humans delays the deployment of robots in the near future. In addition, for reasons discussed elsewhere (Section 2.3.3), human-likeness may interfere with the main function of social robots by confusing human users.

The ethorobotic approach emphasises that humans share many aspects of their expressive behaviours with other biological agents, such as mammals. Since the technology is close to build any simpler and robust agents that may be eventually considered as being similar to animals, deviating from the track of replicating humans, would provide a less confusing scenario for interaction, and offer a closer point in time for utilising the service of social robots.

Based on suggestions by Rawal and Stock-Homburg (2022), there could be an extended agenda for roboticists supporting the functionality of artificial agents by improving communication skills:

1 Most HRI investigations are carried out under very unnatural, laboratory conditions, robots with expressive behaviours should be tested in the field, allowing them to interact spontaneously with all kinds of people. Quantitative data should be collected to measure the effectiveness of robot expressive behaviour when it is displayed in real (unscheduled) interactions.
2 The whole body needs to be involved in the display of expressive behaviours just like in the case of typical biological agents. The rules of physical changes should be applied locally to each body part if possible. More versatile robots (e.g., higher DoF) have a better chance of varying expressive behaviours.
3 The input for displaying expressive behaviours should be based on sensor fusion, that is, visual, acoustic, haptic, etc. Broad range of behaviours should be considered for evaluating the inner mental state of the human partner.
4 The robot's reaction should be made flexible and depending on learning. Both the display of expressive behaviours and the reaction to them are influenced by several intrinsic (e.g., personality) and extrinsic (e.g., familiarity to the actual location) factors.
5 The display of expressive behaviour should be controlled by analogue models (e.g., emotional space model), allowing the gradual, fine-tunned changes in latency, intensity, and endurance. Two-dimensional models (valence, intensity) could be extended to additional dimensions to make expressions more fine tunned.
6 There could be a need to introduce more specific variations in the expressive behaviour of robots. Recent focus on six basic emotions may be a too narrow base for social interactions. Moreover, it should be tested in the field whether people are able to learn about embodiment-specific forms of robotic behaviour over repeated encounters if it is generated along the established rules for expressive behaviour.

3.5.9 Energy sources for autonomous robots

Just like in the case of electric cars, the usefulness of autonomous social robots is determined by how long they can function without the need to be recharged. Although, there are different kind of technical solutions, most today's robots rely on electricity supplied by rechargeable batteries, which they need to carry with themselves (see also Mikolajczyk et al., 2022).

Social robots interacting actively with humans need to be well powered to function for several hours without the need of being recharged. Continuous activity (over a day) could be maintained if the robots can use short breaks for recharging, but this may not be always practical, and recharging in this case would need to be very efficient. In many commercially available robots, it is assumed that the dead battery is swapped for a charged one by a human operator but this is typically not a viable solution under many field conditions. In any case, batteries should continuously support the operations of a robot for at least 4–6 hours (calculated for maximum activity) to be useful in practice.

Some researchers aim the develop measures for calculating the energy efficiency for a robot. This could also help in designing and choosing the best energy source, as well as making robots comparable in this respect (e.g., Sakagami et al., 2002; Tucker, 1975) suggested specific resistance (SR) as a measure calculated as SR = Energy consumption (E)/(mass of the robot (M)*distance of the move (d) * standard gravitation (g)). The problem of these measures is that they calculate efficiency based on horizontal movements while social robots use electricity for various other activities also, and very often other activities like gesturing or manipulating objects use more energy than expected. But even horizontal moving can use more energy if the robot has to stop and start more often or slow down and speed up repeatedly. Many robots

Table 3.4 Energetics comparison of biological and robotic systems (based on Kashiri et al., 2018). Importantly, the values in the table are indicative and heavily reliant on the specific systems chosen for comparison. Additionally, advancements in robotics may have occurred since 2017. Nevertheless, these conceptual comparisons play a vital role in gaining an objective understanding of current technology and projecting potential advancements in the future. Establishing criteria for valid comparisons between biological and robotic systems is also imperative

Criteria	Biological systems	Robotic systems (2017)
Actuator power density	500 W/kg (muscles)	200 W/kg (brushed direct current motor and harmonic drive)
Actuator energy efficiency	20% muscle	40%–50%
Computation power	20 W (human brain)	60 W (regular notebook)
Power consumption at rest	60–80 W (basic metabolism)	150–400 W
Energy storage	17 MJ/kg (complicated digestion system)	0.87 MJ/kg (Li-ion) (efficient lightweight power converters)

advertised as night guards cannot be active autonomously for a few hours because their underpowered battery capacities.

As cited by Kashiri et al. (2018), a horse trots with a SR of 0.2. Calculations reveal a similar value for a walking human. In contrast, the humanoid ASIMO robot exhibits a SR of 2, whereas walking robots, leveraging passive dynamics and employing kinematic and actuator optimisations, can enhance performance, achieving a SR of 0.7. The biological realisation of motion still presents numerous opportunities to enhance robotic functioning (Table 3.4). Increased emphasis should be placed on gathering proprioceptive data (based on IMUs and force/torque sensing) for precise movement control. Furthermore, there is a need to shift away from overly simplistic models of legged movement (see Kashiri et al., 2018).

The need to consider the energy consumption of robots because batteries not only demand significant space but also contribute substantial weight to the robot. These factors can significantly impact the construction and design of the robot. Enhancing the robot's efficiency enables a reliance on smaller batteries, which is also essential in minimising environmental pollution.

3.5.10 Conclusions on movement and actions for ethorobotics

Riener et al. (2023) provided the first in-depth comparison between typical human and robot performance. Comparative studies are always difficult to carry out because one needs to ensure that the objects of the comparison differ only with respect to measure to be taken. Animal behaviour has a long history and experience with similar problems. Comparison of humans with apes, wolves with dogs, or cats with dogs often suffers from the same theoretical and methodological issues (see Miklósi, 2007). Thus, the results or statements of comparative investigations should always be considered with caution. When comparing typical human performance, Riener et al. (2023) looked at the skills of 27 different robots (existing until 2021) that were constructed with the aim and promise of replacing partly or wholly specific human tasks. Robust and very simplified technology can easily surpass human performance if there is a close fit between the agent actions and the environment, such as achieved in industrial robots.

Importantly, it is frequently overlooked that anthropogenic environments are deliberately designed to accommodate human activities, and they also exhibit an unpredictable nature. Predicting or preparing for future events within such environments is challenging. Consequently, social robots must maintain a constant state of specific attention (monitoring) to promptly respond to potential changes. Hence, even if a robot is adequately equipped to perform typical tasks, environmental disturbances, primarily stemming from human activities such as decreased compliance with the robot, can lead to the failure of the agent in practice (Figure 3.36).

A further weak point of human and robot comparison is that robots may appear to display better specific performance only because other factors have not been taken into account. For their comparisons, Riener et al. (2023) determined many variables that were used to normalise the performance of robots and humans (e.g., mass, size, power consumption, operation time, cost of transport). With regard to specific

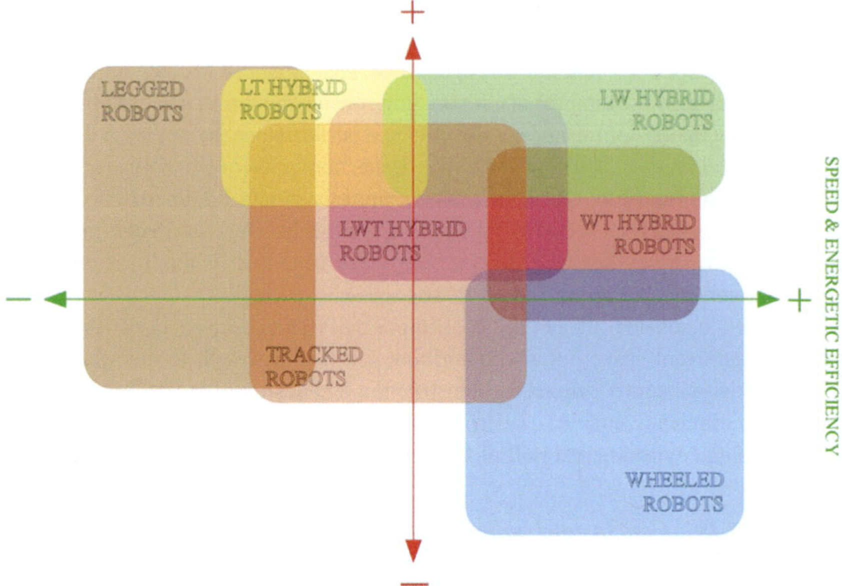

Figure 3.36 The comparison of terrestrial robots based on mobility in unstructured environments (Y axis) and speed and energetic efficiency (X axis). Legged, tracked, and wheeled systems are positioned diagonally on the graph, where mobility in unstructured environments is inversely proportional to speed and energetic efficiency. Wheeled robots prioritise speed and efficiency, while legged robots excel in unstructured environments but sacrifice speed and efficiency. Tracked robots fall in the middle, well-suited for soft terrains due to a large ground contact area. Hybrid systems occupy the right upper zone, combining advantages but are constrained by increased mechanical complexity and performance compromises (figure from Bruzzone & Quaglia, 2012).

functionalities, the authors made the following main conclusions (for details, see Riener et al., 2023):

1 No robot has surpassed humans in terms of gait speed when considering factors like cost of transport or operational time. Even wheeled-legged robots fail to match the sprinting speed of a human.
2 The speeds achieved by the top climbing robots, particularly in ascending tasks, are comparable to those of an average human relative to body mass or size. However, these robots fall short when compared to the maximal possible operation time achieved by a human.
3 Robots excel in performing postures and tracking movements with higher repetition accuracy. Depending on their actuation and transmission type, they can maintain a posture almost indefinitely without consuming additional power.

4. Certain humanoid robots exhibit capabilities such as jumping, crawling, or swimming, closely aligning with the performance levels of the average human population.
5 In straightforward pick-and-place tasks, robots significantly outpace humans. Nevertheless, when it comes to dynamic manipulation tasks and handling deformable materials, robots still lag far behind human performance.

Despite the inferior performance of robots in this comparison, the technical advances over the last 10–20 years give hope for optimism. Riener et al. (2023) also identified some specific aspects where further development would be especially important:

1 There is a need to improve sensor fusion approaches merging full-body, high-resolution sensing of tactile, proprioceptive, visual, and auditory information for environment perception.
2 Robots should be equipped with ability of multi-contact interactions which enable more efficient and versatile manipulations.
3 Robots should achieve that the control of actuators occurs in similar temporal and spatial resolutions as in humans. More preference should be given to feedforward-control modes to perform movements similarly to humans.
4 Present-day electric motors used as actuators seem to have achieved their maximum performance, and thus alternative mechanical devices should be looked for that mimic closer the performance of muscles and joints.
5 It is important to work on more integrated actuation, sensing, and power supply as well as high-resolution soft tactile sensors to achieve much better manipulation capabilities.

Developing operational social robots capable of reliably performing tasks has posed a significant challenge for the field of social robotics. Despite numerous efforts and promises for widely useful practical applications, the majority of social robots find themselves confined to university laboratories or the research and development departments of companies. The reasons for this limitation are diverse, but a central issue is the lack of a comprehensive theory underpinning social robotics. Many researchers set ambitious goals (typically mimicking humans, see above) that currently surpass the technological capabilities at hand.

While the pursuit of challenging objectives in science is commendable, there is an argument for adopting a more pragmatic approach. One could consider that instead of designing one complex robot, specific functions could be shared by two or more simpler agents. Evolution has also experimented with this kind of task sharing. The hive of ants could be considered as a functional unit ('super-organism') while the tasks, reproduction, food collection, territory protection, etc. are shared by individual ants belonging to a specific cast (Hölldobler & Wilson, 2009).

Simpler utilisation of presently available robust technology may provide an alternative for creating functionally useful social robots tailored for specific tasks. In this context, ethorobotics offers valuable insights by drawing from the behaviours of biological organisms and their adaptations to specific environments. This second option could offer a safer and more attainable path to the development of practical social robots.

In summary, based on its strong ties with biology, ethorobotics may offer many ideas to be included in the development of mobile, interactive, and autonomous robots.

3.6 Modelling problem-solving architectures for ethorobots

3.6.1 Introduction

In the context of ethorobotics, a robot is an autonomous, mobile, embodied, and interactive problem-solving agent that generally behaves in a way that is reminiscent of an animal species. This task has been approached from very different directions, from robotics engineering's high-precision industrial robots to philosophical insights based on the human mind. Some of this expertise was brought under a common umbrella of cognitive science while also maintaining their independence in terms of theory and methodology, including AI, cognitive psychology, neuroscience, and robotics. This means that there is a plethora of different conceptualisations on problem-solving architectures especially when the goal is to build autonomous, embodied, and mobile agents. The actual solutions seem to look very different when one has to build a robust functional robot for the market from developing a general concept for some kind of general problem-solving architecture.

There has also been a trend that has differentiated the development of robots from the evolution of living agents. The anthropogenic environment could be changed in a way that makes the solving of problems easier from the robot's perspective (e.g., building roads for cars). This diminishes the pressure for developing sophisticated architectures (and embodiments) that are not the case in living agents.

As laid out in Chapter 2, the task at hand is not to build a structure that can be labelled as less or more cognitive, because of the very anthropocentric use of this term, that is compared to some kind of imagined and idealised human performance, but rather to arrive at a concept of architectures that are successfully solving a wide range of problems. Thus, from an ethorobotic perspective, any architecture should be based on the following criteria (Miklósi et al., 2017):

1 It has to be functional, meaning being able to survive in a specific environment;
2 It has to be able to account for all capacities that are minimally necessary for an autonomous, mobile, and autopoietic system;
3 It should have the potential to increase its complexity to accommodate new environmental challenges or to develop novel competencies;
4 It should aim to distinguish itself from all existing living creatures in form and structure, representing rather a new 'species', and it should not strive to mimic any specific living being, such as humans, nor should it have a structure similar or comparable to living systems, such as the brain.

In general, architectures consist of two types of basic elements, modules, and connections between them. Modules (labelled also as 'hubs' or 'nodes') can be envisioned as well-defined parts of a system (software and/or hardware) that are characterised by specific types of operations. Such modules emerge (or are constructed) because the type and amount of interaction within some units of the system is or becomes different in comparison to other units (see also Section 2.4.6). Modules do not need to be physical entities but a network of units being active in synchrony at specific times. Thus, one unit may be part of different modules at different time points. Modules may be organised in a hierarchical system, consisting of other (sub) modules or being part of a larger network of modules. The architecture functions via the interaction of modules by providing regular and one- or two-way input and/or output for each other, indicating the actual inner state(s) of the module. A module may provide single or multiple input/output by its connections. The problem-solving capacity of the architecture depends on its modular structure and their connections as well as the depth or complexity of the modules and the system as a whole. All architectures rely on some kind of hierarchical processing that is related to a decision-making activity (module) ensuring the stability of the agent because the embodiment provides a critical physical limitation in carrying out actions.

Based on the above, the aim of developing a theory for a general problem-solving architecture may gain some support if the following insights were to be taken on board.

1. At present, there are many types of architectures based on different assumptions depending on the origin of the scientific input (e.g., robotics, neuroscience, etc.). The trend for designing architectures, which correspond to the (human) mind, is in contrast to arriving at a generalised model (see also below).
2. While the actual and functional (human) mind is the product of evolutionary (population-level selective processes over extended time span, 100s of millions of years) and developmental (selective processes at the level of neuronal connections over shorter time scale, days/years) processes, this is only rarely taken into account in the conceptualisation of general models. Especially, on the evolutionary scale, the presence of modules and their connections may show a dynamic variation which may be at the heart of organising a problem-solving architecture which may face different challenges under short and long-term unpredictable environmental conditions.
3. Selective processes may push the general architecture towards optimal functioning. These trends may follow Ashby's (1958) insight suggesting that for a system to be stable, the number of states that its control mechanism can attain (its variety) must be greater than or equal to the number of states in the system being controlled (typically, the environment in the case of biological agents). This required complexity of the architecture may be determined by inputs and outputs, that is, more complex environments (c.f., niches) may be selected for more complex architectures. General architectures may also have basic modules (present in all/most systems) and more specific modules (operating only in specific systems).

4 For many general architectures, the human mind provides the etalon. Early experimental psychologists also emphasised the generality of the human mind. This anthropocentric view has concealed the fact that the human (primate) mind could also not escape the forces of evolution, and thus it is as specific as other minds. This notion of generality originates from the subjective experience that the human mind is able to deal with a broad range of problems in its specific environment. Doing away with this view could give a huge impulse to the development of general architectures for the next generation of ethorobots.
5 Despite the success of algorithms developed in the field of AI, it has been emphasised many times that these are typically not human-like (Chemero, 2023; Mitchell, 2019) and are far from reaching human sophistication which does not mean that they cannot show a higher performance in specific tasks.
6 General architectures should be scalable because they may operate in physically less or more complex environments (spatial scaling) and may exist for different durations (temporal scaling). Both dimensions may affect the nature and complexity of operations in the system.

3.6.2 The Standard Model of the Mind (SMoM)

Following a series of discussions among leading experts of the field, Laird et al. (2017) provided a common computational framework for a Standard Model of the Mind (SMoM; Figure 3.37). Their commendable goal was to distil a model that can become a common reference point for all, and for which there is a general agreement among experts. Their aims included (1) identifying the minimum set of modules/connections/processes for an architecture, (2) pinpointing fundamental types of modules, and (3) revealing main levels of complexity (if possible). Such standard models may not only act to connect diverse fields of science but also serve to develop a common nomenclature and perhaps also to distancing them from the anthropocentric trends (for a more direct application of this concept, see Section 3.8.5 on Soar).

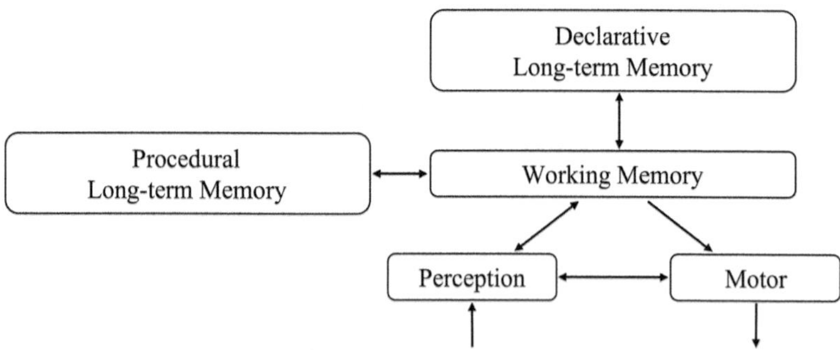

Figure 3.37 The Standard Model of the Mind (SMoM) was introduced by Laird et al. (2017) with the primary objective of creating a foundational model unanimously agreed upon by experts in the field of artificial intelligence (AI) (redrawn based on Laird et al., 2017).

The SMoM follows the triadic partition of perception, action and central mechanisms, and postulates that the modules and submodules of the system have specific functions, and the system's final performance depends on the functioning of these modules and their connections.

SMoM is able to use two types of data (representations): symbolic (non-numerical) and non-symbolic (numerical). The concept of the former was introduced by Newell and Simon (1976) (Physical Symbol System Hypothesis) and has the following key features:

1 *Representation*: They may refer to any physical or abstract entities in the external world or within the architecture itself;
2 *Manipulation*: Symbols can be manipulated or processed. The operations or transformations performed on symbols are analogous to processes in the mind;
3 *Compositionality*: Complex symbols consist of simpler symbols, and their function can be traced to their constituents;
4 *Specificity*: Symbols are system-specific representations, their nature and emergence depend on their native architecture;
5 *Interpretation*: Symbols are relational constructs with differential functions depending on their inner structure and the problem at hand;
6 *Compatibility*: The same problem can be solved by a different set of symbols depending on the actual nature of the architecture.

In the SMoM, the symbols are primitive elements that provide the necessary functionality to represent and manipulate relational structures. The system is capable of performing operations on symbol structures to compose new symbol structures.

The inclusion of non-symbolic data in SMoM makes it a hybrid system which can utilise various forms of statistical learning. Non-symbolic data has two distinct functions. They can represent explicit quantities related to the inner state of the system or outer state of the environment, which can give rise to emergent representations (e.g., neural networks). Alternatively, they act as metadata as an annotation for symbolic and non-symbolic data supporting and structuring the hierarchical processing within the architecture. This also means that the processing of a symbol typically involves the inclusion of non-symbolic data associated with that symbol.

The SMoM consists of the following main modules (core elements) (Figure 3.37) (Laird et al., 2017):

3.6.2.1 Central mechanisms

Working memory has a buffering function (temporal global space) bringing together inputs from all other modules needed to arrive at an optimal decision in the process of problem solving. These inputs arrive from the procedural or declarative memory and perception. This buffering function is important because the collection of inputs may depend on the state of other modules, and thus the working memory ensures that all data is there for a short time window before the action is decided and the next round of data collection starts. For example, data from actual perception should be co-analysed with the feedback on the just completed action, and this data should be there to calculate the best motor output of the system in terms of its future goals.

Procedural memory is concerned with actions including environmental context and selection and hierarchical plans for execution. Procedural memory contains rules that are pattern-directed activation of action sequences. This module provides input for the working memory where the conditions of the rules determine symbolic patterns over the contents and rule actions modify working memory. This buffering offers the combined inputs from declarative memory and motor actions in which metadata associated with specific rules affect the final selection of the motor output.

Declarative memory is the main long-term store for symbols and is represented as a graph of symbolic relations (see also Box 2.18). It also contains metadata that reflects various aspects of past symbol use. It may be useful to divide this module into a semantic and episodic submodule which may differ in dealing with more general experiences obtained over larger time periods ('facts') and representations of contextualised experiential experiences ('unique episodic impressions').

3.6.2.2 Perception

This module converts sensor inputs to symbols and relations and endows them with some metadata. The working memory provides the buffer for further evaluation of perceptual input where spatial and temporal limitations act as a selective process that determines the amount and nature of input to be retained (and lost). Perception has a bi-directional connection with the working memory making it an active process. This supports the manifestation of perceptual learning and also that actual perceptual input reaches the buffer ahead of the execution of actions.

3.6.2.3 Motor

This module transforms the symbol structures and their metadata obtained from the working memory into actions by utilising the low-level motor algorithms for the smooth execution of the behaviour.

Learning is the main procedure for creating new symbols and symbolic systems and takes place whenever the agent is actively interacting with its environment. This process also updates metadata associated with the old and new symbolic structures that are fed into the long-term memory system (both procedural and declarative). Learning also accommodates perception and motor skills; however, the long-term effects of these changes are also subject to a selective process by buffering working memory. This bottleneck can be circumvented by repeated learning events which increase the likelihood that experience reaches the respective long-term memory of the architecture. The process of learning is controlled also by the existing symbolic systems by providing some kind of feedback to the working memory. It is assumed that procedural learning may involve learning of new rules or changing the annotations of existing rules by making one more accessible under particular conditions. Similarly, learning may contribute to novel structures in declarative memory or modify the associated metadata.

3.6.3 *Extended Standard Problem-Solving Architecture (ESPSA)*

SMoM represents an important step forward in describing the modules and the general structure of an artificial architecture; however, it also has some important limitations from the ethorobotic perspective. These limitations originate from the starting notion that the SMoM is based on a consensus of expert scientists and uses the human mind as an etalon. In contrast, there are many fully functional non-human minds that may utilise divergent architectures, and the SMoM is dominated by 'cognitive' structures responsible for rational action, and less focus is given to other needs of the system to maintain integrity.

While going with the original idea put forward by Laird et al. (2017) for developing a general model, it may be useful to set up an extended version of SMoM, incorporating the insights of LeCun's (2022) system architecture for autonomous intelligence. Rather than referring to the complex notion of intelligence the model is named here as Extended Standard Problem-Solving Architecture (ESPSA) (see Figure 3.38) acknowledging the SMoM as a forerunner.

While the SMoM strives to mimic the human mind, LeCun's model does not differentiate human and non-human minds for constructing a general architecture. Furthermore, it contains a module that represents the preferences of the system itself that is especially important for embodied autonomous agents but lacks an explicit long-term memory. Another difference is that while the SMoM argues that a hybrid system of symbolic and non-symbolic parts is how 'cognitive' architectures for robots should be built, the central theme of LeCun's position paper is creating the architecture in such a way that it is as non-symbolic as possible, in order for the system to be differentiable and thus learnable with methods that have made neural networks such a success. In fact, it proposes very specific machine learning (ML) implementation details to this end, which we will not reproduce here.

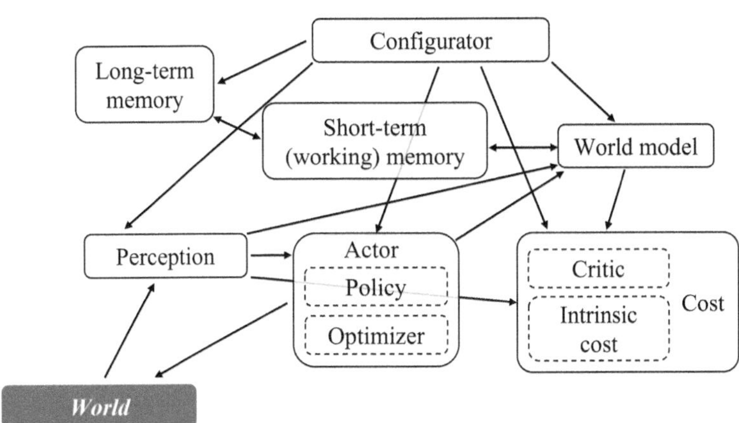

Figure 3.38 LeCun's system architecture for autonomous intelligence (redrawn after LeCun, 2022), extended with insights from Laird et al.'s mental model and thus renamed Extended Standard Problem-Solving Architecture (see text for details).

ESPSA combines the SMoM and LeCun's model but also elaborates on them, in particular the importance of emotions. ESPSA is also a high-level concept as SMoM and LeCun's model (although the former has actual implementations of varying complexity, see the Soar architecture in Section 3.8.5), and that in any actual realisation of ESPSA modules could have various submodules or may be missing. Possible variations within modules and their connections are also noted to indicate how complexity could change when agents (species) are exposed to different challenges in their respective environments. These differences are referred to as 'levels' which can emerge or disappear in specific architectures. Levels in different modules may form a strong linkage, meaning that one functioning of one level in a module may presuppose the existence of the other. Alternatively, levels in modules may emerge independently from each other. Finally, most models do not take into account that problem solving may occur at different levels of complexity. In the natural scenario, some animals or individuals have restricted mental capacities based on few alternative mental operations. This can come about because of their evolutionary or developmental status. Thus, it may be useful to differentiate tasks, which are short-term problem-solving activities, and projects, which are based on hierarchically organised task sequences with long-term outcomes.

3.6.3.1 Central mechanisms

The *configurator module* actively asks for input from all other modules in order to make a specific decision. Based on actual inputs from these modules several tasks are possible but, in each case, there is a need to run updates in order to decide which outcome provides the highest benefit in the actual context (see later). The configurator module is typically concerned with tasks rather than projects, but this may change if the complexity of the world model increases, and generation of projects becomes more feasible. Since the configurator modulates the entire system, it is inextricably tied to motivation but also emotions. If the system is under the effect of strong emotions, it is the configurator that aligns all modules to this emotion and it is also responsible for emotion regulation. For example, when the system is terrified, the configurator can limit perception to look only for signs of danger, the horizon of future state predictions to the immediate future, and selectable actions to ones congruent with a flight or fight situation, in order to focus compute power to immediate survival.

The *perception module* provides input on the current state of the world received from various sensors. Perception can be envisioned as a filter and a mixer of different kinds working in two directions. It filters out the majority of input because of processing constraints but also because of specific preferences of the world model and the configurator. The former is a relative standard and passive process depending on the hardware of the sensor. The latter, however, involves either environmental features with some significance obtained by experience or may be built-in preferences. Alternatively, the configurator may command specific environmental input and initiate attentive processes.

The *world model module* builds and updates a representation (a model) of the inner and outer environment that becomes a database (knowledge) needed for all kinds of problem-solving tasks. This model is then used to predict possible future world states as they may come about either by events in the environment or by the consequences of the agent's action. Just as in other such cases, the reliability of predictions is negatively associated with the prediction's time scale. This may also mean that differences among future choices are more elusive than differences for more closer alternatives. Importantly, much of the uncertainty originates from not knowing the state of the agent (in the future) when it has just finished the last tasks and starts to undertake the next one. This may also mean that very rapid functioning and decision making 'on the fly' may actually be seen (phenomenologically) as planning ahead. Long-term predictions need a lot of input data and computing space to maintain them as possible outcomes thus this kind of future modelling emerges only from larger architectures. It is more likely that in less complex environments the modelling of simple task sequences may suffice, while more complex environments can be overcome by generating projects (see the actor module). An optimally functioning system should ensure that these projects are flexible enough to be changed or merged at decision points if necessary. At a very high level of functioning, complex data (knowledge) and inputs allow the world model to generate future projections in a form of 'fantasies' that are simulations within an artificially created virtual submodule.

The world model works on several time scales in similar ways. It makes predictions for immediate tasks (scale of milliseconds) and also makes predictions for longer time frames for projects (scale of hours or more). Both kinds of predictions, which may be computed in separate submodules, rely on annotated long-term memory content as input data.

As the world model grows (as a result of development and experience), more and more latent variables may be involved which increases the instability of the system. Thus, the world model should have its own motivational state, expressed through the cost module, to decrease this uncertainty by gathering more information, for improving prediction. At the behavioural level, this is manifested in exploratory behaviour involving also attentive and monitoring processes (see Section 1.11.6). Exploratory behaviour may be initiated not only by discrepancy between perception and the world model but also it could be an inbuilt feature activated regularly while monitoring the environment. This kind of proactivity is a crucial feature in biological agents which, however, depends on the complexity of their ecological environment, and possibility that hazards endanger the stability of the agent. This kind of tendency of the world model could be supported by genetic predisposition to learn continuously. This may provide the mental basis of the 'curiosity loop' (Gordon, 2019) that collects information both about the environment and the self. These kinds of processes play a significant role during development (Oudeyer et al., 2007) (see Box 3.20).

From a practical point of view, the world model may use preferentially self-supervising algorithms (see below) for improving its predictive abilities (see Section 1.11.7) not excluding other types of procedures. The latter is achieved by creating a supervisory signal from the input data without relying on external feedback. The model's task is to predict some parts of the input data from other parts. The emergence of these supervisory signals could be supported by regularities which can be recognised relatively easily. For example, measuring the contingencies between

Box 3.20 Exploratory behaviour and curiosity play an important role in gaining new skills in social robots

Gordon (2019) presents an overview of the possibilities of using embodied curiosity in robotics for the creation of a hierarchical structure of behaviours autonomously. In the presented approach, a learning algorithm features a computational model of curiosity, the goal of which is to choose actions that maximise learning. This algorithm drives different actions depending on whether the agent is placed in a social or a non-social environment.

In a non-social environment, it was found that the agent learns about 'sensorimotor correlations', such as the relationship between motor commands and the image of a camera moved by that motor. On higher levels of learning, the agent was able to track its arm on the camera image, then learn the connection between the visual field and its arm's position, and be able to perform a reaching motion for an object. This involves multiple layers of learned behaviours, starting from motion detection, to mapping this motion to the camera's visual field, and being able to deliberately move the robot arm towards an object.

In a social environment, the learning agent attempted to predict the next image frame from the current one and started focusing on faces in the environment, since they provided the most unpredictable changes in the visual field, and thus the most learning opportunities.

A DragonBot robotic platform was used in an experiment to test the emerging behaviours in a situation where the robot was able to interact with people in the environment. This platform was able to perform a range of facial expressions corresponding to emotions, and its goal was to maximise input information for its learning. As faces convey the most information, the robot tried to have people's faces in its visual field for the longest time possible. This resulted in the robot presenting the 'sad' and 'crying' emotion for most of the time, as this kept people interacting with it for the longest (Figure 3.39a, b).

Oudeyer et al. (2007) created a similar, intrinsic motivation-based system for active learning of robots, named 'Intrinsic Adaptive Curiosity' (IAC). They describe the IAC as a system that tries to maximise the learning progress via driving the robot into novel situations, and steering it away from situations that are either too predictable or too unpredictable; this, in turn, results in preferred situations changing over time as the agent's learning progresses. To demonstrate the abilities of the IAC, a Sony AIBO robot was placed in a playing environment for infants along with a set of toys that could be bitten, bashed, or visually detected (Figure 3.39c). The robot's action space consisted of head movements, and bashing and biting movements that could be used to interact with the toys accordingly. During the learning process, the robot learned its sensorimotor correlations starting with the easier behaviours (looking around), and gradually proceeding in complexity to the more complex ones (bashing objects, which requires a correct orientation of the head as well as an angle and strength parameter) while

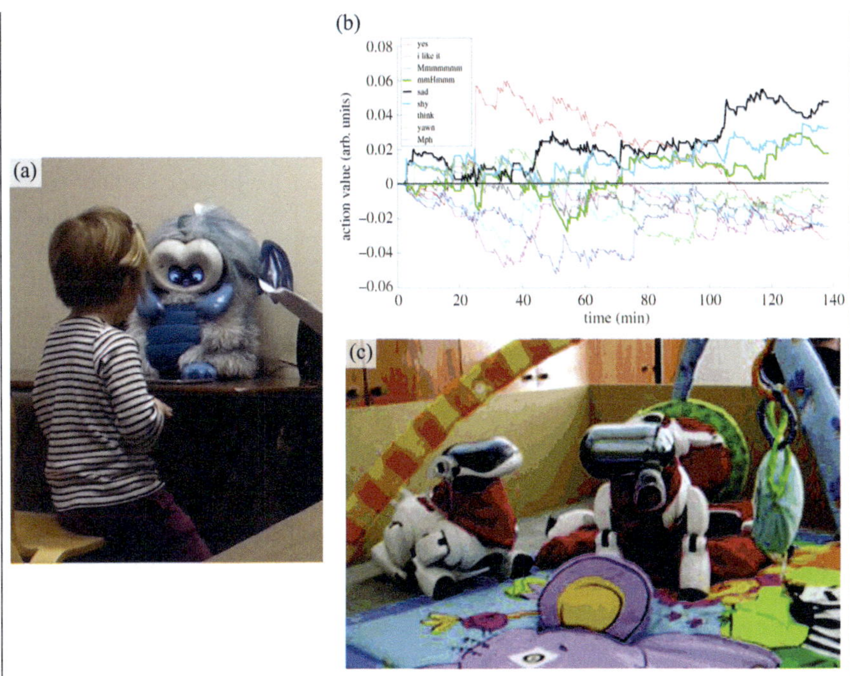

Figure 3.39 (a) DragonBot robotic platform is able to interact with its social environment (photo from Kory Westlund et al., 2017). (b) The robot's reward policy, aimed at piquing people's interest, has resulted in a preference for displaying a 'sad' facial expression (figure from Gordon, 2019). (c) The playground for AIBO shows exploratory behaviours towards the objects (photo from Oudeyer & Kaplan, 2006).

focusing only on one type of behaviour at a time. It has also been noted that the robot performed the more complex behaviours first in a so-called 'non-affordant' version, that is, trying to bite objects, which cannot be bitten, and only after that it learned to perform its behaviours with the correct targets, requiring a more specific set of parameters to be used.

The main contribution of Oudeyer et al. (2007) is that they were able to recreate two aspects of human development using their curiosity algorithm: progressive and incremental development, and an autonomous learning process.

actions and consequences may differ for 'own body' and 'other bodies' since the correlation is close to 100% in the case of the former and much lower in the case of the latter, allowing the discrimination of self and other (Bahrick & Watson, 1985). In the case of development, these learning processes could be repeated over and over again by refining the supervisory signal based on the new data collected (Fehér et al., 2017).

The *cost module*'s main output is a single scalar that measures the level of current deviation from a predetermined homeostatic level. The cost module's goal is to influence behaviour in a way that the system spends less energy to maintain its integrity over longer time scale, making it the main driver of behaviour. Following the models on motivation (Section 2.5.4) this module has two submodules.

The *intrinsic cost submodule* computes the value for biological motivational states that are involved in maintaining the short- and long-term homeostasis of the agent to maintain fitness. Biological agents typically share some of these states, like hunger or thirst, while others are more specific, e.g., sociability or exploration. Calculations regarding different motivational states should be made comparable similarly to the common currency model (Section 2.5.4). Specific parameters influencing the calculations in this submodule could refer to the temperament or personality (individual variation) of the agent (Section 2.4.10). The deviations from the homeostatic level in specific states can provide a possible initiation for specific tasks to be carried out.

The *trainable critic submodule* predicts expected values of the intrinsic cost in the future (1) to anticipate long-term outcomes, and (2) to allow the configurator to make the agent focus on accomplishing subgoals with a learned cost. Like the intrinsic cost, its input is either the current state of the world or possible states predicted by the world model. For training, the critic retrieves past states and subsequent intrinsic costs stored in the associative memory module and trains itself to predict the latter from the former.

Based on insights of living systems, the ESPSA needs *emotion states* to behave in a natural way (Section 2.5.5). LeCun (2022) envisions this as emerging from the cost module, as the product of the intrinsic cost and critic's estimate of future outcomes. Emotions may represent the indirect costs of actions usually referred to as positive or negative valence which are important features in action selection. Emotions associated with tasks or projects are important factors for preferring or avoiding situations in which the associated actions are typically carried out. For example, the function of negative emotions is to label situations that should be avoided. Actions with known negative valence are typically elicited by environmental changes, and not initiated. In contrast, agents seek actions or projects with known positive valence. Different forms of exploration serve to survey as early as possible (in the case of living organisms) the valences associated with actions executed. This procedure maps the tasks and projects to an area in an emotional space rather than to specific emotional categories. We also argue that the role of emotions goes beyond just costs driving behaviour but may also explicitly drive reconfiguration of the whole system (see configurator module).

The *short-term (working) memory module* keeps track of the current world state and also relevant past states and predicted future states. It also stores the intrinsic costs of these states. The short-term memory can be queried by the world model, which can also manipulate the contents of this memory. The architecture of the short-term memory must be such that retrieval and storage is efficient and flexible and is modulated by context. A candidate architecture could be something like a Key-Value Memory

Network. This was designed to be able to answer questions about documents, using a database of symbols and using an embedding of the queries and responses along with the database to answer questions (Miller et al., 2016).

While the world model also incorporates the history of the system through learning, its job is the modelling of the current state and the prediction of the future. In contrast, the *long-term memory module* stores the explicit history of the system. While the short-term memory is basically what is required for the system to operate in the current situation and is rapidly accessed and updated continuously, the long-term memory is a sort of cold storage for short-term memory snapshots in 'important' situations. If needed, these snapshots may be retrieved into the short-term memory.

The *actor module* is responsible for computing the next action or sequence of actions that will be carried out by the robot. At lower levels, the actor module may generate only a few simple tasks repeated over time based on small amounts of data and simple input from other modules. As the complexity of inputs and the world models increases the actor module may shift to generating projects (composed of tasks, see above). This change typically takes place in developing minds, and while it is currently missing from most architectures, it could play an important role in the emergence of a world model that is well-accommodated to solve problems in its specific environment. The actor module can have two submodules, corresponding to two modes it can take: a policy module and an optimiser module.

The *policy submodule* serves as low-cost reactive behaviour control, which corresponds to carrying out tasks that do not require much planning. The policy submodule takes as its input the current state of the world as estimated by the world model and produces a single action. The world module then predicts the state of the world in the next step based on the action, while the action is carried out by the robot. The difference between prediction and outcome can be used to adjust the world model itself.

Operating the *optimiser submodule* is more costly, but can be used to train the policy module for specific tasks, as an analogue of learning new skills with practice. In this mode, the optimiser module outputs a sequence of actions based on the world model's estimate of the current state. The world model then produces a simulation of the world state(s) as the consequences of the proposed actions. The cost of these simulated states is then estimated by the cost module. Based on this outcome, the optimiser submodule proposes a new set of actions, now with a lower cost and this is iterated over until the minimum-cost (or sufficiently low-cost) action sequence is found. From this minimum-cost action sequence, the first action is carried out and the world model (through perception) estimates the current world state after the action. The difference between simulated and current state and cost can be used to train both the critic and the world model, while learning the new skill is essentially running the policy submodule in parallel and updating its policy submodule, so it learns the optimal choices produced by the optimiser.

The actual functioning of the architecture may be driven by a framework designed on the principles introduced by 'ideomotor' (IM) (Shin et al., 2010) and test-operate-test-execute (TOTE) (Miller et al., 1960) paradigms. Pezzulo et al. (2006) suggested that the combination of these two paradigms could provide a possibility to design embodied agents solving problems efficiently. IM assumes that actions are driven by inner representations, and that there is a close matching between systems' representations of goals and internal representations activated by perception. This match is strengthened by experience gained through actions and that the achievement of goals is perceived by the system. This experience provides the basis for action selection. The TOTE mechanism is activated by the system noting a difference between the state of the environment and a goal state and then measures whether the selected action decreases this difference or not.

The combination of some aspects of both systems could be one possible non-exclusive mechanism that may explain interactions among the world model, the actor, and the cost modules (Box 3.21). The basic mechanism of TOTE, which is important to support the necessary exploratory (active) behaviour of the architecture, is complemented by IM's feature where perceptual goal representations, established on the basis of actions, trigger the associated actions in the future.

In summary, LeCun's model (2022) seems to provide a good general structure for including most modules that were postulated or actually developed by researchers working on the architectures of the mind. This or similar models may lay the ground for arriving at a common nomenclature that can be used from ethology to AI and also decide the acceptable level of anthropomorphism that does not constrain research.

Box 3.21 The main features of a Test-Operate-Test-Exit (TOTE) system (based on Pezzulo et al., 2006)

The main characteristics of a Test-Operate-Test-Exit (TOTE) system are outlined based on the work of Pezzulo et al. (2006). (a) The system functions by selecting an action when a disparity (represented by a double-headed arrow) arises between its current and desired states. It halts operation (exits) once this disparity is resolved (Figure 3.40a). (b) According to the Ideomotor Principle (IMP), action selection is guided by the anticipated sensory feedback. The initiation of an action is driven by a mental simulation of the expected future state of the environment (Figure 3.40b). (c) Pezzulo et al. (2006) proposed a potential integration of these two systems (Figure 3.40c). They suggest that both TOTE and IMP are somewhat underspecified, and a unified system may lead to improved practical operation. In this integrated system, goals are represented perceptually, and these perceptual representations of goals trigger associated action commands. The activated goal is continuously compared to the current perceptual input, facilitating the recognition of goal achievement. Achieving this requires the distinction between goal-related perceptual codes and actual perceptual codes, possibly through a tag-based mechanism.

The TOTE system can utilise action-effect rules akin to the IMP while still incorporating the test component and utilising mismatches for selection and triggering.

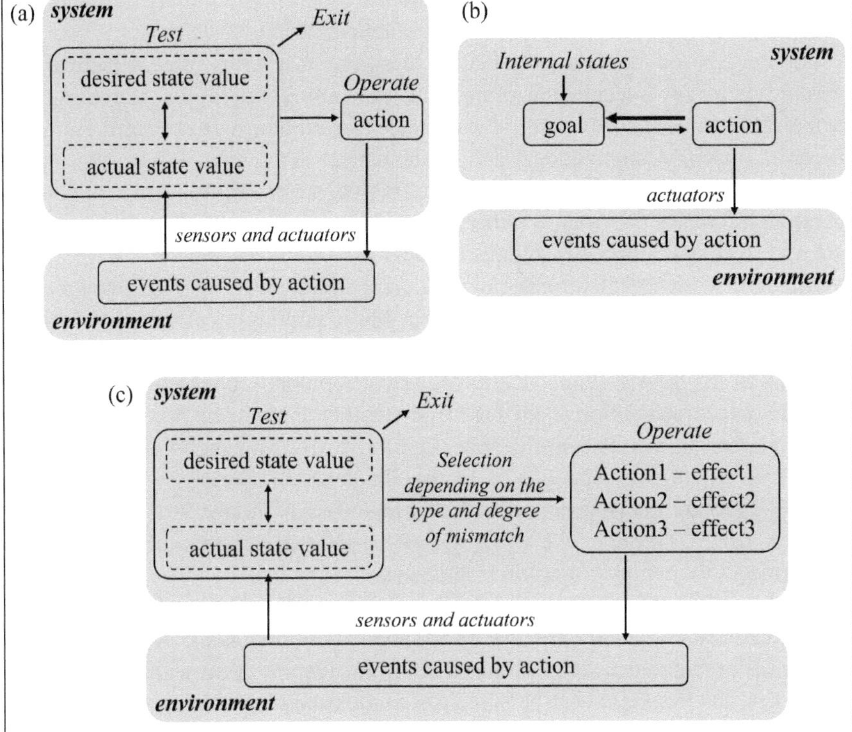

Figure 3.40 Schematic depiction of the (a) TOTE, the (b) IMP, and the (c) integration of the two systems (redrawn based on Pezzulo et al., 2007).

Although the specific processes of matching, selection, and triggering are not detailed, they can be implemented in various ways. This integrated approach was tested in tasks involving visual search or reaching actions.

3.7 Artificial intelligence and machine learning

3.7.1 Introduction

The term AI is commonly used to describe the creation of machines or systems capable of performing tasks that typically require an array of human-like cognitive skills. However, contemporary AI has the potential to surpass human abilities in specific functions, indicating a broader trend towards developing artificial agents with advanced problem-solving capacities (Russell, 2021). This perspective frames AI as a collection of theoretical and practical tools employed to construct specific robotic mental architectures (Section 2.1.1). In contrast, ML represents a distinct field in computer science dedicated to developing algorithms and statistical models that empower computers to execute tasks without explicit programming –an approach facilitating the realisation of AI (Jordan & Mitchell, 2015).

For the ethoroboticist, it is crucial to recognise that terms like 'X [something] learning' (e.g., reinforcement learning, transfer learning) within the AI context may evoke concepts rooted in cognitive psychology or ethology. However, these terms represent specific computational and mathematical techniques. Given our current limited understanding of how to replicate biological problem-solving abilities or learning, it remains uncertain whether today's ML concepts will play a role in future attempts to emulate these capabilities or not.

ML is the science of turning raw data into a mathematical (statistical) model. Some people even go as far as to claim that ML is nothing more than glorified statistics: although this is a gross exaggeration, it does help underline that we should not think of ML as something akin to cognition. A modern robot relies on ML in its AI (e.g., object recognition applied to camera outputs), but many parts, for example, any symbolic logic in its behaviour are not part of ML. This section is not aimed to provide an exhaustive catalogue of all the things called learning in this discipline because there are many text books on this topic (e.g., Burkov, 2019; Deep, 2019; James et al., 2021); rather, we aim to provide here only a short overview, in hopes that it dispels thoughts of 'magical black boxes'.

All forms of ML can be broken down into three categories: paradigms, models, and techniques. *Paradigms* are how information about our data is available, *models* are particular mathematical constructs that fit the given problem within one or more paradigms, and *techniques* can be thought of as tricks helping to enhance performance or make a solution feasible. Although this oversimplifies the process (not everything pairs with everything), given a situation, one chooses a paradigm, then a model and may or may not apply a specific technique to solve the problem (Shalev-Shwartz & Ben-David, 2014).

From a practical standpoint, every learning paradigm finds its origins in biology. Psychologists and biologists began studying learning in animals, including humans, by the end of the 19th century, and the formulation of key concepts such as associative and operant learning, habituation, and others began to take shape. As machines started to exhibit similar performance, adopting these terminologies seemed natural. However, especially at that time, little was understood about the exact biological (physiological and molecular) mechanisms underlying learning phenomena (Burkhardt, 2005). Since then, significant progress has been made through the deployment of sophisticated biological technology, enabling researchers to trace changes down to single cells (e.g., Ortega-de San Luis & Ryan, 2022). Despite this, biologists still have a long way to go in fully comprehending all the intricacies of the learning process. In contrast, while computers and robots may emulate the performance of biological organisms, the mechanisms they employ are often only superficially similar to those found in living beings. For example, as their name may imply, neural networks in AI could be similar to the neural systems of animals, in reality, they serve as distant inspirations. There exist significant qualitative differences between these systems, both in terms of the units involved and the nature of their connections. Nevertheless, it could be interesting to see the parallels in biological and artificial systems, so the relevant biological phenomena are mentioned as appropriate.

3.7.2 Paradigms in machine learning

ML can be roughly categorised into three basic paradigms, supervised learning, unsupervised learning, and reinforcement learning, depending on how information is available about our dataset (e.g., Emmert-Streib & Dehmer, 2022).

Box 3.22 Training and validating machine learning models

The goal of ML is to capture the statistical properties of how the data set relates to the labels, so that the model will be able to predict the label (in the case of classification) of novel data, i.e., data that the model has not 'seen' during training (Burkov, 2019).

To achieve this, the available data is split into two parts, a training set and a test set. The model is trained on the training set and its performance is evaluated on the test set. Training means that with an algorithm suitable for the model in question, the parameters of the model are tuned to minimise the distance of the models' outputs from the correct results. The simplest example of this is the ordinary least square method for linear models. The performance of the model trained like this can be measured in several ways, depending on the problem. An easy-to-grasp example for classification is accuracy, which is the ratio of correct outputs from all the outputs of the model. This splitting ensures that our model does not over-learn the particularities of our data sample, thus degrading its performance on novel data.

Usually, this train-test splitting is not done once, but randomly k times, and the result of the k splits is evaluated together. This method is called k-fold cross-validation that increases the robustness of the metric. This needs to be taken one step further, if we want to choose from several contesting models as the best model, since if we compare the accuracies of several models and choose the best from that, now our test sets are also involved in the choosing. In this case, before starting the training process (and any train-test splitting), we create a validation or 'hold-out' set, which is only used to evaluate the final chosen model's performance but is not actually used in choosing or training the models.

This is a straightforward concept, but there are subtle ways where one can go wrong. One way to go astray is by choosing the wrong metric for evaluating performance. Accuracy is a good metric in cases where the amount of data to each label is approximately equal. But this is not the case when the majority of the data belong to the same label. In such scenarios, the model always predicts the majority label with very high accuracy, rendering accuracy alone insufficient as an evaluation metric. Therefore, it becomes necessary to consider alternative metrics that are more appropriate for assessing performance in imbalanced datasets.

Another way to fool ourselves is inadvertently mixing our training data with the test data. For example, one is training a face recognition model. If every picture one has is from a different individual, one should go ahead and split the data randomly between train and test. But if there are multiple pictures of an individual, splitting randomly would leak the training data into the test data, instead of testing on novel data. In this case, during splitting between train and test, we need to make sure that the split is not among pictures, but individuals.

3.7.2.1 Supervised learning

In supervised learning, each data point is labelled by an outside actor (a teacher) as 'golden standard'. This actor is usually, but not always a human. An example problem is object recognition: we have a set of images, where each image is a data point (consisting of millions of numbers, depending on the resolution and whether it has colours) and labelled with whether that image is a dog, a cat, a human or a microwave, etc. This type of supervised learning is called a classification problem because the data is split amongst two or more, but a finite number of classes. The other basic type of supervised learning is regression, where instead of a label, a numeric value is assigned to each data point, essentially making the number of labels or classes infinite. Technically, labels in classification are also translated into numeric values, but they will come from a finite set of integers, while in regression they will come from an infinite set of numbers, like real numbers between 0 and 1 (Box 3.22).

The goal of supervised learning is to create a mathematical model of the association between the data and the corresponding label/numeric values. In the archetypical realisation, the model is initialised, which produces guesses for the labels/values of the data set, this is evaluated against the 'golden standard' via a cost function, which is a mathematical function that calculates a sort of distance between the current prediction and the 'golden standard'. The model's parameters are then changed with various strategies and evaluated again in order to find the parameter set with which the cost function returns the smallest distance. It is important to note, that although each iteration may inform the next iteration about the optimal parameters, the evaluation of a specific set of parameters is independent of any other set (Figure 3.41a).

In the laboratory model of classical conditioning, the organism learns the association between a neutral stimulus (e.g., sound) with a meaningful stimulus (reinforcement: e.g., food). In supervised learning, this process is akin to the model discerning whether a choice is correct or not based on the feedback (reinforcement) provided by the system. While the formation of an association is pre-established within the artificial system, organisms must learn about the possibility of such associations themselves, acknowledging their existence. Another difference lies in the robustness and definitiveness of the classification process in AI, which consistently provides binary feedback (yes/no), contrasting with the probabilistic nature of stimulus-reinforcement relationships in nature.

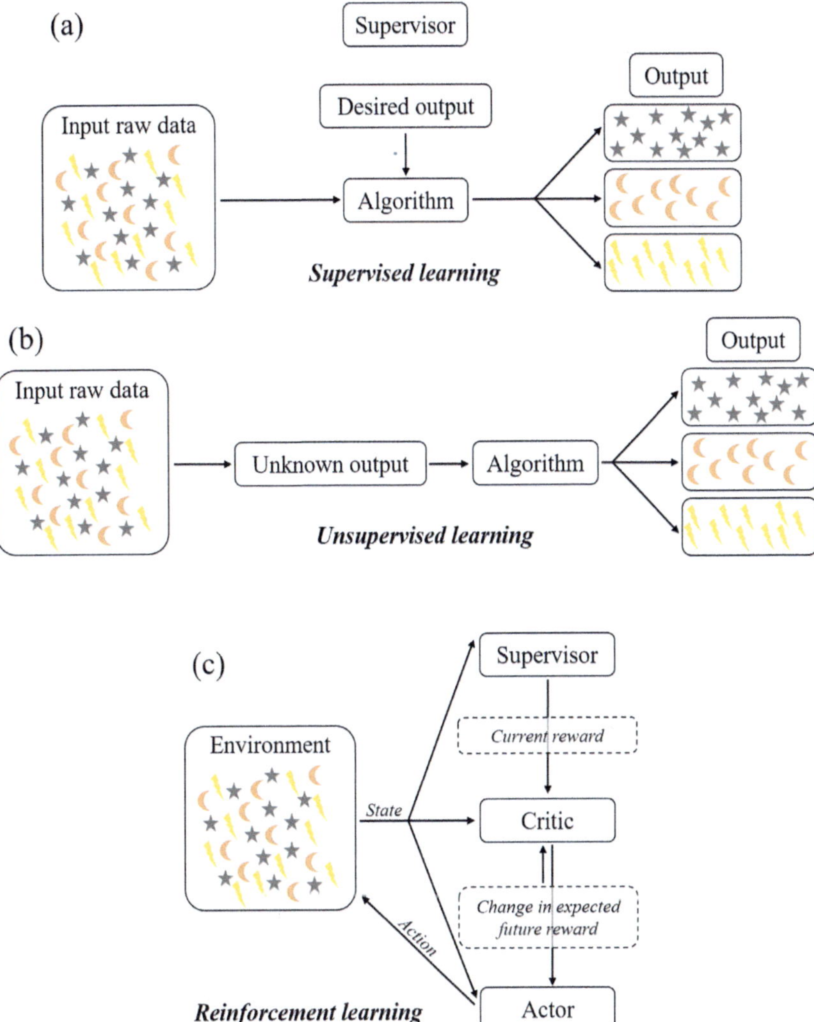

Figure 3.41 Simplified overview of the three machine learning paradigms (drawn based on IntelliPaat.com, 2023). (a) *Supervised learning*: the system tries to create a model of the relationship between the input data and the desired output, which is presented to the system by a supervisor. (b) *Unsupervised learning*: the labels or model outputs are not known, and therefore the task is to find patterns in the input dataset. (c) *Reinforcement learning*: focuses on states and possible behaviours of an agent. In each step, the agent takes an action based on its internal policy, the result of which gets evaluated by a supervisor giving reward or punishment scores. A critic evaluates the change in expected future rewards caused by a change in the state of the system, which changes the policy of the agent and the critic itself.

In both cases, the objective is to establish a specific relationship between a stimulus and its consequence, requiring the development of discriminative capabilities. However, the Pavlovian model only partially replicates natural scenarios, as its true purpose is not merely to associate, for instance, 'a sound' with 'some food', but rather for a fox to learn that a specific sound (e.g., mouse chirping) predicts a specific outcome (mice meat). In supervised learning, researchers typically determine the training sets, and to some extent, this is also the practice in animal training (by excluding all non-relevant stimuli). However, in nature, training sets are less defined, with fewer 'good' examples (stimuli) available for learning. To address this challenge in artificial learning mechanisms, additional mechanisms (discussed below) have been implemented to accommodate more realistic scenarios.

3.7.2.2 Unsupervised learning

In supervised learning, there is no outside feedback about the data, and the goal is to find inherent patterns, structures, or relationships within the data set. Typical use cases are clustering when the aim is to group similar data points together into clusters based on some similarity metric (i.e., trying to find n groups the data can be split into). Clustering is often used for tasks such as customer segmentation, anomaly detection, and image segmentation. Dimension reduction aims to reduce the number of features or dimensions in the data while preserving its important structure or relationships, i.e., trying to find the variables that characterise the data better than others, in order to eliminate some of the less useful variables. Dimension reduction is commonly applied for visualisation, feature extraction, and noise reduction. Association rule learning algorithms discover interesting relationships or patterns between variables in large datasets. These algorithms typically find sets of items that frequently occur together in transactions or observations. This approach is useful in market basket analysis, recommendation systems, and bioinformatics (Figure 3.41b).

From a biological perspective, unsupervised learning appears to mirror perceptual learning in animals, wherein they refine or enhance sensory processing mechanisms through repeated exposure to specific stimuli, thereby improving their perceptual abilities through experience and practice (Hall, 2009). In typical scenarios, animals extract relevant information during mere exposure to various sensory inputs that lead to better discrimination performance. It is assumed that perceptual mechanisms distinguish coincidental features unique to one type of input from features shared by the same inputs (McLaren & Mackintosh, 2000), establishing excitatory links in the former case and inhibitory links in the latter.

3.7.2.3 Reinforcement learning

Reinforcement learning operates with system states and possible actions of an agent, which can take the system from one state to another (the system meaning the environment of the agent). In the archetypical realisation, an agent, given the current state of the system (which includes both the agent and its environment), takes an action, which changes the system's state. It is important to note that the agent may have access to the full description of the current state of the system, which can happen in a simulation, but a robot mostly has incomplete information about the current state of the system. The agent gets feedback (reinforcement) as a consequence of the action and this is repeated in discrete timesteps. Reinforcements, in the form of scores, can be positive ('reward') or negative ('punishment'), which is used to update the agent's behaviour.

An example problem is learning how to play a computer game. This could be a classic arcade game, where the agent literally plays the game: it sees all the information a human player would (i.e., the state of the system, but only partially) and the reward is the current score. The possible actions it can take are the same as the human player could, for example, move left/right, shoot, etc. Learning would mean replaying the game as many times as needed, in each iteration the agent playing better and better.

In contrast with supervised learning, which can be thought of as a static problem, the goal in reinforcement learning is to maximise the long-term reward over several steps, instead of just a single evaluation (Figure 3.41c). There are many formulations of reinforcement learning to do this; however, here we refer to the actor-critic formulation as these are the terms that we have already seen in the ESPSA model (see Section 3.6.3) and indeed, they come from there. It was introduced to overcome issues in systems where the states and possible actions are infinite, which is exactly the kind of system a realistic robot is in. In this formulation, there are actually two separate learnings happening, in parallel. The actor is learning the so-called policy function, which is a mapping describing the probability of taking an action given the current state of the system, while the critic is learning the so-called value function, which estimates the maximal cumulative reward reachable from the current state. At each step, the actor chooses an action based on the policy function, which changes the state and the reward/punishment score is received by the critic. The critic then calculates how this change of state has impacted the estimated cumulative reward, based on the value function which is then used to update both the value and the policy function. The agent learns through trial and error, exploring different actions and learning from the rewards obtained. Exploration of new actions to learn is typically combined with known actions to maximise immediate rewards, with the balance between exploration and exploitation being a very important aspect: it is unfeasible to explore all types of state and action combinations, so the agent must optimise looking for better behavioural patterns versus using already known 'good enough' ones, to maximise the long-term cumulative reward.

Reinforcement learning is a useful paradigm to be deployed in a task of navigating a maze to reach a target location. The robot, equipped with a set of sensors (e.g., cameras), can move forward, turn left or right, or stop. Each action has consequences. Getting forward brings reward, while hitting obstacles or colliding with walls results in punishment. The algorithm iteratively updates the robot's policy based on the rewards received and the observed states and actions. Learning starts with the robot exploring the maze by taking random actions to discover the layout and potential paths to the target. As the robot gains experience and learns from rewards, it shifts towards exploiting the learned policy to navigate the maze more efficiently. During the training process, the robot repeatedly explores the maze, taking actions and receiving feedback in the form of rewards. The reinforcement learning algorithm updates the robot's policy based on the observed rewards and states, gradually improving its navigation strategy. Finally, the robot's policy may be fine-tuned or further optimised through additional training iterations, and for being able to accommodate changing environmental conditions.

The biologically corresponding system to the reinforcement learning paradigm in AI is called instrumental learning or operant conditioning (Rescorla & Solomon, 1967; Shettleworth, 2010). It has been argued that this fundamental learning mechanism serves to teach the organism the consequences of specific actions, which can be positive or negative (or neutral, that is, non-negative in practice). Unlike typical robots, animals possess a wide range of actions, making it crucial for them to discern when to act and when not to. Instrumental learning also aids in acquiring complex actions. Action sequences can be learned incrementally until the goal is achieved (forward chaining). Alternatively, the animal may first acquire the final goal-directed action and then learn additional preceding actions leading to the successful outcome (backward chaining). In natural settings, especially during development, backward chaining is likely more prevalent, as organisms have a higher chance of receiving reinforcement. Forward chaining carries greater risk (in terms of energy cost), as many attempts (trial and error behaviour) to gain reinforcement may prove futile.

The study of foraging behaviour in animals resembles reinforcement learning described above for robots (Parker & Maynard Smith, 1990). The notion is that behaviour evolves to minimise costs and maximise energy intake. These basic assumptions impose significant constraints on organisms, but there are numerous behavioural solutions depending on species (c.f., behavioural repertoire) and actual environment (c.f., food distribution). Thus, it is assumed that on average animals behave 'optimally' when searching for food and they are inherently designed to maximise the cost (value function evaluated by the critic, see above). In parallel, they are also learning specific behavioural policies, that is, the actor learns to carry out the actions that have the highest probability to lead to the goal under specific conditions.

It is important to recognise that reinforcement learning mechanisms form the foundation for goal-oriented behaviour (Section 2.5.6). Similar to living organisms, social robots may also possess multiple goals that they need to achieve at specific points in time. Therefore, information acquired through reinforcement learning mechanisms must be integrated into a hierarchical architecture (see below) that addresses the temporal and spatial organisation of behaviour. This is essential because, in most scenarios, an agent can only be engaged in one action at a time.

3.7.3 Models

Models are specific mathematical formulations of ML problems (e.g., Burkov, 2019). Some models only make sense for one kind of paradigm, while others can be used for multiple purposes. This section does not intend to be a complete and thorough list of models; rather, we have chosen to showcase some models either as an enlightening example or because they are widely used in robotics contexts as well.

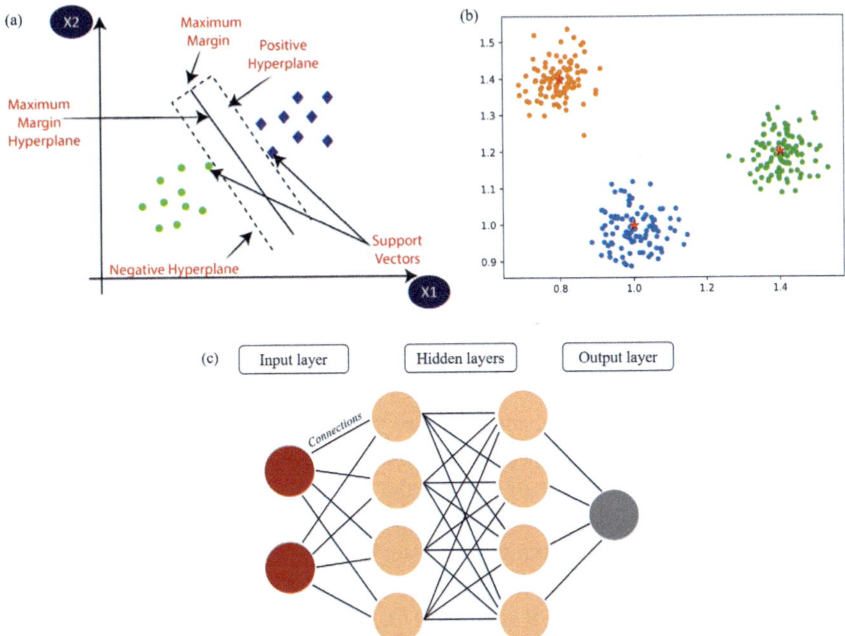

Figure 3.42 (a) The support vector machine model in supervised learning tries to find the hyperplane that separates different classes of data points the best (figure from Das et al., 2023). (b) The K-means clustering model tries to find k number of clusters, by minimising the total within cluster distances (the coloured dots are the actual data point, while the red stars are the calculated cluster centres, showing a k=3 scenario). (c) Neural networks are a highly flexible set of models, as the number of nodes and layers, and connections between nodes can all be changed according to the exact modelling needs.

3.7.3.1 Linear regression (ordinary least squares regression)

When comparing ML to statistics, linear regression is a frequently chosen model solution. The model – well known from statistics of fitting $y=a*x+b$ with ordinary least squares for a trend line – can also be utilised for supervised learning (regression), although the exact usage (and goal) differs slightly. In statistics, one fits the line on all of the data, evaluates the goodness-of-fit on statistical parameters, and aims to draw conclusions about the population the data comes from. In ML, one fits the model on multiple subsets of the data, evaluates the predictive power on other subsets (known as cross-validation, see Box 3.22), and aims to find the parameters maximising predictive power on future (yet unknown) data from the population.

3.7.3.2 Support vector machines (SVMs)

Although SVMs are usually used for supervised learning, they can also be used for unsupervised learning and reinforcement learning as well and are often used when the data has a high dimensionality (i.e., data is described by a large amount of variables). For the sake of the argument, let's imagine one wants to train a binary classifier on 2-dimensional data (Figure 3.42a). The data can be plotted on a 2-dimensional graph and each coloured according to their class. The idea behind the SVM is that it looks for the line between the points that maximises the margin between the two groups. Since each data point is represented by a 2-dimensional vector, the points closest to the line (on the margin) are called the support vectors. In 3 dimensions, the line becomes a plane, and in more than 3 dimensions one calls the same concept a hyperplane: the underlying mathematics is the same regardless of the number of dimensions.

3.7.3.3 K-mean clustering

K-means clustering is a model for unsupervised learning, with the goal of finding clusters (non-overlapping groups) in the data (Figure 3.42b). The 'K' means the user needs to predefine the number of clusters the algorithm is looking for. The model training then entails assigning each datapoint to one of the k clusters in a way that the total distance of points from their cluster's centre is minimal over all clusters (technically, the square of the distances is minimised). Unless we have a solid theoretical backing for the number of clusters to find in the data, we must also explore the number of clusters. This is usually done by training the algorithm multiple times and in each iteration increasing the number of clusters and stopping when adding a new cluster does not drastically decrease the average distances of the points from their cluster's centre.

3.7.3.4 Neural networks and deep learning

Although neural networks were already investigated in the 1960s, their resurgence happened much later, when the computational power required to use them efficiently became available. Neural networks draw inspiration from how nervous systems are organised (Libedinsky, 2023), but it is important to note that only in the most vague manner, i.e., they are usually not in any sense a model of a nervous system.

A neural network consists of layers of nodes ('neurons') where nodes between subsequent layers are connected (Figure 3.42c). The connection means that a node takes input numbers from other nodes, applies a function to these inputs and the output number is then passed onto a subsequent node or nodes. The first layer is called the input layer, the last layer is the output layer, and any layers in between the two are called hidden layers. For example, a neural network designed to decide if grayscale images contain dogs or cats (so two possible outputs) will have the input layer consist of as many nodes as there are pixels in an input image and an output layer consist of a single node outputting a number between 0 and 1. If the output is closer to 0, it is a cat and if closer to 1, it is a dog. The 'magic' happens in the hidden layers where the number of layers, the number of nodes in a specific layer and the exact topology of the connections can get extremely varied depending on the goals and available computation power. This gives great flexibility to building models for various tasks.

Many types of ML one reads about fall into the category of neural networks. For example, deep learning means that the neural network used is deep, that is, there are more than two hidden layers, with some networks having over 100 layers. Large Language Models (LLMs) (e.g., ChatGPT or LLama) and Large Multimodal Models (LMM) (similar to LLM-s, but they can handle other types of media besides text, like images and sound together with text) are also neural networks, albeit substantially more complicated. The size of these models is usually measured in the amount of parameters they have. For example, a state-of-the-art image classifier has parameters on the order of 10–100 million, while LLMs have 50–100 billion parameters (Box 3.23).

Box 3.23 Explainable and interpretable AI?

Unlike statistics or modelling as classically understood in natural sciences, ML in general is not concerned with either the explainability or interpretability of a model. A model is explainable if given an input, we can explain how the model arrived at a given output, and it is interpretable if it allows for extracting relevant knowledge from the model about our data and in consequence, the world. Although the interpretability and explainability of ML models vary, the goal of ML is to build models that perform well on novel data, essentially treating models as black boxes.

Consider for example an experiment where an object is subjected to a series of different forces (F) and its acceleration (a) is measured. We fit a linear regression model on the data with ordinary least squares regression: $F = m * a$. The resulting model is explainable, given any F, it is now easy to see what a will be. The model is also interpretable, the parameter m can be interpreted as a physical property (the mass) of the object. A linear regression model only has two parameters ($y = a * x + b$) and we know that the constant b must be 0, since if there is no force, there is no acceleration, making both explaining and interpreting easier. In contrast, YOLOV's deep learning model for object recognition has 70 million parameters! Explaining or interpreting models of this size is no easy feat and an active area of research unto itself.

An interesting use of neural networks that may be of interest to the ethoroboticist are encoders, which are used to compress information in the input into a more condensed, 'relevant' representation. The easiest to picture this is with autoencoders, where a neural network is thought to reproduce the input as close as possible. Although at first sight, this is a trivial task (just don't do anything with the input), the trick is, that although the input and the output layers has the same amount of nodes (the necessary amount to represent the entire input), in the middle between the two there is a layer with a much smaller amount of nodes, called the encoder layer or bottleneck layer. Let's imagine that we are teaching such a network to reproduce faces of people. The input layer is say 3 million nodes (not particularly high-resolution pictures) and we have the bottleneck layer with ten times less neurons. Since the bottleneck layer does not have enough nodes to represent the original picture in a copy-paste manner, the network must learn to represent the image in a more condensed way. If we give enough faces to this network and train it so the output images (also 3 million nodes) are fairly accurate, then this means that the condensed representation in the bottleneck layer accurately captures the meaningful details. Now if we give a new face to the network and we calculate the layers up to the bottleneck layer, we have a condensed representation of the face, which takes up less space, and can be used to check how close two images of faces are to each other. This encoding or bottleneck layer is now a non-symbolic representation of that given face.

As mentioned above, as appealing as it may look at first glance, such an encoding or in general the information encoded in a neural network, is mostly alien compared to a biological nervous system. One way to showcase this is by adversarial inputs. For example, adding a small amount of linear noise to an input image, a neural network trained to classify objects is completely fooled, while a human observer will not even notice the difference. It is also possible to train adversarial models, whose goal is to fool other models. In an image classification scenario, it could be trained to produce input images that are classified as dogs by the target network, while it clearly looks like a car to a human observer, or generate a seemingly nonsensical image that is still classified as a dog with a high probability.

3.7.4 Techniques

Techniques are neither paradigms nor specific models; rather, they are 'tricks' to solve specific technical issues. Many of them are related to problems such as not enough data, not enough labelled data or how expensive it would be to obtain them (Figure 3.43). Since these techniques are mostly inconsequential regarding our general topic, we only cover a select few, to show how techniques are different from models and paradigms.

3.7.4.1 Self-supervised learning

In self-supervised learning, the learning system generates the labels for itself in some way, so we have a supervised learning paradigm, but the target does not have to be

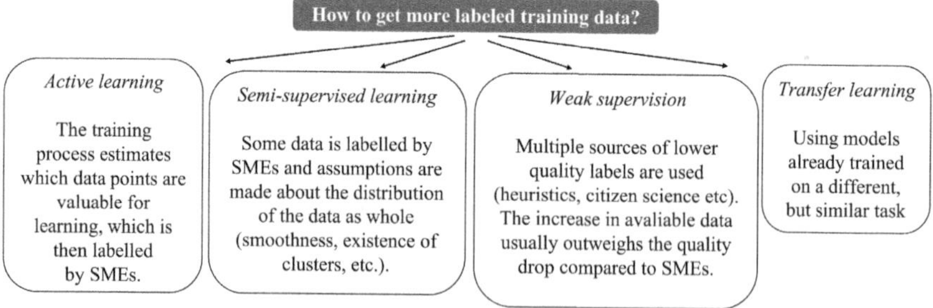

Figure 3.43 In traditional supervised learning all the data is labelled by subject matter experts (SMEs) (drawn based on Stanford AI Lab, 2019). Depending on the problem, this can be very expensive and time consuming, or maybe even downright impossible. Fortunately, there are some techniques to get around this issue.

created by humans. One example of this is how LLMs or in-painting AIs are taught. Given a corpus of text or a set of images, one can delete a short part or piece and use the rest as input. The task of the algorithm is to fill in the blanks thus created and the output can be directly compared against the original. Similarly, in autoencoders, the input and the target are the same.

Self-supervised learning can also take the form where various objects are explicitly labelled by an automated system. This makes sense if we have data in multiple modalities and for one of the modalities, there is already an acceptable classifier. For example, a robot's lidar sensor data could be labelled by using a camera simultaneously to look at the scene. Using a top-tier object recognition algorithm the lidar data can be approximately labelled.

3.7.4.2 Active learning

In active learning, only a subset of the data is labelled by humans manually, with the help from the training system, in order to reduce the cost of labelling (in both time and money potentially). The training starts with a small amount of already labelled data and the training procedure itself actively suggests data points to be labelled by subject matter experts. This allows for channelling the labelling efforts towards data points that are more valuable for the learning process. The idea here is that not all data points are equally important for training, if we think back to how SVM-s work, in theory, labelling only the support vectors themselves could be enough (although, obviously it is not trivial to know which these are beforehand). In practice, this could mean that from among millions of images, the system would actively flag edge-cases for a human to evaluate, e.g., in an autonomous driving scenario, it would flag images where reflections on vehicles also show vehicles and the human would pick out actual obstacles and reflections.

3.7.4.3 Transfer learning

In transfer learning, an already trained model is reused on a similar problem. For example, an already existing image classifier trained on a large amount of general images (e.g., recognising everyday household objects) is reused in order to classify a much smaller, specific dataset, which has labels that were not trained for in the original model (e.g., discriminating different types of apples). In neural networks, this can be achieved by retraining the model's last few layers and changing the output layer, the idea being that the layers before these have some general encoding of images. In this way, even though the smaller dataset (of apples in our example) in itself would not have been enough to produce a good-performing model, one can get good results.

3.8 Architecture implementations

In the previous sections, we have covered theories on how AI in high-functioning robots should be organised, without addressing implementations. This section showcases some of the most widely used implementations in AI and robotics including an example implementation of the SMoM. The origins of these architectures are as varied as the fields that are involved in robotics, thus showing slightly different facets for very similar things, with similar concepts cropping up. For more details, the reader is advised to look for more technical references (e.g., Correll et al., 2021; Siciliano & Khatib, 2008).

For a robot to be functional it requires a hierarchy of architectures: it needs power for moving, sensing and computing parts, microchips and computers to control everything, all of which need to be able to communicate with each other in an engineering sense, before any kind of high-level problem solving can commence. Hierarchies of abstractions are extremely useful: once complicated motor control is solved, any downstream application/module only needs to handle motor control to, e.g., go forward with 1 m/s. Just as most software engineers do not write the drivers (software responsible for communication between the hardware and the operating system) for their computer, roboticists usually also do not want to design motors from scratch and write their drivers but rather look for pre-made software. This also means that a single robot may also have a multitude of the below architectures running in its software each controlling a different sub-component.

From a purely engineering viewpoint, one can imagine most robots as essentially a computer (or computers) with various peripherals plugged into it. The computer must run an operating system, which allows running the software constituting the robot's behaviour and also all the drivers necessary to interact with the peripherals. Thus, the practising roboticist should either be prepared to write a lot of software or should choose an already existing solution that handles most of the low-level control. If extremely precise control of timing is required, the operating system must be a real-time operating system. Most operating systems are not like this and for an everyday social robot this may not be necessary, but for example in autonomous cars, that are expected to function at high speeds, this would be important.

A currently rather popular solution is the Robot Operating System (ROS). ROS is not an operating system, but rather it is better called a middleware as it runs atop an actual operating system (usually Ubuntu Linux, which is not a real-time operating system). It is free (even for commercial use), open source, has wide driver support for hardware often used in robotics, and comes with many utility programmes that are useful in any type of robotics problem. ROS is mostly programmed in either python or C++, with the former widely considered an easy language to pick up. Obviously, any specific system chosen also comes with constraints, but ROS can be used as a basis for any of the architectures discussed here.

3.8.1 Emergence and hierarchy as a property of complex systems

There is a popular paraphrasing of something Aristotle wrote in his *Metaphysics*: the whole is more than the sum of its parts and is often cited when explaining emergent behaviour in complex systems. But what are complex systems and what is emergence (see also Section 2.5.6)?

To understand, it is best to contrast a complex system (not to be confused with a complicated system: a complex need not be complicated) with a simple one. In a simple system, a single cause has a single effect and a small change in the cause always leads to a small change in the effect, hence a simple system is predictable. In a complex system, a single cause may lead to multiple effects, small changes in cause may lead to dramatic changes in effect, the system may have self-reference and self-similarity making cause and effect harder to untangle, and all these lead to unpredictability and emergence: the phenomenon where the rules describing how individual components of the system behave lead to new, collective behaviours of the system as a whole. A very aesthetic example of this is how starlings flock: a simple algorithm of each bird trying to match their few nearest neighbours' velocity reproduces the beautiful aerial display, which looks more like a single flying creature than thousands of individual birds (Ballerini et al., 2008). In this example, there is no hierarchy within the constituents, but often complex systems are also hierarchical: interaction of cells make up the body, but these cells are organised into organs within the body. Indeed, it seems organising systems hierarchically naturally makes them more efficient and is thus likely to appear in complex systems (Zafeiris & Vicsek, 2018). Neural networks are also good examples of complex, hierarchical systems with emergent properties. Nodes are locally connected to each other by very simple mathematical functions, yet the system as a whole is capable of deciding whether a picture shows a cat or dog.

Unpredictable and emergence here does not mean that it is unknowable, or unstructured; rather, it means that looking at the level of individual constituents, predicting the exact behaviour of the constituent, and seeing how the entire system behaves is very hard. But looking at the system as a whole it may still produce easily describable patterns. Just as in a balloon, the individual particles of air move randomly according to their energy, the air as a whole in the balloon has very clear and deterministic properties of pressure, volume, and temperature.

For the robot builder, this also means that as their robots get more and more complex, the robot will exhibit emergent behaviours that were neither planned for nor anticipated (bugs). Even a simple reactive architecture (see below) can produce emergent actions, if it has enough rules and modules running in parallel.

3.8.2 Architectures from computer science

The origin of these architectures was the need to organise and control software behaviour in general and are widespread in software development (Correll et al., 2021; Siciliano & Khatib, 2008). As such, they can also be readily used to organise and control a robot's software.

3.8.2.1 State machines

A state machine is a simple architecture for modelling or defining the dynamic behaviour of a system (e.g., Correll et al., 2021). As the name implies, it consists of discrete states, which refer to the internal status of the system. In a state machine architecture, a system must be in exactly one state at any given time during its operation. Between the states, so-called state transitions can be defined, which declare how the system

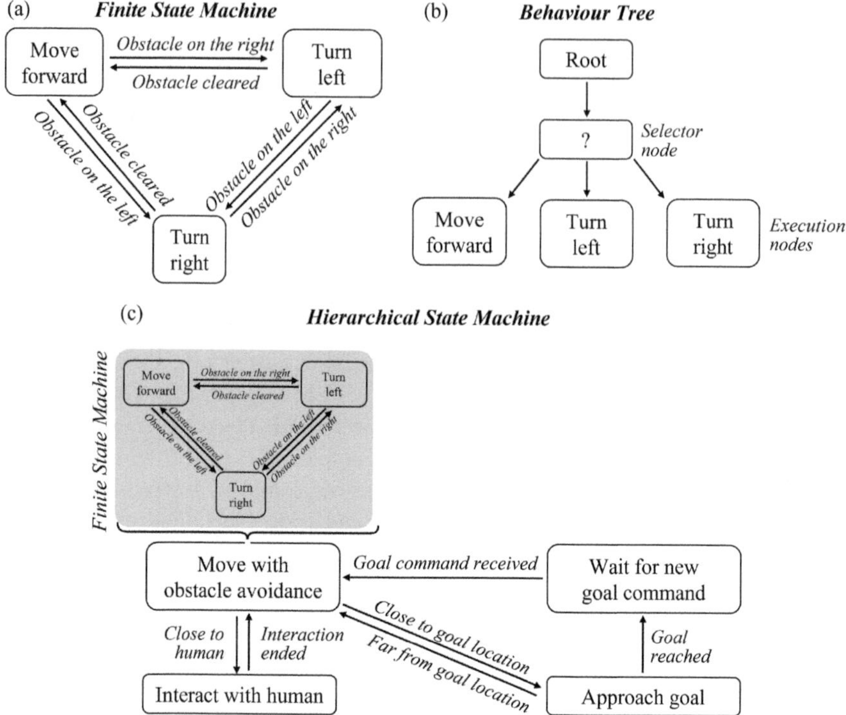

Figure 3.44 Architectures from computer science: (a) finite state machines, (b) behavioural tree, and (c) hierarchical state machines (see also text).

gets from one state to another, and what action is performed when it does so. Each transition is dependent on the currently received inputs and the current state of the system, meaning that the same system with a different state can react differently to the same input. It also means that states can pass memory to each other, even if it is just the identity of the previous state. The transitions can be visualised in a state-transition table containing the relationships between the inputs, current states, and following states. Another common way for visualisation is a so-called state diagram, in which nodes represent states, and the connecting arrows represent the possible state transitions.

An example of using a state machine in defining robotic behaviour is a simple algorithm for obstacle avoidance (Figure 3.44a). In this case, the state machine can have a state for moving forward, and two states for turning its body left and right, respectively. A robot with such a state machine would be in the moving forward state until it encounters an obstacle, when it would transition to one of the turning states, depending on which side the obstacle was detected. After the obstacle is cleared, the state machine would transition back to the moving forward state to continue moving along.

In practical implementations, the term 'state machine' usually refers to a finite state machine, which has an arbitrary but finite number of states. State machines can be used for designing systems in which each distinct state is well defined. Their main advantage is that their core concept is intuitive and easy to understand, and therefore can usually be easily designed and implemented. However, as the modelled system grows in complexity, it quickly increases both the required number of states and the number of state transitions, and adding or changing functionality requires all related transitions to be re-considered, causing scalability issues. Due to this, creating complex decision-making logics using a simple state machine architecture can be rather difficult. This architecture also puts limits on the reusability of the created structures, as states may have each transition specifically designed for the actual use-case.

Models on animal motivational systems may represent an analogy to finite state machines described above. Here the states are represented by specific needs such as 'hunger' or 'thirst'. External stimuli, such as the sight or smell of food or water, as well as internal cues like hormone levels, can trigger transitions between these states. Once a certain state is reached, the animal may exhibit corresponding behaviours, such as seeking food or water or switch to different actions (see Section 2.5.4).

A drawback of the simple finite state machine architecture is that it suffers from what is called 'state explosion'. This term is used to refer to the phenomenon occurring when a system becomes more complex, and that results in a high and unmanageable increase in the number of states and state transitions to the point when the architecture becomes too difficult to construct and control.

Hierarchical state machines propose to solve some of the shortcomings of regular finite state machines by increasing modularity (Correll et al., 2021). They introduce hierarchical relationships known from software development into the field of state machines, thus creating higher and lower-level state machines. The core idea is that in this architecture, a lower-level state machine can be nested into another, high-level one, and be used as a state in the high-level state machine. It is possible then to visualise the whole architecture on different levels, and abstract away the inner workings of lower-level state machines, which makes the changing and extending of functionalities easier. In this structure, low-level state machines can be treated as standalone state machines, and be reused in other contexts, offering better modularity and reusability. With hierarchical state machines, the designer of the robot behaviour only has to consider the state transitions at a given level in the hierarchy, as low-level state machines appear as single states from the higher levels in the hierarchy.

Using the previous example from finite state machines, the whole state machine of obstacle avoidance could be placed into a low-level state machine and used by higher-level components as a package. For example, the high-level navigation of the robot could have a state machine on its own, and one if its states could be called 'obstacle avoidance', containing the simple three-state obstacle avoidance state machine described in the previous section (Figure 3.44b). This approach reduces the number of states on a given level of abstraction, (1 vs 3), and allows for separate development of this module without concerning other parts of the system.

The modelling of animal motivational systems faces similar problems because of the presence of complex and intertwined needs that may also reflect different levels of significance with regard to integrity of the animal. Maslow (1943) suggested that motivations are structured in a hierarchical way, and predicted that higher-order needs (e.g., self-esteem) are fulfilled only if low-order needs (e.g., hunger) are already met.

3.8.2.2 Behaviour trees

Behaviour trees originate from the video game industry, where they were initially created for modelling the behaviour of non-player characters (NPCs) (Correll et al., 2021; Iovino et al., 2022). When contrasted with state machines, they present a different approach for defining behaviours of agents: while the state machine architecture is designed based on the states of a given system, behaviour trees focus on actions or tasks instead, although it is important to note that hierarchical state machines are obtainable as a special case of behaviour trees.

The name of behaviour trees comes from their visual representation: they are usually drawn as a tree having a root and two main types of nodes: control flow nodes and execution nodes (Figure 3.44c). Relationships between nodes can be referred to as parent-child relationships, and nodes as parent or child nodes. The tree is structured such that the root of the tree has no parent, and every other node has exactly one, while the execution nodes have no children, and the control flow nodes have at least one each. Execution nodes therefore are placed at the ends of the tree, and they represent the tasks or actions done by the agent. These nodes can return with *success, failure,* and *running* outcomes, representing the status of the execution of the task.

Control flow nodes control the execution of their children nodes and allow for structured execution of tasks and the creation of decision-making logic. The implementations of the behaviour tree can differ in what kind of control flow nodes are defined, but the two most common types are the selector or fallback node and the sequence node. Both nodes execute their children nodes one by one from left to right until a certain condition is met. The selector node executes until it finds a node that does not fail (i.e., returns *success* or *running*), while the sequence node executes until it finds a node that does not succeed (i.e., returns *running* or *failure*). In practice, this means that the selector node chooses the first non-failing node with priority given to the nodes to the left, while the sequence node executes its children from left to right, one after the other, until all the nodes have succeeded or one of them fails.

The execution of the behaviour tree is coordinated by so-called ticks, which are signals allowing the nodes to execute and are sent from the root periodically at a predefined frequency. The nodes return their outcome values on receiving a tick, and the control flow nodes direct the ticks to the child node they want to execute.

A main advantage of behaviour trees is their inherently modular design. Since transitions do not have to be defined for the nodes themselves, nodes can be added, deleted, or moved to different locations in the tree easily. This also applies to larger subtrees that can implement complex functionality and can also be used to extend a behaviour tree without having to reconsider several other nodes with all their transitions. This allows modules to be easily reused in different contexts, while the remaining parts of the tree can be left unchanged.

As a behaviour tree can be decomposed into smaller subtrees, it has a hierarchical structure by default, scales better than finite state machines, and offers greater readability of the created behaviour. The drawback of behaviour trees is that they require more initial planning and setup compared to state machines, as the developers need to ensure a correctly structured tree even for a relatively simple behaviour. However, behaviour trees are more scalable and modular than state machines, they are gaining popularity in robotics, with some implementation libraries being openly available.

The simple obstacle avoidance algorithm introduced in the previous examples could be implemented as a behaviour tree having a selector node, with children nodes of 'move forward', 'turn left', and 'turn right' in this order. First, the selector node directs its ticks to the move forward node which gets executed as long as it does not fail. If it fails (e.g., because of an obstacle), the selector node would direct the ticks to the next child node, which would be the 'turn left' node in this case. Should this node fail (because of the obstacle being on the left side of the robot), the last node would start executing, thus rotating the robot to the right, until it performs a complete turn and clears the obstacle. As the ticks are sent from the root of the tree with a given frequency, the execution nodes are evaluated periodically, and the selector node can choose the first child (move forward) as soon as it returns with a 'running' value instead of failing.

There exists some parallel between the behavioural tree architecture and an early model of behavioural organisation proposed by Tinbergen (1951). The hierarchical model assumes nested levels of behavioural organisation with specific actions forming the lowest level of actions. The highest level is typically a specific behaviour function (e.g., reproductive behaviour) that controls sub-functions or local problem-solving behaviours (e.g., territorial behaviour, nest building, protecting territory, etc.). Finally, each problem-solving behaviour is composed of several actions., such as nest building involves boring, material collection, etc. Activities at one level are typically mutually exclusive, the animal can do only one of them at a time. All levels and activities are under inhibition, and specific environmental inputs (typically sign stimuli) trigger the appropriate action if the higher level has also been activated. The main concepts of Tinbergen's model have been incorporated also in robotic architectures for mobile and interactive robots (Arkin, 1989; Cañas Plaza & Matellan Oliveira, 2005).

3.8.3 Architectures from control theory

Control theory is an engineering field dealing primarily with the control of machines and thus the physical movements of robots (Matarić, 2007). It is an extensive field in itself, but here we only briefly cover some of those architectures that tie in more closely with what has been discussed so far.

3.8.3.1 Reactive control

Reactive control is a very simple, albeit very useful concept (Figure 3.45a). In this setup, a reactive module directly ties sensory input to specific actions carried out by the robot, e.g., if the distance sensors show something closer than 3 cm-s, stop.

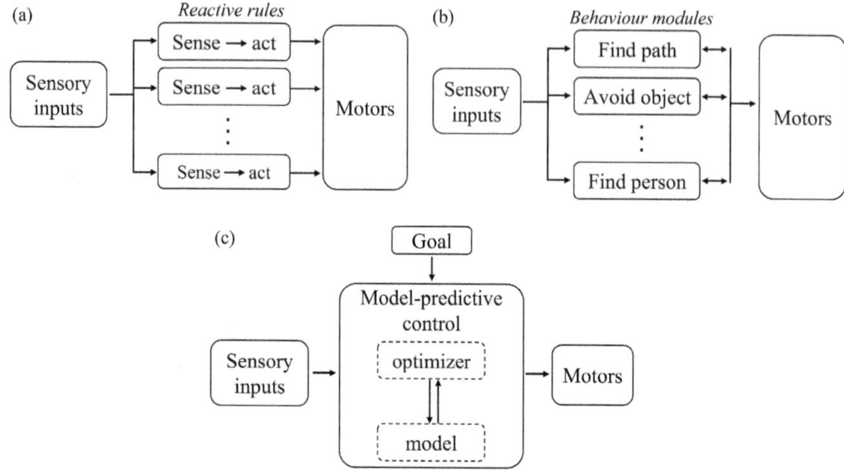

Figure 3.45 Architectures from control theory: (a) reactive architecture, (b) behaviour-based architecture, and (c) model-based architecture (see also the text) (based on Matarić, 2007).

If all the possible states of the system can be mapped out, these if-then rules can be used to completely describe the appropriate behaviour of the robot in the system. The modules are triggered by the inputs and run in parallel, allowing the robot to do multiple things at once. A very simple navigation can be implemented with two reactions: if there is no obstacle detected go forward, if there is an obstacle detected, turn the other way.

This architecture can be further complicated by creating a hierarchy of reactive modules, where modules higher up in the hierarchy control which modules lower in the hierarchy are active. This allows the robot to behave differently under different circumstances, for example, a module responsible for stopping the robot when it senses something too close could shut down all other reactions.

A very important aspect of reactive architecture is that there is a direct and basically memoryless reaction to the environment. The modules can also thus be thought of as short-lived: an input (or lack of one) triggers a module with an immediate action which then terminates. This makes it cheap in memory, complexity, and computational power compared to a more sophisticated system. It also makes it easier to reason about what the robot would do in a given situation. This comes with the drawback of being hardcoded and unable to accommodate, and not having any representation of the world around it.

Similar mechanisms are also present in animals for situations requiring rapid reactions (Elefteriou, 2018). Such reflexes are essential for animal survival and have been naturally selected because there is typically only one 'correct' reaction in these situations. For example, any stimulus that may harm the eye can only be avoided by activating the eye-blink reflex. In such cases, there is no need for decision-making processes. Reflexes also play important roles in behaviours with regular patterns, such as swimming or walking. It is crucial to note that these reflexes, which are carried out by specific neural pathways, should not be confused with 'habitual' behaviours, where learned activities become reflex-like, such as riding a bicycle. The spontaneous reaction of animals to so-called sign stimuli (e.g., a newly hatched chick pecks at a small dot on the ground, see also Figure 1.3) are functionally similar to reflexes but they are based on a specific configuration of the stimulus and are under different kinds of neural control.

Although complex architectures relying solely on reactive control are not feasible for building an ethorobot, such features can serve specific purposes in certain parts of the ethorobot's architectural hierarchy, especially concerning safety. For instance, the simplest way to ensure our robot does not collide with anyone is to implement reactive rules that stop the robot before obstacles.

3.8.3.2 Behaviour-based control

Behaviour-based control is organised around so-called 'behaviours', which are basically longer-running modules, that take sensory input, may communicate with other behaviour modules during their lifetime, and have goals they are trying to fulfil (e.g., Graves & Czarnecki, 2000). As with the reactive control, the behaviour modules are executed in parallel, but the important difference is that due to their memory, and long-livedness, the sensory input is not directly tied to actions; rather, they modulate the behaviour of the module (Figure 3.45b). For example, the robot could be running two behaviours in parallel, one trying to move the robot from A to B and another trying to interact with humans detected. The interactive behaviour will modulate (turn off) the navigation behaviour, when detecting a human.

Unlike in reactive control, where the modules can only interact with each other through the environment itself, here there is a direct connection between them, leading to a network of behaviour modules that may even be hierarchical, which easily lends itself to emergent behaviour. In this sense, calling these modules as behaviours can be very misleading, since they may or may not be something that the robot explicitly manifests as observable behaviour.

With a large number of behaviours, each with their own goals, running in parallel behaviour coordination becomes a problem (Pirjanian & Matarić, 2001). There are two main categories of behaviour coordination mechanisms in behaviour-based control, arbitration, where one behaviour is selected and given control over the robot, and command fusion where multiple behaviours are allowed to control the robot in a kind of consensus. In arbitration, the selection can be based on for example priority, or an internal competition. A more interesting scenario is when some form of command fusion is used to coordinate behaviours.

Several types of mechanisms have been proposed for command fusion (Pirjanian, 2000). Voting mechanisms allow behaviours to vote for actions and the action with the most vote is taken. Superposition mechanisms create a superposition of behaviours via their linear combinations. The multiple objective mechanism uses a formalism from multiple objective decision theory, where the goal is to optimise a decision along multiple criteria at once, where in this case, the criteria are the goals determined by the various behaviours. Finally, there is the fuzzy behaviour coordination, based on fuzzy logic. Fuzzy logic operates with so-called fuzzy sets, where membership in a set is not a binary variable (either in the set or not in the set), but rather a number between 0 and 1. As such in fuzzy logic the truth value of a variable is also not binary, but between 0 and 1. This allows defining fuzzy rules, e.g., 'IF the obstacle is close THEN robot speed is slow', rather than 'IF the obstacle is closer than 0.5 m THEN robot speed is 0.1 m/s'. Multiple fuzzy rules like this can be created for controlling the robot speed that can be evaluated together to reach a decision on the value. The process is that first the 'crisp' inputs (the exact numbers, e.g., 0.5 m obstacle distance) are fuzzified (the 0.5 m is turned into 'close'), the rules are evaluated and aggregated using fuzzy logic and interpolation and finally the output is defuzzified into a crisp value (e.g., 0.12 m/s for robot speed) for control output. The benefit of this over crisp rules is that the multiple, vague, and slightly contradicting rules can be evaluated together in a smoother and more natural manner. Sticking to our example, instead of the robot abruptly slowing down when nearing the obstacle it will do so more gradually, as the various rules concerning its speed are handled together (see also Box 3.24).

Box 3.24 Attachment in robots?

The knowledge representation of an ethological model comprises a series of observations of various facts and action-reaction rules. In principle, such inputs can be used to construct models that describe various kinds of behavioural interactions. Such ethologically inspired behavioural models could be implemented in future ethorobots. The interaction between owner and dog in the Strange Situation Test (SST) offered a possibility to model the behaviour of a virtual dog in a simulated environment (Kovács et al., 2011; Numakunai et al., 2012). The input for such models came from observing dog and owner dyads in such scenarios that provided support for the emergence of attachment between them. Such models have the potential to be implemented in ethorobots when attachment-like behavioural patterns could prove beneficial.

For mathematical modelling of such a system, rule-based knowledge representation is straightforward. Furthermore, the need to follow observed sequences necessitates a model structure resembling that of a state machine. A fuzzy state machine may offer a possible solution, where the state is represented as a vector of membership values, state transitions are controlled by a fuzzy rule base, and observations and conclusions are continuous values. However, a major challenge arises in the practical implementation of such a fuzzy state machine, as implementing the ethologically necessary rules leads to an overly large state-transition rule base.

To address the exponential complexity problem inherent in the required fuzzy automation state-transition rule base, the application of Fuzzy Rule Interpolation (FRI) methods is proposed. The knowledge representation style of the FRI method aligns well with the conceptually sparse, rule-based structure of existing descriptive verbal ethological models. This model can function effectively without relying on a comprehensive and ethologically complete rule base, which considers all possible scenarios.

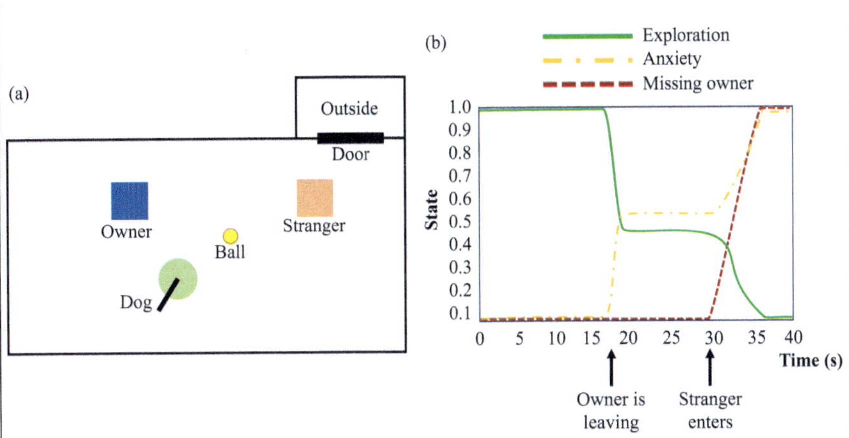

Figure 3.46 (a) Schematic depiction of the visualisation of the simulation, with the owner, the stranger, the dog, and a ball inside the room. (b) Data from a simulation (see the text for details; redrawn based on Kovács et al., 2011).

Examples:

1 State-transition rule related to the room 'Exploration level': If ThePlaceIsUnknown=Low Then ExplorationLevel=Low;
2 State-transition rule related to the 'Missing owner level': If OwnerInTheRoom=False Then MissTheOwner=Increasing
3 State-transition rule related to the 'Anxiety level': If OwnerToDogDistance=Small And StrangerToDogDistance=High Then AnxietyLevel=Decreasing

Figure 3.46b presents data from a simulation. At the onset of the scenario, the Owner occupies the room while the Stranger remains outside, and the location is unfamiliar to the virtual dog (VD). Following the state-transition rules, VD initiates exploration of the room. Subsequently, the Owner exits the room, and after a period, the Stranger enters and remains inside. Consequently, both the 'Anxiety' and 'Missing the owner' levels of VD escalate, leading VD to cease exploration of the room.

The attachment system can be modelled by different architectures. Hiolle et al. (2014) relied on neural networks to construct a behavioural model for a NAO robot displaying attachment towards a human.

3.8.3.3 Model-predictive control

In the previous control architectures, there was no explicit model of the behaviour of the system as a whole, although implicitly, of course it was coded into the modules, but no systematic approach was taken. In MPC, a central part of the control architecture is a mathematical model of the system as whole (e.g., Camacho & Bordons, 2007). In such a model, some of the independent variables can be directly controlled, some of them can be treated as outside perturbations and it aims to predict dependent variables that can be measured. The idea is that given a particular state of the system as a goal and a series of planned actions, the model is used to predict the future states of the system for a given amount of timesteps (with diminishing importance, called a receding horizon). This is then iterated back-and-forth with the optimiser to find an optimal set of control actions in the near future, to reach the desired state (Figure 3.45c).

The question is how does the agent obtain the model? This method is mostly used in industrial applications, where the system that needs to be controlled is sufficiently closed and is closer to basic natural laws than biological behaviour. In general, one could use a purely first-principles approach, where using the laws of physics, one derives a model of the system, called a white box. However, this problem is generally intractable, so typically either a grey-box or a black-box approach is used. In a grey-box model, some first-principles constraints are introduced and measurements of the system's behaviour fill out the gaps, while in a black-box model, the entire model is constructed via measurements of the system.

This method has been successfully used for controlling the movement of flying robots (which is arguably much harder than controlling one on wheels, but the problem is very close to physics to lend itself to tractable modelling) and can likely be used for the more mechanistic parts of a robot's control, while it is unlikely to be successful in controlling social aspects of a robot in such a direct manner. Nonetheless, in the ESPSA's actor module, the optimiser submodule's interaction with the world model is based on this idea (Figure 3.35).

In the context of various movements, animals have the opportunity to predict changes in posture and anticipate how these changes may affect the position of their CoM. Research has uncovered the existence of specific feedforward mechanisms that anticipate the new state and orchestrate compensatory adjustments accordingly (Gordon et al., 2020; Leonard et al., 2009). This ability enables animals to respond more swiftly to changing situations, thereby minimising the need for additional compensatory actions once feedback about the actual state is received from proprioceptors (see also Section 2.4.9).

3.8.4 Deep reinforcement learning architecture

Deep reinforcement learning is doing reinforcement learning with deep neural networks (see neural networks in Section 3.7.3). Although other types of deep learning are also used in robotics extensively (e.g., object classification using supervised learning for processing visual inputs), given that reinforcement learning deals with an agent in an environment, it is easy to conceptualise it as an overarching architecture for robots (Pierson & Gashler, 2017). As we have seen, the ESPSA builds extensively on deep reinforcement learning with its use of an actor and critic, clearly reminiscent of an actor-critic reinforcement learning paradigm.

Indeed, reinforcement learning has been successfully used in many cases for replacing low-level control of robotic bodies and manipulators, but nothing much higher level (Liu et al., 2021). For example, grabbing an object from the right angle and employing just the right amount of pressure is a non-trivial engineering task, if there is a large variety of objects. Even human children take a few years to master this with precise movements. Similarly, a non-trivial complex engineering problem is bipedal walking (Box 3.13). In deep reinforcement learning, deep neural networks are used to approximate the value functions or policies that guide the agent's decision-making process. These networks are capable of learning high-dimensional and abstract representations of states, actions, and rewards, enabling the agent to generalise across different situations and make informed decisions.

The key to teaching such an agent is of course providing it with enough input. For this, either the possible scenarios must be few and easily reproducible, or one must use simulation. This is precisely why it is hard to do higher-level functions with such a method: complex scenarios are out of the question in real life and as the simulations move further away from reality, the further away the trained agent will operate from reality. Since simulations are the easier, the closer we are to physics directly (unlike as for social interactions, we know the precise mathematical rules), the applications are also tied to low-level, close-to-physics problems.

3.8.5 Soar: a 'cognitive' architecture

Although a few other architectures could have been chosen, the Soar architecture is an example of the SMoM, which has been extensively used in virtual environments and some mobile robots as well (Laird, 2012). The 'cognitive' in their own words refers to human-level intelligence but there is no specific reason to exclude the possibility that some analogous versions of the Soar may operate in non-humans. The Soar architecture is based on the concept of the problem-space computational model (PSCM) (Newell, 1990).

In the PSCM, an agent can be in given states and can choose from actions called operators. The selected operation may act on the environment or within the agent and ultimately change the current state. A problem is defined as an initial state and a set of desired states that must be achieved (see also Section 1.5). A problem space is the set of states the agent can move through using the available operators, and thus to solve a problem, a search in the problem state must be carried out, which is done in a processing cycle. In general, in each cycle, the agent considers input from the environment, proposes and evaluates operators, and then applies the selected operator (Figure 3.47).

The default mode of running this cycle is the knowledge-search mode. Knowledge is a collection of rules stored in memory about the operators: whether they can be considered for application given the current state and whether they should be considered, given the state and the goal. Thus, in such a cycle, based on the current state and environmental inputs, the knowledge base is searched for suitable operators and once one has been selected it is applied.

If in a cycle the knowledge search does not succeed or the selected operator cannot be applied an impasse is reached. To resolve the impasse a substate is generated, where instead of a knowledge search, a full problem-space search is carried out, which is somewhat analogous with switching from executing a usual routine to

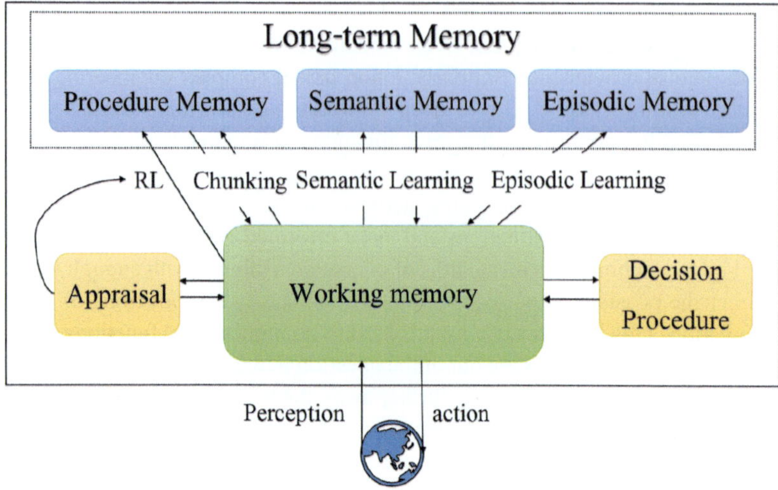

Figure 3.47 Overview of the Soar architecture (figure from Luo et al., 2022). This architecture is an implementation of the SMoM, that has been applied to actual systems of varying complexity.

thinking and generating alternative solutions to a novel, specific problem. Successfully solving an impasse in a substate can also be used to generate new knowledge via an operation called chunking. In essence, this extracts the operator(s) that allowed the impasse to be resolved and stores them away with the state where the impasse occurred.

Soar elaborates this basic scheme with many additional details. The current state of the agent and the environment and its immediate past is encoded in the short-term working memory with a graph structure. A similar, but larger graph structure in the semantic memory holds long-term information, that can be retrieved by knowledge search, and is built up by using pieces from the working memory. In contrast, the episodic memory is a series of snapshots of the working memory, including temporal information. Lastly, the procedural memory holds all the rules learned or precoded for operator selection.

The Soar is deeply symbolic: it assumes that perception creates symbolic descriptions, and its working memory is based on a graph of symbols and the operators and described by its own symbolic (programming) language. However, Soar is not entirely symbolic. There are two important non-symbolic additions to Soar. The first one is applying reinforcement learning to procedural memory. Unlike the new rules generated by a substate after an impasse, reinforcement learning allows for the tweaking of the existing rules, allowing the agent to be more adaptable. The other important non-symbolic addition is spatial-visual memory. Unlike the other two long-term memories that were built up from the working memory, the spatial-visual memory feeds directly from perception and motor functions and its purpose is to aid in spatial reasoning, which is less suited to symbolic representations.

3.8.6 Conclusions for ethorobotics

From the ethological point of view, robotic architectures make an advance over the traditional mind models by not just identifying the components of the system but also inventing their actual way(s) of operation. This is not surprising because while some of these models remain at the theoretical level, the majority has been invented to solve actual problems. This problem-centred approach of robotics is the key not only to building functional robots but also to providing a way of understanding living organisms.

The main goal of the ethorobotic approach is to unify top-down mental models originating from ethology and psychology with bottom-up models of neurobiology, computer science, and robotics. During the last 50 years, engineers have developed a myriad of mental architectures for controlling robot behaviour. Some were based on logical optimisation of machine behaviour, others were inspired by the operation of the nervous system. Despite improvements in both hardware and software capacities, and impressive performance of the most modern technologies, these developments are still very far from competing with animals on the whole. Thus, technology can be of huge help in specific tasks but is unable to capture the entire organism. For ethorobotics, this situation is not uncomfortable because it contrasts with those who aim to 'build humans': this approach takes evolutionary insight to solve problems by constructing agents for specific functions. In fact, the ethorobotics approach actively pushes back against notions of treating Artificial General Intelligence (defined as human-like intelligence) as the holy grail of AI. Humans are also subject to evolutionary constraints and thus however (and however well and flexible) the human mind works, it may not be the most appropriate solution for a set of specific functions. Not to mention that 'human-like' is an ill-defined term, prone to over-stretched analogies, like calling a car 'horse-like'.

While the focus of the research interest is at the higher cognitive processes (e.g., language), the development of ethorobots is also halted without solving 'simpler' problems (e.g., walking). Thus, architectures with different origins may have to be combined to achieve robust, autonomous, and interactive behaviour in robots, just as it is the case in living organisms.

The jury is still out on the question of whether explicit symbolic logic in artificial mental architectures is needed or not. Although at least humans definitely engage in symbolic reasoning, whether there is an explicit and abstract symbolic representation in the human mind is not clear (Landy et al., 2014). Laird (2012) argues for using hybrid systems, with both symbolic and non-symbolic parts. Indeed, achieving symbolic reasoning, by explicitly incorporating it, especially for smaller systems might be technically more feasible and robust, than achieving it via emergence. In contrast, LeCun (2022) argues the contrary via another technical argument, namely, that if reasoning is kept continuous, the toolbox of modern ML can be brought to bear for learning. In favour of emergence, Webb et al. (2021) showed that deep neural networks with an external memory (such as there is in ESPSA) can produce emergent symbolism. It can also be conceptualised, that at sufficiently high levels of aggregation the continuous space of representations breaks down into essentially discrete elements or that appropriate projections of continuous representations lead to discrete elements. The latter is actually congruent with how learning of symbols should progress: the symbol 'fruit' does not disappear, when we learn to differentiate between different kinds of fruits, and restricting ourselves to the subspace of apples allows for the existence of different types of apples.

A further critical question is how artificial mental architectures deal with the issue of learning. One has the impression that most learning models are mimicking 'adult learning' but not 'infant learning'. This is important because living systems with learning abilities collect the to-be-learnt information as early as possible, and later (adult) learning is mainly concerned with minor adjustments. Thus, one important aspect of learning is to increase the problem space (see above), increasing the dimensionality of the mental space. There is also evidence that such mental processes are paralleled by the development of the nervous system. This kind of growing mental 'power' is less reflected in present-day architectures that control robot behaviour.

References

Ache, B. W., & Young, J. M. (2005). Olfaction: Diverse species, conserved principles. *Neuron*, *48*(3), 417–430. https://doi.org/10.1016/j.neuron.2005.10.022

Adarsh, S., Kaleemuddin, S. M., Bose, D., & Ramachandran, K. I. (2016). Performance comparison of Infrared and Ultrasonic sensors for obstacles of different materials in vehicle/robot navigation applications. *IOP Conference Series: Materials Science and Engineering*, *149*(1), 012141. https://doi.org/10.1088/1757-899X/149/1/012141

Albustanji, R. N., Elmanaseer, S., & Alkhatib, A. A. A. (2023). Robotics: Five senses plus one – An overview. *Robotics*, *12*(3), 68. https://doi.org/10.3390/robotics12030068

Altes, R. A., Dankiewicz, L. A., Moore, P. W., & Helweg, D. A. (2003). Multiecho processing by an echolocating dolphin. *The Journal of the Acoustical Society of America*, *114*(2), 1155–1166. https://doi.org/10.1121/1.1590969

Arkin, R. C. (1989). Motor schema - Based mobile robot navigation. *The International Journal of Robotics Research*, *8*(4), 92–112. https://doi.org/10.1177/027836498900800406

Arkin, R. C. (1998). *Behavior-based robotics*. MIT Press.

Arkin, R. C., Fujita, M., Takagi, T., & Hasegawa, R. (2003). An ethological and emotional basis for human-robot interaction. *Robotics and Autonomous Systems*, *42*(3–4), 191–201. https://doi.org/10.1016/S0921-8890(02)00375-5

Ashby, W. R. (1958). Requisite variety and its implications for the control of complex systems. *Cybernetica*, *1*(2), 83–99. https://doi.org/10.1007/978-1-4899-0718-9_28

Au, W. W. L. (1997). Echolocation in dolphins with a dolphin-bat comparison. *Bioacoustics*, *8*(1–2), 137–162. https://doi.org/10.1080/09524622.1997.9753357

Au, W. W. L., & Simmons, J. A. (2007). Echolocation in dolphins and bats. *Physics Today*, *60*(9), 40–45. https://doi.org/10.1063/1.2784683

Bahrick, L. E., & Watson, J. S. (1985). Detection of intermodal proprioceptive-visual contingency as a potential basis of self-perception in infancy. *Developmental Psychology*, *21*(6), 963–973. https://doi.org/10.1037/0012-1649.21.6.963

Baines, R., Patiballa, S. K., Booth, J., Ramirez, L., Sipple, T., Garcia, A., Fish, F., & Kramer-Bottiglio, R. (2022). Multi-environment robotic transitions through adaptive morphogenesis. *Nature*, *610*(7931), 283–289. https://doi.org/10.1038/s41586-022-05188-w

Ballerini, M., Cabibbo, N., Candelier, R., Cavagna, A., Cisbani, E., Giardina, I., Orlandi, A., Parisi, G., Procaccini, A., Viale, M., & Zdravkovic, V. (2008). Empirical investigation of starling flocks: A benchmark study in collective animal behaviour. *Animal Behaviour*, *76*(1), 201–215. https://doi.org/10.1016/j.anbehav.2008.02.004

Beck, A., Hiolle, A., Mazel, A., & Cañamero, L. (2010). Interpretation of emotional body language displayed by robots. *AFFINE '10: Proceedings of the 3rd International Workshop on Affective Interaction in Natural Environments*, 37–42. https://doi.org/10.1145/1877826.1877837

Bee, M. A., & Micheyl, C. (2008). The 'cocktail party problem': What is it? How can it be solved? And why should animal behaviorists study it? *Journal of Comparative Psychology*, *122*(3), 235–251. https://doi.org/10.1037/0735-7036.122.3.235

Bett, N. N., & Hinch, S. G. (2016). Olfactory navigation during spawning migrations: A review and introduction of the Hierarchical Navigation Hypothesis. *Biological Reviews*, *91*(3), 728–759. https://doi.org/10.1111/brv.12191

Betts, J. G., Young, K. A., Wise, J. A., Johnson, E., Poe, B., Kruse, D. H., Korol, O., Johnson, J. E., Womble, M., & DeSaix, P. (2013). *Anatomy and physiology*. OpenStax. https://openstax.org/books/anatomy-and-physiology/pages/1-introduction

Bjelonic, M., Grandia, R., Geilinger, M., Harley, O., Medeiros, V. S., Pajovic, V., Jelavic, E., Coros, S., & Hutter, M. (2022). Offline motion libraries and online MPC for advanced mobility skills. *International Journal of Robotics Research*, *41*(9–10), 903–924. https://doi.org/10.1177/02783649221102473

Bjelonic, M., Grandia, R., Harley, O., Galliard, C., Zimmermann, S., & Hutter, M. (2021). Whole-body MPC and online gait sequence generation for wheeled-legged robots. *2021 IEEE/RSJ International Conference on Intelligent Robots and Systems (IROS)*, 8388–8395. https://doi.org/10.1109/IROS51168.2021.9636371

Blausen.com staff. (2014). Medical gallery of Blausen Medical 2014. *WikiJournal of Medicine*, *1*(2).

Bleckmann, H., Schmitz, H., & von der Emde, G. (2004). Nature as a model for technical sensors. *Journal of Comparative Physiology A: Neuroethology, Sensory, Neural, and Behavioral Physiology*, *190*(12), 971–981. https://doi.org/10.1007/s00359-004-0563-y

Block, A. E., & Kuchenbecker, K. J. (2019). Softness, warmth, and responsiveness improve robot hugs. *International Journal of Social Robotics*, *11*(1), 49–64. https://doi.org/10.1007/s12369-018-0495-2

Bodenhagen, L., Suvei, S. D., Juel, W. K., Brander, E., & Krüger, N. (2019). Robot technology for future welfare: Meeting upcoming societal challenges – an outlook with offset in the development in Scandinavia. *Health and Technology*, *9*(3), 197–218. https://doi.org/10.1007/s12553-019-00302-x

Bradbury, J. W. (2011). *Principles of animal communication*. Sinauer Associates.

Bretan, M., Hoffman, G., & Weinberg, G. (2015). Emotionally expressive dynamic physical behaviors in robots. *International Journal of Human-Computer Studies*, *78*, 1–16. https://doi.org/10.1016/j.ijhcs.2015.01.006

Brosque, C., & Fischer, M. (2022). Safety, quality, schedule, and cost impacts of ten construction robots. *Construction Robotics*, *6*(2), 163–186. https://doi.org/10.1007/s41693-022-00072-5

Brown, A., Lamb, E., Deo, A., Pasin, D., Liu, T., Zhang, W., Su, S., & Ueland, M. (2023). The use of novel electronic nose technology to locate missing persons for criminal investigations. *IScience*, *26*(4), 106353. https://doi.org/10.1016/j.isci.2023.106353

Bruzzone, L., Nodehi, S. E., & Fanghella, P. (2022). Tracked locomotion systems for ground mobile robots: A review. *Machines*, *10*(8), 648. https://doi.org/10.3390/machines10080648

Bruzzone, L., & Quaglia, G. (2012). Review article: Locomotion systems for ground mobile robots in unstructured environments. *Mechanical Sciences*, *3*(2), 49–62. https://doi.org/10.5194/ms-3-49-2012

Burkhardt, R. W. (2005). *Patterns of behavior: Konrad Lorenz, Niko Tinbergen, and the founding of ethology*. The University of Chicago Press.

Burkov, A. (2019). *The hundred page machine learning book*.

Burmeister, S. S. (2017). Neurobiology of female mate choice in frogs: Auditory filtering and valuation. *Integrative and Comparative Biology*, *57*(4), 857–864. https://doi.org/10.1093/icb/icx098

Buschmann, T., Favot, V., Schwienbacher, M., Ewald, A., & Ulbrich, H. (2013). Dynamics and control of the biped robot Lola. In *Multibody system dynamics, robotics and control* (pp. 161–173). Springer. https://doi.org/10.1007/978-3-7091-1289-2_10

Byosiere, S. E., Espinosa, J., & Smuts, B. (2016). Investigating the function of play bows in adult pet dogs (*Canis lupus familiaris*). *Behavioural Processes*, *125*, 106–113. https://doi.org/10.1016/j.beproc.2016.02.007

Camacho, E. F., & Bordons, C. (2007). *Model predictive control*. Springer. https://doi.org/10.1007/978-0-85729-398-5

Campion, G., & Chung, W. (2008). Wheeled robots. In B. Siciliano & O. Khatib (Eds.), *Springer handbook of robotics* (pp. 391–410). Springer. https://doi.org/10.1007/978-3-540-30301-5_18

Cañas Plaza, J. M., & Matellan Oliveira, V. (2005). Integrating behaviors for mobile robots: An ethological approach. In *Cutting edge robotics* (pp. 311–330). IntechOpen.

Capdevila, J., & Belmonte, J. C. I. (2000). Perspectives on the evolutionary origin of tetrapod limbs. *Journal of Experimental Zoology*, *288*(4), 287–303. https://doi.org/10.1002/1097-010X(20001215)288:4<287::AID-JEZ2>3.0.CO;2-5

Carelli, R., & Freire, E. O. (2003). Corridor navigation and wall-following stable control for sonar-based mobile robots. *Robotics and Autonomous Systems*, *45*(3–4), 235–247. https://doi.org/10.1016/j.robot.2003.09.005

Carmo, J. P., & Ribeiro, J. E. (2013). Optical fibers on medical instrumentation. *International Journal of Biomedical and Clinical Engineering*, *2*(2), 23–36. https://doi.org/10.4018/ijbce.2013070103

Chandrashekar, J., Hoon, M. A., Ryba, N. J. P., & Zuker, C. S. (2006). The receptors and cells for mammalian taste. *Nature*, *444*, 288–294. https://doi.org/10.1038/nature05401

Chappell, D. R., & Speiser, D. I. (2023). Polarization sensitivity and decentralized visual processing in an animal with a distributed visual system. *Journal of Experimental Biology*, *226*(4), jeb244710. https://doi.org/10.1242/jeb.244710

Chemero, A. (2023). LLMs differ from human cognition because they are not embodied. *Nature Human Behaviour*, *7*(11), 1828–1829. https://doi.org/10.1038/s41562-023-01723-5

Chen, C. P., Chen, J. Y., Huang, C. K., Lu, J. C., & Lin, P. C. (2015). Sensor data fusion for body state estimation in a bipedal robot and its feedback control application for stable walking. *Sensors*, *15*(3), 4925–4946. https://doi.org/10.3390/s150304925

Cheng, L., Meng, Q. H., Lilienthal, A. J., & Qi, P. F. (2021). Development of compact electronic noses: A review. *Measurement Science and Technology*, *32*(6), 062002. https://doi.org/10.1088/1361-6501/abef3b

Christ, R. D., & Wernli, R. L. (2014). Sonar. In *The ROV manual* (pp. 387–424). Elsevier. https://doi.org/10.1016/b978-0-08-098288-5.00015-4

Cooperrider, K., & Mesh, K. (2022). Pointing in gesture and sign. In A. Morgenstern & S. Goldin-Meadow (Eds.), *Gesture in language: Development across the lifespan* (pp. 21–46). De Gruyter Mouton. https://doi.org/10.1037/0000269-002

Correll, N., Hayes, B., Heckman, C., & Roncone, A. (2021). *Introduction to autonomous robots: Mechanisms, sensors, actuators, and algorithms*. MIT Press.

Craig, J. J. (2018). *Introduction to robotics: Mechanics and control* (4th ed.). Pearson.

Dalton, P. (2000). Psychophysical and behavioral characteristics of olfactory adaptation. *Chemical Senses*, *25*(4), 487–492. https://doi.org/10.1093/chemse/25.4.487

Das, R., Mishra, J., Pattnaik, P. K., & Bhatti, M. M. (2023). Prediction of heatwave using advanced soft computing technique. *Information*, *14*(8), 447. https://doi.org/10.3390/info14080447

Daw, N. (2012). *How vision works: The physiological mechanisms behind what we see*. Oxford University Press.

Deep, J. (2019). *Machine learning for beginners: An introductory guide to learn and understand artificial intelligence, neural networks and machine learning*.

Diogo, R., & Santana, S. E. (2017). Evolution of facial musculature. In J.-M. Fernández-Dols & J. A. Russell (Eds.), *The science of facial expression* (pp. 133–152). Oxford University Press.
Dreyer, D., Frost, B., Mouritsen, H., Günther, A., Green, K., Whitehouse, M., Johnsen, S., Heinze, S., & Warrant, E. (2018). The Earth's magnetic field and Visual landmarks steer migratory flight behavior in the nocturnal Australian bogong moth. *Current Biology, 28*(13), 2160–2166.e1-e5. https://doi.org/10.1016/j.cub.2018.05.030
Durini, D. (2014). *High performance silicon imaging: Fundamentals and applications of CMOS and CCD sensors*. Elsevier.
Dymerski, T. (2017). Electronic noses and electronic tongues. In S. Yurish (Ed.), *Advances in biosensors: Reviews: Vol. 1* (pp. 153–193). International Frequency Sensor Association Publishing.
Eakin, R. M., & Brandenburger, J. L. (1975). Understanding a snail's eye at a snail's pace. *American Zoologist, 15*(4), 851–863. https://doi.org/10.1093/icb/15.4.851
Eargle, J. (2005). *The microphone book*. Elsevier.
Eckstein, M., Mamaev, I., Ditzen, B., & Sailer, U. (2020). Calming effects of touch in human, animal, and robotic interaction - Scientific state-of-the-art and technical advances. *Frontiers in Psychiatry, 11*, 555058. https://doi.org/10.3389/fpsyt.2020.555058
Eguíluz, V. M., Ospeck, M., Choe, Y., Hudspeth, A. J., & Magnasco, M. O. (2000). Essential nonlinearities in hearing. *Physical Review Letters, 84*(22), 5232–5235. https://doi.org/10.1103/PhysRevLett.84.5232
Elefteriou, F. (2018). Impact of the autonomic nervous system on the skeleton. *Physiological Reviews, 98*, 1083–1112. https://doi.org/10.1152/physrev.00014.2017.-It
Emmert-Streib, F., & Dehmer, M. (2022). Taxonomy of machine learning paradigms: A data-centric perspective. *Wiley Interdisciplinary Reviews: Data Mining and Knowledge Discovery, 12*(5), e1470. https://doi.org/10.1002/widm.1470
Fant, G. (1960). *Acoustic theory of speech production: With calculations based on X-ray studies of Russian articulations*. Mouton.
Faraj, Z., Selamet, M., Morales, C., Torres, P., Hossain, M., Chen, B., & Lipson, H. (2021). Facially expressive humanoid robotic face. *HardwareX, 9*, e00117. https://doi.org/10.5281/zenodo.3539723
Fehér, O., Ljubičić, I., Suzuki, K., Okanoya, K., & Tchernichovski, O. (2017). Statistical learning in songbirds: From selftutoring to song culture. *Philosophical Transactions of the Royal Society B: Biological Sciences, 372*(1711), 20160053. https://doi.org/10.1098/rstb.2016.0053
Fletcher, M. D., Verschuur, C. A., & Perry, S. W. (2023). Improving speech perception for hearing-impaired listeners using audio-to-tactile sensory substitution with multiple frequency channels. *Scientific Reports, 13*, 13336. https://doi.org/10.1038/s41598-023-40509-7
Freeman, S., Sichel, J.-Y., & Sohmer, H. (2000). Bone conduction experiments in animals - evidence for a non-osseous mechanism. *Hearing Research, 146*, 72–80. https://doi.org/10.1016/S0378-5955(00)00098-8
Freitas-Magalhães, A. (2012). Microexpression and macroexpression. In V. S. Ramachandran (Ed.), *Encyclopedia of human behavior: Vol. 2*. Elsevier/Academic Press.
Frontera, W. R., & Ochala, J. (2015). Skeletal muscle: A brief review of structure and function. *Calcified Tissue International, 96*, 183–195. https://doi.org/10.1007/s00223-014-9915-y
Fukuhara, A., Gunji, M., & Masuda, Y. (2022). Comparative anatomy of quadruped robots and animals: A review. *Advanced Robotics, 36*(13), 612–630. https://doi.org/10.1080/01691864.2022.2086018
Fukushima, T., Siddall, R., Schwab, F., Toussaint, S. L. D., Byrnes, G., Nyakatura, J. A., & Jusufi, A. (2021). Inertial tail effects during righting of squirrels in unexpected falls: From behavior to robotics. *Integrative and Comparative Biology, 61*(2), 589–602. https://doi.org/10.1093/icb/icab023

Garwood, R. J., & Edgecombe, G. D. (2011). Early terrestrial animals, evolution, and uncertainty. *Evolution: Education and Outreach*, *4*(3), 489–501. https://doi.org/10.1007/s12052-011-0357-y

Gehring, W. J. (2014). The evolution of vision. *Wiley Interdisciplinary Reviews: Developmental Biology*, *3*(1), 1–40. https://doi.org/10.1002/wdev.96

Gong, Y., Sun, G., Nair, A., Bidwai, A., CS, R., Grezmak, J., Sartoretti, G., & Daltorio, K. A. (2023). Legged robots for object manipulation: A review. *Frontiers in Mechanical Engineering*, *9*, 1142421. https://doi.org/10.3389/fmech.2023.1142421

Gordon, G. (2019). Social behaviour as an emergent property of embodied curiosity: A robotics perspective. *Philosophical Transactions of the Royal Society B: Biological Sciences*, *374*(1771), 20180029. https://doi.org/10.1098/rstb.2018.0029

Gordon, J. C., Holt, N. C., Biewener, A., & Daley, M. A. (2020). Tuning of feedforward control enables stable muscle force-length dynamics after loss of autogenic proprioceptive feedback. *ELife*, *9*, e53908. https://doi.org/10.7554/eLife.53908

Granger, J., Walkowicz, L., Fitak, R., & Johnsen, S. (2020). Gray whales strand more often on days with increased levels of atmospheric radio-frequency noise. *Current Biology*, *30*(4), R155–R156. https://doi.org/10.1016/j.cub.2020.01.028

Graves, A. R., & Czarnecki, C. (2000). Design patterns for behavior-based robotics. *IEEE Transactions on Systems, Man, and Cybernetics - Part A: Systems and Humans*, *30*(1), 36–41. https://doi.org/10.1109/3468.823479

Gravina, N., Nastasi, J. A., Sleiman, A. A., Matey, N., & Simmons, D. E. (2020). Behavioral strategies for reducing disease transmission in the workplace. *Journal of Applied Behavior Analysis*, *53*(4), 1935–1954. https://doi.org/10.1002/jaba.779

Green, D. M., & Barber, D. L. (1988). The ventral adhesive disc of the clingfish *Gobiesox maeandricus*: Integumental structure and adhesive mechanisms. *Canadian Journal of Zoology*, *66*(7), 1610–1619. https://doi.org/10.1139/z88-235

Gulletta, G., Erlhagen, W., & Bicho, E. (2020). Human-like arm motion generation: A review. *Robotics*, *9*(4), 102. https://doi.org/10.3390/robotics9040102

Han, C., O'sullivan, J., Luo, Y., Herrero, J., Mehta, A. D., & Mesgarani, N. (2019). Speaker-independent auditory attention decoding without access to clean speech sources. *Science Advances*, *5*, eaav6134. https://doi.org/10.1126/sciadv.aav6134

Häne, C., Heng, L., Lee, G. H., Fraundorfer, F., Furgale, P., Sattler, T., & Pollefeys, M. (2017). 3D visual perception for self-driving cars using a multi-camera system: Calibration, mapping, localization, and obstacle detection. *Image and Vision Computing*, *68*, 14–27. https://doi.org/10.1016/j.imavis.2017.07.003

Hansson, B. S., & Stensmyr, M. C. (2011). Evolution of insect olfaction. *Neuron*, *72*(5), 698–711. https://doi.org/10.1016/j.neuron.2011.11.003

Harley, C. M., & Asplen, M. K. (2018). Annelid vision. In *Oxford Research Encyclopedia of Neuroscience*. Oxford University Press. https://doi.org/10.1093/acrefore/9780190264086.013.177

Harley, H. E., & DeLong, C. M. (2008). Echoic object recognition by the bottlenose dolphin. *Comparative Cognition & Behavior Reviews*, *3*, 46–65. https://doi.org/10.3819/ccbr.2008.30003

Hashimoto, H. (2003). Intelligent space: Interaction and intelligence. *Artificial Life and Robotics*, *7*, 79–85. https://doi.org/10.1007/s10015-003-0270-8

Hashizume, J., Huh, T. M., Suresh, S. A., & Cutkosky, M. R. (2019). Capacitive sensing for a gripper with gecko-inspired adhesive film. *IEEE Robotics and Automation Letters*, *4*(2), 677–683. https://doi.org/10.1109/LRA.2019.2893154

Heckman, C. (2022). *Introduction to autonomous robots*. MIT Press.

Hedayati, H., Suzuki, R., Rees, W., Leithinger, D., & Szafir, D. (2022). Designing expandable-structure robots for human-robot interaction. *Frontiers in Robotics and AI*, *9*, 719639. https://doi.org/10.3389/frobt.2022.719639

Hepper, P. G., & Wells, D. L. (2005). How many footsteps do dogs need to determine the direction of an odour trail? *Chemical Senses*, *30*(4), 291–298. https://doi.org/10.1093/chemse/bji023

Hernandez, J., Sunny, M. S. H., Sanjuan, J., Rulik, I., Zarif, M. I. I., Ahamed, S. I., Ahmed, H. U., & Rahman, M. H. (2023). Current designs of robotic arm grippers: A comprehensive systematic review. *Robotics*, *12*(1), 5. https://doi.org/10.3390/robotics12010005

Higueras-Ruiz, D. R., Nishikawa, K., Feigenbaum, H., & Shafer, M. (2022). What is an artificial muscle? A comparison of soft actuators to biological muscles. *Bioinspiration and Biomimetics*, *17*(1), 011001. https://doi.org/10.1088/1748-3190/ac3adf

Hiolle, A., Lewis, M., & Cañamero, L. (2014). Arousal regulation and affective adaptation to human responsiveness by a robot that explores and learns a novel environment. *Frontiers in Neurorobotics*, *8*, 17. https://doi.org/10.3389/fnbot.2014.00017

Hoffmann, L., & Krämer, N. C. (2021). The persuasive power of robot touch. Behavioral and evaluative consequences of non-functional touch from a robot. *PLoS One*, *16*(5), e0249554. https://doi.org/10.1371/journal.pone.0249554

Hölldobler, B., & Wilson, E. O. (2009). *The superorganism: The beauty, elegance and strangeness of insect societies*. W. W. Norton.

Huggi, A. S., Nilavar, A. C., Bali, J., Giriyapur, A., & Ashwini, G. K. (2021). Implementation of sensor fusion for a mobile robot application. *2021 6th International Conference on Recent Trends on Electronics, Information, Communication and Technology, RTEICT 2021*, 15–20. https://doi.org/10.1109/RTEICT52294.2021.9573730

Hurot, C., Scaramozzino, N., Buhot, A., & Hou, Y. (2020). Bio-inspired strategies for improving the selectivity and sensitivity of artificial noses: A review. *Sensors*, *20*(6), 1803. https://doi.org/10.3390/s20061803

Hwang, S., Park, Y., & Park, Y. S. (2011). Sound direction estimation using an artificial ear for robots. *Robotics and Autonomous Systems*, *59*(3–4), 208–217. https://doi.org/10.1016/j.robot.2010.12.005

Imai, T., Sakano, H., & Vosshall, L. B. (2010). Topographic mapping - the olfactory system. *Cold Spring Harbor Perspectives in Biology*, *2*(8), a001776. https://doi.org/10.1101/cshperspect.a001776

IntelliPaat.com. (2023). *Supervised learning vs unsupervised learning vs reinforcement learning*. https://intellipaat.com/blog/supervised-learning-vs-unsupervised-learning-vs-reinforcement-learning/

Iovino, M., Scukins, E., Styrud, J., Ögren, P., & Smith, C. (2022). A survey of Behavior Trees in robotics and AI. *Robotics and Autonomous Systems*, *154*, 104096. https://doi.org/10.1016/j.robot.2022.104096

James, G., Witten, D., Hastie, T., & Tibshirani, R. (2021). *An introduction to statistical learning: With applications in R* (2nd ed.). Springer.

Johnsen, S., Lohmann, K. J., & Warrant, E. J. (2020). Animal navigation: A noisy magnetic sense? *Journal of Experimental Biology*, *223*(18), jeb164921. https://doi.org/10.1242/JEB.164921

Jordan, M. I., & Mitchell, T. M. (2015). Machine learning: Trends, perspectives, and prospects. *Science*, *349*(6245), 255–260. https://doi.org/10.1126/science.aac4520

Joselevitch, C. (2008). Human retinal circuitry and physiology. *Psychology & Neuroscience*, *1*(2), 141–165. https://doi.org/10.3922/j.psns.2008.2.008

Kaissling, K.-E., & Priesner, E. (1970). Die Riechschwelle des Seidenspinners. *Naturwissenschaften*, *57*, 23–28. https://doi.org/10.1007/BF00593550

Kalita, P., Saikia, M. P., & Singh, N. H. (2015). Electronic-nose technology and its application - A systematic survey. *International Journal of Innovative Research in Electrical, Electronics, Instrumentation and Control Engineering*, *3*(1), 123–128. https://doi.org/10.17148/ijireeice.2015.3126

Karakaya, D., Ulucan, O., & Turkan, M. (2020). Electronic nose and its applications: A survey. *International Journal of Automation and Computing*, *17*(2), 179–209. https://doi.org/10.1007/s11633-019-1212-9

Kashiri, N., Abate, A., Abram, S. J., Albu-Schaffer, A., Clary, P. J., Daley, M., Faraji, S., Furnemont, R., Garabini, M., Geyer, H., Grabowski, A. M., Hurst, J., Malzahn, J., Mathijssen, G., Remy, D., Roozing, W., Shahbazi, M., Simha, S. N., Song, J. B., Smit-Anseeuw, N., Stramigioli, S., Vanderborght, B., Yesilevskiy, Y., & Tsagarakis, N. (2018). An overview on principles for energy efficient robot locomotion. *Frontiers Robotics AI, 5*, 129. https://doi.org/10.3389/frobt.2018.00129

Kavalieros, D., Kapothanasis, E., Kakarountas, A., & Loukopoulos, T. (2022). Methodology for selecting the appropriate electric motor for robotic modular systems for lower extremities. *Healthcare, 10*(10), 2054. https://doi.org/10.3390/healthcare10102054

Kędzierski, J., Muszyński, R., Zoll, C., Oleksy, A., & Frontkiewicz, M. (2013). EMYS-Emotive Head of a Social Robot. *International Journal of Social Robotics, 5*(2), 237–249. https://doi.org/10.1007/s12369-013-0183-1

Khan, M. A., Sun, J., Li, B., Przybysz, A., & Kosel, J. (2021). Magnetic sensors - A review and recent technologies. *Engineering Research Express, 3*(2), 022005. https://doi.org/10.1088/2631-8695/ac0838

Khan, S., Rahmani, H., Shah, S. A. A., & Bennamoun, M. (2022). *A guide to convolutional neural networks for computer vision*. Springer. https://doi.org/10.1007/978-3-031-01821-3

King, B. H., & Gunathunga, P. B. (2023). Gustation in insects: Taste qualities and types of evidence used to show taste function of specific body parts. *Journal of Insect Science, 23*(2), 1–18. https://doi.org/10.1093/jisesa/iead018

Kishi, T., Kojima, T., Endo, N., Destephe, M., Otani, T., Jamone, L., Kryczka, P., Trovato, G., Hashimoto, K., Cosentino, S., & Takanishi, A. (2013). Impression survey of the emotion expression humanoid robot with mental model based dynamic emotions. *2013 IEEE International Conference on Robotics and Automation (ICRA)*, 1663–1668. https://doi.org/10.1109/ICRA.2013.6630793

Kita, S. (2003). *Pointing: Where language, culture, and cognition meet*. Lawrence Erlbaum.

Klemm, V., Morra, A., Salzmann, C., Tschopp, F., Bodie, K., Gulich, L., Küng, N., Mannhart, D., Pfister, C., Vierneisel, M., Weber, F., Deuber, R., & Siegwart, R. (2019). Ascento: A two-wheeled jumping robot. *2019 International Conference on Robotics and Automation (ICRA)*, 7515–7521. https://doi.org/10.1109/ICRA.2019.8793792

Kok, K. Y., & Rajendran, P. (2019). A review on stereo vision algorithms: Challenges and solutions. *ECTI Transactions on Computer and Information Technology, 13*(2), 112–128.

Kolb, H., Fernandez, E., & Nelson, R. (1995). *Webvision: The organization of the retina and visual system [Internet]*. University of Utah Health Sciences Center.

Korcsok, B., Konok, V., Persa, G., Faragó, T., Niitsuma, M., Miklósi, Á., Korondi, P., Baranyi, P., & Gácsi, M. (2018). Biologically inspired emotional expressions for artificial agents. *Frontiers in Psychology, 9*, 1191. https://doi.org/10.3389/fpsyg.2018.01191

Kory Westlund, J. M., Dickens, L., Jeong, S., Harris, P. L., DeSteno, D., & Breazeal, C. L. (2017). Children use non-verbal cues to learn new words from robots as well as people. *International Journal of Child-Computer Interaction, 13*, 1–9. https://doi.org/10.1016/j.ijcci.2017.04.001

Kovács, S., Gácsi, M., Vincze, D., Korondi, P., & Miklósi, Á. (2011). A novel, ethologically inspired HRI model implementation: Simulating dog-human attachment. *2nd International Conference on Cognitive Infocommunications*.

Kunold, L., & Onnasch, L. (2022). A framework to study and design communication with social robots. *Robotics, 11*(6), 129. https://doi.org/10.3390/robotics11060129

Ladich, F., & Winkler, H. (2017). Acoustic communication in terrestrial and aquatic vertebrates. *Journal of Experimental Biology, 220*(13), 2306–2317. https://doi.org/10.1242/jeb.132944

Lagnado, L. (1998). Retinal processing: Amacrine cells keep it short and sweet. *Current Biology, 8*(17), R598–R600. https://doi.org/10.1016/S0960-9822(98)70385-9

Laird, J. E. (2012). *The Soar cognitive architecture*. MIT Press.

Laird, J. E., Lebiere, C., & Rosenbloom, P. S. (2017). A standard model of the mind: Toward a common computational framework across artificial intelligence, cognitive science, neuroscience, and robotics. *AI Magazine*, *38*(4), 13–26. https://doi.org/10.1609/aimag.v38i4.2744

Lakshmanan, V., Görner, M., & Gillard, R. (2021). *Practical machine learning for computer vision*. O'Reilly Media, Inc.

Land, M. F., & Fernald, R. D. (1992). The evolution of eyes. *Annual Review of Neuroscience*, *15*, 1–29. https://doi.org/10.1146/annurev.ne.15.030192.000245

Landy, D., Allen, C., & Zednik, C. (2014). A perceptual account of symbolic reasoning. *Frontiers in Psychology*, *5*, 275. https://doi.org/10.3389/fpsyg.2014.00275

Laska, M. (2017). Human and animal olfactory capabilities compared. In *Springer handbook of odor* (pp. 81–82). Springer International Publishing. https://doi.org/10.1007/978-3-319-26932-0_32

Lavoie, J., Gassó Astorga, P., Segal-Gavish, H., Wu, Y. W. C., Chung, Y., Cascella, N. G., Sawa, A., & Ishizuka, K. (2017). The olfactory neural epithelium as a tool in neuroscience. *Trends in Molecular Medicine*, *23*(2), 100–103. https://doi.org/10.1016/j.molmed.2016.12.010

Lawson, M. J., Craven, B. A., Paterson, E. G., & Settles, G. S. (2012). A computational study of odorant transport and deposition in the canine nasal cavity: Implications for olfaction. *Chemical Senses*, *37*(6), 553–566. https://doi.org/10.1093/chemse/bjs039

LeCun, Y. (2022). A path towards autonomous machine intelligence. *Preprint Posted on OpenReview*. https://openreview.net/pdf?id=BZ5a1r-kVsf

Lee, D. H., Park, H., Park, J. H., Baeg, M. H., & Bae, J. H. (2017). Design of an anthropomorphic dual-arm robot with biologically inspired 8-DOF arms. *Intelligent Service Robotics*, *10*(2), 137–148. https://doi.org/10.1007/s11370-017-0215-z

Lenz, J., & Edelstein, A. S. (2006). Magnetic sensors and their applications. *IEEE Sensors Journal*, *6*(3), 631–649. https://doi.org/10.1109/JSEN.2006.874493

Leonard, J. A., Brown, R. H., & Stapley, P. J. (2009). Reaching to multiple targets when standing: The spatial organization of feed-forward postural adjustments. *Journal of Neurophysiology*, *101*(4), 2120–2133. https://doi.org/10.1152/jn.91135.2008

Li, Y., Zhu, L., Zhang, Z., Guo, M., Li, Z., Li, Y., & Hashimoto, M. (2023). Humanoid robot heads for human-robot interaction: A review. *Science China Technological Sciences*. https://doi.org/10.1007/s11431-023-2493-y

Liang, W., Liu, H., Wang, K., Qian, Z., Ren, L., & Ren, L. (2020). Comparative study of robotic artificial actuators and biological muscle. *Advances in Mechanical Engineering*, *12*(6), 1–25. https://doi.org/10.1177/1687814020933409

Libedinsky, C. (2023). Comparing representations and computations in single neurons versus neural networks. *Trends in Cognitive Sciences*, *27*(6), 517–527. https://doi.org/10.1016/j.tics.2023.03.002

Libersat, F., & Camhi, J. M. (1988). Control of cercal position during flight in the cockroach: A mechanism for regulating sensory feedback. *Journal of Experimental Biology*, *136*(1), 483–488. https://doi.org/10.1242/jeb.136.1.483

Liu, R., Nageotte, F., Zanne, P., de Mathelin, M., & Dresp-Langley, B. (2021). Deep reinforcement learning for the control of robotic manipulation: A focussed mini-review. *Robotics*, *10*(1), 22. https://doi.org/10.3390/robotics10010022

Lohmann, K. J. (2010). Magnetic-field perception. *Nature*, *464*, 1140–1142.

Lohmann, K. J., Goforth, K. M., Mackiewicz, A. G., Lim, D. S., & Lohmann, C. M. F. (2022). Magnetic maps in animal navigation. *Journal of Comparative Physiology A: Neuroethology, Sensory, Neural, and Behavioral Physiology*, *208*(1), 41–67. https://doi.org/10.1007/s00359-021-01529-8

Lohr, C., Grosche, A., Reichenbach, A., & Hirnet, D. (2014). Purinergic neuron-glia interactions in sensory systems. *Pflugers Archiv - European Journal of Physiology*, *466*(10), 1859–1872. https://doi.org/10.1007/s00424-014-1510-6

Luo, F., Zhou, Q., Fuentes, J., Ding, W., & Gu, C. (2022). A Soar-based space exploration algorithm for mobile robots. *Entropy*, *24*(3), 426. https://doi.org/10.3390/e24030426

Lurton, X. (2002). *An introduction to underwater acoustics: Principles and applications*. Springer.
Macías, M. M., Manso, A. G., Orellana, C. J. G., Velasco, H. M. G., Caballero, R. G., & Chamizo, J. C. P. (2013). Acetic acid detection threshold in synthetic wine samples of a portable electronic nose. *Sensors*, *13*(1), 208–220. https://doi.org/10.3390/s130100208
Mainland, J., & Sobel, N. (2006). The sniff is part of the olfactory percept. *Chemical Senses*, *31*(2), 181–196. https://doi.org/10.1093/chemse/bjj012
Manley, G. A. (2017). Comparative auditory neuroscience: Understanding the evolution and function of ears. *JARO - Journal of the Association for Research in Otolaryngology*, *18*(1), 1–24. https://doi.org/10.1007/s10162-016-0579-3
Manzini, I., & Korsching, S. (2011). The peripheral olfactory system of vertebrates: Molecular, structural and functional basics of the sense of smell. *E-Neuroforum*, *17*(3), 68–77. https://doi.org/10.1007/s13295-011-0021-6
Marzke, M. W., & Marzke, R. F. (2000). Evolution of the human hand: Approaches to acquiring, analysing and interpreting the anatomical evidence. *Journal of Anatomy*, *197*(1), 121–140. https://doi.org/10.1046/j.1469-7580.2000.19710121.x
Maslow, A. H. (1943). A theory of human motivation. *Psychological Review*, *50*(4), 370–396.
Matarić, M. J. (2007). *The robotics primer*. MIT Press.
McDermott, J. H. (2009). The cocktail party problem. *Current Biology*, *19*(22), R1024–R1027. https://doi.org/10.1016/j.cub.2009.09.005
McKinnon, P. (2016). *Robotics: Everything you need to know about robotics from beginner to expert*. CreateSpace Independent Publishing Platform.
McLaren, I. P. L., & Mackintosh, N. J. (2000). An elemental model of associative learning: I. Latent inhibition and perceptual learning. *Animal Learning & Behavior*, *28*(3), 211–246. https://doi.org/10.3758/BF03200258
Megat Hasnan, M. M. I., Mohd Sabri, M. F., Mohd Said, S., & Nik Ghazali, N. N. (2014). Modeling of a high force density fishbone shaped electrostatic comb drive microactuator. *Scientific World Journal*, *2014*, 912683. https://doi.org/10.1155/2014/912683
Miklósi, Á. (2007). *Dog behaviour, evolution, and cognition* (1st ed.). Oxford University Press.
Miklósi, Á., Korondi, P., Matellán, V., & Gácsi, M. (2017). Ethorobotics: A new approach to human-robot relationship. *Frontiers in Psychology*, *8*, 958. https://doi.org/10.3389/fpsyg.2017.00958
Mikolajczyk, T., Mikołajewska, E., Al-Shuka, H. F. N., Malinowski, T., Kłodowski, A., Pimenov, D. Y., Paczkowski, T., Hu, F., Giasin, K., Mikołajewski, D., & Macko, M. (2022). Recent advances in bipedal walking robots: Review of gait, drive, sensors and control systems. *Sensors*, *22*(12), 4440. https://doi.org/10.3390/s22124440
Miller, A., Fisch, A., Dodge, J., Karimi, A.-H., Bordes, A., & Weston, J. (2016). Key-value memory networks for directly reading documents. *ArXiv*. https://doi.org/10.48550/arXiv.1606.03126
Miller, G. A., Galanter, E., & Pribram, K. A. (1960). *Plans and the structure of behavior*. Holt, Rhinehart, & Winston.
Mirvakili, S. M., & Hunter, I. W. (2018). Artificial muscles: Mechanisms, applications, and challenges. *Advanced Materials*, *30*(6), 1704407. https://doi.org/10.1002/adma.201704407
Mitchell, M. (2019). Artificial intelligence hits the barrier of meaning. *Information*, *10*(2), 51. https://doi.org/10.3390/info10020051
Newell, A. (1990). *Unified theories of cognition*. Harvard University Press.
Newell, A., & Simon, H. A. (1976). Computer science as empirical inquiry: Symbols and search. *Communications of the ACM*, *19*(3), 113–126.
Nhu, T. V., & Sawada, H. (2017). Singing performance of the talking robot with newly redesigned artificial vocal cords. *2017 IEEE International Conference on Mechatronics and Automation (ICMA)*, 665–670. https://doi.org/10.1109/ICMA.2017.8015895
Nimpf, S., & Keays, D. A. (2022). Myths in magnetosensation. *IScience*, *25*, 104454. https://doi.org/10.1016/j.isci.2022.104454

Nityananda, V., & Read, J. C. A. (2017). Stereopsis in animals: Evolution, function and mechanisms. *Journal of Experimental Biology, 220*(14), 2502–2512. https://doi.org/10.1242/jeb.143883
Nobel Prize Organization (2004). Press release. NobelPrize.org. Nobel Prize Outreach AB 2024. Retrieved: 16 Feb 2024. https://www.nobelprize.org/prizes/medicine/2004/press-release/
Numakunai, R., Ichikawa, T., Gácsi, M., Korondi, P., Hashimoto, H., & Niitsuma, M. (2012). Exploratory behavior in ethologically inspired robot behavioral model. *2012 IEEE RO-MAN: The 21st IEEE International Symposium on Robot and Human Interactive Communication*, 577–582.
Oosterhof, N. N., & Todorov, A. (2008). The functional basis of face evaluation. *Proceedings of the National Academy of Sciences of the United States of America, 105*(32), 11087–11092. https://doi.org/10.1073/pnas.0805664105
Ortega-de San Luis, C., & Ryan, T. J. (2022). Understanding the physical basis of memory: Molecular mechanisms of the engram. *Journal of Biological Chemistry, 298*(5), 101866. https://doi.org/10.1016/j.jbc.2022.101866
Oudeyer, P.-Y., & Kaplan, F. (2006). Discovering communication. *Connection Science, 18*(2), 189–206. https://doi.org/10.1080/09540090600768567
Oudeyer, P.-Y., Kaplan, F., & Hafner, V. V. (2007). Intrinsic motivation systems for autonomous mental development. *IEEE Transactions on Evolutionary Computation, 11*(2), 265–286. https://doi.org/10.1109/TEVC.2006.890271
Parker, G. A., & Maynard Smith, J. (1990). Optimality theory in evolutionary biology. *Nature, 348*(6296), 27–33. https://doi.org/10.1038/348027a0
Patricelli, G. L., & Hebets, E. A. (2016). New dimensions in animal communication: The case for complexity. *Current Opinion in Behavioral Sciences, 12*, 80–89. https://doi.org/10.1016/j.cobeha.2016.09.011
Peckre, L., Kappeler, P. M., & Fichtel, C. (2019). Clarifying and expanding the social complexity hypothesis for communicative complexity. *Behavioral Ecology and Sociobiology, 73*(1), 11. https://doi.org/10.1007/s00265-018-2605-4
Petre, I. (2012). Pneumatic muscle diameter evolution under compressed air action. *Journal of Electrical and Electronics Engineering, 5*(1), 185–190.
Pezzulo, G., Baldassarre, G., Butz, M., Castelfranchi, C., Butz, M. V, & Hoffmann, J. (2006). An analysis of the ideomotor principle and TOTE. In M. V Butz, O. Sigaud, G. Pezzulo, & G. Baldassarre (Eds.), *Proceedings of the Third Workshop on Anticipatory Behavior in Adaptive Learning Systems (ABiALS 2006)*. https://www.researchgate.net/publication/249781497
Pezzulo, G., Baldassarre, G., Butz, M., Castelfranchi, C., & Hoffmann, J. (2007). From actions to goals and vice-versa: Theoretical analysis and models of the ideomotor principle and TOTE. In M. V Butz, O. Sigaud, G. Pezzulo, & G. Baldassarre (Eds.), *Anticipatory Behavior in Adaptive Learning Systems: From Brains to Individual and Social Behavior* (pp. 73–93). Springer. https://www.researchgate.net/publication/249781497
Pickles, J. (2008). *An introduction to the physiology of hearing*. Emerald Group Publishing.
Pierson, H. A., & Gashler, M. S. (2017). Deep learning in robotics: A review of recent research. *Advanced Robotics, 31*(16), 821–835. https://doi.org/10.1080/01691864.2017.1365009
Pirjanian, P. (2000). Multiple objective behavior-based control. *Robotics and Autonomous Systems, 31*, 53–60. https://doi.org/10.1016/S0921-8890(99)00081-0
Pirjanian, P., & Matarić, M. (2001). Multiple objective vs. fuzzy behavior coordination. In D. Driankov & A. Saffiotti (Eds.), *Fuzzy logic techniques for autonomous vehicle navigation. Studies in fuzziness and soft computing: Vol. 61*. Physica.
Prasad, V., Stock-Homburg, R., & Peters, J. (2022). Human-robot handshaking: A review. *International Journal of Social Robotics, 14*(1), 277–293. https://doi.org/10.1007/s12369-021-00763-z
Proske, U., & Gandevia, S. C. (2012). The proprioceptive senses: Their roles in signaling body shape, body position and movement, and muscle force. *Physiological Reviews, 92*, 1651–1697. https://doi.org/10.1152/physrev.00048.2011.-This

Raphan, T., & Cohen, B. (2002). The vestibulo-ocular reflex in three dimensions. *Experimental Brain Research, 145*(1), 1–27. https://doi.org/10.1007/s00221-002-1067-z

Rawal, N., & Stock-Homburg, R. M. (2022). Facial emotion expressions in human–robot interaction: A survey. *International Journal of Social Robotics, 14*(7), 1583–1604. https://doi.org/10.1007/s12369-022-00867-0

Renaud, D. L., & Popper, A. N. (1975). Sound localization by the bottlenose porpoise *Tursiops truncatus*. *Journal of Experimental Biology, 63*(3), 569–585. https://doi.org/10.1242/jeb.63.3.569

Rescorla, R. A., & Solomon, R. L. (1967). Two-process learning theory: Relationships between Pavlovian conditioning and instrumental learning. *Psychological Review, 74*(3), 151–182. https://doi.org/10.1037/h0024475

Riener, R., Rabezzana, L., & Zimmermann, Y. (2023). Do robots outperform humans in human-centered domains? *Frontiers in Robotics and AI, 10*, 1223946. https://doi.org/10.3389/frobt.2023.1223946

Rigo, A., Chen, Y., Gupta, S. K., & Nguyen, Q. (2023). Contact optimization for non-prehensile loco-manipulation via hierarchical model predictive control. *2023 IEEE International Conference on Robotics and Automation, 2023-May*, 9945–9951. https://doi.org/10.1109/ICRA48891.2023.10160507

Romanova-Bolshakova, I. K., & Poleschikov, E. V. (2023). Reconfigurable wheeled walking robot with adaptive control. *AIP Conference Proceedings, 2833*(1). https://doi.org/10.1063/5.0151970

Rubio, F., Valero, F., & Llopis-Albert, C. (2019). A review of mobile robots: Concepts, methods, theoretical framework, and applications. *International Journal of Advanced Robotic Systems, 16*(2), 1–22. https://doi.org/10.1177/1729881419839596

Ruppert, F., & Badri-Spröwitz, A. (2019). Series elastic behavior of biarticular muscle-tendon structure in a robotic leg. *Frontiers in Neurorobotics, 13*, 64. https://doi.org/10.3389/fnbot.2019.00064

Russell, S. (2021). The history and future of AI. *Oxford Review of Economic Policy, 37*(3), 509–520. https://doi.org/10.1093/oxrep/grab013

Sakagami, Y., Watanabe, R., Aoyama, C., Matsunaga, S., Higaki, N., & Fujimura, K. (2002). The intelligent ASIMO: System overview and integration. *Proceedings of the 2002 IEEE/RSJ International Conference on Intelligent Rotots and Systems*, 2478–2483.

Sandoval, J. A., Jadhav, S., Quan, H., Deheyn, D. D., & Tolley, M. T. (2019). Reversible adhesion to rough surfaces both in and out of water, inspired by the clingfish suction disc. *Bioinspiration and Biomimetics, 14*(6), 066016. https://doi.org/10.1088/1748-3190/ab47d1

Sandoval, J. A., Sommers, J., Peddireddy, K. R., Robertson-Anderson, R. M., Tolley, M. T., & Deheyn, D. D. (2020). Toward bioinspired wet adhesives: Lessons from assessing surface structures of the suction disc of intertidal clingfish. *ACS Applied Materials and Interfaces, 12*(40), 45460–45475. https://doi.org/10.1021/acsami.0c10749

Schmitz, H., & Bleckmann, H. (1998). The photomechanic infrared receptor for the detection of forest fires in the beetle Melanophila acuminata (Coleoptera: Buprestidae). *Journal of Comparative Physiology A, 182*, 647–657. https://doi.org/10.1007/s003590050210

Schwaner, M. J., Hsieh, S. T., Swalla, B. J., & McGowan, C. P. (2021). An introduction to an evolutionary tail: EvoDevo, structure, and function of post-anal appendages. *Integrative and Comparative Biology, 61*(2), 352–357. https://doi.org/10.1093/icb/icab134

Seybold, J., Bülau, A., Fritz, K. P., Frank, A., Scherjon, C., Burghartz, J., & Zimmermann, A. (2019). Miniaturized optical encoder with micro structured encoder disc. *Applied Sciences, 9*(3), 452. https://doi.org/10.3390/app9030452

Shailer, M. J., & Moore, B. C. J. (1983). Gap detection as a function of frequency, bandwidth, and level. *The Journal of the Acoustical Society of America, 74*(2), 467–473. https://doi.org/10.1121/1.389812

Shalev-Shwartz, S., & Ben-David, S. (2014). *Understanding machine learning: From theory to algorithms*. Cambridge University Press.

Shangari, T. A., Shams, V., Azari, B., Shamshirdar, F., Baltes, J., & Sadeghnejad, S. (2017). Inter-humanoid robot interaction with emphasis on detection: A comparison study. *Knowledge Engineering Review*, *32*, e8. https://doi.org/10.1017/S0269888916000321

Shettleworth, S. J. (2010). *Cognition, evolution, and behavior* (2nd ed.). Oxford University Press.

Shield, S., Jericevich, R., Patel, A., & Jusufi, A. (2021). Tails, flails, and sails: How appendages improve terrestrial maneuverability by improving stability. *Integrative and Comparative Biology*, *61*(2), 506–520. https://doi.org/10.1093/icb/icab108

Shin, Y. K., Proctor, R. W., & Capaldi, E. J. (2010). A review of contemporary ideomotor theory. *Psychological Bulletin*, *136*(6), 943–974. https://doi.org/10.1037/a0020541

Siciliano, B., & Khatib, O. (2008). *Springer handbook of robotics*. Springer Verlag.

Skoyles, J. R. (2006). Human balance, the evolution of bipedalism and dysequilibrium syndrome. *Medical Hypotheses*, *66*(6), 1060–1068. https://doi.org/10.1016/j.mehy.2006.01.042

Snijders, L., Thierij, N. M., Appleby, R., St. Clair, C. C., & Tobajas, J. (2021). Conditioned taste aversion as a tool for mitigating human-wildlife conflicts. *Frontiers in Conservation Science*, *2*, 744704. https://doi.org/10.3389/fcosc.2021.744704

Spence, C. (2015). Just how much of what we taste derives from the sense of smell? *Flavour*, *4*, 30. https://doi.org/10.1186/s13411-015-0040-2

Stanford AI Lab. (2019). *Weak supervision: A new programming paradigm for machine learning*. https://ai.stanford.edu/blog/weak-supervision/

Stanton, L. A., Sullivan, M. S., & Fazio, J. M. (2015). A standardized ethogram for the felidae: A tool for behavioral researchers. *Applied Animal Behaviour Science*, *173*, 3–16. https://doi.org/10.1016/j.applanim.2015.04.001

Stenfelt, S., & Goode, R. L. (2005). Bone-conducted sound: Physiological and clinical aspects. *Otology and Neurotology*, *26*(6), 1245–1261. https://doi.org/10.1097/01.mao.0000187236.10842.d5

Stevens, M. (2013). *Sensory ecology, behaviour, and evolution*. Oxford University Press.

Szeliski, R. (2022). *Computer vision: Algorithms and applications*. Springer.

Tang, Z., Sekine, M., Tamura, T., Tanaka, N., Yoshida, M., & Chen, W. (2015). Measurement and estimation of 3D orientation using magnetic and inertial sensors. *Advanced Biomedical Engineering*, *4*, 135–143. https://doi.org/10.14326/abe.4.135

Tavakoli, M., Marques, L., & de Almeida, A. T. (2008). A comparison study on pneumatic muscles and electrical motors. *Proceedings of the 2008 IEEE International Conference on Robotics and Biomimetics*, 1590–1594. https://doi.org/10.1109/ROBIO.2009.4913238

Tinbergen, N. (1951). *The study of instinct*. Oxford University Press.

Tucker, V. A. (1975). The energetic cost of moving about. *American Scientist*, *63*(4), 413–419.

Tzschaschel, E., Brooks, K. R., & Stephen, I. D. (2022). The valence-dominance model applies to body perception. *Royal Society Open Science*, *9*(9), 220594. https://doi.org/10.1098/rsos.220594

Van Wanrooij, M. M., & Van Opstal, A. J. (2004). Contribution of head shadow and pinna cues to chronic monaural sound localization. *Journal of Neuroscience*, *24*(17), 4163–4171. https://doi.org/10.1523/JNEUROSCI.0048-04.2004

Vega, J. A., & Cobo, J. (2021). Structural and biological basis for proprioception. In *Proprioception*. IntechOpen. https://doi.org/10.5772/intechopen.96787

Wackermannová, M., Pinc, L., & Jebavý, L. (2016). Olfactory sensitivity in mammalian species. *Physiological Research*, *65*(3), 369–390. https://doi.org/10.33549/physiolres.932955

Ward, D., & Stapleton, M. (2012). Es are good: Cognition as enacted, embodied, embedded, affective and extended. In F. Paglieri (Ed.), *Consciousness in interaction: The role of the natural and social context in shaping consciousness* (pp. 89–104). John Benjamins Publishing. https://doi.org/10.1075/aicr.86.06war

Warren, B., & Nowotny, M. (2021). Bridging the gap between mammal and insect ears – A comparative and evolutionary view of sound-reception. *Frontiers in Ecology and Evolution*, *9*, : 667218. https://doi.org/10.3389/fevo.2021.667218

Webb, T. W., Sinha, I., & Cohen, J. D. (2021, December 28). Emergent symbols through binding in external memory. *9th International Conference on Learning Representations, ICLR*. http://arxiv.org/abs/2012.14601

Wiltschko, R., & Wiltschko, W. (2015). Avian navigation: A combination of innate and learned mechanisms. *Advances in the Study of Behavior*, *47*, 229–310. https://doi.org/10.1016/bs.asb.2014.12.002

Wiltschko, W., & Wiltschko, R. (2005). Magnetic orientation and magnetoreception in birds and other animals. *Journal of Comparative Physiology A: Neuroethology, Sensory, Neural, and Behavioral Physiology*, *191*(8), 675–693. https://doi.org/10.1007/s00359-005-0627-7

Xitco, M. J., & Roitblat, H. L. (1996). Object recognition through eavesdropping: Passive echolocation in bottlenose dolphins. *Animal Learning and Behavior*, *24*(4), 355–365. https://doi.org/10.3758/BF03199007

Yasa, O., Toshimitsu, Y., Michelis, M. Y., Jones, L. S., Filippi, M., Buchner, T., & Katzschmann, R. K. (2023). An overview of soft robotics. *Annual Review of Control, Robotics, and Autonomous Systems*, *6*, 1–29. https://doi.org/10.1146/annurev-control-062322

Zabala, U., Rodriguez, I., Martínez-Otzeta, J. M., & Lazkano, E. (2021). Expressing robot personality through talking body language. *Applied Sciences*, *11*(10), 4639. https://doi.org/10.3390/app11104639

Zafeiris, A., & Vicsek, T. (2018). *Why we live in hierarchies? A quantitative treatise*. Springer. http://www.springer.com/series/8907

Zamani, N., Moolchandani, P., Fitter, N. T., & Culbertson, H. (2020). Effects of motion parameters on acceptability of human-robot patting touch. *IEEE Haptics Symposium, HAPTICS, 2020-March*, 664–670. https://doi.org/10.1109/HAPTICS45997.2020.ras.HAP20.36.d8bb0c58

Zecca, M., Mizoguchi, Y., Endo, K., Iida, F., Kawabata, Y., Endo, N., Itoh, K., & Takanishi, A. (2009). Whole body emotion expressions for KOBIAN humanoid robot – preliminary experiments with different emotional patterns. *The 18th IEEE International Symposium on Robot and Human Interactive Communication*, 381–386. https://doi.org/10.1109/ROMAN.2009.5326184

Zhang, X., Gong, J., & Yao, Y. (2016). Effects of head and tail as swinging appendages on the dynamic walking performance of a quadruped robot. *Robotica*, *34*(12), 2878–2891. https://doi.org/10.1017/S0263574716000011

Zhang, Z., Wang, S., Liu, C., Xie, R., Hu, W., & Zhou, P. (2022). All-in-one two-dimensional retinomorphic hardware device for motion detection and recognition. *Nature Nanotechnology*, *17*(1), 27–32. https://doi.org/10.1038/s41565-021-01003-1

Zheng, C., Sane, S., Lee, K., Kalyanram, V., & Lee, K. (2023). α-WaLTR: Adaptive wheel-and-leg transformable robot for versatile multiterrain locomotion. *IEEE Transactions on Robotics*, *39*(2), 941–958. https://doi.org/10.1109/TRO.2022.3226114

Zheng, X., Shiomi, M., Minato, T., & Ishiguro, H. (2020). How can robots make people feel intimacy through touch? *Journal of Robotics and Mechatronics*, *32*(1), 51–58. https://doi.org/10.20965/jrm.2020.p0051

Zhu, H., & Thomas, U. (2023). Mechanical design of a biped robot FORREST and an extended capture-point-based walking pattern generator. *Robotics*, *12*(3), 82. https://doi.org/10.3390/robotics12030082

Zufall, F., & Leinders-Zufall, T. (2000). The cellular and molecular basis of odor adaptation. *Chemical Senses*, *25*, 473–481. https://doi.org/10.1093/chemse/25.4.473

4

ETHOROBOTICS IN PRACTICE

Past, present, and future – case studies

Beáta Korcsok, Ádám Miklósi and Judit Abdai

4.1 Introduction

In this chapter, we review the functional capabilities of several popular social robots, highlighting how they have been tailored to suit specific niches over the course of their technological development. We outline the software and hardware attributes that form the foundation of these capabilities, while also addressing the inherent limitations of these robots. Additionally, we discuss various studies in human–robot interaction (HRI) to shed light on areas of interest within this domain.

Each example of a robot is examined in detail from an ethorobotic perspective. We consider aspects such as embodiment, sensory capabilities, autonomous functionality, locomotion, object manipulation, and interactive skills, providing a critical analysis of their accomplishments.

4.2 NAO

4.2.1 Overview

NAO is a small humanoid robot (Figure 4.1). It was developed with the goal to serve as an easy-to-program robot with a friendly appearance, originally aimed at improving the life of its users (Gelin, 2019), which together with other features of this easy to transport and cost-effective humanoid robot platform with comprehensive interactive capabilities prompted its widespread use in various research applications (Amirova et al., 2021).

The development of NAO started in 2005 by the founder of Aldebaran Robotics (SoftBank Robotics since 2015), and the RoboCup started using NAO robots with its first release in 2008 (Gelin, 2019). The latest 6th version of the robot was released in 2018 and according to company data, currently there are more than 15,000 NAO robots in use all over the world.

 DOI: 10.4324/9781003182931-5

Figure 4.1 Photo of NAO, version 6. The robot shown in the picture is used for conducting social psychology HRI experiments by the Digital Interactions Research Group at the University of Debrecen, Hungary.

4.2.2 Embodiment

Apart from the various versions of the robot, different body types also exist. The robot has a total of 25 degrees of freedom (DoFs) in case of the H25 body type, and 23 DoFs in the H21 body type, which lacks the wrist yaw joints on both sides. The NAO T2 only features the head and the upper torso, and the T14 contains the head, upper torso and arms. Compared to the H25 body type (see Table 4.1), the Academics Edition of NAO has two more microphones and two infrared sensors in the eyes of the robot (SoftBank Robotics, 2024). A different head version is also available with a laser sensor for object avoidance and SLAM (Simultaneous Localization and Mapping). The runtime of the robot is between 45 and 90 mins with fully charged batteries. The detailed kinematic description of the robot together with technical guides and actuator and sensor datasheets are provided online by the manufacturer.

Table 4.1 Hardware overview of NAO (H25 body type, V5 version), mainly based on the NAO website (SoftBank Robotics, 2024). The ABOT score is from the Anthropomorphic Robot Database (Phillips et al., 2018), which measures the human-likeness of robots, providing a score between 0 and 100.

Height		57.4 cm
Weight		5.4 kg
Body plan		Humanoid
Locomotion		2 legs
Body surface		Plastic
DoFs	Total	25
	Head	–
	Neck	2
	Torso	1
	Legs	10 (2×5)
	Arms	12 (2×6, from which one-one is for closing/opening the fingers)
	Fingers	2×3 fingers: one thumb with 2 phalanges and two longer fingers with 3 phalanges per hand; fingers are curled or straightened via cables
Interactive hardware	LEDs	51 in total: 12 around touch sensor of the head, 2×10 around loudspeakers on the side of the head, 2×8 in eyes, 1 in chest button, 2×1 in feet
	Loudspeaker	2 (placed as ears)
	Screen	–
Sensors	Head	1 touch sensor (3 capacitive sensors), 4 microphones, 2 infra-red sensors placed as eyes (its presence varies between versions), 2 cameras (placed in the mouth and on forehead, with 60.9° horizontal and 47.6° vertical field of view)
	Neck	–
	Torso	1 touch sensor (switch), 3-axis accelerometer, 3-axis gyroscope, 2 ultrasonic sensors
	Legs	1 touch sensor (2-2 switch) per leg, 4 force-sensitive resistors per feet
	Arms	1 touch sensor per arm (3-3 capacitive sensors)
Operation time with charged battery		45–90 mins
Human-likeness score (ABOT)		45.92

The robot is commercially available for approximately $10,000–13,000 (in 2023). NAO received a score of 45.92 in human likeness by The Anthropomorphic Robot Database (ABOT) (Phillips et al., 2018). The ABOT human-likeness score is based on four aspects of robot appearance: body and manipulators, surface look, facial features, and mechanical locomotion.

4.2.3 Further developments and improvements

The sensorisation of NAO has been improved with newer models from 2008 to the most recent, V6 version with better resolution cameras (up to 2560 × 1920 px for the top, and 1280 × 960 px for the bottom camera) and microphones (from 10 mV/Pa sensitivity and 20 Hz–20 kHz frequency range, to omnidirectional microphones with 250 mV/Pa ± 3dB sensitivity and 100 Hz to 10 kHz frequency range). The overall structure of the robot was also improved by changing various plastic parts to metal and the robot received magnetic shielding, resulting in a 0.5 kg additional weight in the current version (Gelin, 2019).

However, the vast array of software developments could have had a greater impact on the robot's widespread use. The original robot programming framework, NAOqi (SoftBank Robotics, 2024) (run via the NAOqi OS, formerly on the OpenNAO, operating system, a GNU/Linux distribution), provides graphical programming (via the Choregraphe desktop application) and feedback monitoring options, as well as SDKs (software development kits) for Python and C++. The NAOqi has been frequently updated since the release of the robot, improving the features and programmability of NAO. Additionally, with the spread of NAO's use, first a ROS (Robot Operating System) driver, and later a full ROS stack has been released. Other software options were released as well, for example, NAOsim (Brown et al., 2013), a virtual world simulator for testing algorithms before implementing them on the robot, MONITOR, for receiving data from the sensors and cameras (Amirova et al., 2021), AskNao Blockly, a graphical programming application for informatics education, and AskNao Tablet for special-needs education, for the purpose of helping children with autism spectrum disorder (ASD) with, for example, various interactive games, imitation, body schema learning.

4.2.4 Autonomy

The robot can be controlled remotely or can function autonomously. Autonomous operation is supported by the Autonomous Life mode (see also the description of the *ALAutonomousLife* module in Section 4.2.6) together with the Basic Channels, various downloadable applications, and the potential of user-created activities. Although NAO has been used autonomously already in studies from 2010, the majority of studies were conducted via teleoperation, Wizard of Oz methods, or via pre-scripted sequences of behaviour elements (Amirova et al., 2021). Technical advancements moved forward the use of autonomous functioning in studies (e.g., Pulido et al., 2017), especially in technological research, while in user studies in healthcare and assistive robot roles for children with ASD, the individual requirements made the continued use of the Wizard of Oz method of control popular. NAO's autonomous functionality has been broadened over the years with the continued development of NAOqi, as well as by multiple research projects, in which they developed various cognitive architectures for the robot (Amirova et al., 2021).

4.2.5 Sensory and motor capabilities

The specific interactive capabilities of NAO are based on several different modality inputs and outputs. The functional capabilities are described based mainly on the NAO and NAOqi websites (SoftBank Robotics, 2024).

4.2.5.1 Vision

The visual inputs of NAO are provided by the two cameras on the robot's head which are placed at the centre of the forehead and in place of its mouth. NAO's recognition and visual processing capabilities can be used among others for navigation, detecting specifically coloured areas, landmarks, movement, learning and recognising objects and locations, and people detection. The vision-based interactive skills of NAO consist of tracking and orienting towards humans; detecting in which of the three adjustable engagement zones the human is in (based on the closeness of the stimulus to the robot), enabling the use of specific behaviours in the zones; face detection with a limit of 2–4 m distance (with face position and angular coordinates of facial features); facial characteristics detection (age, gender, smile, and emotion estimation); gaze direction detection; detection of sitting/standing position and waving; and recognition of pre-specified individuals. HRI studies have readily utilised these visual capabilities for gesture recognition, joint attention, and imitation tasks; however, due to e.g., hardware limitations, these studies also frequently utilise additional external sensors.

4.2.5.2 Hearing

The robot can locate sound sources, react to them, and can recognise verbal commands in multiple languages from a list of pre-specified phrases according to manufacturer specifications. The number of recognised languages has been increased from 9 to 19, then 22, however, most studies seem to use additional custom hardware or software for speech recognition instead of only using NAO's own systems (e.g., Sphinx-4 speech recogniser (Mubin et al., 2014), Pocketsphinx automatic speech recognition system with a multipass decoder (Heinrich & Wermter, 2011), and Baidu API for Chinese speech recognition (Han et al., 2018)).

A study by Kennedy et al. (2017) used the speech recognition system Nuance VoCon 4.7, originally implemented on NAO, to investigate various effects of, for example, different microphones, background noise, and test setups (distance and orientation to the robot) on the recognition of children's speech in English. The speech was produced by 11 children with a mean age of 4.9 years (±0.3 SD). The speech recognition was performed on recorded audio files. The distance and orientation of the sound source greatly influenced the speech recognition using NAO's built-in microphones and the Nuance engine, with recognition rates decreasing steeply after 50 cm distance from the robot: in case of relatively clean audio samples (i.e., minimal background noise) the recognition rate dropped from 73% at 25 cm distance to 65% at 50 cm, 27% at 75 cm, and finally to 4% recognition at 1 m. The recognition rate of utterances from noisy sound samples dropped more at 50 cm to 44%. Regarding the angle of the sound source, when the source is at more than 45° angle from the front of the robot recognition rate decreases greatly, and reaches 0 at 90° (which could be improved, as off-the-shelf only the frontal 2 out of 4 audio channels are used for speech recognition).

The authors also compared four speech recognition engines, the Nuance, the Google and the Microsoft Speech APIs, and the Pocketsphinx with audio files recorded with a studio microphone. The Google Speech API performed best with a mean average normalised Levenshtein Distance (mean LD) value of 0.34, and the Nuance was the worst with a mean LD value of 0.76 in case of fixed utterances (recorded by repeating pre-defined sentences). The LD indicates from 1 to 0 how well the transcript produced by the recognition engine matches the actual utterance on the level of letters, 0 indicating total match and 1 indicating no matched letter. The LD is normalised for utterance length in order to be usable for spontaneous speech utterances (Kennedy et al., 2017). In spontaneous speech, again the Google API had the best mean LD value (0.39), and the Nuance and PocketSphinx both had the worst 0.8 mean LD. Even though the Google Speech API was found to be the best from the four speech recognition engines, it was still considered insufficient to adequately recognise the speech of children in HRI, as if the researchers also considered semantic closeness of the original utterances produced during realistic spontaneous speech and their transcripts, the recognition rate was only 18%. This is in part explained by the frequent grammatical errors in the speech of children, but recognition was also only 38% in case of grammatically correct, fixed utterances. Both of these recognition rates were 0 with the PocketSphinx and NAO's Nuance engine. With the Google API if the confidence threshold of the engine was set to 0.8, a close to 50% semantically correct recognition rate could be achieved with spontaneous speech, but the engine would not process the majority of the utterances at all (only 36 utterances from 222).

While children's speech seems to be not recognised adequately by NAO, and its microphones are harshly criticised (Mubin et al., 2014), based on the official documentation of NAO the robot still has some speech recognition capabilities, which we assume works better in highly optimised circumstances. The low-level speech recognition (via the *ALSpeechRecognition* module) enables the robot to go through the pre-specified phrases when a human voice is perceived and list out the phrases it recognised with the probability of correct recognition. Word spotting is also possible, in which the robot looks only for a specific word, regardless of the surrounding speech. More complex speech recognition is provided by the *ALDialog* module.

The robot has the option to distinguish emotions from human speech, selecting from unknown, sorrow, joy, anger, or calm. NAO is also capable of detecting the source location of sounds via the Interaural Time Differences measured by its four microphones (the amount of time difference between detecting the sound with the sensors, based on the direction of the sound). In optimal circumstances, the localisation accuracy can in theory reach 10° (in contrast to 1–2° in humans); however, background noise greatly decreases it. NAO can only locate one sound source at a time and cannot filter out human voices from other sources using the standard NAOqi software.

4.2.5.3 Touch

NAO provides the possibility of creating various reactions to tactile stimuli; however, this feature is less frequently utilised in HRI studies. The *ALTouch* module of the NAOqi framework raises events when a capacitive sensor is being touched, or if the touch has finished, while in body parts without capacitive sensors touch detection is based on the comparison of commands sent to joints, and the feedback via joint sensors. The *ALTactileGesture* module enables the robot to recognise a list of 18 pre-defined tactile motions (e.g., single tap, double tap) performed by a user on the robot's three capacitive head sensors, to which new tactile motions can also be added. Similarly, the *ALChestButton* module is responsible for raising events if the button is pressed. The robot can react to various touch events by the means of the *ALMemory* module, as defined by developers.

4.2.5.4 Other sensing capabilities

NAO has four force-sensitive resistors (FSRs) in each foot, sensing pressure in a 0–25 N range. The robot also has an inertial measurement unit (IMU), comprised of a 3-axis accelerometer (measuring acceleration in m/s^2) and a 3-axis gyroscope (providing rotation speed values in rad/s), from which the torso angle data (in rads) is also computed. Both the FSRs and the IMU aid the robot in maintaining its balance. The two ultrasonic sensors located in NAO's torso can be turned on to sense obstacles (including the arms of the robot) in front of the chest in a 60° cone, with distance measurements in a 20–80 cm range and with 1–4 cm resolution. There are also joint position sensors in NAO, enabling a sort of proprioception: the magnetic rotary encoders provide the angular positions of the joints. All the motors are equipped with a current sensor, measured in amperes. This enables the control of stiffness of the joints, contributing to, e.g., fall protection and gives the basis of computed temperature values, with the goal of protecting motors from overheating.

4.2.5.5 Speech

NAO can provide verbal outputs, based on different text-to-speech engines, with the option of changing some acoustic parameters of the emitted voice (e.g., pitch, speed) in general, or at specific words/texts via tags. Tags also enable the robot to synchronise particular actions with bookmarked verbal outputs. NAO comes with two installed languages, English and the local or requested language, while two additional languages can be acquired from the set of available languages.

4.2.5.6 Actions

There are predefined positions that NAO can use (three positions for standing, two for sitting, lying on back and stomach, and crouching), and a library of more than 200 default behaviour elements (called animations) it can exhibit, with the option of creating new behaviour elements as well, with, namely, infinite posture variations.

4.2.5.7 Gestures/signals

NAO is able to perform numerous hand gestures due to its high degrees of freedom. It can also express light signals produced with the LEDs located on its body and head.

4.2.6 Software architecture and related system capabilities

The NAOqi is the primary framework running on the robot (or on a computer with a simulated NAO), providing a cross-platform and cross-language environment with introspective application programming interface (API) for robot programming (SoftBank Robotics, 2024). The NAOqi contains various remote or local modules, in multiple functional categories (in version 2.8 these are the core, emotion, interaction engines, motion, audio, vision, people perception, and sensors and LEDs categories), as well as the LoLa (formerly DCM, Device Communication Manager) module.

In NAOqi 2.8, the core modules and services establish the main base functions of the system (SoftBank Robotics, 2024). The modules and their main functions are presented in Table 4.2. In the NAOqi core, we can also find extractor modules, which serve as base classes for modules in other functional categories, belonging to vision and perception.

Vision modules provide video and picture recording, barcode reading, managing video input settings and optimising processing resources, detecting lighting conditions, keeping tabs on previous head positions (and therefore the past visually perceived areas of the environment), as well as further modules for various vision-related tasks and cognitive capabilities (SoftBank Robotics, 2024). For example, the detection of movement in the environment when the robot itself is not moving (*ALMovementDetection*); detection of visual markers called Naomarks, which can be placed in the environment of the robot, aiding localisation (*ALLandMarkDetection*); localisation based on the comparison of a reference image or a composite 360° panorama image to the current visual inputs (*ALVisualCompass* and *ALLocalization*); separate modules for the fast detection of blobs with specified colours (*ALColorBlobDetection*) and for the detection of red, circular objects (*ALRedBallDetection*), and even the recognition of other previously learned specific objects/visual features, based on keypoint matching (*ALVisionRecognition*).

Audio modules manage all audio inputs and outputs, audio recording and playback, sound detection and the localisation of its source, and low-level speech-related audio functions (SoftBank Robotics, 2024). The basic speech recognition via the *ALSpeechRecognition* module uses the speech recognition software of Nuance (see also Section 4.2.5 Hearing), while speech synthesis is provided via the *ALTextToSpeech* module, using the ACAPELA, Nuance or microAITalk speech synthesisers, depending on the language (see also Section 4.2.5 Speech). As an enrichment, the *ALAnimatedSpeech* module provides randomised or contextual movements during speech synthesis, with the help of the *ALSpeakingMovement* module (belonging to the interactive engines of NAO). Annotated text can be accompanied by contextually relevant downloadable or user-created animations (movement patterns).

Table 4.2 Overview of NAOqi Core modules and services with their functions. Table adapted and modified based on the NAOqi website (SoftBank Robotics, 2024).

Module name	Function
ALBehaviorManager	Managing e.g., what behaviours should be started automatically with NAOqi
ALMemory	Serving as centralised memory on hardware configurations, current states of actuators and sensors, serving as an event notification hub
ALResourceManager	Managing resources hierarchically, such as actuators, stiffness parameters, CPU, etc., for optimal behaviour synchronisation
ALExpressionWatcher	Creating complex events based on multiple other events, e.g., if conditions are fulfilled due to the presence of multiple stimuli
ALUserSession	Managing active users as a state machine of people currently perceived and one person tracked by the robot, identifying known individuals with the help of the *ALAutonomousLife* module, creating temporary profiles for new users
ALUserInfo	Managing persistent data of users (each user has to be created previously by the *ALUserSession* module)
ALModule	Handling methods of user modules, creating custom modules
ALKnowledge	Service, managing persistent data as triples (defining subject, predicate and object relations)
ALWorldRepresentation	Storing persistent data in a spatial graph structure
ALNotificationManager	Managing event and state notifications for the end-user
ALDiagnosis	Managing hardware diagnostics
ALConnectionManager	Managing network settings
ALPreferenceManager	Managing robot settings
ALSystem	Managing system configurations (e.g., robot name, time zone) and operations (e.g., rebooting, shutting down)
PackageManager	Managing application packages
ServiceManager	Managing previously installed services

Various modules are responsible for managing the sensors of NAO and controlling its LEDs (SoftBank Robotics, 2024). There are designated modules dealing with internal stimuli, e.g., processing sensor data from the battery, monitoring temperature status, and diagnosing overheating.

Other modules are responsible for processing environmental stimuli, such as data from the FSRs in the robot's feet, which indicate whether at least one foot is in contact with the ground. These modules also handle tactile stimuli from the chest button and other tactile sensors, as well as simple or complex tactile gestures applied to the robot's head. Additionally, they process ultrasonic sensor data to facilitate object avoidance (in case of NAO, object avoidance is limited to stopping when an object is detected at one of the pre-set security distances). While most modules in this category deal with inputs, the *ALLeds* module enables the control and management of single or multiple LEDs of the robot.

The people perception modules of the NAOqi system contribute to the interactive capabilities of NAO by providing recognition of various human features and related spatial aspects (SoftBank Robotics, 2024). The *ALPeoplePerception* module tries to detect people in the robot's environment, and maintains lists of currently visible and not visible, but recently detected people with the respective aspects (e.g., position, time since detection), while another module estimates if the people are sitting or standing. Waving motions performed from a short distance from the robot can also be detected by the *ALWavingDetection*. The *ALFaceDetection* module enables the detection of faces, providing their positions and the angular coordinates of the main facial structures, as well as face recognition based on the 2D camera images of individuals whose face the robot has learned beforehand. The *ALFaceCharacteristics* module estimates age, gender, and facial gestures of the detected faces, collecting input for e.g., emotion modules, while the *ALGazeAnalysis* module detects human gaze orientation and the closing of eyelids. The *ALEngagementZones* module handles spatial zones defined around the robot, dividing it into 3 separate zones in which if objects or people are detected, the robot can react differently.

The LoLa (Low-Level Abstraction) is a key component of NAOqi, which controls the hardware of NAO, while incorporating the ALMotion, ALTouch, ALRobotModel, and ALNavigation modules. The HAL (Hardware Abstract Layer) is responsible for ensuring the communication between LoLa and the electronic boards of hardware components (SoftBank Robotics, 2024) (Figure 4.2).

Motion modules control the movements of NAO (SoftBank Robotics, 2024). The *ALMotion* module is responsible for most movement functionalities, including stiffness control of all or specific joints, positions of joints (either via blocking calls or reactive control), locomotion control (navigation goal, velocity, footstep planning, etc.), Cartesian control of NAO's effectors (the endpoints of kinematic chains, e.g., right hand, head; or all joints) and its whole body via inverse kinematics, reflexes such as balance, self-collision and fall protection, external collision avoidance (via stopping at security distances; only works with ALNavigation), Smart Stiffness to optimise energy saving by limiting joint torque, safety measures based on reactions to joint, actuator and sensor malfunctions, idle behaviour control (maintaining a posture or imitating breathing motions), programming tools aiding motion development, and motion task handling. NAO can exhibit many default animations (behaviour elements) via the *ALAnimationPlayer* service, which aids the *ALBehaviorManager* module (SoftBank Robotics, 2024). NAO can also take up predefined postures via the *ALRobotPosture* module, and its navigation is controlled by *ALNavigation*. NAO, using NAOqi, is not yet able to avoid obstacles during navigation and stops instead when an obstacle is perceived. Finally, the *ALTracker* module controls NAO's ability to follow faces, a red ball, landmarks, specified people, and sound locations, which positions the robot can track and can turn towards with only its head, its whole body or can also move towards in order to maintain a given distance from it. The *ALTracker* module uses various tracking modules for the different types of targets, belonging to vision, audio, and people perception module categories.

The interaction engines of NAO give the backbone of its autonomous functions and interactive communication with humans (SoftBank Robotics, 2024). Verbal

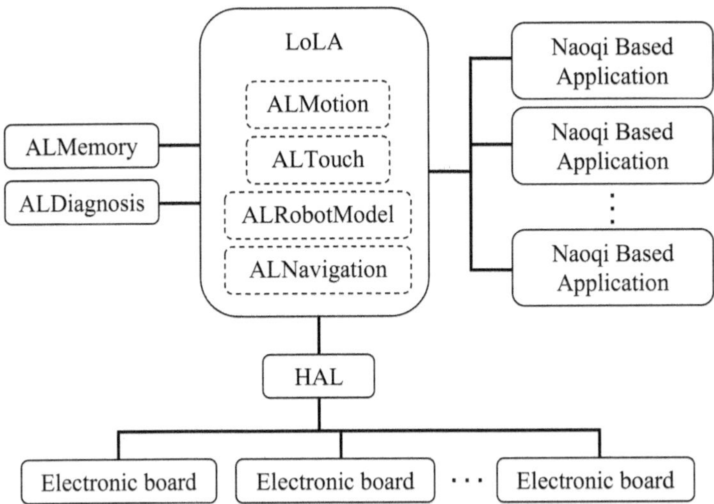

Figure 4.2 The structure of the hardware control architecture of NAO. The LoLa (Low-Level Abstraction) incorporates the ALMotion, ALTouch, ALRobotModel, and ALNavigation modules, while the HAL (Hardware Abstract Layer) ensures communication between LoLa and the electronic boards of hardware components from SoftBank Robotics, 2024.

communication is handled by the *ALDialog* module, in which various rules, categorised into functional topic groups, can be formulated. With the rules, the robot can respond to specified verbal inputs of the user with its own verbal outputs or can say things with no prior user speech. Applications created by the user can also be started verbally on the robot. The autonomous behaviours of NAO are mainly organised into the *ALAutonomousLife* module, and some more specialised, smaller modules that are activated depending on *ALAutonomousLife* states. These modules provide so-called *autonomous abilities,* e.g., blinking with LEDs, small autonomous movements during idle time and when the robot is listening or speaking, as well as behaviours related to awareness about others. The *ALBasicAwareness* module uses information of multiple sensory modules to enable the robot to focus its attention on various modalities of incoming stimuli by looking towards them, and if a person is detected, tracking them (Figure 4.3). The level of robot engagement can be specified at three levels, making it more or less distractable from the tracked person.

The *ALAutonomousLife* module works as a state machine (Section 3.8.2), with four states called solitary, interactive, safeguard, and disabled. The solitary state has two main functions (SoftBank Robotics, 2024): searching for people to interact with, and performing actions that can increase its perceived aliveness and might attract the attention of people. In solitary state, the robot can also perform activities such as exploration and learning. The solitary state is considered as the default state, from which the robot switches to the interactive state when an interactive activity is started. The interactive state is focused on robot–human interaction. The safeguard state is activated in case of critical hardware failure or environmental risks to the robot, if the

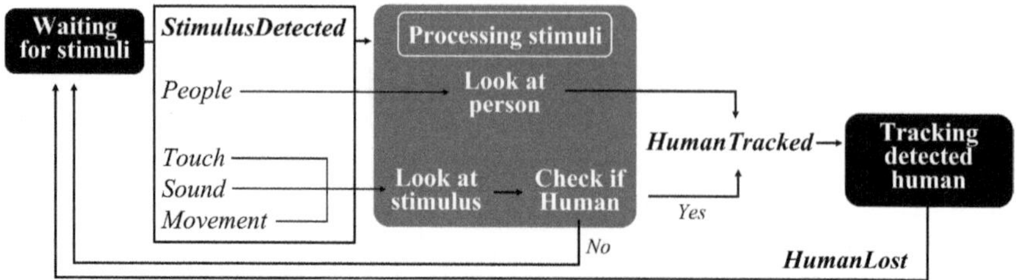

Figure 4.3 Diagram showing the functioning of the ALBasicAwareness module. The module utilises multiple types of stimuli provided via other modules to detect and track important environmental stimuli, e.g., humans. The considered stimuli for NAO are people (detected human based on camera image, *ALPeoplePerception* module); touch (touch detected on the robot's arm or head, *ALTouch* module); sound (any sound registered by the robot, *ALSoundLocalization* module); movement (detected movements, *ALMovementDetection* module) (Based on the NAOqi website figure, modified to exclude stimuli specific to the Pepper robot; SoftBank Robotics, 2024).

currently running activity is not prepared to handle the event. Finally, in the disabled state, the activities and autonomous abilities are inactive.

The *ALAutonomousLife* module manages both built-in *autonomous abilities* and user-created or downloadable *activities*. It uses perception and vision modules as extractors to gain information about the environment, contributing to the outer and internal *context* which the robot is in. The context limits what activity can be running, while also enables the use of *launch trigger conditions* with which activities can be started autonomously if their specified conditions are met. Only one activity can be running at a time, but autonomous abilities continue to run, unless specifically disabled by the user or by higher priority events (e.g., activities). Activities that can be started by *ALAutonomousLife* have to be categorised as either solitary or interactive, corresponding to the module's related state machine states, with interactive activities having priority over solitary ones.

Emotion modules provide two emotion engines for NAO (SoftBank Robotics, 2024), the *ALMood* and the *ALRobotMood. ALMood* deals with the affective stimuli surrounding the robot: when the *ALAutonomousLife* module is active, *ALMood* estimates the emotional state and attention of a person the robot is focused on (which is determined by the *ALUserSession*), by incorporating the degree of smiling and facial expressions interpreted as neutral, sad, happy or angry, expressed as a probability score between 0 and 1 for each expression (via the *ALFaceCharacteristics* module). For the attention estimation, *ALMood* takes into account the head and gaze orientation (via *ALGazeAnalysis*), resulting in the user being characterised as fully engaged, semi-engaged, or unengaged. The direction of gaze is also estimated. The *ALMood* module also gauges a general affective ambience surrounding the robot, based on the sound (noise energy level) and amount of detected movement in the environment, forming a calm or excited environment estimation. *ALMood* determines both high-level and low-level representations of these stimuli that can be accessed and utilised by developers: high-level information keys (valence and attention) also contain previously estimated emotional states and environmental (ambience) and contextual (user profile) information, while low-level information keys provide simple, current observations (e.g., smile).

The other emotion engine, the *ALRobotMood* module takes into account internal and external stimuli to create an emotional state for the robot (SoftBank Robotics, 2024): the evaluation of the physical state of the robot is based on the battery level, contributing to a pleasure component (this component can be negative, positive or neutral, on a scale of −1 to 1). External stimuli also influence the pleasure scale, while providing the basis for the other component as well, called excitement. These external stimuli are procured via the *ALMood* module, and are based on the perceived emotional state of the people in the robot's environment, and the interpretation of the calm/excited level of environmental ambiance. The emotional state of the robot usually intends to reflect the emotional state of the people, and the excitement level of the environment around it.

These features are available in the original NAOqi framework, but a long line of research projects have dealt with improving the capabilities of NAO, developing or adapting a range of novel software components.

4.2.7 Utilisation in HRI research

4.2.7.1 Assistive role

NAO has been utilised in multiple research studies investigating the applicability of Socially Assistive Robots (SAR) in therapies. Pulido et al. (2017) have developed and evaluated an autonomous system called NAOTherapist that provides movement therapy with NAO to children in need of upper-limb rehabilitation. The system aims to motivate children to perform exercises through social interaction via an imitation game in which they are encouraged to recreate poses shown by the robot, and it also provides verbal and visual feedback in order to facilitate the correct replication of poses.

The autonomy of the NAOTherapist is achieved with a complex system architecture, divided into three levels. The high-level planning selects various exercises from the system's database to be included in a therapy session and recommends novel exercises if there are no available ones in the database.

The medium-level planning (Figure 4.4) is responsible for executing the actual therapy session and handles the interaction with the children. The environment of the robot is modelled via functions and environmental (e.g., detection of a human) and internal predicates (e.g., changes due to planned actions). The system supports the replanning of actions by the processes of the Executive component, which deals with the representation of the environment, and the Decision Support component, which generates an expected state of it. Replanning occurs when the two components differ in their predicates, or in case of unexpected events. The medium-level planning also deals with the pose comparison between what the system perceives as the current body position of the child (based on video frames), and what is the position that the child should recreate. Poses are considered correct based on an experimentally

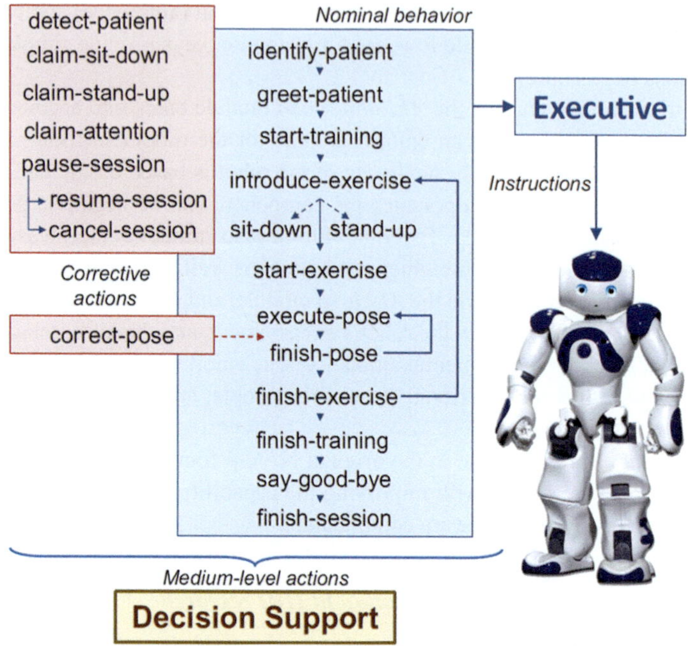

Figure 4.4 Diagram of the initial planning and corrective actions for replanning in the Nao-Therapist system (figure from Pulido et al., 2017).

determined, dynamic threshold of comparison, the time percentage of failure or success, and the amount of time the pose is maintained for. In case of an incorrect pose, the system can determine the error in the imitation and the robot signals to the child what to change: after a first failed attempt NAO signals verbally and by twisting its own wrists of the affected arms, then after a second failed attempt by repeating the demonstration of the required movement. The medium-level planning also controls the actions that the robot performs autonomously during the therapy sessions, for example, detecting and greeting the patient, introducing exercises, executing robot poses and correcting the poses of the children, pausing sessions, finishing poses, exercises or the session, or saying goodbye. Finally, the low-level planning controls subtasks (e.g., planning movement trajectories, changing the colours of LEDs, moving an actuator to a specific position) that the robot has to perform to be able to exhibit medium-level actions. The system also gets input data from a Microsoft Kinect sensor placed on a table behind the robot, for measuring and tracking exact positions of the children's body parts, integrated via a RoboComp component utilising Microsoft's Software Development Kit (SDK).

The NAOTherapist system was tested in two main steps. First, the system was set up in two schools, in which a total of 117 healthy school children, between the ages of 5 and 9, interacted with NAO for approximately 5 mins/child, performing two types of exercises. This stage of the experiments was used to analyse various aspects of HRI with children to discover possible technical issues and to further develop the system architecture. Furthermore, in both this and the next stage, the researchers wanted to investigate multiple viewpoints. Regarding the children, they examined how well were the children engaged in the exercises, if they could follow the instructions of the robot and if they liked NAO and considered it a social agent; from the robot's perspective, whether the robot could adequately complete the sessions autonomously; and finally, whether observing experts considered the robot useful.

During the sessions NAO autonomously interacted with the children, guiding them throughout the exercises during the imitation game, complete with various socially interactive elements (e.g., greeting, goodbye), as well as corrective actions based on replanning (e.g., pose correction, attention attracting, asking for the child to sit down/stand up). The researchers used a simple questionnaire to ask children about the behaviour of the robot, its perceived capabilities, aliveness, how much they liked various aspects of the robot and the session, etc. The sessions were recorded on video, and 50 of the sessions were analysed to measure the behaviour of the children. Finally, they also scored the performance of the children regarding correct pose execution. The children could attempt to produce each pose three times before it was considered a failed pose. Healthy school children had to use on average 24% of the possible attempts and failed the execution of 9.65% of poses when following NAO's instructions. As these children were not familiar with physiotherapy exercises, misunderstandings could occur about how precisely they should imitate the robot. Additionally, the children sometimes incorrectly assumed that the robot could hear them.

In the second set of experiments, three children between the ages of seven and nine with cerebral or brachial plexus palsy participated. Their sessions were longer, taking about 15–20 mins and containing four exercise types. An added questionnaire was also introduced for measuring the opinions of experts (e.g., physicians) and the children's families who observed the sessions about the usefulness of NAO. The performance of the NaoTherapy system and its autonomous functioning was evaluated positively and was considered useful by therapists and other experts as well. Some limitations were observed in certain poses due to the sensing capabilities of the Kinect. The children readily interacted with the robot, oriented towards it most of the time and followed its instructions. The paediatric patient children used 61% of the allowed attempts to achieve the correct poses and failed between 22.7% and 31.8% of the poses. All three children could realise during the longer, 15 mins sessions that the robot was not able to hear them.

4.2.7.2 Guide role

The number of experiments exploring HRI in real-world scenarios where NAO is employed as a guide is relatively small, despite several studies showcasing robot development tailored for such applications (e.g., development of a system with which NAO can play an interactive game with visitors, helping them discover exhibitions; Mondou et al., 2017).

Gehle et al. (2014) shed light on the difficulties faced by robots aimed at functioning in museum environments and by the visitors trying to interact with them. They conducted two consecutive studies in which a NAO robot provided information about the exhibited artworks to the visitors and tried to attract them with closed (yes/no answers, Study 1), and open and closed questions (Study 2). The goal was to investigate the behaviour of the visitors regarding their engagement and their communicative behaviours in case of problems during the interaction.

The verbal communication and gestures of the robot were mostly pre-defined, with timed silent intervals for the visitors' answers; however, the robot could orient towards the visitors' faces during, for example, greeting and at set moments during the interaction. The system itself was also enhanced between the studies: in the first study the visitors had to wear helmets, which was recognised by a VICON motion capture system to provide information for the robot. NAO did not have acoustic inputs and stayed at a fixed location. In the second study, the sensors of the robot were used to establish gaze behaviours, and NAO's speech recognition was utilised to detect yes/no answers (however, these answers did not modify the behaviour of the robot), and the robot was also moving around on top of a table autonomously. In both setups, the robot was displayed for multiple days at the museum, and while participants were aware of being recorded for an experiment (and also had to wear helmets in the first study, and microphones in the second), they were not familiar with the goal of the studies or with NAO. The interactions were recorded on video.

Two types of analysis were conducted for both studies. Quantitative analyses of the video recordings of 75 and 37 interactions were used with descriptive statistics to explore the prevalence and type of answers/clarifying questions of visitors (interacting with the robot as individuals or in groups). Second, qualitative case evaluation of verbal communication, gaze orientation, and body posture changes were conducted using conversation analysis methods in 1-1 example interactions to gain some insight into how visitors signal problems during the interaction.

Visitors readily answered the questions of the robot in the majority of cases, providing both yes/no and elaborate answers, and some were also willing to repeat/rephrase their answers if the robot asked two similar questions after each other (e.g., 'How does it look like?' and 'What does it represent?'). The main problems during the interaction originated from the robot's inability to time turn-taking in the conversation according to the behaviour of visitors, and from when the visitors asked clarifying questions or provided answers that the robot then ignored. Visitors whose answers the robot ignored or interrupted tended to look at other members of their group and disengaged from the interaction for a short time. The study emphasises the difficulty of museum guide robots when verbally interacting with visitors or groups of visitors in noisy environments and proposes the possibility of using body posture and gaze behaviours as an indication of failure in communication for the robotic systems.

4.2.8 Ethorobotic perspectives

4.2.8.1 Embodiment

NAO should be considered as a small adult human model. Its size is smaller than children capable of walking (its height is equivalent to that of two-four-month-old babies), which can lead to confusion (Alenljung et al., 2018). Not surprisingly, many HRI studies with NAO place the robot on top of a desk to provide a better view. The small size has also been raised as an issue of perceiving the robot as fragile.

NAO's body proportions are a mixture of that of adults and children. Human body proportions change drastically during ontogeny (Burdi et al., 1969). The head length of infants is around 25% of their body length, with a circumference narrowly exceeding the circumference of the chest cavity, and the upper body with the shoulders is slightly bigger than the head. By adulthood, the head is only about 14% of the body height, and much smaller in circumference than the chest cavity. The cranium of newborns features a big forehead and has a more elongated shape from the side view, while being relatively circular from the front. In infants, the cranium is also much larger than the facial area, only constituting 12.5%, which is 40% in adults, making the skull more vertically elongated from the front view. The proportion of limb lengths also changes, as in babies the upper limbs are longer, but by adulthood the legs are almost 17% longer than the arms. In NAO, the vertical length of the head is around 20% of its body length; the head is elongated horizontally instead of vertically, matching the width of the chest area. Its eyes are placed far apart and are relatively big, creating a childlike impression. NAO's legs are around 13%–14% longer than its arms, and its broad shoulder and small waist also resemble that of adults, similarly to its trunk-leg proportion (Figure 4.5).

The placement of the cameras (on the forehead and in the 'mouth') can be misleading, as people interacting with the robot could reasonably assume that the visual sensors of the robot are placed in its eyes and might show objects or perform hand gestures specifically in front of the eyes of NAO.

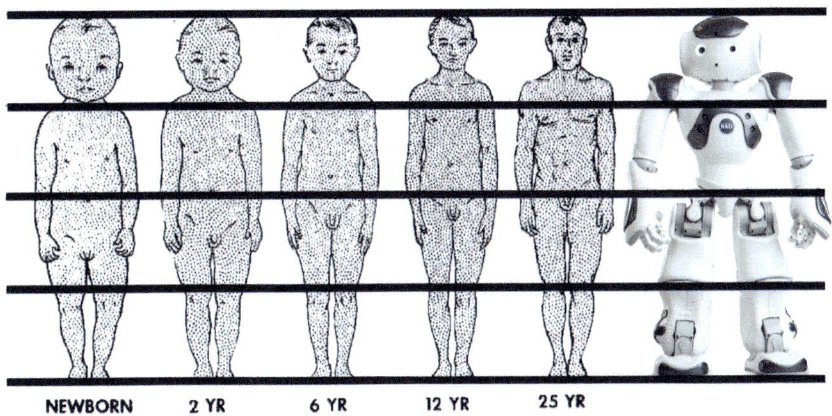

Figure 4.5 Body proportions of humans during ontogeny (figure from Burdi et al., 1969), compared with the body proportions of NAO. In NAO, the proportions do not expressly follow adult or child characteristics. Thus, overall NAO has an adult–child hybrid appearance, because the ratios of the legs to the arms, the trunk to the legs, and the shoulder to the waist are more similar to that of adults, while the big, set apart eyes are somewhat indicative of childlike features.

4.2.8.2 Sensing

Even though NAO features several sensors in multiple modalities, the quality of some sensors is not ideal for varied interactive scenarios and communicative capabilities. Speech recognition is greatly influenced by environmental noise, distance of the speaker from the robot (with a frequently used limit of 2 m distance), the noise caused by the head movements of the robot, the functioning of the electrical drive of the laser scanner if the robot is equipped with one on its head, as well as the restricting quality of the built-in microphones (Mubin et al., 2014). Similarly, the vision-based skills of the robot are restricted by the cameras, as it is greatly influenced by, for example, the lighting conditions (Albani et al., 2016), its field of view, and by the limitation of only being able to use one of them at a time. Most vision-based skills are optimised for office lighting, and their robustness generally decreases in relation to lighting, tilt, pan, and rotation of e.g., an object to be recognised, and its size in the camera image (i.e., distance from the robot). Due to the placement of the cameras in the head, the fields of vision have no overlap. In applications where stereo vision is needed, researchers used a modified head setup in which the cameras were placed in the position of the eyes, and the additional ability of using both cameras at the same time, providing the necessary visual field overlap (Müller et al., 2014).

4.2.8.3 Autonomous operation

The usability of the NAO robot outside of research and education is limited by various factors. The runtime of the robot is shortcut by the battery's capacity, and while batteries can be switched, this requires turning off the robot and screwing off a back panel. To achieve autonomous charging, researchers have modified the robot's hardware creating electrical contacts on its back, enabling the robot to dock into a docking station e.g., with the help of a reinforcement learning algorithm (Navarro et al., 2011). Chair-shaped docking stations are commercially available for the robot, intended for autonomous charging. However, even though the NAOqi system contains a module responsible for automatic charging (*ALRecharge*), as of now this function is not available for NAO. NAO can be used with constant charging via cable, but this restricts the working area of the robot, and is warned against for damage concerns. The lack of object avoidance is also limiting the array of complex interactive scenarios in which NAO could participate.

4.2.8.4 Locomotion

The general use is limited by the locomotive capabilities of NAO. While the robot is capable of autonomous walking and navigation, including climbing stairs, the speed of locomotion (maximum walking speed of NAO is 0.4 m/s) is low for HRI (Gelin, 2019). Researchers have developed a new walking pattern generation method for increased stability, with which the robot achieved only 0.24 m/s walking speed with body posture control (Liu & Urbann, 2016).

The Whole Body Balancer (via the *ALMotion* module) uses generalised inverse kinematics to optimise NAO's motions, considering all joints of the robot to provide more stability, redundancy, and safety, but it is not usable during walking. The robustness of walking is not ideal, especially on soft surfaces.

To accommodate potential falls, multiple solutions are implemented in NAO. The structure of the robot was developed to withstand a high number of falls, aided by the flexibility of its printed circuits in its neck, as well as the specially created cylindrospheric actuation modules designed for the neck, elbows, and shoulders of NAO, also enabling backdrivability (meaning that a force acting on these limbs of the robot can be transferred to the motors, increasing shock resistance). NAO detects falls as they occur and moves its head and hands to a protected position and lowers joint stiffness, as to ensure that during impact the joints are not damaged. When the Autonomous Life and Basic Channel options are active and the robot falls, it tries to get up to a standing or sitting position and restore the last stable position it was in. If this fails, the robot turns off its joint stiffness, as to enable a human to put the robot in a stable sitting/crouching position.

4.2.8.5 Object manipulation

The three fingers on NAO's hands enable the robot to grasp and manipulate small objects in a limited fashion. The grasping mechanism of the fingers is operated via metal cables, with which the robot can curl or straighten its fingers, with the stiffness of fingers being maximal in the closed position. NAO can only grasp objects firmly with one hand if they can be enclosed by the fingers and the palm, creating a form-closure type grasp (Müller et al., 2012).

In an object-handover task, NAO had to autonomously recognise cups or pens, be able to take the object from a human holding it, then give it back if asked to (Müller et al., 2014). The time necessary for NAO to grasp the object was on average around 7 s. For NAO to hand over a lightweight, empty cup, at least two conditions had to be fulfilled: (1) it had to hear a verbal request from the human, (2) it had to see the human's hand or feel a touch on the back of the robot hand. Following this, the robot would release the object if it sensed it being pulled (measured indirectly by the traction force acting on the robot's arm). Although in the study they greatly increased the manipulation capacities of NAO, the system necessitated the use of a modified NAO head for stereovision. As of now, the off-the-shelf manipulation skills of NAO are not robust enough to be considered ideal for functional use.

4.2.8.6 Interactive skills

A recurring shortcoming in the interactive skills of NAO expressed by the participants of HRI studies is the limited expressiveness, naturalness of the robot, including the ability to show attentive and turn-taking behaviour. The lack of facial expressions also limits emotion expression to sounds and gestures, while participants would prefer the sounds and gestures of NAO to be more natural (as reviewed by Albani et al., 2016).

The size of the robot poses other limitations. While enabling better portability, the small size of NAO makes many manipulations and cooperative tasks impractical which narrows the scope of HRI.

4.2.9 Future utility

We expect that the primary use of NAO and other robots with similar embodiments and hardware limitations will remain in the application areas of research concerned with education and socially assistive robotics. Based on its high DoFs and well-programmable movement capabilities, there could be a time to re-think its embodiment and adjust the hardware (especially sensors) and software capabilities to the needs of more practical applications.

4.2.10 Similar robots

Mobile robots with similar humanoid body structures (Figure 4.6) are anticipated to fulfil more elaborate roles in the future, also indicated by the announcement of the development of Tesla's humanoid and human-sized robot, Optimus in 2021, with its first prototype introduction in 2022, and a demonstration of Gen 2 in December of 2023. According to Malik et al. (2023), the Optimus is designed to be 173 cm tall, weighing 57 kgs with a carrying capacity between 20 and 10 kgs (depending on arm

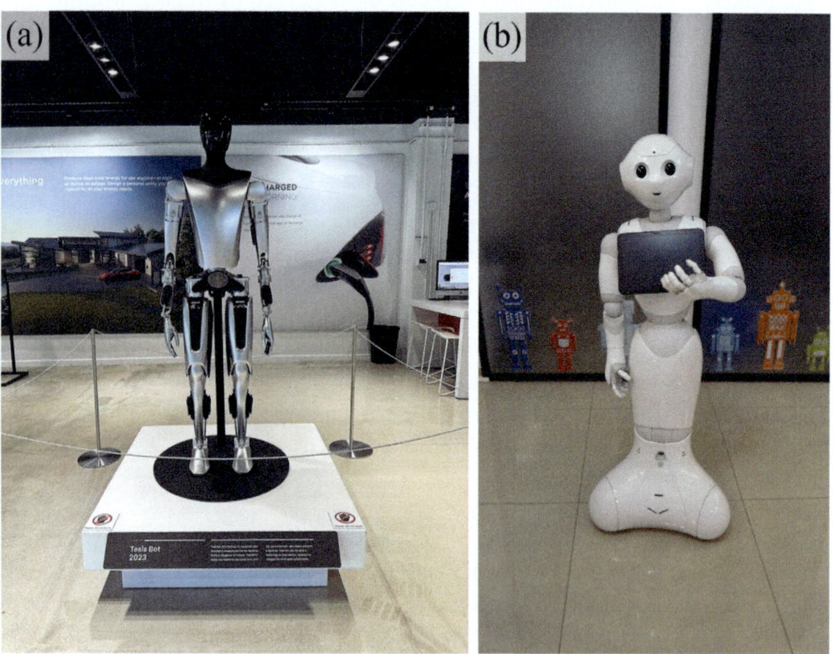

Figure 4.6 Examples of mobile humanoid robots: (a) Optimus (Tesla) (photo by Phillip Pessar, Flickr), (b) Pepper (SoftBank Robotics).

posture) and 8 km/h locomotion speed. As not much detail is available at this time, the expectations and reactions towards the Optimus are mixed. The robot promises good object manipulation capabilities due to its humanoid hand structure, which could be an advantage over most humanoid social robots.

A widely used robot is Pepper, which can be considered the conversation-focused counterpart of NAO. Pepper is produced by the same company as NAO, also utilising the NAOqi software with some elaborated functionality (e.g., tracking an object and maintaining a specified distance from it, by navigating with object avoidance). Pepper is larger with its 121 cm height and uses wheels as a means of locomotion instead of legs, but it shows an overall similarity to NAO in many aspects of its embodiment.

4.3 Kaspar

4.3.1 Overview

The humanoid robot Kaspar was created with the goal of aiding children living with ASD, both by furthering the research area and by serving as a specifically designed tool and interactive partner for such children in therapeutic settings, helping them in the development of their social skills. Importantly, the robot is aimed at serving as a learning tool and as a mediator to improve children's interactions with others, not as an artificial replacement of social partners.

Kaspar has been developed by Blow, Dautenhahn, and others in 2005 (see Dautenhahn et al., 2009) and has been continuously refined ever since in the Adaptive Systems Research Group, resulting in 5 main generations of Kaspar robots by 2019 (Wood et al., 2021).

4.3.2 Embodiment

The first version of Kaspar (K1) was the size of a two-year-old, with a 16 DoFs embodiment that ensured a wide variety of facial gestures, had black and white cameras in its eyes, and 2 hours of operation time. The robot construction was optimised to cost less than 2,000 euros. The embodiment of the Kaspar robot received a score of 75.95 out of 100 in the 2019 version of ABOT (Phillips et al., 2018) in overall human-likeness. Details of the embodiment of Kaspar are listed in Table 4.3.

4.3.3 Further developments and improvements

An alternate version of Kaspar was constructed (K1-L) with the aim of serving as a development platform, with added laser depth camera and feedback sensors in its joints. The K1-L version, in contrast with later generations, was not optimised to be portable, as it was mounted on a PC and was also bigger in size, resembling a six-year-old. The DoFs of the robot were increased to 18, with the addition of two DoFs in its arms. Kaspar was used to investigate whether the success of cued imitation of waving gestures was affected by the robot's congruent or incongruent waving. Unlike interactions with human partners, Kaspar's demonstration of congruent or

Table 4.3 Hardware overview of Kaspar (version K5.5) and its ABOT score measuring human-likeness between 0-100 (Phillips et al., 2018).

Height		56 cm
Weight		N/A
Body plan		Humanoid (seated)
Locomotion		No locomotion
Body surface		Silicone skin and 3D printed NinjaFlex TPU for joint covering
DoFs	Total	22
	Head	8 (2 in mouth, 2×3 in eyes)
	Neck	3
	Torso	1
	Legs	-
	Arms	10 (2×5) + neodymium magnets in hands for holding objects
	Fingers	5 fingers, not movable
Interactive hardware	LEDs	-
	Loudspeaker	Mounted on head
	Screen	-
Sensors	Head	4 force sensitive resistors on the face, 2 cameras in the eyes
	Neck	-
	Torso	1 force-sensitive resistor on chest area, RFID reader
	Legs	4 force-sensitive resistors (2×1 on each leg, 2×1 on each foot)
	Arms	6 force-sensitive resistors (2×1 on each arm, and 2×2 on each hand: 1 on the palm, 1 on the back of the hand)
Operation time with charged battery		7 hours
Human-likeness score (ABOT)		75.95

incongruent actions did not influence the actions carried out by the human partner (Shen et al., 2009). The consecutive Kaspar robots were the improved versions of K1, with each generation augmented with better sensing capabilities, the optimalisation of the level of autonomy, and by the use of newer fabrication techniques (see Figure 4.7). While only single units existed from the first four versions of Kaspar, the number of robots increased to 7 with the K4, and then to 20 with the K5 versions, with the aim of providing robots for use in nursery schools, schools, and eventually in homes.

4.3.4 Autonomy

The research studies that utilise Kaspar for child-robot interactions divide the development of the system's (including the robot itself and the necessary external devices) autonomous capabilities into two main phases, with phase one being remote-controlled (Wizard of Oz) functioning, and phase two semi-autonomous/

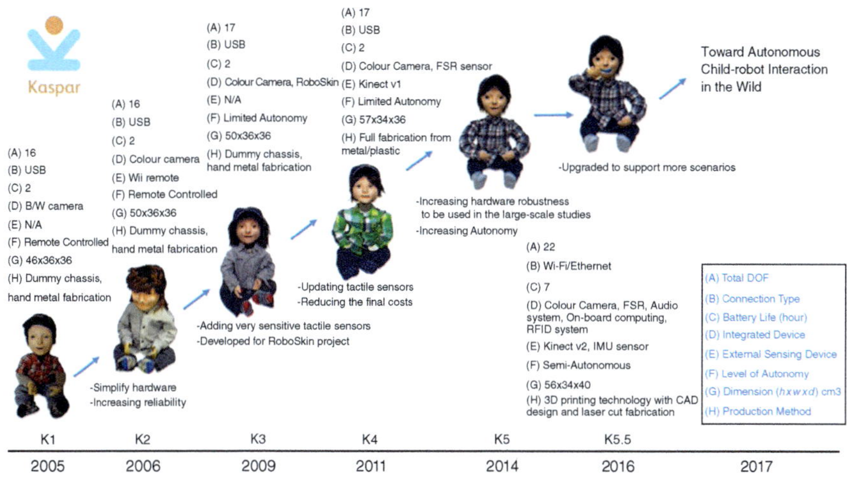

Figure 4.7 Comparing the different versions of Kaspar from 2005, based on (A) the total number of DoFs, (B) connection types, (C) the hours of battery life, (D) integrated devices, (E) external sensing devices, (F) level of autonomy, (G) their dimensions, and (H) the method of production (figure from Wood et al., 2021).

autonomous operation (see also Zaraki et al., 2017). The first two versions of the Kaspar robot were remote-controlled, by the means of a 20-key wireless keyboard. Specific pre-programmed movements were initiated with one button presses, also used in collaborative interactions. Remote control and the programming of novel behaviours were also available via a graphical user interface designed to be used with Kaspar (Kaspar Wizard of Oz (KWOZ); Dautenhahn et al., 2009). Additionally, the Kaspar system was also endowed with the capability of autonomous gameplay. While the system worked autonomously, Kaspar as a separate embodied agent did not: the robot's behaviour was controlled via an external laptop (Kose-Bagci et al., 2009). As the researchers intended to use the K2 robot to investigate dyadic and triadic interactions between the robot and children, Nintendo Wii remotes (wearable devices on the children's arms) were integrated with the robot's system, enabling autonomous operation in collaborative games. Autonomous operation was developed using the software Yet Another Robot Programme (YARP). However, the full autonomy during gameplay and children's difficulty with the wearable Wii sensors resulted in decreased flexibility in the games and the exclusion of low-functioning children with ASD. To ensure that the system is ideal for a wide range of children, further developments of Kaspar robots aimed at achieving a level of semi-autonomy instead of full autonomy, maintaining flexible robot operation in which the human operator is still included in the high-level control of the robot, but the immense cognitive load of fully remote controlling a robot in complex interactive scenarios is lowered (Wood et al., 2021).

A later version of the Kaspar (K5.5) system was equipped with a new additional software architecture with which semi-autonomous functions can be achieved (see

Section 4.3.6). The software architecture is responsible for autonomously processing environmental inputs (*Sense layer*), integrating these with additional high-level information and selecting the appropriate action (*Think layer*), and performing the action and providing feedback on the success of execution (*Act layer*). However, the action is only exhibited by the robot if an operator approves it. The system controlling the robot's autonomous behaviour is operated on a separate laptop, not on the robot itself.

4.3.5 Sensory and motor capabilities

4.3.5.1 Vision

The first version of Kaspar had two monochrome (black and white) miniature cameras embedded in its eyes, and later visual sensing was enhanced with colour cameras in the K2 version. Despite this, much of the vision-based capabilities of Kaspar rely on the use of sensors, such as a Kinect and a full-HD camera exterior to the robot's embodiment. The vision system of the newer Kaspar models, as part of the Sense layer of the Sense-Think-Act software architecture, is capable of object recognition and tracking (including humans) via the use of these exterior sensors (Zaraki et al., 2017). With the Microsoft Kinect SDK, the Sense layer can detect 5–6 people with 25 joints for each, enabling tracking, as well as gesture and head position detection.

For the detection, tracking, and orientation estimation of other objects, e.g., toys, the object analysis module of the Sense layer uses various filters and methods (e.g., blob detection) to find colour regions in the image acquired via an external wide-angle RGB camera. The robot operator can view the visualised colour regions in Kaspar's user interface and can select a point on a target object, which the system then uses as the basis of further image filtering, followed by object locating and tracking (Zaraki et al., 2017). The same object filtering and tracking capabilities are based on the image provided by the built-in cameras in Kaspar's eyes in the K5.5 version (Wood et al., 2021) and are applicable for the tracking of multiple objects at the same time (Zaraki et al., 2018).

4.3.5.2 Hearing

Most versions of the Kaspar robot were not equipped with microphones. Several studies have been conducted in which hearing capability was provided via exterior microphones, enabling, for example, drumming-based interactive games (Dautenhahn et al., 2009). More complex audio-based capabilities were also achieved with such microphones, for example, researchers used the microphone array of a Microsoft Kinect for tracking a speaking individual, electing from multiple children. A beamforming method is used for localisation (beamforming methods are based on simultaneous data collection of sound pressure values with multiple microphones; Gombots et al., 2017), during which the system compares the sound source location (the beam angle) with the 3D position of the children in the robot's field of view, also measured by the Kinect (Zaraki et al., 2017). By the means of the Windows speech recognition platform, Zaraki et al. (2017) implemented a speech recognition feature into the software system of Kaspar, also via the use of a Kinect. However, Kaspar only recognised predefined phrases, and the recognition was optimised for adult voices. An auditory system has been mentioned in connection to the newer (K5-K5.5) versions, indicated as an integrated device on a figure (see Figure 4.7), but we did not find further information on its capabilities or the level of integration (Wood et al., 2021).

4.3.5.3 Touch

The initial Kaspar robots did not have tactile sensors. The third Kaspar robot, K3, was equipped with spots of ROBOSKIN tactile sensors, containing distributed capacitive pressure sensors (taxels) applied to the arms, feet, and chest of the robot in big patches consisting of smaller triangular patches (12 sensors/triangular patches, resulting in a total number of 1,128 taxels). The sensors could detect varying pressures (soft/hard touch, hitting/stroking), due to the elastic effect of an upper foam layer covering the sensors. With these sensors, it was possible to detect and classify touch patterns occurring during HRI, independently of the overall location of the touch on the robot's body. The method was optimised to be used with relatively low amount of training data and was based on the creation of a codebook containing the local structures of taxels and the occurring pressure distributions, which were then classified via a histogram-based approach (Ji et al., 2011).

In later versions, the ROBOSKIN prototype sensors were replaced with ten FSRs, with a smaller type of FSR sensor for Kaspar's head, hands, and shoulders, and larger ones for other areas. The FSR sensors were covered with beige skin-coloured neoprene foam in the K4 (Barbadillo et al., 2011). The K5 version contained 15 FSR sensors in total, covered with silicone. It is unclear whether these sensors were also able to distinguish between soft and hard touch.

4.3.5.4 Other sensing capabilities

The K5.5 version was improved with the inclusion of a Radio Frequency Identification (RFID) reader used for easily changing between game modes, and with the intention to be later included in autonomous functions. The software architecture controlling the Kaspar K5.5 also supports the addition of environmental sensors. For example, during interactive games, the system is made aware of the exact position and orientation of specific objects (e.g., a toy cube) via IMUs embedded in them, which are also modelled in a virtual visualisation, aiding robot operators. The 3D positioning of the toy cube enables the system to detect which side of it is turned towards the robot and which one towards the child. This information is used for developing the Visual Perspective Taking skills of children: understanding the line-of-sight of others, and that different features of an object might be visible to them (Wood et al., 2021).

Joint position sensors were incorporated in the K1-L version of Kaspar to measure the current positions of the arm joints via rotary potentiometers (Dautenhahn et al., 2009). The servos used in Kaspar 5 and newer robots have integrated position sensors, providing the appropriate feedback for Kaspar's software system. According to the manufacturer's specifications, the main servos of the motors also have feedback capabilities for speed, voltage, temperature, and load; however, it is unclear whether these are utilised by Kaspar's software.

4.3.5.5 Speech

The earlier versions of Kaspar were used with exterior loudspeakers, which were in later versions integrated to the head of the robot. As the initial research goals with Kaspar focused on non-verbal communication, the development of a speech system was not a priority. Kaspar can provide verbal outputs with pre-recorded audio files, frequently deployed in a Wizard of Oz framework (Gou et al., 2023), with the possible use of some autonomous phrases triggered by sensor inputs (e.g., Kaspar says '*That's nice, it tickles me*' when its leg is tickled; Syrdal et al., 2020).

4.3.5.6 Actions

Kaspar has no locomotive or other leg movement capabilities and is fixed in a seated position. The robot can execute a set of pre-defined actions. For example, in a 2008 study by Mirza et al. (2008), the robot exhibited 17 actions in a peekaboo game. The actions belonged to three groups: nine movement actions (e.g., hide head with hands, wave left arm), three facial expressions (smile, neutral, frown), and five resetting actions (e.g., head to forward position). With the inclusion of an added DoF, the K3 version was able to turn left or right with its torso, which has been used, for example, for the robot to turn away when a tactile sensor detects the robot being hit by a child, enabling children to learn about appropriate and inappropriate tactile interactions. The K5.5 robot also has some limited object manipulation skills, realised via the neodymium magnets embedded in its hands, with which it can hold (and with its overall arm movements, move) various objects to teach/encourage children with ASD for their uses, for example, a spoon or a toothbrush (Wood et al., 2021).

The pre-programmed actions that the robot can perform during remote-controlled operation or semi-autonomously via the Interaction Architecture (see in Section 4.3.6) are recorded in external files. The behaviour sequences are played via the Act layer which runs on an external computer (Zaraki et al., 2018). The files determine all the servo position values the robot has to take up, together with the option of playing specific audio files.

4.3.5.7 Gestures/signals

Kaspar is capable of showing simple facial gestures, as well as emotional behaviour via face, arm, and torso movements. Facial gestures are based on the DoFs allowing movement for the eyes, eyelids, and the mouth. The mouth can be moved to create a smile or a frown and can be opened and closed as well (e.g., for a yawn, see Huijnen et al., 2021). The eyes are also movable up-down and left-right, while the eyelids can be closed for blinking. The three DoFs of the neck enable head movements, such as, for example, shaking its head or tilting it to the side.

4.3.6 Software architecture and related system capabilities

The software architecture of the Kaspar robots used only for remote control is less published. Remote controlling the robot was aided by a graphical user interface (KWOZ) in case the human operator wanted to control Kaspar via a laptop, and not via the custom remote-control keyboard which could also be utilised by the children. In early studies (e.g., turn-taking during drumming game; Kose-Bagci et al., 2010), autonomous operation was achieved using the YARP framework (Dautenhahn et al., 2009), which was used, for example, as a communication layer between custom software modules and as an interface towards the robot hardware, aiding more complex software systems with which Kaspar was able to learn and select its actions based on the rewarding environmental signals in its interaction history (Mirza et al., 2008).

A new software architecture was implemented for the semi-autonomous functioning of the latest Kaspar models (Wood et al., 2021). The Interactive Architecture (also called as a hybrid Deliberative-Reactive Control Architecture and the Sense-Think-Act architecture) runs on an external computer. It has been developed to both decrease the cognitive effort needed from the human operators and to provide a way of autonomously recording the events and sensory information that occur during the interactions (Zaraki et al., 2018). The goal of the system is to collect and analyse environmental inputs and to select and exhibit the most adequate robot behaviours based on these and other high-level information (e.g., game status) available to the robot. The Interactive Architecture uses *objects* to define salient items with specified characteristics, such as the robot itself, its environment, children, and the toys used during the HRI studies. Each object has its defined features, conceptualised by the sensory information the system can collect about the object (Zaraki et al., 2017). The architecture consists of three interconnected layers with separate functions, the *Sense*, the *Think*, and the *Act* layers (Figure 4.8). The layers have specific modules for processing different sensory and high-level information. Data handling between the layers (e.g., between the *Think* and *Act* layer) was in the past established via YARP ports (Zaraki et al., 2017). Later the architecture was implemented using the Intelligent Real-time Interactive Systems Toolkit (IrisTK), designed to provide an easy-to-adapt system and modules for multimodal and multi-party HRI (Skantze & Al Moubayed, 2012), in which the IrisTK Broker (a TCP/IP network) is responsible for relaying information between the modules of the different layers. Information is sent as Java Script Object Notation data packets called *events*, which function either as *sense*, *action,* or *monitor* events, which are available for all layers. The layers or specific modules do not have to receive input about all events, the modules can subscribe to specific events needed for their operation to optimise the computational costs of the system (Zaraki et al., 2018). Further reduction of the computational cost is possible, as the system is capable of functioning in a distributed manner across multiple processors (Wood et al., 2021). Additionally, it is also platform and

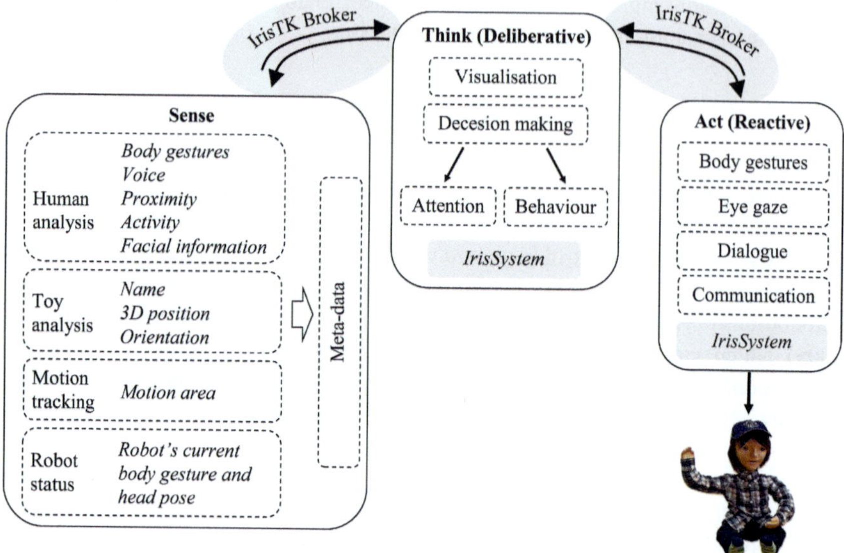

Figure 4.8 Figure showing the Sense-Think-Act software architecture of Kaspar. The Sense layer collects sensory information and processes them in designated modules (Human analysis, Toy analysis, Motion tracking, and Robot status modules) and provides the resulting meta-data to the Think layer, relayed by the IrisTK Broker. The Think layer is responsible for decision-making processes that elect two possible next actions for the robot, one of which is then enacted via the Act layer after a human operator selects it (drawn based on Wood et al., 2021; Zaraki et al., 2017).

programming language independent, enabling the easy addition of new modules and hardware.

The *Sense* layer is responsible for collecting data both from internal and external sensors, providing information for the *Think* layer about the robot and its environment. The perception modules of this layer process the incoming information and produce the necessary data for the *Think* layer via *sense events*. The main modules of the *Sense* layer are the Human analysis, Toy analysis (also referred to as Object analysis), Motion tracking and Robot status modules, with most available descriptions focusing on the first two. The Human analysis module processes all the sensory information needed for detecting and tracking humans, recognising their body and head posture, localising and tracking a speaking person during multi-person interactions, and recognising some pre-defined phrases in speech (for more details see Section 4.3.5). As the input stimuli are provided by a Kinect as an exterior sensor, the Microsoft Kinect Software Development Kit v2.0 has been integrated into the *Sense* layer (Zaraki et al., 2018). The Object analysis module (sometimes referred to as *Toy analysis module*, or two separate modules: *3D Orientation of Cube and Turn Table Tracking* and *Object Recognition*) processes incoming data about specific physical objects (toys and their accessories) used during child–robot interactions. The 3D orientation is detected via IMUs with Bluetooth connection embedded into the objects (a toy cube and a turntable), with the sensors' own API (see also Section 4.3.5). The 3D orientation data is relayed to the *Think* layer as an orientation vector. Object recognition is based on camera images as inputs, either from external cameras or the cameras in Kaspar's eyes. It uses the AForge.Net image processing library, and filters colours automatically based on an area selected by the robot operator on the GUI (see also Section 4.3.5) (Zaraki et al., 2018).

The *deliberative* functions of the software architecture are fulfilled by the *Think* layer. In a previous iteration, the layer used rule-based action selection (Zaraki et al., 2017), but in later system versions the *Think* layer decision making is conducted via an IrisTK subsystem, the IrisFlow. The IrisFlow was developed as a tool for dialog flow control for multi-party HRI scenarios, using a statechart method. IrisFlow can handle hierarchically structured states and multiple levels of sub-states. Event handlers are used for transitioning between states (Skantze & Al Moubayed, 2012). All the games in Kaspar's system are implemented as states of one interaction flow. Going from an initial state at the beginning of the child–robot interaction, the system transitions between states and substates with the *goto* and *return* commands, according to the information coming from the *Sense* and *Act* layers. The IrisFlow system also considers high-level pre-defined information, for example, the rules of the games, how to interpret sense events and how to react to specific behaviours of children (Zaraki et al., 2018). The endpoint of the decision process is the action event that the *Think* layer should send to the *Act* layer.

The system also contains a component that enables a human operator to approve or modify the action/behaviour sequence that the robot should display next. The system offers two options to choose from, showing them on the system's GUI: one is the result of the *Think* layer deciding based upon the inputs it received from the *Sense* layer, while the other is determined by the current status of the interaction and the Interaction Flow (e.g., the progression of the game) based on the IrisFlow module (Zaraki et al., 2018).

The *Act* layer serves as the *reactive* part of the software architecture, responsible for the control signals of Kaspar, by which the robot can display a sequence of movements/audio playback (see also Section 4.3.5). The *Act* layer is only subscribed to the *Think* layer, which sends it action events. The modules of the *Act* layer are the Gaze, Gesture, Dialogue and Data communication modules. After the robot displays an action, the *Act* layer sends a monitor (feedback) event to the other layers (Zaraki et al., 2018).

4.3.7 Utilisation in HRI research

4.3.7.1 Assistive role

Assistive robotics is one of the most prominent areas in which HRIs are studied, but there is still more focus on short-term, laboratory-based experiments.

In contrast, Syrdal et al. (2020) have conducted the (to our knowledge) longest running *in situ* assistive robotics study, in which two–six years old children participated in frequent interactions with Kaspar for 16.53 months on average. The importance of this study is signified by the way it was conducted: the robot was used in a nursery (TRACKS autism) for children with ASD and was operated by the staff of the nursery according to their own considerations, without hands-on researcher oversight, providing a major stepping stone for the development of assistive robots. As the authors emphasise, assistive robots are developed and most frequently used in laboratory environments, or at least under the direct control of researchers.

The ultimate goal of Kaspar, however, is to be operated in various child-centric environments in schools, nurseries, and homes, by users other than trained engineers and researchers. Developing a robot prototype for laboratory experiments and 'field use' is very different however, the latter presenting more diverse and unexpected challenges and problems. To face such challenges, Syrdal et al. (2020) investigated the use and acceptance of Kaspar as a new technology on an individual and organisation level, and aimed to identify the points of improvement for Kaspar by the insights of the end users. The research loosely followed a method of Davis' Technology Acceptance Model (Davis et al., 1992) and considered factors such as extrinsic benefits of using the robot (e.g., therapy outcomes), intrinsic benefits (e.g., interaction quality), ease of use (e.g., technical difficulties), intention to use Kaspar, and actual use of the robot. It is unclear which exact model was used in the study but based on the specifications (17 DoFs and ROBOSKIN tactile sensors) we assume that it was the K3 version.

The robot was used with limited autonomy, with some simpler autonomous actions (e.g., reacting to tactile stimuli). The primary participants of the study were three members of the staff of the nursery, while 19 children were secondary participants. The staff was already somewhat familiar with Kaspar, as previous research studies were also conducted in the nursery. In addition, prior to the start of this study, the staff received two training sessions, one for detailed instructions, and one for practice of use with the help of researchers present. Other than these demonstrations, the actual manner of use during the following months (frequency, types of games played, etc.) was decided by the staff.

Due to the longevity of the study, data collection had to be optimised with some trade-offs; therefore, data on the primary participants was collected via questionnaires with open-ended questions and in-detail interviews in four occasions, in different stages of robot deployment. The data of children was also collected with questionnaires, in which the staff made an initial assessment when a child entered the study, and a final one at the end. These questionnaires measured five areas of mental functioning (perception, cognitive, psychomotor, social-emotional, and communication domains). The results revealed that none of the children experienced negative effects in these domains during the study, and correlation analysis between the number of interactions with Kaspar and the changes in various domains demonstrated significant development in the perception skills when controlled for total duration of participation.

The staff questionnaire data was handled together with information from the semi-structured interviews and was described in chronological order to show the progression of Kaspar's use according to the aspects of the Technology Acceptance Model. The first set of interviews preceded the training of the staff. Originally, their main fears were about the technological operation of the robot, setting up the robot, and handling hardware breakdowns during child-robot interactions. They hoped, however, that the use of Kaspar will have positive effects on children, and that the research on Kaspar will progress. By the second visit, the staff had a one-month experience with the robot and could highlight multiple points of development (e.g., easier way to change between games), as well as frequent hardware problems (e.g., overheating of the neck servos) to be solved, and recommended solutions to avoid the issues from occurring. The staff also described so-called 'wow-moments', in which they experienced a child interacting with Kaspar showing unexpected novel behaviour, indicative of positive developments (e.g., a mediating effect of the robot between the child and the adults). By the third visit, the use of the robot became ordinary part of the nursery's activities, and the staff was more confident and proficient in handling technical issues. However, they also described some bottlenecks in the use of Kaspar (e.g., the place to use the robot in, currently available staff to use the robot with, the current desire for children to interact with the robot). The importance of intrinsic

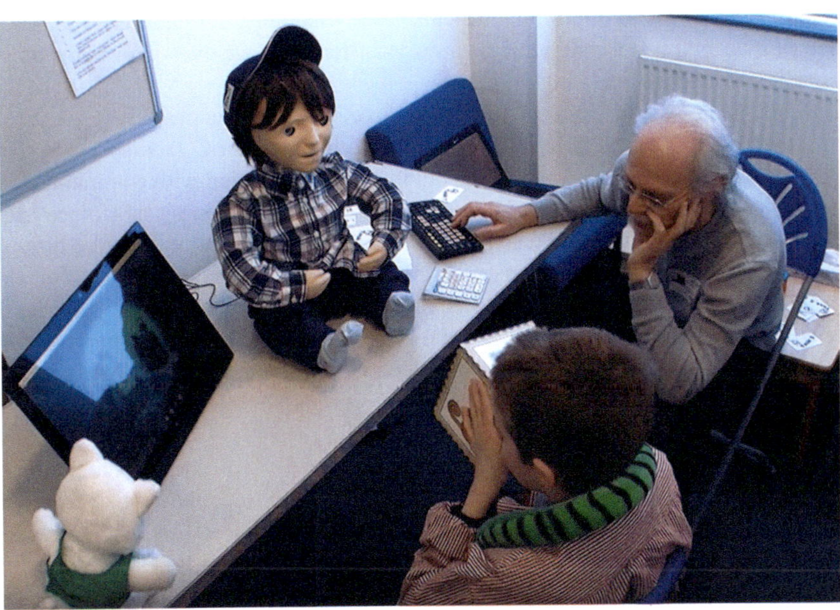

Figure 4.9 An example of the use of Kaspar. In the study of Wood et al. (2017), Kaspar was applied to teach children with ASD that another person has a different perspective as themselves, both regarding the line of sight and the details they can see. In the different games, the children had different tasks. For example, either Kaspar's head was fixed and the child had to move a toy in a way that it would enter the robot's field of vision; Kaspar asked for a specific animal that was on one of the sides of a cube, thus the child had to move the correct side of the cube towards Kaspar; the child had to move Kaspar's head towards a toy in the room; or two children had to cooperate to direct Kaspar's gaze towards a toy (one controlling up/down, and the other the left/right motion of the head by using a joystick) (Photo from Wood et al., 2021).

factors became more apparent with time (e.g., mediating effect). By the fourth visit (more than a year later than the start of the study), the staff better integrated the robot with the other activities in the nursery and also advised the implementation of a picture-based communication system (PECS cards used in the nursery) to be included, enabling better child-led interactions. After the study has ended, the robot remained at the TRACKS nursery with no further data collection, with seven years of total use (at the time of the 2020 publication).

Kaspar's usefulness for research and therapy for children with ASD is well-established (Figure 4.9), but recently the Kaspar robots have also been used in increasingly wide applications in assistive robotics. Lakatos et al. (2023) investigated whether a humanoid robot such as Kaspar can be utilised in Speech, Language and Communication (SLC) therapy (Lakatos et al., 2023) to aid children with a variety of speech production and comprehension difficulties and improve their related social skills. Additionally, they also investigated what are the measurable effects of interacting with the robot. As children in need of SLC therapy have problems with both understanding verbal communication and partaking in it, a robotic interaction partner could be useful (in addition to its benefits seen in therapies for children with ASD) due to their ability to communicate in a consistent manner over high numbers of repetitions (Lakatos et al., 2023).

In this study, the researchers developed three interactive games for Kaspar, each of which addressed multiple objectives of the children's SLC therapy goals. The development of the games was carried out in close cooperation with communication experts and with the special needs schools in which the HRI testing took place later. Based on their insights and on a frequency analysis of their data about the therapy needs of children, the researchers identified the most important objectives for the games. The first one was a feeding game, in which Kaspar asked the children to 'cut up' a specific toy food (composed of easy-to-separate parts) and put it by the robot's mouth. The game had nine SLC therapy objectives, for example, to practice understanding simple sentences, the use of objects, and to follow relatively complex instructions.

The second game consisted of multiple rounds of two episodes: in the first one, the child was asked to point at an animal picture out of three, based on the verbal request of the robot. If the child succeeded, he/she was praised by the robot. In the second episode, the roles were reversed, and the children asked Kaspar to point at a picture. The game tackled six goals, including joint attention skills and turn taking. In the third type of game, the robot prompted the children to say sentences describing actions in different tenses, based on pictures. The eight goals of the game contained, for example, creating full sentences, using tenses, and explaining events. The control method of the robot is not detailed, but a Wizard of Oz approach was likely used as, for example, in the first game a researcher selected remotely which toy food should Kaspar request, and additionally, there are no mentions of autonomy in the robot's behaviour.

In the testing phase of the study, 20 children, between the ages of 6 and 14 years, participated by playing the games with Kaspar in nine sessions during a three-week period. The children were students in special needs education schools and had speech difficulties related to their diagnoses (e.g., attention deficit hyperactivity disorder, Down syndrome, ASD). The sessions with Kaspar were recorded on video, and the behaviour of the children was analysed in their first two and last two sessions by the means of a scoring system. The children received separate scores in the specific games for their speech production and speech comprehension skills. For example, in the third game for the speech production, children received a score 0 if they did not vocalise, 1 if they vocalised unintelligibly, 2 if they said meaningless words, etc., up to score 5 in which children said a correct sentence in the necessary tense. The results of the statistical analysis showed significant improvement in comprehension and production between the beginning and the end of the intervention with Kaspar.

4.3.8 Ethorobotic perspectives

As we will see throughout this chapter, if measured in the same frame of reference as the NAO robot, or as a general-purpose robot, Kaspar would fall behind due to its relatively low sensorisation, movement capabilities, and autonomy. However, Kaspar has been designed and refined in every step of its development to serve a specialised assistive role, to fulfil its functions in a particular niche.

4.3.8.1 Embodiment

The developers of Kaspar have paid great attention to the embodiment of the robot in order to create an ideal physical appearance for interacting with children with ASD, and to make a robust and portable robot at a low cost (Dautenhahn et al., 2009). The hands and feet of the robot are covered with a beige silicone skin for better tactile experience. The childlike body, sitting pose, and size of the robot are not only the result of the practical decision of using a commercial child mannequin as the chassis of the old versions of Kaspar but was a deliberate design choice to make the robot seem less potentially threatening, and to lower the expectations that users might have about the capabilities of this toy-like doll. Although the researchers wanted to make it visible that Kaspar is indeed a robot by leaving the mechanical parts of the neck visible (in old versions by not covering it with clothing and since the redesign for the K5 version by covering the neck with transparent plastic), they did use old children's clothing to dress Kaspar up, with the intention of making the robot more relatable (Dautenhahn et al., 2009). Not hiding all mechanical parts of Kaspar and introducing it to children as a robot instead of a child enables the use of the robot in more therapy scenarios, as children can practice appropriate tactile interactions and understand inappropriate ones (e.g., hitting) in a calm and non-judgemental manner. The robot was designed to be adequately sturdy as well for tactile interactions with children (Syrdal et al., 2020). The head structure of the robot was simplified to increase robustness in K2. The number of DoFs increased with the newer versions of Kaspar, and the quality of the servos was improved as well.

The head of Kaspar is larger than it would be expected based on children's body proportions to make it less threatening and to emphasise the iconicity of Kaspar's face (Dautenhahn et al., 2009). The iconicity of the face (emphasising salient features and simplifying details) was a main design consideration based on multiple factors: maintaining a similar level of complexity throughout the robot's embodiment, focusing on more salient facial gestures to avoid overwhelming children with ASD, and for the robot to be able to represent itself as less person-specific with a more general, mostly genderless and ageless looking face. However, Kaspar is frequently dressed more similarly to boys (as Kaspar interacts with more male children; Dautenhahn et al., 2009), and its body size reflects childlike features as well.

Safety during child–robot interactions is crucial for the embodiment of Kaspar, for which safety features were incorporated as intrinsic parts of its design. As the plans for the K5.5 version were for them to be operated in e.g., schools with no researchers present, their safety features were updated to ensure that there are no sharp edges on the robot, as well as no pinch-points nor a danger of electric shock (Wood et al., 2021). Therefore, all wires and metal parts were covered (at the arms and neck with a transparent, bendable plastic). The K5 model also uses Wi-Fi for connecting to the external parts of its system, and as such no longer needs to be connected with a cable, limiting the tripping hazard (Wood et al., 2021).

In most use cases of social robots, ethorobotics does not consider such a highly humanoid embodiment necessary or ideal; however, this is one of the niches where the human-like appearance is advantageous for Kaspar's functions. This is indicated by the (anecdotally) different reactions of adults and children when first meeting Kaspar: while children (both typically developing and children with ASD) tend to react positively to Kaspar and try to engage in playful interaction or physical contact with it, adults tend to be critical towards the appearance and capabilities of the robot (Dautenhahn et al., 2009). Videos of Kaspar had been used in short questionnaire studies on the Uncanny Valley hypothesis, with results indicating an effect prompted by the human-like face of the robot (Gray & Wegner, 2012; Yam et al., 2021). In the videos in which the robot is visible with its humanoid face, the metallic parts on the back of its head are also visible instead of a wig that is usually used for Kaspar, which might enhance the Uncanny Valley effect. On the other hand, using morphed virtual faces showed that children with ASD are not affected by the Uncanny Valley, in contrast to typically developing children (Feng et al., 2018).

The simplicity of the robot's appearance mostly follows the notion that it should not promise more than the robot is actually capable of (although e.g., the nose as a facial feature was kept without olfaction), an explicit design choice of the research team (Dautenhahn et al., 2009). Similarly to keeping the nose, Kaspar's legs could logically indicate the capability to locomote or at least move them (as it can move its upper limbs). For example, Kaspar can move its head and hands to cover its face if it is hit by a child, but it cannot move its feet away from harm (which would be a natural reaction of a human). We assume that this limitation is explained to children before interacting with the robot.

4.3.8.2 Sensing

The built-in sensing capabilities of Kaspar are limited to tactile and visual modalities, an RFID reader and some proprioception via position sensors of servo motors, while external sensors are used to augment its perception of the environment with 3D depth image data and microphones usually provided by a Kinect, IMUs for measuring toy orientation and in some cases additional visual images via an external camera. While some functions (e.g., autonomous reactions to tactile stimuli) can be achieved with only the built-in sensors, the more complex interactive skills of Kaspar require the use of exterior sensors as well. This practice is not aligned with the ethorobotics perspective, and it probably reflects an optimisation towards robustness. While much research deals with perception based on built-in sensors, it can be assumed that the reliability and field of vision are better in a commercially available Kinect than what could be reasonably achieved with the built-in sensorisation of Kaspar in a cost-effective manner. The larger K1-L version of Kaspar had an opening on the chest to accommodate the inclusion of a 3D camera (Dautenhahn et al., 2009). In other Kaspar models this is not included, we assume due to space limitations. The IMUs embedded in toys that are used to measure their 3D orientation compared to Kaspar need to be calibrated before each session, and the system can lose them as well if the toys are moved too fast, necessitating a reset. Even with this, the researchers consider the use of IMUs easier than vision-based orientation estimation (Zaraki et al., 2018).

Interestingly, while the number of tactile sensors (as whole patches) increased slightly in the newer robot versions, the current sensors are simpler but more reliable than what was initially developed and used (see Section 4.3.5), showing another optimisation towards robustness over complexity. It is unclear whether the simpler sensors can still differentiate between soft and hard touch as ROBOSKIN could.

In a similar vein, the visual perception provided by the built-in cameras in Kaspar's eyes and the related 2D vision module in the Interactive Architecture face difficulties due to sensitivity to lighting conditions, and the research team has been considering the use of RFID tags for object (toy) recognition instead (Zaraki et al., 2018).

While the use of external sensors is understandable with the above reasons, we expect that their use complicates the setup process of the Kaspar system for on-site experiments or therapeutic uses.

4.3.8.3 Autonomous operation

Kaspar has a very low level of autonomy because it is optimised for its assistive applications for children with ASD, which necessitates the active inclusion of a human operator in the control of the robot (Cabibihan et al., 2013). Some basic actions of the robot can be used both autonomously or be initiated via remote-control, such as actions performed in case of a tactile stimulus (e.g., turning away, covering the face, and saying 'Ouch that hurts' by playing an audio file when being hit; Syrdal et al., 2020). However, the movements (and audio playbacks) themselves follow a pre-programmed form and sequence, with specified motor positions for all servos. Although the lack of variability might make a robot less animate, the repetitiveness (and the perceived predictability) is considered as a positive aspect for children with ASD (Cabibihan et al., 2013).

In most cases, Kaspar is used with a Wizard of Oz method, or semi-autonomously with the Sense-Think-Act interactive architecture (see also Section 4.3.5). The semi-autonomous operation has to function in an environment much different from in which robots are developed and frequently used in research, as Kaspar has to work in uncontrolled, child-focused environments such as schools or nurseries instead of laboratories (Zaraki et al., 2018). In addition, to accommodate the needs of children with ASD, the robot's semi-autonomous operation needs to be stopped if the child loses interest or becomes frustrated. In this case, the therapist can engage the child in some free play and then re-establish the interaction between the child and robot (Zaraki et al., 2018).

As discussed by Wood et al. (2021), autonomous perception and computing capabilities are not sufficiently reliable yet, especially if one considers the individual needs of children. This can lead to situations when Kaspar's inadequate behaviour may upset children who may react with inappropriate behaviours towards the robot (Wood et al., 2021). An example was observed in Syrdal et al. (2020): Kaspar played audio files with increased complexity when reacting to being hit by a child compared to when reacting to stimuli resulting from appropriate interactions. This resulted in some children hitting the robot more often. As the audio output was not under user control, a solution was proposed to increase the complexity of what Kaspar says in case of appropriate interactions as well.

In spite of such potential difficulties, a general direction is to slowly increase the autonomy of the robot, with the inclusion of learning capabilities, proposing that the robot could learn during the interactions from the individual child and the therapist about the reactions it should display (Zaraki et al., 2018). Similarly to what we have seen regarding the sensors used by Kaspar's system, the Sense-Think-Act software architecture also relies on exterior devices, as the architecture itself is run on an external laptop.

4.3.8.4 Locomotion

Kaspar has no locomotive capabilities, which for a general-use robot would be quite limiting. In the niche in which Kaspar operates, a sedentary robot might prevail as the focus is on social skills such as e.g., turn-taking, emotion recognition, and tactile interactions, for which locomotion is not needed. The sedentary setup of Kaspar can also contribute to the robot being perceived as non-threatening and predictable, both important aspects for child–robot interactions. However, children with ASD are generally more interested in moving things, and therefore a robot with locomotion might better maintain their interest (Cabibihan et al., 2013).

4.3.8.5 Object manipulation

Kaspar has limited object manipulation capabilities, restricted to the 5 DoFs movements of its arms with which it can produce rhythmic beats on a drum, and to the use of small accessories with magnets in them, via the neodymium magnets placed in the hands of the robot. The drumming has been used in various research studies on e.g., turn-taking and non-verbal communication, in which the robot imitated rhythms played by adult humans (Kose-Bagci et al., 2010). The joints of the robot determined the minimum amount of time between beats with which it could replicate the perceived rhythm at 0.3 s, and a few seconds were also needed before the start of a set of beats in order to position its arms accordingly. The speed of its movements and the forces it can exert are relatively low, which makes the child–robot interaction inherently safer. Similarly, its movements can be easily stopped by children by hand (Dautenhahn et al., 2009), with no damage to the servos in the newer models, as they have elastic response to forces acting on them (Wood et al., 2021).

The robot's hands were kept as rigid bodies even in the newer versions of Kaspar with no articulation, which restricts the further development of object manipulation skills. The reason for using the dummy-like hands was to maintain a similar level of complexity throughout the embodiment of the robot and to encourage children for tactile interaction (Dautenhahn et al., 2009) (together with its silicone cover in later models). We assume that avoiding the implementation of articulated fingers could help keep down the costs of production and reduce the possibility of hardware failures. The limited ability to move/lift small objects with the help of magnets had been considered useful for demonstrations of e.g., eating with spoons or using a toothbrush, motivating children to do so as well (Wood et al., 2021).

As we can see in case of assistance dogs, a human-like hand structure is not needed for simple object manipulation tasks, as holding objects in their mouths can be considered similar to the working of a gripper. The inclusion of a 1 DoF articulation in the thumb of Kaspar could create a hand structure that could be used to hold onto more objects placed in Kaspar's hands without magnets, or even to perform simple handover tasks during interactions.

4.3.8.6 Interactive skills

The communicative signals of Kaspar are based on visual and auditory modalities. Kaspar can show various non-verbal signals by moving its neck, mouth, and eyes, including simple, minimalistic emotion expressions created by the mouth actuators (resulting in a smile, neutral face, and a frown). The robot can change its gaze direction and can lower its eyelids to close its eyes or to blink. The head actuators of the robot move at a speed which, according to the authors, could help create a less artificial transition between facial expressions (about 2 s from the neutral expression to a smile) (Dautenhahn et al., 2009). Blow et al. (2006) conducted a preliminary study on people's reaction to Kaspar's smile by showing video recordings of static and dynamic smiles, different degrees of smiles and also by manipulating the transition time between the neutral and smiling expression by leaving it at its original 2 s length, or removing the transitional frames of the video, resulting in an instantaneous change from the neutral face to the smiling one. While the authors conclude that the participants rated the smiles with the 2 s transition as happier and more appealing, the results are presented only with rudimentary descriptive data and no statistical analysis. According to Ekman and Friesen (1982), felt (also called enjoyment, genuine, or Duchenne) smiles that are produced due to positive inner states tend to last from two-thirds of a second to four seconds (depending also on the context), while false smiles (including phony smiles without positive inner states, and masking smiles covering negative inner states) are usually longer than four seconds, but their onset and offset is short compared to genuine smiles. These notions are also reflected by the experiments of Krumhuber and Kappas (2005), in which they found smiles with onsets lasting 528 ms to be perceived as more genuine than smiles with more abrupt (132 ms) onsets; however, longer onsets than 528 ms were not considered in this study, and as the authors note, the perceived naturalness of a smile with a longer offset (as is the case with Kaspar's) is unclear. On the other hand, Ekman has defined a high number of smile types other than felt and false smiles, with various characteristics and occurrences (Ekman, 1992). Dautenhan et al. (2009) describe Kaspar's smile as similar to the Duchenne smile: the use of a silicone face mask (in the older models a resuscitation mask) and the placement of its attachment points to the actuators of the head create a smile similar to a genuine smile, as the face mask deforms around the eyes as well (Dautenhahn et al., 2009). However, we are unaware of an empirical categorisation of Kaspar's smile based on e.g., the Facial Action Coding System (Ekman et al., 2002). Emotion expressions may be accompanied with body movements (e.g., turning away) and audio outputs. While in earlier versions an external speaker placed by the robot provided the sound output, a speaker has been integrated into Kaspar, first in its body (K3–K4) then in its head (K5–K5.5) (Wood et al., 2021). The interactive skills of Kaspar reflect simplicity and repeatability, but its limitations are seen as adequate for the children its designed to interact with.

4.3.9 Future utility

As presented in the above chapter, Kaspar has undergone some technological development during the last almost 20 years, with continual optimisations for its functional niche. The Kaspar robots have already interacted with more than 500 children (Lakatos et al., 2023), but their use so far has been mostly in the framework of research studies (see Section 4.3.7 and Syrdal et al., 2020). The study by Syrdal et al. (2020) provides some evidence that Kaspar is making its (literary) steps towards in situ use, with decreasing need for human supervision. This will most likely necessitate and result in newer adaptations to Kaspar and might decide its 'survival' in the long run.

4.3.10 Similar robots

The Ameca robot developed by Engineered Art (first released in 2022) has a mixture of humanoid and mechanical features: while it has a humanoid body plan (standing 187 cm tall) and expressive, human-like face and hands covered with grey silicone, the rest of its body has a metallic cover with highly visible mechanical parts (Figure 4.10). Similarly to Kaspar, Ameca is incapable of locomotion but can move its upper body, neck, head, arms, and face. The facial gestures of Ameca are designed to closely imitate that of humans. The robot serves an entertainment function, interacting with humans during events and exhibitions with the help of its two cameras and microphones, enabling face and voice recognition. Additionally, a desktop

Figure 4.10 Ameca robot (Engineered Arts) (Photo by Steve Jurvetson, Flickr).

version (with the upper chest and head) is offered for academic research, presented as a modular platform that uses a bespoke software framework called Tritium, with a Python-integrated development environment.

4.4 AIBO

4.4.1 Overview

AIBO is a small quadruped zoomorph robot, developed for entertainment purposes by the Sony Corporation (Figure 4.11). Various prototypes (e.g., the MUTANT; Fujita, 2001) have preceded the first commercially available AIBO model, the ERS-110, which was developed by researchers at the Digital Creatures Laboratory of Sony. The ERS-110 was sold in 1999, representing the first four-legged entertainment robot to be used autonomously in home environments (Fujita, 2004). The entertainment role of the robot was a decision based on contemporary limitations of autonomous robotics, as users were expected to be more forgiving towards the mistakes and shortcomings of such robot compared to ones with more practical uses, where reliability is imperative (Fujita, 2004). Fujita et al. designed the AIBO robots to maximise their behavioural complexity, which they intended to achieve with hardware and software features such as a high degrees of freedom embodiment, various sources of motivation in the behaviour generation of the robot, and by avoiding repetitive behaviour (Fujita, 2004) (see also Section 4.4.6).

Following the release of the first AIBO, multiple new models have been introduced in the next few years (Fujita, 2004) (see Figure 4.12). The ERS-111, which was very similar to the original AIBO, was followed by the redesigned ERS-210. The next generation of robots arrived in 2001 with the ERS-31X series and the

Figure 4.11 The most recent version of Sony's AIBO, ERS-1000 (Photo by KKPCW, Wikimedia Commons).

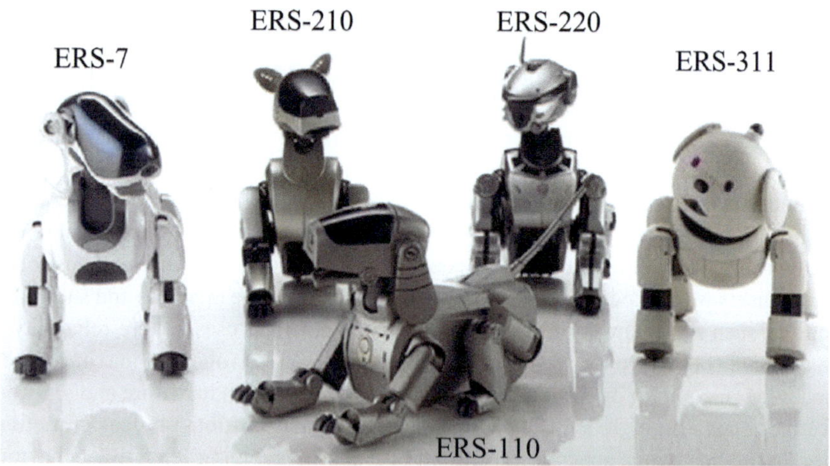

Figure 4.12 The figure shows the most important AIBO models from the old generations. The first commercially available AIBO, the ERS-110 is in the foreground (release date: 1999), while in the background from left to right we can see the ERS-7 (release date: 2003), the ERS-210 (2000), the ERS-220 (2001) and the ERS-311 (a part of the ERS-31X series, 2001) (Photo from Kertész & Turunen, 2019).

remote-controlled ERS-220 (Fujita, 2004). The last generation of old AIBOs was the ERS-7, which became available in 2003–2005. All main models had variants, mostly differing in colouration or small improvements.

More than 150,000 units of the older AIBO models were sold over the years (Knox & Watanabe, 2018), but the AIBO line was still deemed unsuccessful and was discontinued in 2006 (Schellin et al., 2020). While support was terminated for the old models in 2014, Sony announced the development of a new line of AIBOs in 2017, and the new generation, the ERS-1000 was released in 2018 (Knox & Watanabe, 2018). The ERS-1000 is available for $2,900 in 2023.

4.4.2 Embodiment

AIBO has a tetrapod body structure (Table 4.4), complete with ear-shaped appendages and a tail. In the ERS-110 model, the ears moved passively when the robot moved its head (Fujita, 2004), while in the new ERS-1000 models, they have dedicated motors. The appearance of AIBO always resembled dogs to a point, with varying levels of curved (see ERS-31X series, the most dog-like appearance from the old generations, created to appeal to women) or angular designs (ERS-220 model, created to appeal to men; Fujita, 2004). However, until the release of the ERS-1000, the resemblance to dogs was less pronounced, as the older models maintained an overall robotic or stylistic look. The third generation ERS-7 robots presented a step towards the currently used more curved design, but the overall appearance tended towards a futuristic, robotic aesthetic. In contrast, the new ERS-1000 is designed to look more like a dog puppy, which is aided by the new type of eyes featured on the robot's head. While older models had a single central 'visor' as eyes (apart from the ERS-31X series models that had separate eyes) with LED displays showing simple forms or icons (seen most prominently in the ERS-7), the ERS-1000 has OLED displays, showing actual (although cartoonish) eye graphics.

4.4.3 Further developments and improvements

The AIBO was the standard hardware robot used in the Four-Legged League of the RoboCup championship until 2008, which sparked a wide range of research projects related for example to locomotion (Chernova & Veloso, 2004), object manipulation via kicks (Hausknecht & Stone, 2011), sound source localisation (Liu & Wang, 2010), and machine learning (Chalup et al., 2007). As the production and later the support of the old AIBO models stopped, the technical research projects also dwindled. However, we can find projects that aim to develop open-source solutions for the continued operation and development of the ERS-7 (AIBO+; Kertész, 2013). As of yet, we do not see a similar rush of development projects with the new ERS-1000 as during the RoboCup years of AIBO. Most of the development of the new model seems to be coming directly from Sony with frequent software updates that expand the action repertoire of the robot. Additional development for the new AIBO by users or researchers is still possible with the designated Developer Program web API (documentation available online; Sony, 2024) and the Visual Programming application. To access the

Table 4.4 Hardware overview of AIBO (version ERS-1000), mostly based on the official manual (Sony, 2018).

Height		29.3 cm
Weight		2.2 kg
Body plan		Quadruped, dog-like
Locomotion		4 articulated legs
Body surface		Plastic
DoFs	Total	22
	Head	6 (3 at the base of the head + 2×1 for the ears + 1 for the mouth)
	Neck	1 (lifting the head at the base of the neck)
	Torso	3 (1 at the waist + 2 for the tail)
	Legs	12 (4×3)
	Arms	–
	Fingers	–
Interactive hardware	LEDs	Status LED on neck, network LEDs (separately for Wi-Fi and for mobile network)
	Loudspeaker	Yes, but placement is unclear
	Screen	2 OLEDs on head as eyes
Sensors	Head	Wide-angle camera (nose), Time-of-Flight sensor (mouth), microphones, 2 capacitive touch sensors (head and jaw), 3-axis accelerometer, 3-axis gyroscope
	Neck	–
	Torso	Motion sensor, 2 ranging sensors (front of torso), SLAM camera (top of back), light sensor (top of back), pressure sensitive capacitive touch sensor (back), 3-axis accelerometer, 3-axis gyroscope
	Legs	4 paw pads with contact sensors (based on information on the ERS-7; (Kertész, 2016)
	Arms	–
Operation time with charged battery		2 hours
Human-likeness score (ABOT)		N/A

development tools an AIBO AI Cloud Plan (AI cloud service) subscription is needed, which is provided freely with every robot for three years, after which it costs $300/year. The cloud service is needed for most advanced functions of the robot.

4.4.4 Autonomy

Apart from one model, the ERS-220 which was created as a remote-controlled robot (Fujita, 2004), all AIBOs have been made for autonomous operation. The robots can perceive their environment, move around in it, and perform actions, autonomously interacting with objects and humans. In case of the ERS-1000, the more complex functions are enabled by the involvement of the AI cloud service, which stores the necessary data (sensory information, learned actions, specified locations, etc.). Autonomous functions also include self-charging, with the option of automatic software updates being performed in the meantime. If enabled by the user, the robot can also take various photos during its active periods without an explicit command to do so. The robot signals the fact that it is taking a photo by performing a specific ear movement pattern. The photos are then uploaded to the cloud service.

An overarching aspect of the AIBOs' autonomy is that the robots also go through a simple form of maturation, in which their behaviour and actions change as they 'grow up' during approximately three years.

4.4.5 Sensory and motor capabilities

4.4.5.1 Vision

The AIBOs have a small wide-angle camera in their nose, which is the basis of their visual perception. Additionally, the new ERS-1000 has a second camera at the rear end of its backs, looking upwards. This camera is used for navigation based on SLAM.

The first commercially released AIBO, the ERS-110 already had a micro-camera unit at the front of its head and utilised a colour detection engine able to recognise eight colours, as part of image processing. The segmentation based on colour is used for functions such as object recognition and tracking. Multiresolution filtering was also used to generate three images with different resolutions (with 60×30, 120×60, and 240×120 pixels) to provide optimal input for processes that have to take place quickly (e.g., object tracking) or that needs more detailed images (e.g., pattern recognition) (Fujita, 2001).

The vision system of the ERS-1000 is not known in detail but there is some information on its capabilities. This AIBO model can scan QR codes (used for establishing Wi-Fi connection) and can stream video as well with its frontal camera, which is watchable in the AIBO applications. The eye animation of the robot changes to indicate that a video is being streamed. It has been stated that the ERS-1000 can recognise about 100 faces and may also recognise smiles, which are utilised as positive reinforcement.

We can find mentions of additional visual sensors, e.g., the light sensor above AIBO's tail measures the ambient lighting level, which influences the appearance (the eye brightness) and behaviour of the robot. Another optical sensor, a ToF (Time-of-Flight sensor) detects obstacles in front of AIBO.

4.4.5.2 Hearing

Most AIBO models had two microphones, usually placed in/near the ears, while the ERS-1000 has four microphones (details unknown).

Sound processing was already a part of the capabilities of the original AIBO, which had to deal with abiotic environmental (e.g., air conditioning) and self-generated (e.g., by the servos of the robot) noises, as well as voice interference from biotic sources (multiple people talking simultaneously) (Fujita & Kitano, 1998). For this reason, the researchers recommended the use of a 'tone-like language' instead of natural human speech, from which noise and voice interference could be separated via a moving average filter (Fujita, 2001; Fujita & Kitano, 1998). The word-learning research conducted by Kaplan in the early years of AIBO (2000; presented in more detail in Section 4.4.5 Speech) utilised an external computer with a speech system for speech recognition, and an external microphone as well (Steels & Kaplan, 2000). By 2004, the hearing-related social skills of AIBO were expanded with the implementation of word-learning. The microphones of various models were also used for sound source localisation in the framework of the RoboCup (Becker & Risler, 2009).

The current ERS-1000 can recognise a set of voice commands, which list grows as the robot matures. A set of phrases can also serve as positive or negative reinforcement by the user (Sony, 2018).

4.4.5.3 Touch

The ERS-1000 has touch sensors on various parts of its body, with varied (but sometimes overlapping) functionality. The touch sensors on the top of the head, on the chin, and on the back are primarily used for tactile interactions with humans: gently stroking these areas provides positive reinforcement to the robot, encouraging it to exhibit the rewarded behaviour more frequently. In contrast, the sensor on the back of the robot can also sense various pressures, and tapping it serves as punishment/negative reinforcement. Petting the robot fulfills other functions as well, e.g., it prompts the robot to 'go to sleep' when it is on its charger.

The type of touch sensors in the paws of the ERS-1000 are not known, but we assume that these are in part used for sensing whether AIBO's legs are on the ground. In the previous ERS-7, the paws contained two-state buttons that provided simple pressed–not pressed inputs (Kertész, 2014).

In the ERS-1000, the paw pads are also used for teaching, with which the robot can learn up to 30 new actions. During this process, the user commands the robot with a phrase to learn the following actions, which prompts the robot to sit down and lift its front legs. If the user then pushes at least one of the front leg paw pads, the joints of the front legs relax, and the user can move around the legs freely for a maximum of 15 s or while the paw is pressed. After release, the robot can be asked to reproduce the action and memorise it. This form of learning resembles the infrequently utilised modelling (moulding) method in animal training used for teaching very simple actions, and in industrial robotics for defining positions (Kaplan et al., 2002).

4.4.5.4 Other sensing capabilities

The internal sensors of the ERS-1000 include two detection systems providing 3 axes of detection based on gyroscope, and 3 axes of accelerometer data each, placed in the head and the torso of the robot (Sony, 2018), respectively.

The motion and ranging sensors in the front chest area of the ERS-1000 contribute important sensory information to AIBO. The two ranging sensors provide distance measurements and edge detection in front of the robot, which can help the robot not to fall off surfaces, while the motion sensor enables AIBO to detect movement in front of it.

We assume that the ERS-1000 contains minor sensors not listed in the manual, e.g., encoders providing proprioceptive data (in the previous ERS-7 each joint had an encoder; Sousa et al., 2010). The ERS-7 also had temperature and vibration sensors (Datcu et al., 2004), and the manual of the ERS-1000 also refers to LED signals indicating that the robot is overheated, so some sort of temperature sensor is also onboard.

4.4.5.5 Speech

AIBO is not capable of speech; however, it can emit a high number of vocalisations. Some of these are vocalisations resembling distorted, 'robotic' barking, various sound effects (sniffing, chewing, etc.), or playing music and/or singing melodies with non-verbal vocalisations. According to Fujita (2004), the robot can also mimic the prosody or tone of words and recreate them in its own non-verbal vocalisation. In the ERS-1000, the volume of the acoustic signals can be changed via physical buttons on the robot, or via the application.

Despite AIBO's lack of speech, this skill has been implemented in the robot in the framework of research. The AIBO has been used to test a system that could recognise, categorise, learn, and name objects (Kaplan, 2000). The system builds upon the off-the-shelf autonomous operation of AIBO and affects the robot's behaviour, acting as an additional cognitive layer. An external computer was used for speech recognition and the cognitive functions necessary for verbal interaction with a human user. The camera image, containing an object shown to the robot, went through segmentation and filtering based on colour detection. The shape and colour features of a chosen segment were determined, representing the object the robot saw. The AIBO learned the name of objects via language game dialogues played with humans, and with repetition the robot could identify objects from multiple angles, forming categories. Following this, the robot could tell the human the name of known objects it saw via verbal interactions (Kaplan, 2000).

4.4.5.6 Actions

All autonomous AIBO models can perform a high number of actions, as more than 1,000 actions (motions, acoustic, and light signals) were implemented as early as 2004 according to the developers (Fujita, 2004). We assume that some of these actions were variations of each other, or behaviour sequences comprised of multiple behaviour elements. The AIBO has actions such as performing various dances, playing music, and vocalising melodies as well.

In case of the new ERS-1000, the number of potential actions is also high and is frequently increased via software updates. Additionally, the users can also teach short movement patterns for their AIBOs that they can remember and reproduce later. New learned actions can also be shared between AIBOs via the My aibo application. AIBO has an automatic reaction to being lifted: its joints get relaxed, then it takes up a lying down position for better in-hand portability (Sony, 2018).

According to the manual, the ERS-1000 can respond to its name given by its owner, can move around, and perform basic tricks. As mentioned before, the more advanced functions of AIBO are only available with the use of the AI cloud service. These include, among others, augmented reality uses (e.g., for the imitation of consuming virtual foods/drinks from a physically present bowl), learning based on positive and negative reinforcement, user settings, e.g., marking territories on the map that the robot should avoid, and complex behaviours like patrolling or greeting. In patrolling the robot walks around a mapped area and looks for specific individuals, then creates a report, while in greeting the user can teach the robot when do they usually arrive home, by which time the robot will go to the door to greet the user. The robot can also play with its designated toys (see Section 4.4.8). The behaviour of AIBO can be set to be 'well-behaved' via a verbal command and petting it while holding its paw, or via the My aibo app. In this mode, the robot will not move around or vocalise/play audio. As to our knowledge, the advanced features like the patrolling or greeting were not yet investigated or used in studies.

4.4.5.7 Gestures/signals

The ERS-1000 can exhibit various gestures and communicational signals in visual and acoustic modalities. Even though most of its body surface is a rigid plastic cover that does not allow deformations, AIBO can display emotional behaviour by moving its mouth and ears and by changing its eye displays on the OLED screens (it can show half-closed and closed eyes, different gaze directions, blinking, etc.). In addition, it can also use body gestures to express emotional states by, for example, lowering/lifting its head, tilting it to the side, changing its body posture, moving/wagging its tail. For details on its acoustic signals, see above.

4.4.6 Software architecture and related system capabilities

The most detailed software architecture description that is accessible is about the first-generation AIBO and its prototype. Here we present a short overview of the software architecture of the earliest model, the ERS-110.

The ERS-110 AIBO had a behaviour-based architecture, in which different behaviour modules were implemented, each of which was stochastic state-machines (Fujita, 2004). This stochastic nature ensured the addition of randomness to the actions of AIBO, as this way the occurrence of a specific action could be defined with a probability parameter. The behaviour modules were activated via an action selection mechanism, while motivations for the modules were provided by the simulation of 'instincts and emotions' (Fujita, 2001). Six basic emotional states were implemented, which were all influenced by environmental stimuli and simulated motivational states (e.g., hunger, tiredness, and curiosity; Fujita, 2001) that were later changed to affection, investigation, exercise, and appetite (Fujita, 2004). The state values were affected by time and environmental stimuli.

The capacity to learn and develop was already a part of the software capabilities of the ERS-110: reinforcement learning was implemented to change the probabilities of occurrence for possible behavioural responses of the state machines to a given stimuli, based on user feedback (e.g., petting the robot). Long-term changes in AIBO's behaviour were created by changing the graph structure of state machines belonging to behaviours (Fujita, 2001), enabling switching between state machines (Fujita, 2004). Interestingly, the various body parts of the robot (head, legs, and tail) also possessed their own distinct motivational states, which could be inhibited by upper-layer commands. For example, a motivational state of the head could be to keep the head horizontally aligned or to track certain environmental stimuli. Some actions of AIBO were designed to be reactive, providing a rapid behavioural response to a stimulus, while the execution of deliberative actions could take longer. Based on the description of the Mutant prototype (which gave the basis for the ERS-110), the reactive behaviours were produced at the lowest layer of the architecture, called the motor command generator. The layer utilised commands from upper layers and environmental stimuli to create motor commands. The next layer (action sequence generator) was responsible for the generation of action sequences, forming behaviours, while taking into account mechanical constraints. The upper layer was the behaviour generator, which considered both environmental stimuli and the inner state of the robot (Fujita, 2001).

The old generation AIBOs' operating system, Aperios (in some publications referred to as Apertos) has been developed by the Computer Science Laboratory of Sony. Aperios was a closed, real-time operating system in which applications were used as Open-R objects. The objects were loaded from an inserted memory stick after the robot was switched on (Kertész, 2013). An Open-R Software Development Kit was released by Sony to enable development for AIBO. Its use has been described and enhanced for example with the ERS-7 by various researchers (Kertész, 2013; Rico et al., 2004). Custom software created for AIBO could be developed on an external computer and utilised with the robot via transferring it to the insertable memory stick.

The software architecture and operating system descriptions of the current ERS-1000 AIBO are not available in such details. The software running on the robot is named My aibo, similarly to its PC/mobile applications. The custom programming of AIBO is possible via APIs (see Section 4.4.3). The software capabilities of the ERS-1000 are discussed in the relevant sections of Sections 4.4.5 and 4.4.8.

4.4.7 Utilisation in HRI research

The AIBO represented a new type of commercial robot at its release, surpassing similar robots with its high mobility, sensorisation, and availability. Not surprisingly, this resulted in a good number of research studies over the years, both regarding technological developments and HRI, also inspired by the robot's participation in the RoboCup championship. The same level of research-driven enthusiasm is yet to be seen for the new generation of AIBO. As examples, we chose to present research utilising the ERS-1000 model.

4.4.7.1 Friendliness or eeriness

As the renewed design of the ERS-1000 emphasises its many embodiment features and behaviour elements resembling that of puppies, its perception by users on a dog-likeness scale comes into question. Thus, Schellin et al. (2020) raised the idea of an '*Uncanine*' valley (see Uncanny Valley hypothesis, Section 2.3.3). They wanted to find out whether the puppy-like look of AIBO increased people's attraction towards the robot or, in contrast, the lower performance of AIBO in comparison to a dog made people rather wary and anxious. They also considered the possible positive effects of zoomorphism on HRI, and the perceived need for the maturation of the robot (i.e., like a young puppy) which might promote caretaking behaviours in users.

In an exploratory HRI experiment, Schellin et al. (2020) measured users' opinions on the AIBO ERS-1000 before and after a short interaction. For different subjects, AIBO was introduced (and later referred to) as a robot or as a puppy, and it had its original plastic or a furry cover. In total, 33 adults participated in one of the four experimental groups, in which, following the robot's introduction by an experimenter, participants interacted with AIBO for ten minutes by asking it to perform a set of pre-defined commands. The questionnaire measuring their opinions on the robot contained items relating to ten characteristics (creepiness, trustworthiness, dog-likeness, friendliness, etc.) with 10-point Likert-scales, and two additional questions on how much connection they felt with the robot and how warm/caring it was (5-point Likert-scales). The researchers also used video recordings of the tests to analyse some behaviour variables of the participants (e.g., number of smiles, laughs, praises, etc.), and the success and failure of AIBO in following commands in 28 interactions. The reported results indicate that participants rated AIBO as less scary if it had fur, both before and after the interaction. Contrary to the expectations of the researchers, they also considered the AIBO that was introduced as a robot to be friendlier, warmer/more caring and noted stronger connection towards it before the actual interaction, than if the robot was introduced as a puppy. The authors conclude that referring to the AIBO as a puppy might have elevated the users' expectations towards its capabilities, leading to a discrepancy.

4.4.7.2 Assistive role

As mentioned before, AIBO was intended and designed to be an entertainment robot. Despite this, some early studies investigated the robot's applicability in assistive roles in robot-assisted therapies for elderly people and children with ASD (Tetsui et al., 2008; François et al., 2009, respectively).

Newer research has been carried out with the ERS-1000 model. For example, Tanaka et al. (2022) investigated the usability of AIBO in an assistive role for children undergoing medical interventions. A pilot study explored the reactions of children and their caretakers who participated in (almost all cases, multiple) group therapy sessions with AIBO during their hospitalisation. The 24 sessions of group robot interventions lasted 30 minutes in hospital playrooms, with multiple children (number of children per session is not disclosed) and their caregivers present, as well as with the oversight of a psychologist or psychiatrist, who also observed and took notes on the interactions. The observations of 127 children (between the ages of 6 months and 13 years) and 116 caregivers were analysed qualitatively, focusing on verbal and non-verbal expressions, which were recorded as 'sentences' in the form of notes. These were then sorted into nine categories in case of children (e.g., 'affinitive/positive/cooperative', 'indifferent', 'defensive/cautious', and 'change'), and four categories for caregivers ('participation, exploration, watch over, encourage'). The 'change' category in children referred to events in which a child first expressed negative feelings towards the robot, then later expressed positive ones.

The majority of verbal or non-verbal expressions of children were positive, with 52.1% categorised as 'affinitive/positive/cooperative' and 11.4% as 'empathy/comforting/healing'; whereas 2.9% were reported to be indifferent, and 10.7% as 'changed'. The remaining expressions belonged to various categories reflecting negative reactions (e.g., crying, aggression, and defence). Children frequently participated in triadic interactions with the robot and another child. In addition, two-thirds of the caregivers also interacted with or explored the robot. Based on this study AIBO may have some potential to be used in paediatrics, but more investigations are needed.

4.4.8 Ethorobotic perspectives

4.4.8.1 Embodiment

The small size and weight of AIBO was decided in part based on technological limitations, as at the time of its creation (and arguably up to this day) the reliability of mobile autonomous robots made the use of robots with big body sizes and weights potentially hazardous in home environments. For this reason, the focus shifted to entertainment robotics, where a smaller size did not hinder the intended function of the robot (Fujita, 2004).

In the design of the earlier AIBO models, the developers considered it important for the robot not to resemble dogs too much, and to maintain a mechanical aesthetic to limit user expectations towards the capabilities of the robot. The emphasis was instead on making the movements of the robot smooth and its behaviour complex in order to increase its perceived animacy (Fujita, 2004). With the complete restart of AIBO development and production for the ERS-1000, this previous notion seems to have been forsaken, as both the embodiment of the new AIBO and its surrounding marketing clearly highlight the dog-likeness of the robot. Although precise measurements are not available, the head of the ERS-1000 seems to be bigger in proportion to its body compared to previous models (apart from the very cartoonish ERS-31X series), making it more similar to a young puppy. This is further accentuated by its big eyes, which is a step towards both looking dog-like (compared to the visors of most previous models) and a young mammal. We expect that these big eyes were probably a deliberate design choice to create the baby schema effect, in which infant-like proportions of facial features that can be observed similarly in various mammalian species (e.g., humans, non-human primates, dogs, and cats) elicit not only a perception of cuteness (Sanefuji et al., 2007) but increases motivation for caretaking as well (Glocker et al., 2009). The general baby schema features shared by human and non-human primate infants (compared to the adult features of their respective species) are big eyes located relatively low on the face, a triangular face shape that is wider towards the top of the head and a relatively rounder and shorter face (Kawaguchi et al., 2023). We can see most of these aspects reflected on the face of the ERS-1000. In accordance, most promotional Sony AIBO materials, as well as the manual, frequently refer to the ERS-1000 as a puppy. We assume that this emphasis (both in design and brand communication) was created with the purpose of making the robot look cuter, to promote feelings of caretaking in users, and to make users more tolerant towards the shortcomings of the robot.

In our opinion, and based on the available videos, the movements of the ERS-1000 are very smooth and fluid, creating a relatively life-like impression due to the inclusion of frequent, small movements.

The salience of dog-like features could raise expectations towards the capabilities of the robot and lead to its comparison to real dogs, as shown by Schellin et al. (2020) as well. Additionally, we consider it a possibility that these relatively life-like movements and the puppy-like embodiment could create a discrepancy between them and the otherwise clearly artificial nature of the AIBO, leading to categorical uncertainty. Categorical uncertainty is one of the possible explanations for the Uncanny Valley effect (Zhang et al., 2020), which, in case it occurs in the new AIBO, could hinder its more widespread adoption.

4.4.8.2 Sensing

The older generations of AIBOs were ahead of their time regarding the number and variety of sensors available for an off-the-shelf social robot. While robot technology greatly improved in the past decade, and the AIBO is not such an outlier anymore, its level of sensorisation compared to its small size is still noteworthy.

The ERS-1000's visual SLAM navigation is a great addition to the previously available sensors on AIBOs, which has the potential to increase the autonomous locomotion of the robot in a familiar environment. The robot also updates its map frequently, adapting to, for example, rearrangements of the furniture. However, based on information from a video on AIBO's official website, the mapping process is quite slow, as in this example the mapping of a ~70 m^2 room takes an hour. A Lidar-based SLAM method might be much faster, although the lack of a Lidar (light detection and ranging) sensor is understandable from a design perspective, as its placement in the robot body would be very challenging.

We can see a similar approach to the placement of the head-camera of AIBO as in case of the NAO: both robots have visual features on their head that look like eyes, but in reality, the cameras are placed elsewhere, and the eyes only serve a communicative function for human interaction partners. While the NAO has cameras on its forehead and where its mouth would be, the AIBO has a camera in its nose (in the ERS-1000 in the tip of the nose, in older models on a plain surface at the front of the head). Interestingly, AIBO performs 'sniff actions' when it approaches objects (e.g., moves its head closer and emits sniffing sounds). Despite this dog-like action, only visual information may be gained, creating a false image of AIBO being similar to a real dog. It is not clear why it is necessary to imitate sniffing, because dogs may also explore objects visually from a close range.

4.4.8.3 Autonomous operation

The autonomous operation of the robot is greatly aided by its capacity of charging itself by going to the charging station and laying down on it (the charging pins of its Lithium-ion batteries can be found at the ventral side of the robot). Full charging takes about 3 hours. This feature also aids the impression of AIBO being more similar to a living creature.

The actions of AIBO include simulations of some biological needs, for example, the robot can imitate eating or drinking with head movements and sound effects. However, not being fed does not affect its abilities or other actions. Another dog-like but seemingly unnecessary behaviour is the 'urinating' action the robot can perform by sniffing an area then lifting its hind leg over it. The robot can be potty trained as well: it can learn what location the owner prefers for this action, which location will be marked on the map available in the mobile/PC application. While these actions resembling behaviours necessary for biological functions were most likely implemented to increase the similarity of AIBO to a living dog, these may rather emphasise the discrepancy between the biological and the artificial agent because the latter performs these actions without the actual need to do so, and thus also highlights the lack of consequences when the behaviours are not carried out. This gives the impression of the modern AIBO of being a toy for children.

An interesting aspect of AIBO's autonomy is that even though the robot can recognise and enact various verbal commands, it does not always do so. According to the manufacturer, the robot's responsiveness depends on its inner state (Sony, 2018); however, it is likely that in many cases lack of response is the consequence of sensory and/or perceptive failure. However, this cannot be verified due to the missing test data.

4.4.8.4 Locomotion

The autonomous locomotion of the first AIBO already utilised the research of Hornby et al. (1999), in which they used an evolutionary algorithm to develop dynamic gait control for a quadruped robot. The evolutionary algorithm was implemented onboard the robot, with the various values of the 20 gait parameters constituting the genes of 'individuals' on which the evolutionary algorithm acted upon. The fitness of the individuals was determined by the success of embodied, real-life locomotion trials. While many of the gait patterns of the AIBO ERS-110 were specified manually, some were the results of the evolutionary algorithm, reaching 10.2 m/min speed (Fujita, 2001). In the following years, various artificial intelligence methods were used to create even faster gaits, mainly for the RoboCup (e.g., Röfer, 2004) (see Figure 4.13). An important functional aspect of AIBO's locomotion is that it can get up if it falls over.

Figure 4.13 AIBO ERS-7 members of two teams participating in the RoboCup Championship, competing in a football game. The robots can be seen without the removable long covers of their ears and tail (Photo by Pablo, Flickr).

AIBO is designed for indoor use, but preferably not on long-thread carpets or slippery surfaces (Sony, 2018). While the tetrapod body structure and the legs with multiple joints make the AIBO more similar to terrestrial vertebrates than many other robotic toys, the locomotive capabilities of the robot do not approximate that of dogs in robustness or in speed. From a general ethorobotics perspective, the locomotion of the AIBO is quite slow, making it unable to interact with real dogs, but children may also get bored or give up when they do not experience some expected minimal reaction speed in AIBO's actions.

4.4.8.5 Object manipulation

AIBO can manipulate objects via its legs and mouth by either kicking or grabbing specially designed objects. All AIBOs come with a pink ball that the robot can recognise based on its colour, and kick with its legs. Additionally, a frame cube toy ('dice') and a bone-shaped toy are also available, which the robot can grab and lift with its mouth. The robot can also throw the cube, or stack two cubes. While the AIBO can perform actions indicating interest towards different objects (by approaching it, moving its head closer, 'sniffing' the object, etc.), it is unclear whether it can manipulate these toys/objects as well. In the RoboCup championship, however, in which the last used model was the ERS-7, approximately 30 types of kicks were developed for the various AIBO models, with which they could move the ball on the field using the legs, the head, or the body (Visser & Burkhard, 2007). As since 2009 the AIBO has not been used in the RoboCup, this avenue of development is unlikely to continue with the ERS-1000.

The object manipulation skills of AIBO clearly focus on the entertainment aspect and have no practical applicability. This is not only due to the limitations of the object holding and carrying skills of the robot (although it is likely that the robot can carry only light objects in its mouth) but also caused by the slow locomotion of the robot. From the entertainment perspective, while the Sony website states that the robot could figure out new ways to play with the dice toys, the possibilities still seem limited.

4.4.8.6 Interactive skills

The AIBO utilises multimodal communication to interact with humans, which is enabled both by the sensorisation of the robot, its high DoFs embodiment, and the devices producing its auditory and visual communicative signals. According to the manual (Sony, 2018), AIBO is able to recognise individuals based on their faces, it can behave differently (show preference) with familiar people and recognise smiles as positive reinforcement (Zhong et al., 2021).

AIBO's interactive skills are influenced by its personalisation via settings defined by the user such as AIBO's name, the language it should recognise (English or Japanese), its voice and eye colour, or limitations on the use of cloud services (e.g., taking and uploading photos automatically, etc). The gender of AIBO has to be selected during the first setup, which affects the pitch of its voice and its movements, but not its personality. Its gender can only be changed later with the initialisation of the robot, during which all settings, recorded data, and learned actions are deleted. The overall personality of the robot changes based on its previous long-term interactions with humans, which in turn influences its interaction style, making the robot 'clingy' or 'wild', as defined by Sony. AIBOs also have individual likes and dislikes, for example in the form of food preferences, expressed by their differential behaviour (e.g., emotionally expressive behaviours) towards them (Sony, 2018). However, due to the lack of available information, the extent and form of these behavioural differences are not clear.

We consider several of AIBO's reactions to environmental stimuli or stimuli coming from a human to be well designed and functionally relevant (e.g., it follows or approaches its toys, looks at human faces, and expresses emotions in accordance with the stimulus, such as reacting to loud sounds with surprise). Less 'natural' actions of AIBO (e.g., dancing or playing music) create a contrast with its otherwise dog-like embodiment and behaviour repertoire, making the overall impression less coherent and more toy-like.

4.4.9 Future utility

Utility in a strict sense has never been a main consideration with AIBO, as its primary goal is to provide entertainment. The niche it is designed to operate in might decrease, or even disappear with time, as the development of robotics technology will likely bring forth robots that can fulfil more practical roles *and* entertain their users. There might be some interest towards artificial home companions due to living conditions and cultural shifts (either long-term, or more temporarily such as the experience of a pandemic lockdown), in which a more controllable pet with no biological needs and an option to suspend its operation might be favoured.

Questions regarding data privacy will have to be considered in detail with the further popularisation of robots functioning in home environments, as even AIBO uses, processes, and stores an array of sensory information via the AI cloud service (see also Lutz et al., 2019).

Independently of what the future holds for AIBO, it already had an interesting cultural impact, highlighting the emotional bonds that users could form towards their robots. With the diminishing support of older models from Sony, hundreds of AIBO funerals were held in Japan to bid farewell to robotic pets too old or broken to repair. After the ceremonies, parts of the robots are donated to other AIBOs in need of repair, which is provided in an increasing number of community-oriented repair shops across the country. The practice emphasises differences between eastern and western cultures in attitudes and customs towards inanimate but personally or culturally relevant objects (Knox & Watanabe, 2018).

4.4.10 Similar robots

Quadruped robots have been present in various applications (Figure 4.14). The dinosaur-like small robot, Pleo from 2006, and its newer version, Pleo reborn (released in 2011) (Causo et al., 2016) were designed to be artificial pets. The original Pleo had two speakers and 14 motors to enable interactive behaviours, as well as various sensors including multiple touch sensors, push buttons in its feet, two microphones, a camera, two infrared sensors, and a tilt sensor (Fernaeus et al., 2010). The Pleo reborn has four life stages from newborn to adult in which it exhibits different behaviours and skills (Kamino, 2023). The robot can slowly walk, react to its environment, and express moods, and its behaviour is influenced by internal drives (e.g., simulated 'hunger') (Heerink et al., 2012). The capabilities of the old version of Pleo regarding e.g., its learning and interactive skills, development, and motion did not meet the expectations of users in a long-term study (Fernaeus et al., 2010). The Pleo reborn has been used in various research studies, for example, investigating child–robot interactions in a paediatric hospital (Moerman & Jansens, 2021) or engagement of elderly people living with dementia (Hendrix et al., 2019). However, it is still unclear whether the Pleo reborn with its augmented sensorisation and interactive skills was able to better match the expectations set for successful robotic toys.

Recently, the quadruped robots that received a lot of limelight are the Spot, developed by Boston Dynamics and the various quadruped robot platforms from Unitree Robotics. These robots are not intended as pets, with developers and researchers envisioning various future applications like the oversight of construction sites (Wetzel et al., 2022), guide roles for people with visual impairments (Due, 2023), and uses in disaster sites or rescue situations (Sombolestan et al., 2021). The expected functions of these quadruped robots are based on their advanced locomotion capabilities,

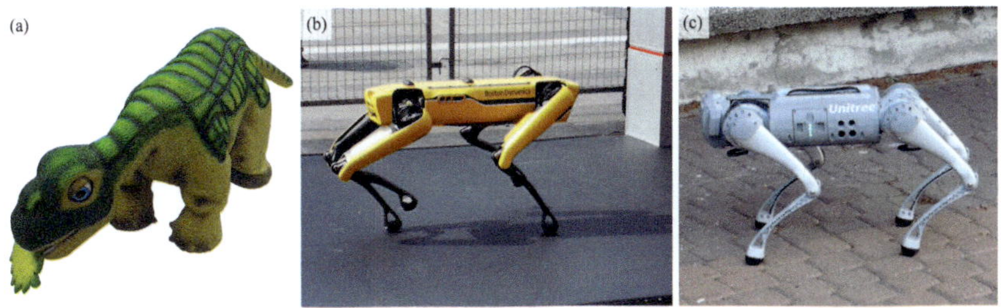

Figure 4.14 Examples of quadruped robots. (A) Pleo (Innvo Labs; photo by Jiuguang Wang, Wikimedia Commons) (B) Spot (Boston Dynamics; photo by Saggittarius A, Wikimedia Commons) (C) Unitree Go1 (Unitree Robotics; photo by DGtal, Wikimedia Commons).

which enable them to move around and navigate in areas where wheeled robots are usually inadequate. So far, the majority of research focused on technical implementations, and not much emphasis had been paid for interaction with humans. The runtime of most of these robots is currently rather short: on average 90 minutes for Spot (Boston Dynamics, 2023), and 1–2 for the standard, or 2–4 hours for the EDU version of the Unitree Go2 (Unitree, 2024), according to the manufacturers. Their payload capacity is a maximum of 14 kg (Spot) and 8–12 kg (Unitree Go2 EDU). On the other hand, the Unitree B2 promises 4–6 hours of runtime and a maximum of 40 kgs payload during walking. The autonomous skills of Spot seem to mainly focus on behaviours relating to locomotion and navigation, with the option of custom developments. While these robots are not inherently designed for HRI, we assume that simple social skills and interactive behaviours could be easily implemented.

4.5 BellaBot

4.5.1 Overview

The BellaBot is a service robot developed for delivery tasks in commercial applications by PUDU Robotics, with its main intended use being as a waiter assistant robot in restaurants. BellaBot is a cylindrically shaped relatively tall robot, with four levels of tray areas incorporated in its torso (Figure 4.15). The upper part of the robot features some design elements (e.g., 'ears', animated face) with zoomorphic aspects, reminiscent of a cat. PUDU Robotics was established in Shenzen (China) in 2016 and had developed various robot platforms before the release of BellaBot in 2020. According to company data around 70,000 units have been sold in more than 60 countries by August 2023 for a retail price of approximately $15,900.

As the BellaBot is a commercial product and not an open-source development or a research platform, there is less publicly available information on its hardware and software components. Interestingly, despite its wide-spread use since 2020, it has not yet appeared in the focus of research studies, regardless of the relatively fast-spaced scientific field of robotics and HRI.

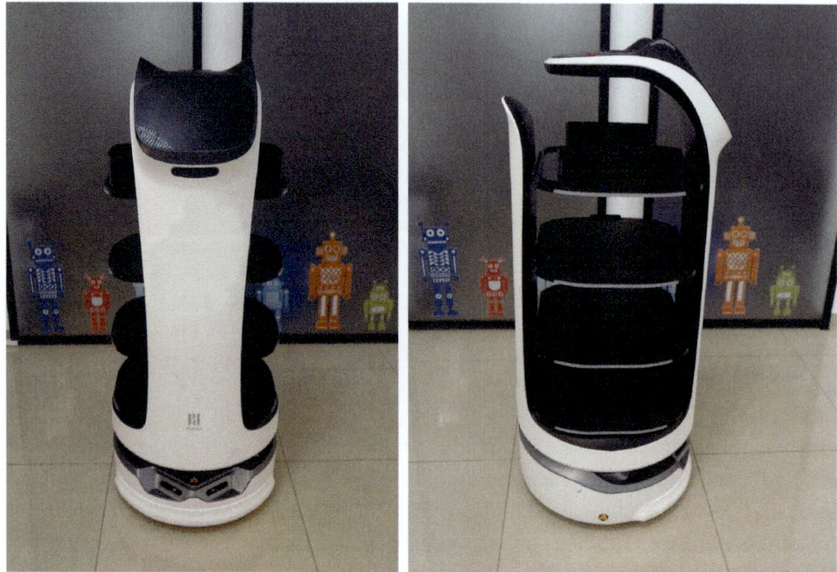

Figure 4.15 The BellaBot service robot by PUDU Robotics.

4.5.2 Embodiment

The embodiment of BellaBot resembles a cylindrical shape with three main sections: the base, the torso, and the head. The base of the robot contains the drive mechanism, the wheels, 2 RGB-D cameras (placed in a slightly upwards angle), and the battery. The base is separated from the torso by an aperture needed by the robot's Lidar sensor. The torso of the robot has a solid, plastic-covered front and back side, while its side is open, providing access to the four trays placed above each other. The head of the robot is a continuation of the torso with no neckline or articulation. The head carries various sensors (see Table 4.5 and Section 4.5.5), with a touchscreen as the main interface. The touchscreen, apart from the displays of the various functions, also shows an animated cat-like face in an iconic style. The cat-like appearance is further emphasised by the two protrusions (placed and shaped similarly to cartoon cat ears) on the top of the head. The robot can carry a total of 40 kgs of load, 10 kg/tray (PUDU, 2023). While the trays are modular, they are not to be taken out by guests, but can be removed to, e.g., change the distance between one above the other.

4.5.3 Further developments and improvements

Although we have not found official information on different hardware versions of BellaBot, the placement of the small frontal camera of the robot (in the same plane as the touchscreen) varies between being above or below the touchscreen. In older videos, it seems to be missing, which is indicative of the continued development of the robot. Software updates have also been provided for the BellaBot.

Table 4.5 Hardware overview of BellaBot, based on the PUDU Robotics website (PUDU, 2023) and uploaded video recordings.

Height		129 cm
Weight		55 kg
Body plan		cylindrical
Locomotion		Wheels: 2 standard wheels, 4 dual caster wheels (based on an official PUDU video)
Body surface		ABS plastic
DoFs	Total	0 (not counting locomotion)
	Head	–
	Neck	–
	Torso	–
	Legs	–
	Arms	–
	Fingers	–
Interactive hardware	LEDs	Colour changing LEDs in 'ears', near the bottom of the torso, at the edge of the trays
	Loudspeaker	Yes (probably two loudspeakers near the bottom of the torso, based on an official PUDU video)
	Screen	Touchscreen on head, dot matrix screen on the backside of the torso
Sensors	Head	Camera facing upwards for visual SLAM, touch sensors in 'ears', touch sensor button near the back of the head, touchscreen, emergency stop button, small camera at the top or bottom of the touchscreen, presumably a microphone
	Neck	RGB-D camera
	Torso	Lidar, 2 RGB-D cameras, infrared sensors for each tray
	Legs	–
	Arms	–
Operation time with charged battery		12–13 hours
Human-likeness score (ABOT)		N/A

Accessories are available for the robot, expanding its functionality. A large delivery box has been designed, which can be put on the torso of the robot, covering the trays and thus protecting the food during delivery. The box has magnetic doors for tray access. A nine-part silicone cup holder can also be put on the trays, making the delivery of beverages safer. Considering IoT accessories, a 4G watch can be used for e.g., calling the robot to a goal location. A door control system is also available for automatic doors and turnstiles, enabling the robot to autonomously open gates and move through them. Another accessory is a pager device with which the robot can be called to the location of the pager (e.g., to a restaurant table).

An opportunity for robot hardware and software customisation and development is offered by the company's project called PUDU Open Platform, which is available for distributor members (PUDU, 2023).

4.5.4 Autonomy

BellaBot has been designed for autonomous operation, with some remote-control options via a 4G watch, an Android mobile app and a pager. With the pager and the 4G watch the robot can be called to a location, but the functionality of the mobile app is not detailed. In the autonomous mode the robot can carry out several functions such as cruise, delivery, escorting etc. (see also Section 4.5.5 Actions). As visible in some demo videos, the path of the robot in cruise mode can be defined via an application that shows the layout of an already mapped area. During its autonomous locomotion, the robot can detect and stop before obstacles with a reaction time of 0.5 s, or it can also avoid them.

4.5.5 Sensory and motor capabilities

Officially available information on BellaBot is very scarce; therefore for this section, we relied on the PUDU website, brochures, and videos available online to find the relevant information or to make observations based on them. We also had the opportunity to observe a BellaBot robot operating at a local restaurant.

4.5.5.1 Vision

BellaBot uses its visual inputs for navigation. Its visual SLAM navigation is based on a camera on top of the robot's head, looking straight up, which is be able to perceive specifically designed visual markers. The visual markers have to be placed on a ceiling that is no higher than 8 m. One of the markers is for indicating the starting location of the robot. When the robot boots up, as well as during navigation, it recognises the individual markers and uses them for rudimentary localisation.

The more refined part of BellaBot's navigation is a laser SLAM method, for which the input data is provided by a Lidar located between the robot base and torso at an approximately 30 cm height, and 3 RGB-D cameras, with all sensors on the forward side of the robot. Two of the cameras can be found at the base of the robot looking slightly upwards, and one below the head of the robot facing slightly downwards. The frontal detection angle is 192.64° horizontally, and the two cameras at the base cover 97° vertically. The robot can detect objects more than 10 m in front of it.

4.5.5.2 Hearing

There is no mention of a microphone, thus the hearing capacities of the robot are unknown. However, the robot probably has a microphone because in some of the available demo videos the robot reacts to voice commands with displaying a specific screen not shown in other situations and answers the person it interacts with both in writing and with synthesised speech (see also Section 4.5.5 Speech).

4.5.5.3 Touch

The tactile interaction between the robot and people is enabled by the touchscreen on the robot's head, touch sensors in its ears, and two buttons on the back of the head. The larger one serves as a pause or sending away button, and the smaller one is the on–off switch. The robot also has an emergency stop button on the top of its head.

The robot reacts to its ears being touched by varying its face animations and by playing pre-defined phrases and cat-like vocalisations (see also Section 4.5.5 Gestures/signals). If the touchscreen or the pause button is touched once during navigation the robot stops, so the user can use the screen to access the settings or to give new tasks to the robot.

4.5.5.4 Other sensing capabilities

Each tray of the robot is equipped with infrared sensors, which enable the robot to sense whether a tray is empty or not.

4.5.5.5 Speech

BellaBot can synthesise speech in different voices and in multiple languages (most likely depending on the country of use/acquisition). The robot usually says pre-defined phrases related to its current tasks (e.g., the robot can tell guests when it arrives near a table to take off their orders from its tray), or to the interaction it is in (e.g., when its ears are touched). The loudness of the speech can be adjusted. The robot also vocalises during speech, making cat-like 'meow' sounds in a human voice. These vocalisations are also used for emotion expression, as they have 'happy' and 'frustrated/angry' versions as well. According to the main Chinese website, BellaBot also received an AI voice module, providing it with more situation-specific dialogues. Some demo videos indicate that the BellaBot can also be asked to tell jokes, the current time, or the weather, which assumes a constant connection to an online source, presumably via the official PUDU cloud service.

4.5.5.6 Actions

The actions of the robot are optimised for simple delivery tasks and mainly consist of locomotion. The robot can move to pre-defined checkpoints (e.g., start location, charging place, pick-up point, and tables), with various modes. In the delivery mode, the robot visits multiple checkpoints (e.g., tables) after each other. It waits by the checkpoints for a person to take away the delivered item, which can be perceived by the infrared sensors right above the trays. The robot moves to the next checkpoint if a specified time has passed, or if the 'done' button on the touchscreen or on the back of its head is pressed. After visiting all checkpoints, the robot returns to the start (e.g., to the kitchen). A different delivery mode is also possible in which the robot moves to a selected destination and stays there until further commands are received. In cruise mode the robot moves around continuously in a known area following a pre-defined path, while in the escorting mode, it goes from a starting point (e.g., reception) to a destination, waits a set number of seconds, then goes back to the starting point. There are also modes for events: a birthday mode and a special event mode, which can be customised. The robot can also play custom music and playlists during the operation modes. The music volume can be changed independently from the speech volume.

4.5.5.7 Gestures/signals

BellaBot relies on visual and acoustic modalities for communication. The touchscreen serves as both a 'classic' GUI solution for giving commands and accessing settings, and as the 'face' of the robot. The animated cat-like face changes based on the current task progress (e.g., it can make a happy face when it starts to move towards a goal) and during interaction with humans: BellaBot reacts to its ears being touched with various emotionally expressive, iconic faces (Figure 4.16). The changes of the face display are accompanied with vocalisations, either as phrases or 'meowing'. If the ears are touched repeatedly, the touchscreen displays angrier faces, with corresponding vocalisations, with the aim of discouraging guests from taking too much time away from the robot. However, the cartoonish, symbolic aesthetic is maintained. The animated face can be switched off in the settings.

The robot uses colourful LEDs for communication as well. The LEDs in the ears, the pause button, and at the sides of the trays serve as status updates. Based on videos, the robot can also signal with the tray LEDs that which of its trays carries the food that belongs to the currently reached table. There are also LEDs in the aperture between the base and the torso of the robot, which can display a wide range of colours. We assume that these mainly have entertainment purposes.

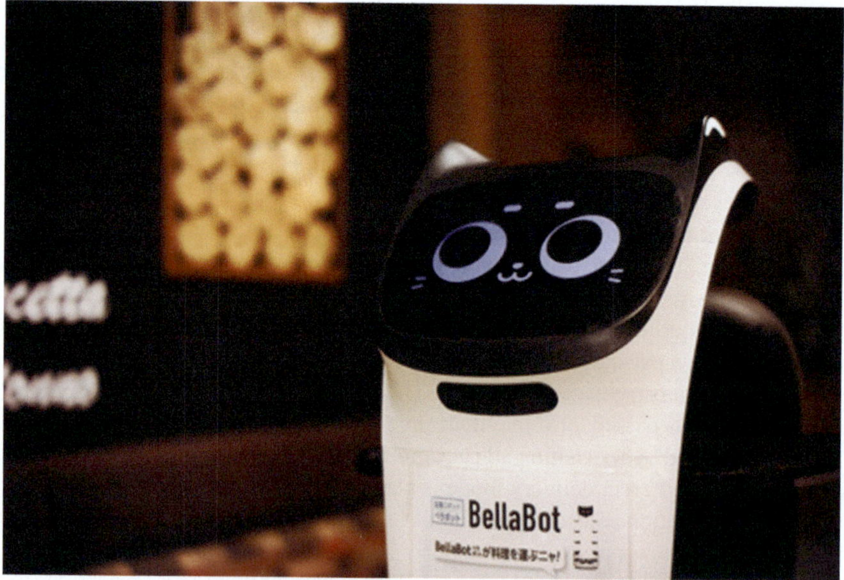

Figure 4.16 The 'basic' cat-like animated face of BellaBot can change into different facial expressions (Photo by MIKI Yoshihito, Flickr).

At the upper back of the robot, we can find a dot matrix screen, on which the robot can display texts and simple symbols. It uses the dot screen to show which direction is it going to turn next and can show e.g., the number of the table it is going to. The text can be modified, and different texts can be displayed during the various modes.

4.5.6 Software architecture and related system capabilities

We have not found publicly available information on the software architecture of BellaBot, apart from that the robot uses an Android operating system. The robot utilises the PUDU cloud service, which functions as a maintenance, scenario data collection, and cloud intelligent service platform, among others. It also contains the Scheduler system, with which the side-by-side operation of up to 20 other PUDU robots can be optimised with direct communication between them.

4.5.7 Utilisation in HRI research

As of now we have not found HRI studies published in reviewed scientific journals with BellaBot. The articles mentioning BellaBot usually only refer to it as an example of a restaurant delivery robot or provide only minimal descriptions (e.g., Sala, 2023). One of the papers by Smirnov and Kovalev (2021) mentions that the BellaBot has IMU sensors; however, we have found no additional information on this. Furthermore, there are research studies with another robot named Bellabot, however, this is a fundamentally different robot developed in the UK to study bioinspired sensing (via whisker-like structures) and adaptive robot control systems (Wilson et al., 2021).

Van Doorn et al. (2023) aimed to investigate the effect of autonomous technologies (AT) in hospitality services on workers and consumers. Based on a new framework (Consumer – Autonomous Technology – Worker (CAW), the authors analysed this three-way relationship for business implications. To collect insights for the CAW framework, the authors conducted interviews with 15 staff members who have been working alongside service robots, as well as 9 customers who interacted with both the human staff and with robots. Four subjects (three waiters and one waiter-manager) worked with BellaBot for less than six months, and three individuals had firsthand experience with the robots as customers. According to the interviews, they suggest that consumer–worker relationships are enhanced when autonomous technologies complement rather than replace human workers. Human supervision plays a crucial role in fostering acceptance of autonomous technologies among both consumers and workers, while the level of anthropomorphism of the technology becomes less critical when there is a human worker present.

4.5.8 Ethorobotic perspectives

4.5.8.1 Embodiment

The structural aspect of the robot's embodiment has been optimised to its functional niche of a delivery robot with multiple, modular trays. The trays are covered in a non-slip silicone material, which provides more stability for the carried items during delivery. The zoomorphic, iconic design elements (the ears and the cat-like face) could ease short-term interactions by making the robot seem less threatening and more inviting for interaction. The choice of using a simplistic design resembling an animal (in contrast to using a human-like design or a realistic rendering) could have some effects on moderating the initial expectations towards the capabilities of the robot.

4.5.8.2 Sensing

The sensorisation at the front of the robot provides some redundancy via the Lidar and the 3 RGB-D cameras, creating a good opportunity for sensor fusion. It is not clear, however, whether there are any sensors for perceiving the environment behind the robot.

According to some videos and descriptions, the robot can receive commands via voice, but the placement of the microphones and the features and limitations of this system are not detailed. The function of the small built-in camera by the touchscreen is unknown.

4.5.8.3 Autonomous operation

There is no adequate information to assess the robustness of BellaBot's autonomous operation, which might be strongly influenced by the particular environment the robot has to function in (e.g., number of people walking around). The autonomous operation is aided by the long runtime, which can reach 12–13 hours with a charged battery. Additionally, while the robot's battery can be charged while inside the robot through a port, the battery can also be easily taken out and switched to an already charged one. This, together with the 4.5 hours charging time, can ensure an all-day-long operation.

4.5.8.4 Locomotion

The main function of BellaBot lies in its ability of locomotion. The food delivery is supported by a vehicle-grade independent suspension dampening system, which makes the movements of the robot smoother in case it has to move on an uneven surface. The speed of its movement can be customised between 0.5 and 1.2 m/s according to an official brochure (PUDU, 2023).

4.5.8.5 Object manipulation

The robot has no movable actuators (apart from its wheels), and no object manipulation capabilities. At present, the BellaBot relies on the cooperative attitude of both customers and workers to assist in placing or removing objects carried by the robot. However, this reliance on human assistance limits the robot's utility. Given the current state of technology, there is no foreseeable realistic and practical prospect of equipping such robots with their own manipulation skills.

4.5.8.6 Interactive skills

The virtual face of the robot facilitates the use of iconic, emoticon-like facial expressions. However, this face is not always visible due to the dual function of the touchscreen, which also serves as an interface for assigning tasks to the robot and adjusting its settings. Consequently, users frequently interact with the robot by tapping or 'poking' its face to access controls, which deviates from natural human and interspecies interactions and disrupts the consistency of the robot having a face. An alternative approach to addressing this issue could involve dividing the screen, with the robot's face displayed in one half and the control panel in the other.

In terms of acoustic communication, the combination of human speech and cat-like vocalisations, albeit produced with a human voice, is quite incoherent. We anticipate that the cat-like vocalisations may become bothersome for human staff working with the robot over the long term.

Furthermore, the turn signal displayed on the dot matrix screen at the back of the robot features arrows pointing in the direction the robot intends to turn. However, for individuals encountering the robot for the first time, such as those walking behind it, it may be confusing whether the robot is indicating its own path or directing people behind it to manoeuvre around the side indicated by the arrows.

4.5.9 Future utility

BellaBot can be viewed as a robot assistant for waiters, although it currently resembles more of a tea trolley equipped with an autonomous navigation system. This relatively straightforward functionality may be sufficient, as tasks such as meal delivery or collecting dirty dishes consume a significant amount of time for waitstaff. Consequently, these robots may represent an instance of disappearing technology, where social skills are not necessarily required.

Alternatively, a more animal-like design could serve as the foundation for certain social interactions. For instance, incorporating a head with a movable neck into the robot's design could enable the display of various emotional states. Social behaviours could foster a customer–robot relationship, particularly in settings like offices or hospitals where the establishment of a more personalised connection is feasible.

4.5.10 Similar robots

Recent years have brought forward the introduction of numerous waiter assistant robots (Figure 4.17), which is not surprising with the predicted rise of service robotics in hospitality roles (Bowen & Morosan, 2018). Besides some other waiter assistant robot models from PUDU like the KettyBot or the PuduBot, robots from other manufacturers are also used in multiple countries, including Europe. The T1 dinerbot from Keenon and the various delivery robots from Reeman are structurally and functionally quite similar to the BellaBot with their wheeled locomotion, cylindrical or rounded non-humanoid body shape with multiple trays, and the use of a small touchscreen display for expressions and interaction (although the Keenon T1 has

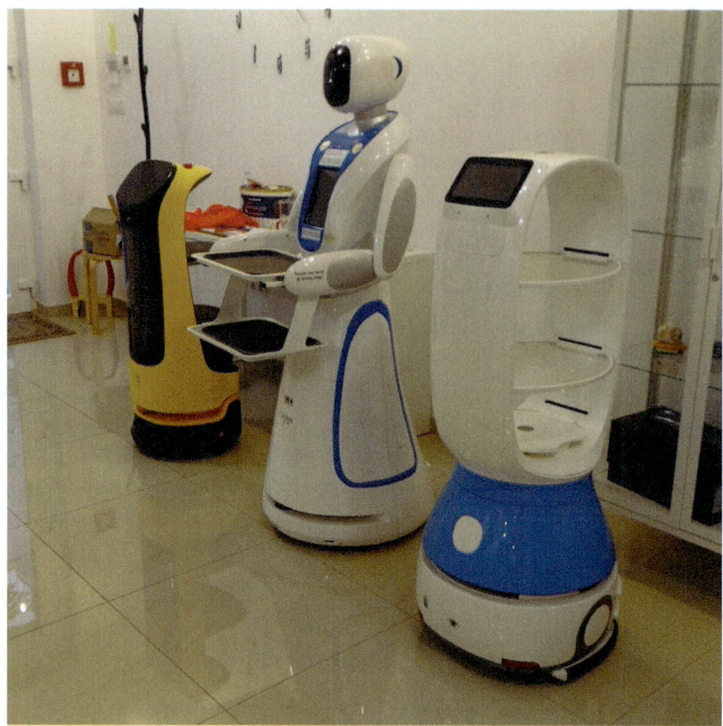

Figure 4.17 Examples of waiter assistant robots in the coffee shop of the Enjoy Robotics Company, Budapest, Hungary. From left to right: KettyBot by PUDU Robotics, Amy by Suzhou Pangolin Robot Corporation (Csjbot), and T1 dinerbot by Keenon.

two separate displays, one for displaying simple, square-like animated eyes and one for user instructions). Another approach is represented by the Amy waiter robot by the Suzhou Pangolin Robot Corporation (also known as Csjbot) which, while also moving on wheels, resembles a humanoid form carrying a two-level tray in its fixed, non-movable hands. Despite their use in coffee shops and restaurants, research with these robots focusing on HRI (instead of technological developments) is basically non-existent for now.

4.6 Conclusions from an ethorobotic perspective

For this chapter, we have selected four distinct types of robots, each emblematic of a specific field of application. Our decision was guided by the aim to showcase functionally different social robots, yet accessibility to scientifically reviewed and published material strongly influenced the inclusion of NAO, Kaspar, and AIBO.

NAO was originally conceived as a companion robot for studying HRI, although it also exhibits limited potential for application in robot-assisted therapy. Similarly, Kaspar may possess similar functionalities, but it has not reached the level of a truly vendable product yet. Indeed, the technical development of the NAO and the Kaspar had an opposite course. While Kaspar underwent continuous development and testing alongside published findings, NAO was introduced to the market without prior publicly available testing data. The AIBO shared the story of NAO with a specific focus on entertainment for people. However, over the years, AIBO has changed in response to user feedback and the company's decision of repositioning it on the market. Thus, AIBO was transitioned from a mere pet-like entertainer to a more nuanced role as a robotic 'pet dog'. Finally, the BellaBot represents an example for a more industrial approach serving primarily as a tea trolley equipped with autonomous navigation capabilities, only showcasing social skills on a level that is not functionally important to its role, and rather serves as added entertainment.

4.6.1 Task description and benchmarking

One could argue that all these robots have been meticulously designed to align with their designated marketing niches, a fact further corroborated by their descriptions in the manuals. However, it remains ambiguous whether the specific abilities outlined are truly effective within the typical parameters of their work environments and whether they satisfy the expectations of the purchasing companies. There's a growing tendency to oppose the use of robots due to the belief that 'they replace human jobs'. Consequently, it is imperative for social robots to fulfil their promises; otherwise, they encounter resistance instead of cooperation from the very workers they aim to assist.

Two important aspects warrant consideration: task analysis and benchmarking (Section 2.2.4). For example, the BellaBot is often promoted as a 'waiter robot'. This characterisation not only understandably generates fear in waitstaff that they may lose their jobs but also misrepresents BellaBot's real skills. A comprehensive task analysis of waiter duties reveals that BellaBot could either supplement or assist 'food runners', whose primary task involves aiding waiters in rapidly delivering orders to customers. Task analysis is also pivotal in determining the optimal morphological and behavioural attributes for the robot. It is essential to initially design a minimally functional system, as premature finalisation may impede future advancements.

Alongside task analysis conducted prior to the design and construction of the robot, it is vital to establish a benchmarking system. Benchmarking should correspond to the expected functionality of the robot and provide feasible and practical data (most of which could be measured automatically by the system). These measurements should encompass primary and secondary variables, including performance parameters and specific behaviours of the robot.

4.6.2 Comparative research

The development of ethorobots is evidently a comparative endeavour. Assessing the performance of any robot necessitates contrasting it with a reference system to highlight the advantages of new solutions. This can be achieved by comparing the newer robot to existing similar models or by using human performance as a baseline. The former approach may pose challenges, as other robots are typically unavailable for testing, and acquiring the necessary expertise to manipulate a robot from another provider entails a significant investment of time and resources.

Comparing robots with human performance presents an alternative, though there is a risk that the robot shows inferior performance. However, such comparisons can still be informative, revealing the extent to which a robot can replace human labour. Nevertheless, such analysis may still indicate that the introduction of the new technology offers distinct advantages. Therefore, it would be beneficial for all publications on ethorobots to include a 'human control' group across all potential implementation fields. This approach would also be valuable in providing measures of human performance and behaviour, including error tolerance, which are currently lacking.

4.6.3 Cutting edge technology and robustness

The situation is comparable to that in the pharmaceutical industry. Chemical compounds show their potential to cure specific diseases in the laboratory, but they need to be developed further and tested for the following five to ten years before they become medications for the human population at large. Robots in the laboratories may show extraordinary performance, recorded on videos, but it could take many years before these features are integrated into a marketable agent. However, robotics, in contrast with the pharmaceutical industry, could progress by smaller steps and thus making the transition from the laboratory to the market faster. Evolution is typically also progressing this way, but various features including robust fundamental functioning, flexibility, and modularity support this mechanism in living systems.

4.6.4 Where are we? Technology Readiness Levels (TRL)

There seem to be two different avenues for developing social robots. Given significant amount of funding, laboratories at universities or research institutes initiate the construction of robot prototypes, which may be improved step by step by many generations of students. These robots are used only within the laboratory, partly because they need continuous engineering support to stay functional for a specific set of experiments. Over the years, many skills of these robots are improved based on the feedback from the experiments but these changes in hardware and software are often not professional, so these robots never go beyond level 3 of technological readiness (TRL), that is, representing a 'proof of concept'. This hinders their deployment in other laboratories to get independent validation of the obtained achievements (replicability).

Robots like the NAO or AIBO had a different history, as both these and many others were introduced by companies promptly as a marketed product, that is they complied with all levels of steps of technological readiness. However, it turned out quite often that while these robots satisfy the expectations of being a product, their features provide very limited possibilities for research or are not useful at all.

This situation suggests a need for much closer cooperation between companies who have the means to produce robots, and researchers who want to test these robots in realistic environments with regard to their usefulness, robustness, interactivity with humans, etc. This collaboration should be initiated from both sides when a specific technology has been validated in the lab (Level 4). This is the stage after which it needs to be ensured that the technology is robust enough to be utilised for the planned function and performs at acceptable levels.

4.6.5 Versatility versus specificity

The pursuit of 'human-like' robots often overlooks a practical consideration regarding single or multi-functionality (Riener et al., 2023). Biological evolution demonstrates that living organisms transitioned from possessing specific (narrow) functions to developing a broader range of skills. This evolutionary trend coincided with slowly increasing complexity in embodiment, including multicellularity and the evolution of the brain, in addition to the increased complexity of mental architecture.

Certainly, the dichotomy between specificity and versatility is contingent upon the definition. However, when applied to ethorobots, it suggests that they should ideally have a narrow range of functions. It could be argued that, similarly to industrial robots, human skills required for solving complex tasks could be distributed among many specific ethorobots. This approach could potentially enhance the robustness of ethorobots by eliminating the need to integrate complex embodiments and architectures.

References

Albani, D., Youssef, A., Suriani, V., Nardi, D., & Bloisi, D. D. (2016). A deep learning approach for object recognition with NAO soccer robots. In S. Behnke, R. Sheh, S. Sarıel, & D. Lee (Eds.), *RoboCup 2016: Robot World Cup XX. RoboCup 2016. Lecture Notes in Computer Science: Vol. 9776*. Springer.

Alenljung, B., Andreasson, R., Lowe, R., Billing, E., & Lindblom, J. (2018). Conveying emotions by touch to the Nao robot: A user experience perspective. *Multimodal Technologies and Interaction, 2*(4), 82. https://doi.org/10.3390/mti2040082

Amırova, A., Rakhymbayeva, N., Yadollahi, E., Sandygulova, A., & Johal, W. (2021). 10 years of human-NAO interaction research: A scoping review. *Frontiers in Robotics and AI, 8*, 744526. https://doi.org/10.3389/frobt.2021.744526

Barbadillo, G., Dautenhahn, K., & Wood, L. (2011). *Using FSR sensors to provide tactile skin to the humanoid robot KASPAR.*

Becker, D., & Risler, M. (2009). Mutual localization in a team of autonomous robots using acoustic robot detection. In L. Iocchi, H. Matsubara, A. Weitzenfeld, & C. Zhou (Eds.), *RoboCup 2008: Robot Soccer World Cup XII* (pp. 37–48). Springer-Verlag. https://doi.org/10.1007/978-3-642-02921-9_4

Blow, M., Dautenhahn, K., Appleby, A., Nehaniv, C. L., & Lee, D. C. (2006). Perception of robot smiles and dimensions for human-robot interaction design. *The 15th IEEE International Symposium on Robot and Human Interactive Communication (RO-MAN06)*, 469–474. https://doi.org/10.1109/ROMAN.2006.314372

Boston Dynamics. (2023). *Spot® - The agile mobile robot*. Https://Bostondynamics.Com/Products/Spot/.

Bowen, J., & Morosan, C. (2018). Beware hospitality industry: The robots are coming. *Worldwide Hospitality and Tourism Themes, 10*(6), 726–733. https://doi.org/10.1108/WHATT-07-2018-0045

Brown, R. L., Helton, H. L., Williams, A. C., Shrove, M. T., Milošević, M., Jovanov, E., Coe, D., & Kulick, J. (2013). Android control application for Nao humanoid robot. *Proceedings of the International Conference on Frontiers in Education: Computer Science and Computer Engineering (FECS), WORLDCOMP 2013*, 1.

Burdi, A. R., Huelke, D. F., Snyder, R. G., & Lowrey, G. H. (1969). Infants and children in the adult world of automobile safety design: Pediatric and anatomical considerations for design of child restraints. *Journal of Biomechanics, 2*, 267–280. https://doi.org/10.1016/0021-9290(69)90083-9

Cabibihan, J.-J., Javed, H., Ang, M., & Aljunied, S. M. (2013). Why robots? A survey on the roles and benefits of social robots in the therapy of children with autism. *International Journal of Social Robotics, 5*(4), 593–618. https://doi.org/10.1007/s12369-013-0202-2

Causo, A., Vo, G. T., Chen, I. M., & Yeo, S. H. (2016). Design of robots used as education companion and tutor. In S. Zeghloul, M. Laribi, & J. P. Gazeau (Eds.), *Robotics and Mechatronics. Mechanisms and Machine Science* (Vol. 37, pp. 75–84). Springer. https://doi.org/10.1007/978-3-319-22368-1_8

Chalup, S. K., Murch, C. L., & Quinlan, M. J. (2007). Machine learning with AIBO robots in the four-legged league of RoboCup. *IEEE Transactions on Systems, Man and Cybernetics - Part C: Applications and Reviews, 37*(3), 297–310. https://doi.org/10.1109/TSMCC.2006.886964

Chernova, S., & Veloso, M. (2004). An evolutionary approach to gait learning for four-legged robots. *Proceedings 01 2004 IEEE/RSJ International Conference on Intelligent Robots and Systems*, 2562–2567. https://doi.org/10.1109/IROS.2004.1389794

Datcu, D., Richert, M., Roberti, T., De Vries, W., & Rothkrantz, L. J. M. (2004). AIBO robot as a soccer and rescue game player. *Proceedings of GAME-ON 2004*, 45–49.

Dautenhahn, K., Nehaniv, C. L., Walters, M. L., Robins, B., Kose-Bagci, H., Mirza, N. A., & Blow, M. (2009). KASPAR - a minimally expressive humanoid robot for human-robot interaction research. *Applied Bionics and Biomechanics, 6*(3–4), 369–397. https://doi.org/10.1080/11762320903123567

Davis, F. D., Bagozzi, R. P., & Warshaw, P. R. (1992). Extrinsic and intrinsic motivation to use computers in the workplace. *Journal of Applied Social Psychology, 22*(14), 1111–1132. https://doi.org/10.1111/j.1559-1816.1992.tb00945.x

Due, B. L. (2023). A walk in the park with robodog: Navigating around pedestrians using a Spot robot as a "guide dog". *Space and Culture, 00*(0), 1–17. https://doi.org/10.1177/12063312231159215

Ekman, P. (1992). *Telling lies: Clues to deceit in the marketplace, politics, and marriage.* W. W. Norton & Company.

Ekman, P., & Friesen, W. V. (1982). Felt, false, and miserable smiles. *Journal of Nonverbal Behavior, 6*(4), 238–252. https://doi.org/10.1007/BF00987191

Ekman, P., Friesen, W. V, & Hager, J. C. (2002). *Facial action coding system: The manual.* Research Nexus division of Network Information Research Corporation.

Feng, S., Wang, X., Wang, Q., Fang, J., Wu, Y., Yi, L., & Wei, K. (2018). The uncanny valley effect in typically developing children and its absence in children with autism spectrum disorders. *PLoS ONE, 13*(11), e0206343. https://doi.org/10.1371/journal.pone.0206343

Fernaeus, Y., Håkansson, M., Jacobsson, M., & Ljungblad, S. (2010). How do you play with a robotic toy animal? A long-term study of Pleo. *IDC '10: Proceedings of the 9th International Conference on Interaction Design and Children*, 39–48. https://doi.org/10.1145/1810543.1810549

François, D., Powell, S., & Dautenhahn, K. (2009). A long-term study of children with autism playing with a robotic pet. *Interaction Studies, 10*(3), 324–373. https://doi.org/10.1075/is.10.3.04fra

Fujita, M. (2001). AIBO: Toward the era of digital creatures. *The International Journal of Robotics Research, 20*(10), 781–794. https://doi.org/10.1177/02783640122068099

Fujita, M. (2004). On activating human communications with pet-type robot AIBO. *Proceedings of the IEEE, 92*(11), 1804–1813. https://doi.org/10.1109/JPROC.2004.835364

Fujita, M., & Kitano, H. (1998). Development of an autonomous quadruped robot for robot entertainment. *Autonomous Robots*, *5*, 7–18. https://doi.org/10.1023/A:1008856824126

Gehle, R., Pitsch, K., & Wrede, S. (2014). Signaling trouble in robot-to-group interaction. Emerging visitor dynamics with a museum guide robot. *HAI 2014- Proceedings of the 2nd International Conference on Human-Agent Interaction*, 361–368. https://doi.org/10.1145/2658861.2658887

Gelin, R. (2019). NAO. In A. Goswami & P. Vadakkepat (Eds.), *Humanoid robotics: A reference* (pp. 147–168). Springer. https://doi.org/10.1007/978-94-007-6046-2_14

Glocker, M. L., Langleben, D. D., Ruparel, K., Loughead, J. W., Gur, R. C., & Sachser, N. (2009). Baby schema in infant faces induces cuteness perception and motivation for caretaking in adults. *Ethology*, *115*(3), 257–263. https://doi.org/10.1111/j.1439-0310.2008.01603.x

Gombots, S., Egner, F., & Kaltenbacher, M. (2017). Beamforming with Kinect V2. *Tagungsband Der 5. Tagung Innovation Messtechnik*, 76–81.

Gou, M. S., Lakatos, G., Holthaus, P., Robins, B., Moros, S., Wood, L., Araujo, H., deGraft-Hanson, C., Mousavi, M. R., & Amirabdollahian, F. (2023). Kaspar explains: The effect of causal explanations on visual perspective taking skills in children with autism spectrum disorder. *2023 32nd IEEE International Conference on Robot and Human Interactive Communication (RO-MAN)*, 1407–1412. https://doi.org/10.1109/ro-man57019.2023.10309464

Gray, K., & Wegner, D. M. (2012). Feeling robots and human zombies: Mind perception and the uncanny valley. *Cognition*, *125*(1), 125–130. https://doi.org/10.1016/j.cognition.2012.06.007

Han, Y., Zhang, M., Li, Q., & Liu, S. (2018). Chinese speech recognition and task analysis of Aldebaran Nao robot. *2018 Chinese Control and Decision Conference (CCDC)*, 4138–4142. https://doi.org/10.1109/CCDC.2018.8407843

Hausknecht, M., & Stone, P. (2011). Learning powerful kicks on the Aibo ERS-7: The quest for a striker. In J. Ruiz-del-Solar, E. Chown, & P. G. Plöger (Eds.), *RoboCup 2010: Robot Soccer World Cup XIV* (pp. 254–265). Springer-Verlag. https://doi.org/10.1007/978-3-642-20217-9

Heerink, M., Díaz, M., Albo-Canals, J., Angulo, C., Barco, A., Casacuberta, J., & Garriga, C. (2012). A field study with primary school children on perception of social presence and interactive behavior with a pet robot. *2012 IEEE RO-MAN: The 21st IEEE International Symposium on Robot and Human Interactive Communication*, 1045–1050. https://doi.org/10.1109/ROMAN.2012.6343887

Heinrich, S., & Wermter, S. (2011). Towards robust speech recognition for human-robot interaction. *Proceedings of the IROS2011 Workshop on Cognitive Neuroscience Robotics (CNR)*, 29–34.

Hendrix, J., Feng, Y., van Otterdijk, M., & Barakova, E. (2019). Adding a context: Will it influence human-robot interaction of people living with dementia? *Lecture Notes in Computer Science (Including Subseries Lecture Notes in Artificial Intelligence and Lecture Notes in Bioinformatics), 11876 LNAI*, 494–504. https://doi.org/10.1007/978-3-030-35888-4_46

Hornby, G. S., Fujita, M., Takamura, S., Yamamoto, T., & Hanagata, O. (1999). Autonomous evolution of gaits with the Sony quadruped robot. In W. Banzhaf, J. Daida, A. E. Eiben, M. H. Garzon, V. Honavar, M. Jakiela, & R. E. Smith (Eds.), *Proceedings of Genetic and Evolutionary Computation Conference* (pp. 1297–1304). Morgan Kauffmann.

Huijnen, C. A. G. J., Verreussel-Willen, H. A. M. D., Lexis, M. A. S., & de Witte, L. P. (2021). Robot KASPAR as mediator in making contact with children with autism: A pilot study. *International Journal of Social Robotics*, *13*(2), 237–249. https://doi.org/10.1007/s12369-020-00633-0

Ji, Z., Amirabdollahian, F., Polani, D., & Dautenhahn, K. (2011). Histogram based classification of tactile patterns on periodically distributed skin sensors for a humanoid robot. *2011 RO-MAN 20th IEEE International Symposium on Robot and Human Interactive Communication*, 433–440. https://doi.org/10.1109/ROMAN.2011.6005261

Kamino, W. (2023). Towards designing companion robots with the end in mind. *HRI '23: Companion of the 2023 ACM/IEEE International Conference on Human-Robot Interaction*, 76–80. https://doi.org/10.1145/3568294.3580046

Kaplan, F. (2000). Talking AIBO: First experimentation of verbal interactions with an autonomous four-legged robot. In A. Nijholt, D. Heylen, & K. Jokinen (Eds.), *Learning to behave: interacting agents CELE-TWENTE workshop on language technology* (pp. 57–63). http://www.csl.sony.fr

Kaplan, F., Oudeyer, P.-Y., Kubinyi, E., & Miklósi, Á. (2002). Robotic clicker training. *Robotics and Autonomous Systems*, *38*, 197–206.

Kawaguchi, Y., Nakamura, K., Tajima, T., & Waller, B. M. (2023). Revisiting the baby schema by a geometric morphometric analysis of infant facial characteristics across great apes. *Scientific Reports*, *13*(1), 5129. https://doi.org/10.1038/s41598-023-31731-4

Kennedy, J., Lemaignan, S., Montassier, C., Lavalade, P., Irfan, B., Papadopoulos, F., Senft, E., & Belpaeme, T. (2017). Child speech recognition in human-robot interaction: Evaluations and recommendations. *ACM/IEEE International Conference on Human-Robot Interaction, Part F127194*, 82–90. https://doi.org/10.1145/2909824.3020229

Kertész, C. (2013). Improvements in the native development environment for Sony AIBO. *International Journal of Interactive Multimedia and Artificial Intelligence*, *2*(3), 50–54. https://doi.org/10.9781/ijimai.2013.237

Kertész, C. (2014). Exploring surface detection for a quadruped robot in households. *2014 IEEE International Conference on Autonomous Robot Systems and Competitions (ICARSC)*, 152–157. https://doi.org/10.1109/ICARSC.2014.6849778

Kertész, C. (2016). Rigidity-based surface recognition for a domestic legged robot. *IEEE Robotics and Automation Letters*, *1*(1), 309–315. https://doi.org/10.1109/LRA.2016.2519949

Kertész, C., & Turunen, M. (2019). Exploratory analysis of Sony AIBO users. *AI and Society*, *34*(3), 625–638. https://doi.org/10.1007/s00146-018-0818-8

Knox, E., & Watanabe, K. (2018). AIBO robot mortuary rites in the Japanese cultural context. *2018 IEEE/RSJ International Conference on Intelligent Robots and Systems (IROS)*, 2020–2025. https://doi.org/10.1109/IROS.2018.8594066

Kose-Bagci, H., Dautenhahn, K., Syrdal, D. S., & Nehaniv, C. L. (2010). Drum-mate: Interaction dynamics and gestures in human-humanoid drumming experiments. *Connection Science*, *22*(2), 103–134. https://doi.org/10.1080/09540090903383189

Kose-Bagci, H., Ferrari, E., Dautenhahn, K., Syrdal, D. S., & Nehaniv, C. L. (2009). Effects of embodiment and gestures on social interaction in drumming games with a humanoid robot. *Advanced Robotics*, *23*(14), 1951–1996. https://doi.org/10.1163/016918609X12518783330360

Krumhuber, E., & Kappas, A. (2005). Moving smiles: The role of dynamic components for the perception of the genuineness of smiles. *Journal of Nonverbal Behavior*, *29*(1), 3–24. https://doi.org/10.1007/s10919-004-0887-x

Lakatos, G., Gou, M. S., Holthaus, P., Wood, L., Moros, S., Litchfield, V., Robins, B., & Amirabdollahian, F. (2023). A feasibility study of using Kaspar, a humanoid robot for speech and language therapy for children with learning disabilities. *2023 32nd IEEE International Conference on Robot and Human Interactive Communication (RO-MAN)*, 1233–1238. https://doi.org/10.1109/ro-man57019.2023.10309615

Liu, J., & Urbann, O. (2016). Bipedal walking with dynamic balance that involves three-dimensional upper body motion. *Robotics and Autonomous Systems*, *77*, 39–54. https://doi.org/10.1016/j.robot.2015.12.002

Liu, R., & Wang, Y. (2010). Azimuthal source localization using interaural coherence in a robotic dog: Modeling and application. *Robotica*, *28*(7), 1013–1020. https://doi.org/10.1017/S0263574709990865

Lutz, C., Schöttler, M., & Hoffmann, C. P. (2019). The privacy implications of social robots: Scoping review and expert interviews. *Mobile Media and Communication*, *7*(3), 412–434. https://doi.org/10.1177/2050157919843961

Malik, A. A., Masood, T., & Brem, A. (2023). Intelligent humanoids in manufacturing to address worker shortage and skill gaps: Case of Tesla's Optimus. *ArXiv Preprint*. https://arxiv.org/abs/2304.04949

Mirza, N. A., Nehaniv, C. L., Dautenhahn, K., & te Boekhorst, R. (2008). Developing social action capabilities in a humanoid robot using an interaction history architecture. *2008 8th IEEE-RAS International Conference on Humanoid Robots*, 609–616. https://doi.org/10.1109/ICHR.2008.4756013

Moerman, C. J., & Jansens, R. M. L. (2021). Using social robot PLEO to enhance the well-being of hospitalised children. *Journal of Child Health Care*, *25*(3), 412–426. https://doi.org/10.1177/1367493520947503

Mondou, D., Prigent, A., & Revel, A. (2017). A dynamic scenario by remote supervision: a serious game in the museum with a Nao robot. *Advances in Computer Entertainment Technology*, 103–116. https://doi.org/10.1007/978-3-319-76270-8_8ï

Mubin, O., Henderson, J., & Bartneck, C. (2014). You just do not understand me! speech recognition in human robot interaction. *The 23rd IEEE International Symposium on Robot and Human Interactive Communication*, 637–642.

Müller, J., Frese, U., & Röfer, T. (2012). Grab a mug - Object detection and grasp motion planning with the Nao robot. *2012 12th IEEE-RAS International Conference on Humanoid Robots*, 349–356. https://doi.org/10.1109/HUMANOIDS.2012.6651543

Müller, J., Frese, U., Röfer, T., Gelin, R., & Mazel, A. (2014). GRASPY - Object manipulation with NAO. In Fv. G. Röhrbein & C. Natale (Eds.), *Gearing up and accelerating cross-fertilization between academic and industrial robotics research in Europe. Springer Tracts in Advanced Robotics: Vol. 94* (pp. 177–195). Springer. https://doi.org/10.1007/978-3-319-03838-4_9

Navarro, N., Weber, C., & Wermter, S. (2011). Real-world reinforcement learning for autonomous humanoid robot charging in a home environment. In R. Groß, L. Alboul, C. Melhuish, M. Witkowski, T. J. Prescott, & J. Penders (Eds.), *Towards Autonomous Robotic Systems. TAROS 2011. Lecture Notes in Computer Science: Vol. 6856* (pp. 231–240). Springer. https://doi.org/10.1007/978-3-642-23232-9_21

Phillips, E., Zhao, X., Ullman, D., & Malle, B. F. (2018). What is human-like?: Decomposing robots' human-like appearance using the Anthropomorphic roBOT (ABOT) database. *HRI '18: Proceedings of the 2018 ACM/IEEE International Conference on Human-Robot Interaction*, 105–113. https://doi.org/10.1145/3171221

PUDU. (2023). *BellaBot*. Https://Www.Pudurobotics.Com/Products/Bellabot.

Pulido, J. C., González, J. C., Suárez-Mejías, C., Bandera, A., Bustos, P., & Fernández, F. (2017). Evaluating the child-robot interaction of the NAOTherapist platform in pediatric rehabilitation. *International Journal of Social Robotics*, *9*(3), 343–358. https://doi.org/10.1007/s12369-017-0402-2

Rico, F. M., González-Careaga, R., María, J., Plaza, C., & Olivera, V. M. (2004). Programming model based on concurrent objects for the AIBO robot. *Actas de Las XII Jornadas de Concurrencia y Sistemas Distribuídos.*

Riener, R., Rabezzana, L., & Zimmermann, Y. (2023). Do robots outperform humans in human-centered domains? *Frontiers in Robotics and AI*, *10*, 1223946. https://doi.org/10.3389/frobt.2023.1223946

Röfer, T. (2004). Evolutionary gait-optimization using a fitness function based on proprioception. *LNCS. RoboCup2004: Robot Soccer World Cup VIII*, 310–322.

Sala, K. (2023). Innovation in hotel enterprises during the Covid-19 pandemic and the Ukraine conflict. Poland: Case study. In F. Moreira & S. Jayantilal (Eds.), *Proceedings of the 18th European Conference on Innovation and Entrepreneurship, ECIE 2023* (pp. 794–802).

Sanefuji, W., Ohgami, H., & Hashiya, K. (2007). Development of preference for baby faces across species in humans (*Homo sapiens*). *Journal of Ethology*, *25*(3), 249–254. https://doi.org/10.1007/s10164-006-0018-8

Schellin, H., Oberley, T., Patterson, K., Kim, B., Haring, K. S., Tossell, C. C., Phillips, E., & de Visser, E. J. (2020). Man's new best friend? Strengthening human-robot dog bonding by enhancing the doglikeness of Sony's Aibo. *2020 Systems and Information Engineering Design Symposium (SIEDS)*, 1–6. https://doi.org/10.1109/SIEDS49339.2020.9106587

Shen, Q., Kose-Bagci, H., Saunders, J., & Dautenhahn, K. (2009). An experimental investigation of interference effects in human-humanoid interaction games. *The 18th IEEE International Symposium on Robot and Human Interactive Communication*, 291–298. https://doi.org/10.1109/ROMAN.2009.5326342

Skantze, G., & Al Moubayed, S. (2012). IrisTK: a statechart-based toolkit for multi-party face-to-face interaction. *ICMI '12: Proceedings of the 14th ACM International Conference on Multimodal Interaction*, 69–76. https://doi.org/10.1145/2388676.2388698

Smirnov, P., & Kovalev, A. (2021). Dedicated payload stabilization system in a service robot for catering. In A. Ronzhin, G. Rigoll, & R. Meshcheryakov (Eds.), *Interactive Collaborative Robotics. ICR 2021. Lecture Notes in Computer Science: Vol. 12998* (pp. 194–204). Springer. https://doi.org/10.1007/978-3-030-87725-5_17

SoftBank Robotics. (2024). *Aldebaran documentation: NAOqi - Developer guide*. Http://Doc.Aldebaran.Com/2-8/Index_dev_guide.Html. http://doc.aldebaran.com/2-8/index_dev_guide.html

Sombolestan, M., Chen, Y., & Nguyen, Q. (2021). Adaptive force-based control for legged robots. *IEEE International Conference on Intelligent Robots and Systems*, 7440–7447. https://doi.org/10.1109/IROS51168.2021.9636393

Sony. (2018). *Aibo ERS-1000: Help guide*. Https://Helpguide.Sony.Net/Aibo/Ers1000/v1/En-Us/Index.Html.

Sony. (2024). *Aibo developer site*. Https://Developer.Aibo.Com/Us/Docs.

Sousa, J., Matos, V., & Peixoto dos Santos, C. (2010). A bio-inspired postural control for a quadruped robot: an attractor-based dynamics. *The 2010 IEEE/RSJ International Conference on Intelligent Robots and Systems*, 5329–5334. https://doi.org/10.1109/IROS.2010.5648945

Steels, L., & Kaplan, F. (2000). AIBO's first words: The social learning of language and meaning. *Evolution of Communication, 4*(1), 3–32. https://doi.org/10.1075/eoc.4.1.03ste

Syrdal, D. S., Dautenhahn, K., Robins, B., Karakosta, E., & Jones, N. C. (2020). Kaspar in the wild: Experiences from deploying a small humanoid robot in a nursery school for children with autism. *Paladyn, Journal of Behavioral Robotics, 11*(1), 301–326. https://doi.org/10.1515/pjbr-2020-0019

Tanaka, K., Makino, H., Nakamura, K., Nakamura, A., Hayakawa, M., Uchida, H., Kasahara, M., Kato, H., & Igarashi, T. (2022). The pilot study of group robot intervention on pediatric inpatients and their caregivers, using 'new aibo.' *European Journal of Pediatrics, 181*(3), 1055–1061. https://doi.org/10.1007/s00431-021-04285-8

Tetsui, T., Ohkubo, E., Murata, H., Kato, Wakabayashi, R., Kimura, M., & Naganuma. (2008). Design and evaluation of simple control console usable by therapist and elderly people in nursing home for robot assisted activity/therapy. *Design and Evaluation of Simple Control Console Usable by Therapist and Elderly People in Nursing Home for Robot Assisted Activity/Therapy*, 720–723.

Unitree. (2024). *Unitree Go2 EDU*. https://shop.unitree.com/.

van Doorn, J., Smailhodzic, E., Puntoni, S., Li, J., Schumann, J. H., & Holthöwer, J. (2023). Organizational frontlines in the digital age: The Consumer-Autonomous Technology-Worker (CAW) framework. *Journal of Business Research, 164*, 114000. https://doi.org/10.1016/j.jbusres.2023.114000

Visser, U., & Burkhard, H.-D. (2007). RoboCup: 10 years of achievements and future challenges. *AI Magazine, 28*(2), 115–132. https://doi.org/10.1609/aimag.v28i2.2044

Wetzel, E. M., Liu, J., Leathem, T., & Sattineni, A. (2022). The use of Boston Dynamics SPOT in support of LiDAR scanning on active construction sites. *Proceedings of the International Symposium on Automation and Robotics in Construction, 2022-July*, 86–92. https://doi.org/10.22260/isarc2022/0014

Wilson, E. D., Assaf, T., Rossiter, J. M., Dean, P., Porrill, J., Anderson, S. R., & Pearson, M. J. (2021). A multizone cerebellar chip for bioinspired adaptive robot control and sensorimotor processing: A multizone cerebellar chip for bioinspired adaptive robot control and sensorimotor processing. *Journal of the Royal Society Interface*, *18*(174), 20200750. https://doi.org/10.1098/rsif.2020.0750

Wood, L., Dautenhahn, K., Robins, B., & Zaraki, A. (2017). Developing child-robot interaction scenarios with a humanoid robot to assist children with autism in developing visual perspective taking skills. *2017 26th IEEE International Symposium on Robot and Human Interactive Communication (RO-MAN)*, 1055–1060. https://doi.org/10.1109/ROMAN.2017.8172434

Wood, L. J., Zaraki, A., Robins, B., & Dautenhahn, K. (2021). Developing Kaspar: A humanoid robot for children with autism. *International Journal of Social Robotics*, *13*(3), 491–508. https://doi.org/10.1007/s12369-019-00563-6

Yam, K. C., Bigman, Y., & Gray, K. (2021). Reducing the uncanny valley by dehumanizing humanoid robots. *Computers in Human Behavior*, *125*, 106945. https://doi.org/10.1016/j.chb.2021.106945

Zaraki, A., Dautenhahn, K., Wood, L., Novanda, O., & Robins, B. (2017). Toward autonomous child-robot interaction: Development of an interactive architecture for the humanoid Kaspar robot. *3rd Workshop on Child-Robot Interaction (CRI 2017), and International Conference on Human Robot Interaction (ACM/IEEE HRI 2017).*

Zaraki, A., Wood, L., Robins, B., & Dautenhahn, K. (2018). Development of a semi-autonomous robotic system to assist children with autism in developing visual perspective taking skills. *Proceedings of the 27th IEEE International Symposium on Robot and Human Interactive Communication*, 969–976. https://doi.org/10.1109/ROMAN.2018.8525681

Zhang, J., Li, S., Zhang, J. Y., Du, F., Qi, Y., & Liu, X. (2020). A literature review of the research on the uncanny valley. *Lecture Notes in Computer Science (Including Subseries Lecture Notes in Artificial Intelligence and Lecture Notes in Bioinformatics)*, *12192 LNCS*, 255–268. https://doi.org/10.1007/978-3-030-49788-0_19

Zhong, J., Ling, C., Cangelosi, A., Lotfi, A., & Liu, X. (2021). On the gap between domestic robotic applications and computational intelligence. *Electronics*, *10*(7), 793. https://doi.org/10.3390/electronics10070793

INDEX

Note: **Bold** page numbers refer to tables and *italic* page numbers refer to figures.